The Economics of Industrial Innovation

The Economics of
Industrial Innovation

Chris Freeman and Luc Soete

The MIT Press
Cambridge, Massachusetts

First MIT Press edition, 1997

Printed and bound in Great Britain.

Library of Congress Cataloging-in-Publication Data

Freeman, Chris
    The economics of industrial innovation / Chris Freeman and
Luc Soete. — 3rd ed.
        p. cm.
    Includes bibliographical references and index.
    ISBN 0-262-56113-1 (pbk: alk. paper). - ISBN 0-262-06195-3
(hardcover: alk. paper).
    1. Technological innovations—Economic aspects. I. Soete, Luc.
II. Title.
HD45.F725   1997
338'.06—dc21   97-8562
                                                                    CIP

CONTENTS

PREFACE

This third edition of *The Economics of Industrial Innovation* has been thoroughly revised and expanded. Ten entirely new chapters have been added and all the other chapters have been extensively rewritten, so that in many respects it is a new book. The main reason for such a thoroughgoing revision was of course the nature and pace of technical change in the world economy in the closing decades of the twentieth century. There are, however, four other reasons for the revision which we have undertaken.

First of all, there has been a very marked upsurge of interest in the economics of innovation, both in the economics profession itself and among a wider public. This has been reflected in the appearance of several new journals in the field in the 1990s, in a proliferation of research papers and numerous new books. Whereas in the 1970s there were relatively few publications in the field, now it is a real problem to keep pace with the expanding literature. Indeed it has been impossible to do justice to all of it and for this reason we have paid special attention to up-to-date literature reviews and to the bibliography. Short lists of review articles, literature surveys and key references are included at the end of each part and a major bibliography at the end of the book.

Second, earlier editions of the book have been widely used as a textbook for university courses. Naturally, we welcome this and we have tried to respond to numerous suggestions and comments for improvement of the book. In particular, we have extended the scope of the book in several directions, but especially in the international dimension. Since many of our readers have been in the Third World, we have included for the first time chapters dealing with underdevelopment, international trade and with globalization. These constitute the new Part Three of the book.

Third, we have also strengthened the historical dimension of our book. This is somewhat unusual for what is widely seen as an economics textbook. However, it is very much in line with the recent developments in economic theory which have put increasing emphasis on path dependence in models of evolutionary change. Moreover, it is also in line with the advice given by Joseph Schumpeter, who did more than any other twentieth-century economist to give innovation pride of place in the theory of economic development. Schumpeter began his own major work on 'Business Cycles' with the Industrial Revolution and believed this to be essential to the understanding of what he called 'successive industrial revolutions' or long waves of technical change. We have followed his example and have included some new chapters in the historical section of our book in Part One, dealing with the rise of mechanization in the Industrial Revolution, electrification in the late nineteenth century and mass production in the twentieth century.

Finally, this does not mean that we have neglected contemporary developments which have been characterized above all by the rise of microelectronics and computerization and the change from mass production to lean production in the automobile industry. We discuss information and communication technology, not only in relation to the history of the computer industry but more extensively in a new chapter on the Information Society in Part Four. Our collaboration over many years on policy issues for innovation has benefited in particular from our joint work for OECD and for the EU, as well as for industry. We have taken advantage of this experience in compiling Part Four which has three additional chapters mainly concerned with problems of science and technology policy.

Despite this very extensive revision we have nevertheless retained the basic theoretical and historical approach of the first edition of the book. Much of the recent literature has shown a surprising lack of historical memory, yet Richard Nelson's concept of 'appreciative theory' is still relevant in the rapidly changing world of current policy concerns. For this reason, it was relatively straightforward to take into account many of the new research findings of the 1980s and 1990s.

Chris Freeman
Luc Soete

ACKNOWLEDGEMENTS

An earlier version of this book was published by Penguin Modern Economics Texts in 1974, and a second edition by Pinter in 1982. This edition has been extensively revised to take account of new research findings in the 1980s and 1990s. The book makes use of extracts from some papers which were previously published in various journals and reports to various organizations. We have made numerous changes from the original versions, for which, of course, these organizations and journals bear no responsibility. Tables 4.1, 4.2 and 4.3 are reprinted from *Petroleum Progress and Profits*, by J. L. Enos, by permission of the MIT Press, copyright © 1962 by the Massachusetts Institute of Technology.

In writing the book, we have been conscious all the time of our debt to colleagues who have worked with us at the Science Policy Research Unit and at the Maastricht Economic Research Institute on Innovation and Technology, and who have assisted us in a great variety of ways. Much of their work is cited, but we are particularly grateful to Anthony Arundel, Keith Pavitt, Roy Rothwell and Bart Verspagen for their help and advice. We are also grateful to colleagues at the OECD, particularly Alison Young, for their help with R&D statistics. The OECD statistics for which they have been responsible have been an invaluable help for all researchers in this field, as have the new European Science and Technology Indicators published by the EU and we are grateful to the OECD and EU for permission to use these statistics. We are also grateful to Bas Ter Weel for his help in updating some tables and figures. International collaboration in this field has always been important and we have gained a great deal from the work and the advice of many fellow researchers in all parts of the world. It would be impossible to do justice to the many individuals from whose work we have benefited, both from their published work and through their visits and other forms of contact.

In an altogether different dimension there are four people without whose help the book could never have been written at all. Jackie Fuller contributed as friend and collaborator to several of the research projects described, both at the NIESR and at SPRU. Wilma Coenegrachts and Karin Kamp contributed in a similar way to much of our joint research. Finally, Susan Lees has steered through all the work on this third edition with extraordinary patience, thoroughness and accuracy. We are deeply grateful to them.

CHAPTER 1

INTRODUCTION

All the improvements in machinery, however, have by no means been the inventions of those who had occasion to use the machines. Many improvements have been made by the ingenuity of the makers of the machines, when to make them became the business of a peculiar trade; and some by that of those who are called philosophers or men of speculation, whose trade is not to do anything but to observe everything; and who, upon that account, are often capable of combining together the powers of the most distant and dissimilar objects. In the progress of society, philosophy or speculation becomes like every other employment, the principal or sole trade and occupation of a particular class of citizens. Like every other employment too, it is subdivided into a great number of different branches, each of which affords occupation to a peculiar tribe or class of philosophers; and this subdivision of employment in philosophy, as well as in every other business, improves dexterity and saves time. Each individual becomes more expert in his own peculiar branch, more work is done upon the whole, and the quantity of science is considerably increased by it.

(Smith, 1776, p. 8)

It is a scientifically based analysis, together with the application of mechanical and chemical laws, that enables the machine to carry out the work formerly done by the worker himself. The development of machinery, however, only follows this path once heavy industry has reached an advanced stage, and the various sciences have been pressed into the service of capital. . . . Invention then becomes a branch of business, and the application of science to immediate production aims at determining the inventions at the same time as it solicits them.

(Marx, 1858, p. 592)

When you adopt a new systematic model of economic principles you comprehend reality in a new and different way.

(Samuelson, 1967, p. 10)

1.1 INTRODUCTION

In the world of microelectronics and genetic engineering, it is unnecessary to belabour the importance of science and technology for the economy. Whether like the sociologist, Marcuse, or the novelist, Simone de Beauvoir, we see technology primarily as a means of human enslavement and destruction, or whether, like Adam Smith and Marx, we see it primarily as a liberating force, we are all involved in its advance. However much we might wish to, we cannot escape its impact on our daily lives, nor the moral, social and economic dilemmas with which it confronts us. We may curse it or bless it, but we cannot ignore it.

Least of all can economists afford to ignore innovation, an essential condition of economic progress and a critical element in the competitive struggle of enterprises and of nation-states. In rejecting modern technology, Simone de Beauvoir was consistent in her deliberate preference for poverty. But most economists have tended to accept with Marshall that poverty is one of the principal causes of the degradation of a large part of mankind. Their preoccupation with problems of economic growth arose from the belief that the mass poverty of Asia, Africa and Latin America and the less severe poverty remaining in Europe and North America, was a preventable evil which could and should be diminished, and perhaps eventually eliminated.

Recently both the desirability and the feasibility of such an objective have been increasingly questioned. However, innovation is of importance not only for increasing the wealth of nations in the narrow sense of increased prosperity, but also in the more fundamental sense of enabling people to do things which have never been done before. It enables the whole quality of life to be changed for better or for worse. It can mean not merely more of the same goods but a pattern of goods and services which has not previously existed, except in the imagination.

Innovation is critical, therefore, not only for those who wish to accelerate or sustain the rate of economic growth in their own and other countries, but also for those who are appalled by narrow preoccupation with the quantity of goods and wish to change the direction of economic advance, or concentrate on improving the quality of life. It is critical for the long-term conservation of resources and improvement of the environment. The prevention of most forms of pollution and the economic recycling of waste products are alike dependent on technological advance, as well as on social innovations.

In the most general sense economists have always recognized the central importance of technological innovation for economic progress. The famous first chapter of Adam Smith's *Wealth of Nations* plunges immediately into discussion of 'improvements in machinery' and the way in which division of labour promotes specialized inventions. Marx's model of the capitalist economy ascribes a central role to technical innovation in capital goods – 'the bourgeoisie cannot exist without constantly revolutionizing the means of production'. Marshall had no hesitation in describing 'knowledge' as the chief engine of progress in the economy. A standard pre-war textbook states in the chapter on economic progress that 'Our brief survey of economic expansion during the last 150 years or so seems to show that the main force was the progress of technique' (Benham, 1938, p. 319). The standard post-war textbook by Samuelson (1967) comes to much the same conclusion.

Yet although most economists have made a deferential nod in the direction of technological change, until recently few have stopped to examine it. Jewkes and his colleagues explained this paradox in terms of three factors: ignorance of natural science and technology on the part of economists; their preoccupation with trade cycle and employment problems; and the lack of usable statistics (Jewkes *et al.*, 1958).

These factors may partly explain the relative neglect of innovation but they cannot be held to justify it, as all of them can be overcome at least to some extent. Jewkes and his colleagues demonstrated this in their study of *The Sources of Invention*, and it has been confirmed by other empirical studies before and since. Indeed, whereas earlier literature reviews (e.g. Kennedy and Thirlwall, 1971) complained of the dearth of studies of innovations and their diffusion, more recent reviews (e.g. Dosi, 1988; Freeman, 1994) pointed to the explosion of interest in the 1980s and 1990s.

The earlier neglect of invention and innovation was not only due to other preoccupations of economists nor to their ignorance of technology; they were also the victims of their own assumptions and commitment to accepted systems of thought. These tended to treat the flow of new knowledge, of inventions and innovations as outside the framework of economic models, or more strictly, as 'exogenous variables'. A large body of economic theory was concerned with short-term analysis of fluctuations in supply and demand for goods and services. Although very useful for many purposes, these models usually excluded changes in the technological and social framework from consideration, under the traditional *ceteris paribus* assumption (other things being equal). Even when, in the 1950s, economists increasingly turned their attention to problems of economic growth, the screening off of 'other things' was largely maintained, and attention was concentrated on the traditional factor inputs of labour and capital, with 'technical change' as a residual factor embracing all other contributions to growth, such as education, management and technological innovation.

It was, of course, always recognized in principle that 'other things' were extremely important, but it was only recently that they began to be the subject of systematic economic analysis. For what they are worth, most of the early econometric studies of growth in industrialized countries attributed the greater part of measured growth to technical progress, rather than to increases in the volume of the traditional inputs of capital and labour. However, technical change remained on the fringe and not at the centre of economic analysis. Yet it would not be unreasonable to regard education, research and experimental development as the basic factors in the process of growth, relegating capital investment to the role of an intermediate factor. This is indeed the tendency of the so-called new growth theory (Romer, 1986; Verspagen, 1992b). It is of course new only in the sense of the belated recognition by modellers of some of the long-held ideas of economic historians and of those economists, such as Schumpeter, who always gave a central place to technical and institutional change. The World Bank (1991) review of development theory also reflected this major shift in thinking about growth mainly in terms of 'intangible investment' (see Chapter 13).

Looked at in this way, the investment process is as much one of the production and distribution of knowledge as the production and use of capital goods, which embody the advance of science and technology.[1] 'Intangible' investment in new knowledge and its dissemination are the critical elements, rather than 'tangible' investment in bricks and machines. Yet our whole apparatus of economic thought, as well as our whole system of statistical indicators, are still largely geared to the 'tangible' goods and services approach.

This will surely change in the coming decades, if only for the reason that the specialized industries concerned with generating and distributing knowledge will employ a large part of the working population. Bernal's model (1958) of the probable patterns of future employment (Figure 1.1)

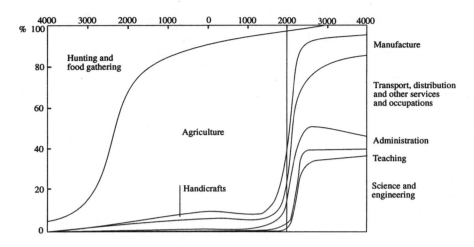

Fig. 1.1 Changes in occupation in the past and future

Source: Bernal (1958).

was speculative. It probably exaggerates the future share of science and engineering and underestimates the future share of 'teaching' but it illustrates the kind of fundamental change which is occurring. Agriculture, which once occupied almost the whole population, now employs less than 10 per cent in the most advanced economies (although still more than 50 per cent in many less developed countries). Not only is the share of manufacturing declining, as services expand their share, but within manufacturing and services an increasing number of people are concerned primarily with generating and disseminating information rather than goods.

Indeed, if a very wide definition of knowledge industries is adopted, then Machlup demonstrated that they already employed a quarter of the United States labour force in 1959. In his book, *The Production and Distribution of Knowledge* (1962), he estimated that over 30 per cent of the US labour force were engaged in occupations essentially concerned with producing and handling information rather than goods. In his definitions he included not only research, development, design and education of all kinds, but also the larger numbers of people employed in printing, publishing, scientific libraries, testing laboratories, design and drawing offices, general statistical services, resource survey organizations, radio, television and other communication industries, as well as computers and information machines of all types, and professional services concerned with analysing and displaying information. All of these activities are important in generating, disseminating and applying advances in technology, although some of them are more important in a broader sense as entertainment. More recently, Porat (1977) estimated that the share of 'information occupations' in the United States economy was already half the total, hence the increasing use of the expression 'information society'. As we shall see in

Chapter 7, when we come to discuss Information and Communication Technology (ICT), the distinction between information and knowledge is an important one. Raw data have to be converted into useful knowledge. The information society can be regarded as the culmination of a long process of the growth of intangible investment in information-based activities.

1.2 THE RESEARCH AND DEVELOPMENT SYSTEM

Research and inventive activities are only a small proportion of this very wide complex of 'information' industries. The professional labour force engaged in research and experimental development is less than 2 per cent of the total working population in the United States, and less than 1 per cent in most other countries. But this Research and Development system is at the heart of the whole complex, for in contemporary society it originates a large proportion of the new and improved materials, products, processes and systems, which are the ultimate source of economic advance. This is not to underestimate the importance of dissemination of knowledge through the education system, industrial training, the mass media, information services and other means. Nor is it to deny the obvious fact that in the short run rapid progress may be made simply by the application of the existing stock of knowledge. Nor yet is it to deny the importance of feedback from production and from markets to R&D and other technical and scientific activities. It is only to assert the fundamental point that for any given technique of production, transport or distribution, there are long-run limitations on the growth of productivity, which are technologically determined. In the most fundamental sense the winning of new knowledge is the basis of human civilization.

Consequently there is ample justification for concentrating attention on the flow of new scientific ideas, inventions and innovations. Efforts to generate discoveries and inventions have been increasingly centred in specialized institutions – the Research and Experimental Development network. This professionalized system is generally known by the abbreviated initials R&D. Its growth was perhaps the most important social and economic change in twentieth-century industry. This book is primarily concerned with the innovations arising from the professional R&D system, and with the allocation of resources to this system. Its interaction with other knowledge industries and with industrial production and marketing are of critical importance for any economy, but it is only recently that it has become the subject of systematic study. The policy adopted for R&D in any country, whether it is implicit in the sense of 'laissez-faire, laissez-innover', or explicit in the sense of national goals and strategies, constitutes the main element of policy for science and technology, or, more briefly, national science policy. A wider spectrum of scientific and technological services (STS) link the R&D system with production and routine technical activities. STS includes such activities as design, quality control, information services, survey and feasibility studies. They are also essential for efficient innovation, and may predominate in the diffusion of technical change in many branches of industry.

Although government and university laboratories had existed earlier, it was only in the 1870s that the first specialized R&D laboratories were established in industry. The professional R&D system was barely recognized at all by economists in the nineteenth century and even in the early part of this century the young Schumpeter (1912), who gave innovation pride of place in his theory of economic development, treated the source of inventions as exogenous to the economy. We owe to Schumpeter the extremely important distinction between inventions and innovations, which has since been generally incorporated into economic theory. An invention is an idea, a sketch or model for a new or improved device, product, process or system. Such inventions may often (not always) be patented but they do not necessarily lead to technical innovations. In fact the majority do not. An innovation in the economic sense is accomplished only with the first commercial transaction involving the new product, process system or device, although the word is used also to describe the whole process. Of course, further inventions often take place during the innovation process and still more inventions and innovations may be made during the diffusion process. Nevertheless, Schumpeter's conceptual distinction is a valuable one.

The chain of events from invention or specification to social application is often long and hazardous. Schumpeter (1912, 1928, 1942) always stressed the crucial role of the entrepreneur in this complex innovative process. But as Almarin Phillips (1971) has pointed out, it was only in his later work that he recognized the 'internalization' of much scientific and inventive activity within the firm. In his 1928 article he pointed out that the 'bureaucratic' management of innovation was replacing individualistic flair and that the large corporation was becoming the main vehicle for technical innovation in the economy. This shift of emphasis from the early Schumpeter ('Mark' 1) to the late Schumpeter ('Mark' 2) will be discussed further in later chapters. It reflected the real change which had taken place in the American economy between the two world wars and the rapid growth of industrial R&D in large corporations during that period.

By the outbreak of the Second World War there was already in existence an extensive network of organized research laboratories and related institutions in government, universities and industry, employing a full-time professional staff. This R&D industry can be subjected to economic analysis like any other although it has some unique characteristics. Its 'output' is a flow of new knowledge, both of a general character (the result of 'fundamental' or 'basic' research) and relating to specific applications ('applied' research). It is also a flow of models, sketches, designs, manuals and prototypes for new products, or of pilot plants and experimental rigs for new processes ('experimental development'). The inputs and outputs of this system are summarized in Table 1.1. But, of course, long before the twentieth century, experimental development work on new or improved products and processes was carried out in ordinary workshops. When Boulton brought Watt's steam engine from the stage of laboratory invention to commercial production model, he most certainly carried out extensive 'research and development' at his Soho works, even if there was no department with that name.

The classical economists were well aware of the critical role of R&D in economic progress even though they used a different terminology. Adam Smith (1776) observed that improvements in machinery came both from

Table 1.1 Inputs and outputs in research, invention, development and innovation

Stage	Input			Output	
	(i) Intangible	(ii) Tangible and human	(iii) Measurable	(iv) Intangible	(v) Measurable
1 'Basic research' (intended output: 'formulas')	Scientific knowledge (old stock and output from 1a) Scientific problems and hunches (old stock and output from 1b, 2b and 3b)	Scientists Technical aides Clerical aides Laboratories Materials, fuel, power	People, hours Payrolls, current and deflated Outlays, current and deflated Outlays per person	a. New scientific knowledge: hypotheses and theories b. New scientific problems and hunches c. New practical problems and ideas	Research papers and memoranda
2 'inventive work' (including minor improvements but excluding further development of inventions) (intended output: 'sketches')	Scientific knowledge (old stock and output from 1a) Technology (old stock and output from 2a and 3a) Practical problems and ideas (old stock and output from 1c, 2c, 3c and 4a)	Scientists Non-scientists Inventors Engineers Technical aides Clerical aides Laboratories Materials, fuel, power	People, hours Payrolls, current and deflated Outlays, current and deflated Outlay per person	a. 'Raw inventions' technological recipes patented inventions patentable inventions, not patented but published patentable inventions, neither patented nor published non-patentable inventions, published non-patentable inventions, not published minor improvements b. New scientific problems and hunches c. New practical problems and ideas, 'bugs'	Patent applications and patents Technological papers and memoranda Papers and memoranda

continued overleaf

Table 1.1 (continued)

Stage	Input			Output	
	(i) Intangible	(ii) Tangible and human	(iii) Measurable	(iv) Intangible	(v) Measurable
3 'Development work' (intended output: 'blueprints and specifications')	Scientific knowledge (old stock and output from 1a) Technology (old stock and output from 3a) Practical problems and ideas (old stock and output from 1c, 2c, 3c and 4a) Raw inventions and improvements (old stock and output from 2a)	Scientists Engineers Technical aides Clerical aides Laboratories Materials, fuel, power Pilot plants Prototypes	People, hours Payrolls, current and deflated Outlays, current and deflated Outlay per person Investment	a. Developed inventions, blueprints, specifications, samples b. New scientific problems and hunches c. New practical problems and ideas, 'bugs'	Blueprints and specification for new and improved products and processes
4 'New-type plant construction' (intended output: 'new-type plant' and new products)	Developed inventions (output from 3a) Business acumen and market forecasts Financial resources Enterprise (venturing)	Entrepreneurs Managers Financiers and bankers Builders and contractors Engineers Building materials Machines and tools	$ investment in new-type plant and products $ investment in new-type plant	a. New practical problems and ideas, 'bugs'	New-type plant or production lines producing novel products, better products, cheaper products, i.e. products and process innovations

Note: Ames (1961) has pointed out that 'bugs' ('persistent irritating obstacles to the completion of scientific and technical work) are an important part of the output, since they may lead to novel results at later stages, including new scientific theories.

Source: Modified slightly from Machlup (1962).

the manufacturers and users of machines and from 'philosophers or men of speculation, whose trade is not to do anything but to observe everything'. Although he had already noted the importance of 'natural philosophers' (the expression 'scientist' only came into use in the nineteenth century), in his day the advance of technology was largely due to the inventiveness of people working directly in the production process or immediately associated with it: 'a great part of the machines made use of in those manufactures in which labour is most subdivided, were originally the inventions of common workmen' (Smith, 1776, p. 8). Technical progress was rapid but the techniques were such that experience and mechanical ingenuity enabled many improvements to be made as a result of direct observation and small-scale experiment. Most of the patents in this period were taken out by 'mechanics' or 'engineers', who did their own 'development' work alongside production or privately.

1.3 THE PROFESSIONALIZATION OF INDUSTRIAL R&D AND ITS GROWTH

What is distinctive about modern industrial R&D is its scale, its scientific content and the extent of its professional specialization. A much greater part of technological progress is now attributable to research and development work performed in specialized laboratories or pilot plants by full-time qualified staff. It is this work which is recorded in R&D statistics. It was not practicable to measure the part-time and amateur inventive work of the eighteenth or nineteenth century. Thus our R&D statistics are really a measure of professionalization of this activity. This professionalization is associated with three main changes:

1. The increasingly scientific character of technology.[2] This applies not only to biological, chemical and electronic processes but often to mechanical processes as well. Even eighteenth-century mechanics actually depended on the formal science of Newton but the combination of mechanical with electronic engineering strengthens this dependence. The Japanese, who are one of the most advanced nations in the applications of electronics to mechanical engineering, have coined the word 'mechatronics', which aptly expresses this transition. A formal body of 'book learning' is usually necessary now for those who wish to advance the state of the art, as well as practical experience.
2. The growing complexity of technology and the partial replacement of 'batch' and 'one-off' systems of production by 'flow' and 'mass' production lines. It is expensive and sometimes almost impossible to use the normal production line for experiments in large-scale plants. The physical separation of experimental development work into specialized institutions was often a necessity in such cases. The sheer number of components in some processes and products has similar effects in prototype and pilot plant work. These are now designated as 'complex systems'.
3. The general trend towards division of labour, noted by Adam Smith, which gave some advantages to the specialized research laboratories,

with their own highly trained people, information services and scientific apparatus. R&D activities are characterized by a very high concentration of engineers and scientists with a relatively small proportion of supporting staff – often only one or two per engineer or scientist.

Starting in the chemical and electrical industries, these laboratories have become increasingly characteristic institutions. Like all changes in the division of labour, the specialization of the R&D function and other STS has given rise to serious social problems, as well as to the benefits, which Adam Smith observed. As we shall see, the departmental separation of R&D from the production line and the marketing function in the firm gives rise to major management co-ordination problems. The rise of a professional 'R&D establishment' as a distinct social group may also lead to even more serious divisions and tensions in society, between those who generate new knowledge and others who may not understand it or may not want to see it applied. The R&D 'establishment' itself becomes a vested interest and political lobby, both in the industrial and in the military field. Some of these problems are discussed in the final section of this book.

The extent of specialization should not be exaggerated. Important inventions are still made by production engineers or private inventors, and with every new process many improvements are made by those who actually operate it. In some firms there are technical or engineering departments or Operations Research (OR) sections, whose function is often intermediate between R&D and production and who may often contribute far more to the technical improvement of an existing process than the formal R&D department, more narrowly defined. But the balance has undoubtedly changed, and it is this specialization of the R&D function which justifies some such expression as the 'research revolution' to describe what has been happening in twentieth-century industry. During this time most large firms in the industrialized countries have set up their own full-time specialized R&D sections or departments. Until the late 1960s, R&D activities were expanding very rapidly in many countries, but during the 1970s and 1980s growth slowed down somewhat, especially in the United Kingdom and the United States. In the 1990s there was a more general slow-down and even some decline, except in some Asian countries where very rapid growth continued. In the former communist countries of Eastern Europe there was a steep decline of formal R&D in the 1990s (Table 1.2 and Figures 1.2 and 1.3).

These contrasting trends are discussed in Part Three. Regular survey publications of the European Union (European Science and Technology Indicators, 1994 onwards) of the US National Science Foundation and of the OECD now provide detailed statistics for many countries of R&D expenditures and of personnel employed. Most of the early surveys confined themselves to these 'input' statistics but the more recent publications make increasing use of 'output' statistics, such as patents, publications and citations (Table 1.1). Some of the problems of measuring output are discussed in Chapter 5.

For the economist, it is obviously desirable to examine the operations of this R&D system from the standpoint of its efficiency in employing scarce

Table 1.2 Trends in gross domestic expenditures on R&D (GERD)

	GERD million current PPP $	Average annual growth rate		Percentage change from preceding year(s)				GERD as a percentage of GDP		
	1993	1981–85	1985–89	1990	1991	1992	1993	1981	1991	1993
USA	169,964	7.3	2.0	3.2	—[9]	1.4	-0.5	2.4	2.8[9]	2.7
Canada	8,320	6.7	2.4	6.0	1.9	0.8	1.3	1.2	1.5	1.5
Mexico	1,964	—	—	—	—	—	—	—	—	0.3
Japan[1]	69,535	8.9	6.5	8.4	3.2	-1.0	-3.0	2.1	2.9	2.7
Australia[2]	3,713	8.2	4.6	5.0	—	—	—	1.0	1.4	—
New Zealand[3]	410	—	—	0.3	-0.8	—	—	—	0.9	—
Austria	2.416	4.0	4.6	8.0	8.8	3.2	3.7	1.2	1.5	1.6
Belgium[3]	2,853	—[9]	—[9]	—	1.6	—	—	—	1.7	—
Denmark	1,786	6.9	7.0	6.4	5.8	3.6	3.6	1.1	1.7	1.8
Finland	1,755	10.5	8.1	4.2	—[9]	1.4	0.3	1.2[9]	2.1[9]	2.2
France	25,984	5.0	4.0	6.1	0.5	0.9	-0.8	2.0[9]	2.4	2.4
Germany[4]	37,265	4.3	—[9]	1.5	—[9]	—[9]	-1.1	2.4	2.6[9]	2.5
Greece	560	—[9]	—[9]	—	1.1	—	15.3	0.2[9]	0.5	0.6
Iceland[5]	65	5.4	12.2	-1.8	18.8	10.8	—	0.6	1.2	1.3
Ireland[5]	504	5.6	5.3	13.4	18.6	10.7	—	0.7	1.0	1.1
Italy	13,220	8.3	5.8	6.7	3.2	-0.3	-1.3	0.9	1.3	1.3
Luxembourg	—	—	—	—	—	—	—	—	—	—
Netherlands[5]	4,965	—[9]	3.6	-0.6	-3.3	-1.3	—	1.9	1.9	1.9
Norway	1,632	—[9]	2.4	—	1.1	—	4.2	1.3	1.8	1.9
Portugal[5,6]	709	5.6	9.8	16.1	—	9.8	—	0.4	0.6	0.7
Spain	4,567	8.7	13.2	16.9	5.1	—[9]	-5.3	0.4	0.9	0.9
Sweden	4,578	8.2	3.0	—	-1.4	—	2.4	2.3[9]	2.9	3.1
Switzerland[5,7]	4,243	—[9]	—[9]	—	—	-1.4	—	2.3	2.9[9]	2.7
Turkey	1,436	—	—	—	64.3	-1.7	—	—	0.5	0.5
United Kingdom	21,584	1.8	3.2	1.9	-4.8	0.3	2.5	2.4[9]	2.2	2.2
North America[8]	180,248	7.3	2.0	3.3	—[9]	1.3	-0.6	2.3	2.6[9]	2.4
EU-15[4]	123,056	4.6	4.4	3.7	—[9]	0.3	-0.3	1.7	2.0[9]	2.0
Total OECD[4,8]	385,495	6.6	3.6	4.3	1.6	0.7	-0.8	2.0	2.3	2.2

1. Adjusted by the Secretariat to improve international comparability.
2. Latest year available 1990. Growth 1981–6 and 1986–90. 1990 for 1991.
3. Latest year available 1991.
4. German totals from 1991 onwards include former East Germany.
5. Latest year available 1992.
6. Growth 1980–84 and 1984–8. 1982 for 1981.
7. 1990 for 1991.
8. Including Mexico from 1991 onwards.
9. Break in series.

Source: OECD, MSTI database, July 1995.

resources. How can the flow of new information, knowledge, inventions and innovations be improved? Could the scientists, engineers and technicians employed in an industrial laboratory or a government research station be more effectively deployed elsewhere? Could the information required be obtained free or at a lower cost from abroad? Are part-time or amateur inventors or scientists sometimes more productive than full-time professionals? What kind of economies of scale are there in research or in development? Can the gestation period for innovations be shortened?

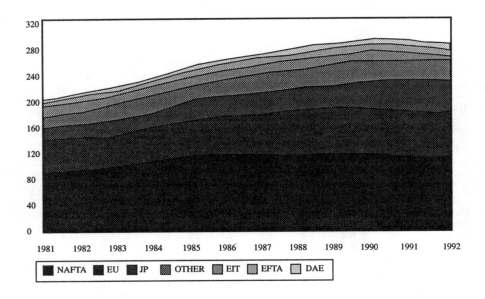

Fig. 1.2 Trends in R&D expenditures in 50 leading countries (1987 US prices Bn. ECU)
NAFTA: USA, Canada, Mexico
EIT: East and Central Europe (Economies in Transition)
DAE: East Asia (Dynamic Asian Economies)
JP: Japan

Source: European report on Science and Technology indicators (1995).

What kind of firms are most likely to innovate and under what market conditions? What type of incentives stimulate invention and innovation most effectively? How are innovations diffused through the economy? In what ways do universities contribute to industrial innovation and how could this contribution be improved? These are the kind of questions which economists ask about the R&D system. They should also ask some more fundamental questions about the relationship of innovations to wider human values. Are the main goals of science and technology the most desirable way of using these resources?

There is a considerable resistance to looking at invention and research in this way. One result has been that many studies of invention and innovation have been written by biographers who tended to concentrate on the personal peculiarities of famous inventors and innovators and memorable anecdotes of their exploits. A mythology has grown up, stressing mainly the random accidental factors in the inventive and innovative process. Sometimes these myths depart altogether from reality as in the case of Watt and the steam from the kettle; in other cases they simply exaggerate the role of chance events as in the case of penicillin.

The treatment of R&D as an exogenous and largely uncontrollable force, operating independently of any policy, has been promoted in the past by

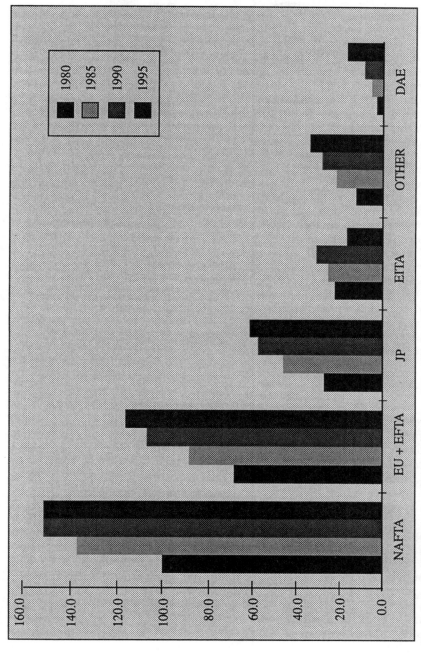

Fig. 1.3 Gross domestic expenditure on R&D (GERD), 1980–1995 (NAFTA 1980 = 100)

Source: MERIT.

both economists and scientists, though for different reasons. In either case it encouraged the 'black box' and 'magic wand' approach to science and technology, which not only discouraged attempts to understand the social process of innovation, but even endangered the whole future relationship between science, technology and society. What is not understood may often be feared, or become the object of hostility.

Polanyi (1962) made an interesting analogy between a free market economy and the basic research system, arguing that in both cases decisions must be completely decentralized to get optimal results. In the one case only firms have the necessary information on which to base good decisions and in the other case only scientists. Like most economists, Polanyi accepted the need for a central government subsidy to basic research because the private market would not finance such an uncertain long-term investment, but he maintained that the scientists should be completely free to pursue whatever projects they thought best. Friedmann took the argument one stage further in maintaining that there was no need for government to finance basic research at all. Kealey (1996) elaborated this position at book length in attempting to establish 'laws' of the economics of science. Like all such arguments, they can be carried to the point of absurdity by over-zealous logic. The market mechanism can be a useful technique for allocating resources in certain rather specific circumstances, but it has its limitations, so that the definition and implementation of social priorities for science and technology cannot be left simply to the free play of market forces (Nelson, 1959, 1977; Pavitt, 1996). The political system is inevitably involved and the full implications of this situation are taken up in Part Four.

1.4   MODERN TECHNOLOGY

The 'research revolution' was not just a question of change in scale, it also involved a fundamental change in the relationship between society and technology. The very use of the word technology usually carries the implication of a change in the way in which we organize our knowledge about productive techniques. If by technology we mean simply that body of knowledge which relates to the production or acquisition of food, clothing, shelter and other human needs, then of course all human societies have used technology. It is perhaps the main characteristic which distinguishes humanity from other forms of animal life. But until recently knowledge of these 'arts and crafts', as they used to be called, was largely based on skills of hand and eye, and on practical experience which was transmitted from generation to generation by some sort of apprenticeship or 'learning by doing'.

The expression technology, with its connotation of a more formal and systematic body of learning, only came into general use when the techniques of production reached a stage of complexity where these traditional methods no longer sufficed.[3] The older arts and crafts (or more primitive technologies) continue to exist side by side with the new 'technology', and it would be ridiculous to suggest that modern industry is now entirely a matter of science rather than craft. The 'heating and ventilating engineer' may still be a plumber, the 'tribologist' may still be a greaser and the 'food technologist' has not yet superseded the cook. They may never do so.

Nevertheless, there has been an extremely important change in the way in which we order our knowledge of the techniques used in producing, distributing and transporting goods. Some people call this change simply 'technology'; others prefer to talk about 'advanced technology', or 'high technology', to distinguish those branches of industry which depend on more formal scientific techniques than the older crafts. Because in a sense human societies have always had technology, some people see little new in modern technology. It will be argued here that this is a profound mistake and that the newer technologies are revolutionizing the relationships between science and society.

Some historians have argued that 'science' and 'technology' are two subsystems which developed autonomously and with a considerable degree of independence from each other. Derek Price (1965) maintained that the two bodies of knowledge were generated by distinct professions in quite different ways and with largely independent traditions. The scientific community was concerned with discovery and with the publication of new knowledge in a form which would meet the professional criteria of their fellow scientists. Application was of secondary importance or not even considered. For the engineers or technologists on the other hand, publication was of secondary or negligible importance. Their first concern was with the practical application and the professional recognition which came from the demonstration of a working device or design. Derek Price did not of course deny that 'science' and 'technology' have interacted very powerfully. He used the simile of two dancing partners who each have their own steps although dancing to the same music. The development of the steam engine obviously influenced thermodynamics (to put it mildly), whereas scientific knowledge of electricity and magnetism was the basis for the electrical engineering industry. But each partner in the dance has his or her own interpretation and moves in a different way.

This simile can be a useful one, but if it is used to argue that nothing has changed since the nineteenth century in the relationship between science and technology, then it can be dangerously misleading. At the very least there are some new 'dances' and some of them are 'cheek to cheek'. The relationship has become very much more intimate, and the professional industrial R&D department is both cause and consequence of this new intimacy. Two very important empirical studies, one British (Gibbons and Johnston, 1972) and the other American (National Science Foundation, 1973) demonstrated in some depth the importance of science and communication with the scientific community for contemporary technical innovation. Since the relationship is one of interaction, the expression 'science-related' technology is usually preferable to the expression 'science-based' technology with its implication of an oversimplified one-way movement of ideas. Marx spoke of the machine as the 'point of entry' of science into the industrial system, but today this expression might be used with more justification about the R&D department.

Walsh et al. (1979), in their study of science and invention in the chemical industry, showed that there was a very close similarity in the patterns of growth of patenting activity by firms and the publication of scientific papers. Liebermann (1978) demonstrated that scientists in the electronics industry actually cited more recent papers from the fundamental physics journals than their colleagues in universities (see also Chapter 7).

Other historians and economists, notably Hessen (1931), Musson and Robinson (1969) and Jewkes *et al.* (1958) have insisted that already in the seventeenth, eighteenth and nineteenth centuries, there was a great deal of interaction between science and industry technology. There is much truth in this contention, but it does not alter the fact that professionalized R&D, carried out within industry itself, has put the relationship on a regular, systematic basis and on a far larger scale.

This change has affected especially the design of new products, but the new science-related technologies also affect the way in which improvements and changes are made in production. As has already been suggested, in the older industries these could be made predominantly 'at the bench' by direct participants in the production process. The subdivision of mechanical process did not remove this possibility. Indeed, as both Adam Smith and Marx noted, the workers themselves were often responsible for inventions leading to further subdivision. But the introduction of flow processes in the chemical industry and of electronic control and automation in other branches of industry mean that improvements and changes now depend increasingly on an understanding of the process as a whole, which usually involves some grasp of theoretical scientific principles. It also means that experiments often have to be made 'off-line' in a separate workshop or pilot plant, rather than 'on-line' by production engineers or operatives. 'Systems analysis' becomes important in its own right. All this has accentuated the relative importance of the specialized R&D group or engineering or technical service department and diminished the relative importance of the 'ingenious mechanic'. In the newer industries R&D personnel, as well as other technical departments and OR sections, often have to spend a good deal of time 'troubleshooting', that is resolving difficulties which arise in the normal production process and are referred back to them for solutions. This is not strictly R&D but it illustrates the changed position of production staff. The use of R&D personnel to start and control new production lines in the semiconductor industry is another indication of this change, as is the trial operation of new instruments and machines first of all by R&D personnel.

This can also be seen from the patent statistics for the various branches of industry. In mechanical engineering, applications from private individuals are still important by comparison with corporate patents, but in electronics and chemicals they are very few. The overall share has been declining since 1900 (OECD, 1982).

The increasingly scientific content of technology and the increased subdivision and specialization within science itself have led to major problems of communication between specialist and non-specialist. These have been accentuated by the divisions within the educational system between the different disciplines and between the arts and the sciences. For many people these tendencies, together with some of the unpleasant features of modern industrialization, have increased the sense of alienation from modern technology to the point where they question the desirability of any further innovation. They feel that the whole system is like an uncontrollable and unpredictable juggernaut which is sweeping human society along in its wake. Instead of technology serving human beings it sometimes seems to be the other way about. The constant reiteration of the stock reply, 'You can't stop technical progress anyway', serves to reinforce rather than to diminish these fears.[4]

As a result, the social mechanisms by which we monitor and control the direction and pace of technical change are one of the most critical problems of contemporary politics. In Part Four of this book it is argued that a more explicit policy for science and technical innovation is increasingly necessary. It is also argued that the market demand mechanism for innovation in consumer goods and services has serious deficiencies. But it is by no means easy to understand or to control this complex system and the high degree of autonomy which it enjoys is partly the result of this difficulty. Socialist societies were not particularly successful either.

This is not to deny that a pure 'laissez-innover' system is unacceptable. Nor is it to deny the paramount importance of human values in deciding whether to promote or to halt particular new technical developments. Technical innovation need not be a purely random or arbitrary process, but control depends upon understanding. An important part of this understanding relates to economic aspects of the process, such as costs, return on investment, market structure, rate of growth and distribution of possible benefits. We still know far too little about these economic aspects of innovation, but slowly we are beginning to build up a body of systematic observations and generalizations, together with explanatory hypotheses which are supported to a varying extent by the empirical data. No doubt some of these hypotheses will be wholly or partly refuted or modified by future observations and experiments. As our knowledge extends so does the possibility of using innovations more satisfactorily.

1.5 SCHUMPETER'S THEORY OF SUCCESSIVE INDUSTRIAL REVOLUTIONS

This book reflects the relatively elementary state of our present knowledge. The generalizations are tentative because they have been insufficiently tested and corroborated by applied research. Although the book describes the results of some of the empirical studies by economists, it also poses some of the principal unsolved problems, in the hope that this will help to stimulate new thinking and research. Finally, the last part of the book raises some of the difficult policy issues which arise from the analysis.

The choice of a historical method of approach in the first part of the book is deliberate. The abstract 'representative firm' is a fictional device which is of little value in understanding the role of industrial R&D. In order to make useful generalizations about R&D in relation to firm behaviour it is essential to place the growth of this phenomenon firmly in a historical context and also in the context of specific industrial sectors. Robinson Crusoe is of little help, and a pure hypothetico-deductive approach is impotent without a preliminary process of observation and description. This is the purpose of Part One. It is designed to illustrate the three basic aspects of the rise of the professionalized industrial R&D system discussed above – growing complexity of technology, increased scale of processes and specialization of scientific work. Such historical description is of course intended to lead to the generation and examination of hypotheses in a systematic manner.

The whole of Part Two is devoted to an examination of the empirical evidence which might be held to support or refute various contemporary theories of innovation, particularly in relation to firm behaviour. The evidence which is used includes both the historical material cited in Part One and additional studies which have a bearing on the problems. The main concern of Part One is with description and historical context, Part Two with micro-level analysis and Part Three with macro-economic aspects of technical change and innovation, i.e. 'national systems of innovation' and international trade and technology flows. Finally, Part Four deals with some issues of public policy. Some readers may wish to skip the historical detail contained in Part One, but they will find that Parts Two and Three sometimes revert to cases cited in Part One for illustration and support.

Part One (Chapters 2–7) deals in a historical–descriptive manner with research, invention and innovation in the waves of technical change which Schumpeter described as 'successive industrial revolutions' (Table 1.3). He followed the Russian economist Kondratieff (1925) in describing these long, roughly half-century phases of development as 'cycles' but most economists have preferred to call them 'waves' or 'phases' of growth. The expression cycle carries too much of a deterministic flavour for what is a rather variable and imprecise phenomenon. Many economists, including Jevons, Pareto and Dupriez had discussed these long-term fluctuations in the economy in terms of price trends or variations in interest rates or trade flows. Schumpeter (1939) and Van Gelderen (1913) were the first to suggest that these long waves were due to the introduction of major new technologies into the economic system.

Table 1.3 illustrates this Schumpeterian conception of long waves based on successive technological transformations but it does not strictly follow Schumpeter's own work. He urged his successors not to follow his scheme precisely but to build on the results of new research and we have followed his advice.

In his major work on *Business Cycles*, Schumpeter (1939) accepted the reality of the phenomenon of 'Kondratieff'[5] long cycles, lasting half a century or so, and offered a novel explanation of them, differing from that of Kondratieff (1925) himself. According to Schumpeter (1939, Chapter 2), each business cycle was unique because of the variety of technical innovations as well as the variety of other historical events such as wars, gold discoveries or harvest failures. But despite his insistence on the specific features of each fluctuation and perturbation, he believed that the task of economic theory was to go beyond a mere catalogue of accidental events, and analyse those features of the system's behaviour which could generate fluctuations irrespective of their specific and variable form. The most important of such features in his view was innovation, which, despite its great specific variety, he saw as the main engine of capitalist growth and the source of entrepreneurial profit.

The ability and initiative of entrepreneurs (who might or might not themselves be inventors but more usually would not be) created new opportunities for profits, which in turn attracted a 'swarm' of imitators and

Table 1.3 Successive waves of technical change

	Long waves or cycles		Key features of dominant infrastructure			
Approx. timing	Kondratieff waves	Science technology, education and training	Transport communication	Energy systems	Universal and cheap key factors
First 1780s–1840s	Industrial revolution: factory production for textiles	Apprenticeship, learning by doing, dissenting academies, scientific societies	Canals, carriage roads	Water power	Cotton
Second 1840s–1890s	Age of steam power and railways	Professional mechanical and civil engineers, institutes of technology, mass primary education	Railways (iron), telegraph	Steam power	Coal, iron
Third 1890s–1940s	Age of electricity and steel	Industrial RD labs, chemicals and electrical, national laboratories, Standards laboratories	Railways (steel), telephone	Electricity	Steel
Fourth 1940s–1990s	Age of mass production ('Fordism') of automobiles and synthetic materials	Large-scale industrial and government RD, mass higher education	Motor highways, radio and TV, airlines	Oil	Oil, plastics
Fifth 1990s–?	Age of microelectronics and computer networks	Data networks, RD global networks, lifetime education and training	Information highways, digital networks	Gas/oil	Microelectronics

improvers to exploit the new opening with a wave of new investment, generating boom conditions. The British industrial revolution was a clear example of this process and was viewed by Schumpeter as the first Kondratieff wave. It is analysed in some detail in Chapter 2 since it was the starting-point of the entire historical process with which we are concerned.

However, in this first period, mechanization was largely based on water power and confined mainly to the textile industries. It was in the second Kondratieff wave that the widespread diffusion of steam power made possible the mechanization of many other industries and the development of the new railway infrastructure. Although these changes required many more engineers and new craft workers and the spread of literacy in the population, it was not until the rise of the electrical industry (Chapter 3) and the technical transformation of the chemical industry (Chapters 4 and 5) that the professional industrial R&D department became a key institution in the development of new products and processes (Table 1.3). Its importance grew still further with the worldwide diffusion of automobiles and petrochemical based products described in Chapters 5 and 6. Finally, the last few decades of the twentieth century have been characterized by the computerization of the economy based on cheap microelectronics (Chapter 7).

In Schumpeter's theory therefore the 'successive industrial revolutions' were based on the qualitative transformation of the economy by new technologies, rather than the simple quantitative growth of individual industries.

Whether or not such a theory offers a plausible explanation of 'long' waves in economic development depends crucially – as Kuznets (1940) pointed out in his review of Business Cycles at the time – on whether some innovations are so large and so discontinuous in their impact as to cause prolonged perturbations or whether they are bunched together in some way. The construction of a national railway network might be the type of innovative investment which would qualify as a 'wave generator' in its own right, but obviously there are thousands of minor inventions and technical changes which are occurring every year in many industries whose effect is far more gradual and which might well adapt to some sort of smooth equilibrium growth path. If these smaller innovations were to be associated with economic fluctuations, then this could only be if they were linked to the growth cycles of new industries and technologies.

Our account differs from Schumpeter's own account in his book on *Business Cycles* in several important respects. First, of course, Schumpeter himself died soon after the Second World War so that he only analysed the first three waves. The notes on the fifth and sixth waves in Table 1.3 are partly speculative although the speculation on the fifth is fairly well grounded. Second, this table is based on the large-scale *diffusion* of new technology systems, not on their first introduction. Schumpeter himself discussed the steam engine mainly in relation to the first Kondratieff wave and steel in relation to the second. In both cases of course the very first innovations came even earlier. However, the standpoint adopted in this book is that what matters for a major upswing and transformation of the economy in terms of new investment and employment is the widespread diffusion of numerous innovations based on a new infrastructure. The

Table 1.4 Boulton and Watt
engines by horsepower, c.1800

Horsepower	cost per HP £
2	89
10	40
20	30

Source: von Tunzelmann (1978, p. 51).

previous gestation period for this new infrastructure and a corresponding cluster of innovations can be several decades. Thus, whereas Schumpeter (quite correctly) spoke of the innovations in automobile production and especially the internal combustion engine in the period from the 1880s to the 1940s, we take the age of mass production and universal use of the automobile as the 'fourth Kondratieff'.

The first steam engines (especially the Newcomen engine) were in use in European coal mines quite early in the eighteenth century but they were confined to pumping applications in the mines. Even with Watt's greatly improved engine towards the end of the century, the number of applications was still very limited as von Tunzelmann (1978) showed in his book on *Steam Power and British Industrialisation*. The mills and factories of the first British industrial revolution mainly used water power not steam. The widespread diffusion of the steam engine in the second Kondratieff wave (Table 1.3) depended on three trajectories:

1. The fall in cost per horsepower with increasing size of steam engine (Table 1.4).
2. The reduction in coal consumption per HP in the new high pressure engines mainly developed in the Cornish mining industry (Table 1.5).
3. The improvements in design of railway locomotives and the rapid increase in their use for the transport of people and goods from 1825 onwards.

This example has been chosen because all of these trends were mutually reinforcing. The huge improvements in transport greatly reduced the price of coal in the key industrial areas where the new industries were growing most rapidly (Table 1.6). The falling costs of steam power facilitated its application in many other industries in addition to cotton (although cotton still accounted for one-third of total UK manufacturing horsepower as late as 1870). An attempt has been made in each of the historical chapters to illustrate this sort of interdependence of technical and economic change and the interdependence of many innovations themselves. Innovations are systemic in nature (Gille, 1978; Hughes, 1982), not isolated events. It is this systematic economic and technological interdependence which gives rise to the 'lock-in' effects of each dominant style in technology.

Each of these technological revolutions was based on clusters of innovations, some of them involving big changes and discontinuities ('radical' innovations) and others involving many small improvements ('incremental'

Table 1.5 Coal consumption in various types of steam engine in manufacturing applications (lbs of coal per hour per HP)

Savery engine (18th century)	30
Newcomen engine (mines) (1700–1750)	20–30
Newcomen engine (1790)	17
Watt low pressure engines (1800–1840)	10–15
High pressure engines (1850)	5

Source: von Tunzelmann (1978, pp. 68–70).

Table 1.6 Coal prices in Britain by region, 1800–1850 (shillings per ton)

	London	Birmingham	Manchester
1800	46	9	16
1810	38	12	13 (1813)
1820	31	13	10 (1823)
1830	26	6 (1832)	10 (1833)
1840	22	8	7 (1841)
1850	16	5	6

Source: von Tunzelmann (1978).

innovations). The selection of innovations which are discussed are not of course more than a small fraction of the total; they have been selected to illustrate some of the main features of each phase of historical development.

1.6 STRUCTURE OF THE BOOK

It is not possible in Part One to describe any of the innovations fully, as each one would merit a book in its own right. Some of the books which have been written are cited in the references. The intention here is to select only some of the most important characteristics of the innovations for discussion, from the standpoint of the economist. The treatment of technical aspects of the innovations is minimal, and so is the treatment of the personal characteristics of the inventors and innovators. Attention is concentrated on such questions as scale of effort, patents, size of firm, marketing and time lags. What kind of firms made the principal innovations? At what stage and in which industries were they made? Were they the result of professionalized R&D? How long did it take to develop and launch the new products and processes? How much did it cost? What were the expectations of management and the pressures which led to the decision to innovate? What are the implications for the theory of the firm?

Although the approach concentrates on the economic aspects, this does not mean that technical, psychological and social aspects of innovation are unimportant. Such an attitude would obviously be absurd. It would be a fair criticism that a more integrated theory of innovation is desirable, but it is beyond the scope of this book. However, some of the wider social issues involved in policy for technical innovation are discussed in the concluding chapters.

The largely descriptive historical treatment of technical innovation in Part One is followed by an analytical treatment of some of the general implications for innovation theory in Parts Two and Three. Chapters 7 to 11 are concerned with problems of the firm in relation to innovation.

In the analytical section it is argued that the professionalization of R&D described in Part One had far-reaching consequences on the nature of the competitive struggle between firms, both on the national and the world market. The factors which lead to success or failure in this new type of competitive struggle are discussed in Chapter 8, and the implications for size of firm in Chapter 9. In general the growth of industrial R&D has favoured the large firm and has contributed to the process of industrial concentration, but small new firms retain an advantage in some types of innovation. The giant international corporation has the great advantage of being able to spread the very high development costs of some kinds of innovation and the associated technical services over a very large sales volume. This is an enormous asset in industries such as telecommunications, turbine generators, refineries, aircraft and drugs. But a high degree of uncertainty remains characteristic of technical innovation whether in large or small firms. The problems for the firm in coping with this high degree of uncertainty in managing innovation are discussed in Chapter 10.

The type of groping and experimental decision-making characteristic of the innovation process is not compatible with theories of the firm which postulate a high degree of accuracy in investment calculations or extensive foreknowledge of the consequences of the firm's behaviour. The uncertainty associated with innovation is such that differences of opinion about the desirability of alternative projects and strategies are the norm rather than the exception. This means that the firm is typically the arena of political debate between the advocates of alternative courses of action, and that power struggles will take place around these issues.

This leads to some reconsideration of the theory of the firm in Chapter 11. The firm attempts to use R&D and other scientific and technical services to reduce the uncertainty which confronts it. But the nature of R&D is such that technical and market uncertainties remain despite its best efforts. Some types of R&D may indeed increase the uncertainty. Consequently, a high degree of instability will remain and decision-making in the firm will continue to resemble a process of 'muddling through' rather than the ordered, rational calculation beloved of neoclassical theory.

The analysis in Part Three moves from the micro to the macro level. The historical analysis in Part One and the theoretical analysis in Part Two show that the performance of firms in their innovative efforts is strongly related to the institutional environment in which they operate. Countries have varied greatly in their rates of economic growth over the last two centuries. The leading countries which forged ahead in the nineteenth and twentieth centuries opened up a huge gap in living standards with the less developed countries of Africa, Asia and Latin America. Many European countries closed this gap in the twentieth century and more recently some East Asian countries have begun to do so. These efforts to catch up and close the gaps in living standards depend heavily on closing the gaps in *technology*. The chapters in Part Three analyse the process of forging ahead, falling behind and catching up in economic growth and the ways in which national performance relates to the transfer of technology, international investment flows and the 'national system of innovation' within each country.

Finally, Part Four discusses the responsibilities of government for science, technology and innovation. During the last half-century governments have increasingly accepted some responsibility, not only for some aspects of R&D and other STS but also for some forms of technology assessment, that is for comprehensive social cost benefit analysis of the probable consequences of technical change. The socialization of some of the risks and uncertainties of technical innovation is difficult to avoid because of the pressures of world competition, externalities and scale factors in R&D, and some of the adverse consequences of '*laissez-innover*'. Such socialization, however, carries with it the responsibility for the development of an explicit rather than an implicit national policy for science and technical innovation. Some problems associated with this major government responsibility are discussed in Part Four.

It is argued there that in the USA, the USSR, France and Britain the priorities of the 1945–89 period were largely determined by the Cold War. Government support for aircraft, nuclear and electronics R&D was both massive and effective. Firms in these industries became part of a special military–industrial complex, in which state-supported innovation was normal. Quite different priorities should be established in the next century and national policy should be concerned to promote other kinds of innovation. A great deal of R&D will be needed to cope with environmental problems, to secure long-term supplies of renewable cheap energy, to deal with natural resource limitations, to promote full employment, to develop much better transport and construction systems and generally to improve the quality of life in industrialized countries. Even more critical is R&D to deal with problems of underdevelopment. This redeployment of scarce R&D resources to meet the most urgent priorities is unlikely to occur solely as the result of short-term market factors. It must therefore be the main concern of national policy for science and technology, and increasingly of international policy.

NOTES

1. Strictly speaking, as the word itself implies, technology is simply a body of knowledge about techniques. But it is frequently used to encompass both the knowledge itself and the tangible embodiment of that knowledge in an operating system using physical production equipment. In this book the expression 'technical innovation' or simply 'innovation' is used to describe the introduction and spread of new and improved products and processes in the economy and technological innovation to describe advances in knowledge.

2. For the changing connotation of the word 'technology' see Ezrahi *et al.* (1995, p. 17).
3. The establishment of the Massachusetts Institute of Technology in 1861 was a landmark in the use of the word.
4. See Ezrahi *et al.* (1995) *Technology, Pessimism and Postmodernism.*
5. As has often been pointed out, Kondratieff was by no means the originator of the long cycle theory and it is in some respects a misnomer that the phenomenon bears his name. The Dutch Marxist van Gelderen could be much more fairly credited with the idea, which he articulated clearly in 1913. At about the same time a variety of economists, including Pareto (1913), had drawn attention to the apparent tendency for long-term price movements, interest rates and trade fluctuations to follow a cyclical movement lasting about half a century. However, during the 1920s while heading the Institute of Economic Research in Moscow, Kondratieff did more to propagate and elaborate the idea than any other economist. For a selection of papers illustrating the controversies surrounding Schumpeterian theories of long waves see Freeman (1996).

REVIEW ARTICLES, LITERATURE SURVEYS AND KEY REFERENCES

These references include some major books which cover the whole subject of the economics of technical change, including innovation. Some of these references are also included at the end of the relevant Parts.

Boyer, R. (1997) *Les Systèmes d'Innovation à l'Ère de la Globalisation*, Paris, Economica.

Chandler, A.D. (1992), 'What is a firm?: a historical perspective', *European Economic Review*, vol. 36, pp. 483–494.

Coombs, R., Saviotti, P. and Walsh, V. (1987) *Economics and Technological Change*, London, Macmillan.

DeBresson, C. (1987) *Understanding Technological Change*, Montreal, Black Rose Books.

DeBresson, C. (1996) *Economic Interdependence and Innovation Activity*, Cheltenham, Elgar.

de la Mothe, J. and Paquet, G. (eds) (1996) *Evolutionary Economics and the New International Political Economy*, London, Cassell.

Dodgson, M. and Rothwell, R. (eds) (1994) *The Handbook of Industrial Innovation*, Aldershot, Elgar.

Dosi, G. (1988) 'Sources, procedures and micro-economic effects of innovation', *Journal of Economic Literature*, vol. 36, pp. 1126–71.

Dosi, G., Freeman, C., Nelson, R., Silverberg, G. and Soete, L. (eds) (1988) *Technical Change and Economic Theory*, London, Pinter.

Edquist, C. (1997), *Systems of Innovation*, London, Pinter.

Foray, D. and Freeman, C. (eds) (1993) *Technology and the Wealth of Nations*, London, Pinter.

Freeman, C. (ed.) (1990) *The Economics of Innovation*, International Library of Critical Writings in Economics, vol. 2, Aldershot, Elgar.

Freeman, C. (1994) 'The economics of technical change: a critical survey', *Cambridge Journal of Economics*, vol. 18, pp. 463–514.

Freeman, C. and Soete, L. (eds) (1992) *New Explorations in the Economics of Technical Change*, London, Pinter.

Gomulka, S. (1990) *The Theory of Technical Change and Economic Growth*, London, Routledge.

Granstrand, O. (ed.) (1994) *Economics of Technology*, Amsterdam, North Holland.

Grupp, H. (1997), *Messung und Erklärung des technischen Wandels: Grundzüge einer empirischen Innovationsökonomik*, Berlin, Springerverlag.

Hodgson, G.M. (ed.) (1995) *Economics and Biology*, The International Library of Critical Writings in Economics, Vol. 50, Aldershot, Elgar.

Kennedy, C. and Thirlwall, A.P. (1971) 'Technical progress', *Surveys in Applied Economics*, 1, pp. 115–77, London, Macmillan.

Landau, R. and Rosenberg, N. (eds) (1986) *The Positive Sum Strategy: Harnessing Technology for Economic Growth*, Washington, National Academy Press.

Mackenzie, D. (1992) *Economic and Sociological Exploration of Technical Change*, in Coombs, R., Saviotti, P. and Walsh, V. (eds), *Technical Change and Company Strategies: Economic and Soliological Perspectives*, London: Academic Press, pp. 25–48.

Mansfield, E. (ed.) (1993) *The Economics of Technical Change*, International Library of Critical Writings in Economics, vol. 31, Aldershot, Elgar.

Mansfield, E. (1995) *Innovation, Technology and the Economy, Selected Essays*, 2 vols, Aldershot, Elgar.

Metcalfe, J.S. (1997) *Evolutionary Economics and Creative Destruction*, London, Routledge.

Nelson, R. (1996) *The Sources of Economic Growth*, Cambridge, MA, Harvard University Press.

OECD (1992) *Technology and the Economy: The Key Relationships*, Paris, OECD.

Rosenberg, N. (1994) *Exploring the Black Box: Technology, Economics and History*, Cambridge, Cambridge University Press.

Saviotti, P. and Metcalfe, J.S. (eds) (1991) *Evolutionary Theories of Economic and Technological Change*, Chur, Harwood.

Stoneman, P. (ed.) (1995) *Handbook of the Economics of Innovation and Technological Change*, Oxford, Blackwell.

Tidd, J., Bessant, J. and Pavitt, K. (eds) (1997) *Managing Innovation: Integrating Technological, Market and Organizational Change*, Chichester, Wiley.

Utterback, J. (1993) *Mastering the Dynamics of Innovation*, Boston, Harvard Business School Press.

Witt, U. (ed.) (1991) *Evolutionary Economics*, The International Library of Critical Writings in Econonomics, Vol. 25, Aldershot, Elgar.

THE RISE OF SCIENCE-RELATED TECHNOLOGY

INTRODUCTORY NOTE

The first part of this book is today still a little unusual for a book on economics, although it was common in the early days of the subject. It starts with history in line with Schumpeter's objective of putting history back into economics. Part One aims to give a condensed historical account of the major waves of technical change which started with the Industrial Revolution in Britain in the late eighteenth and early nineteenth centuries. There are several reasons for this approach.

First of all, recent developments in economic theory have converged in their recognition of path dependence in an evolutionary approach. In many ways, this marks a return to the tradition of the classical economists, including Adam Smith and Marx, who both devoted a great deal of attention to the evolution of the economy and of social institutions.

An evolutionary approach is important above all in the study of technical change. Almost all economists agree that this is one of the main sources, if not the main source, of the dynamism of capitalist economies, of their growth, and of their instability. Technical innovation contributes to the everlasting uncertainty and evolutionary turmoil, which are so characteristic of capitalism. The growth of capitalist firms, industries and nations is not just a matter of the quantitative increase of inputs and outputs, important though this undoubtedly is, but of the qualitative transformation of the structure of the economy through successive waves of technical change.

As already indicated in Chapter 1, our account follows Schumpeter in his theory of 'successive industrial revolutions' or long waves in economic development, although with modifications arising from recent research. The chapters in Part One show that historians of technology have been right to insist on the systemic interdependencies of myriad technical and organizational innovations. Like Hamlet's troubles, they come not singly but in battalions. Process innovations, product innovations, organizational innovations and materials innovations are all interdependent in mechanization, electrification or computerization.

Our historical account is necessarily selective but it seeks to illustrate these systemic features of the innovative process as well as describing other characteristics of important innovations. The account starts with the Industrial Revolution since this marked the transition to an accelerated rate of technical change and of economic growth, at first in Britain and later in many other countries. Chapter 2 describes some of the archetypal entrepreneurs of the Industrial Revolution – Josiah Wedgwood, Richard Arkwright and Samuel Crompton – and the main innovations for which they were responsible. This focus on a few of the most innovative entrepreneurs may be justified in terms of the characteristic pattern of small firm invention and innovation of this period but the chapter also includes a brief analysis of some of the systemic features of the dominant fast-growing industry of the time – the cotton industry.

In Chapter 3, the focus shifts to the United States, which was by far the fastest growing industrial economy of the second half of the nineteenth and early part of the twentieth century. Our account first concentrates on innovation in steel, the essential material for innumerable new industrial products and processes and for improvements in the older ones. Steel was also closely linked to electrification, which transformed many of these processes through its flexibility. Electricity facilitated both the relocation and the reorganization of many industries as well as leading to many changes in the home through new domestic appliances. A major feature of the growth of the American steel industry and of the electrical engineering industry was the link between technical innovation and economies of scale. Giant firms, such as United Steel and General Electric, became characteristic institutions of the twentieth-century American economy. Chapter 3 endeavours to show the links between size of firm and numerous managerial innovations, including the rise of inhouse industrial R&D.

Chapters 4 and 5 show that similar scale economies based on technical process innovations and managerial innovations led also to the emergence of giant firms in the oil and chemical industries, both in the United States and in Germany. These chapters concentrate especially on the new industrial R&D activities of these firms and on the plant design and construction activities of their contractors. They also serve to introduce some of the main problems of measurement of invention and innovation (the outputs of R&D).

Whereas in the industries producing basic materials huge economies were achieved by the scaling up of flow processes, in the production of automobiles and consumer durables this was achieved by the innovation of the moving assembly line. Chapter 6 describes this innovation at Ford's Detroit plant and its subsequent worldwide diffusion. The story concludes with the major modification of the Fordist system by the Japanese automobile industry. The Toyota–Ohno 'lean production' system dramatically increased the productivity and market shares of the Japanese producers in the world market and led to determined efforts by American and European firms to imitate some of its main features.

The final chapter in Part One deals with the most revolutionary innovation of the twentieth century – the electronic computer. Starting with radio, television and radar, the chapter outlines the history of electronic technology, from its modest beginnings to the most pervasive technology of the twentieth century. Underlying the extraordinarily rapid growth of the electronic industry were the successive generations of electronic components, culminating with the integrated circuits combining millions of components on one chip. Scale economies were thus an essential part of this story too.

Although the story in Part One brings out the combined effect of technical innovations, organizational innovations and scale economies in the growth of industrial R&D and of large firms, it also demonstrates the vitality of small firms. Indeed, the account shows that in the early stages of new technologies and industries, small firms have played a key role. Part

Two turns from this historical account to a more systematic analysis of the role of large and small firms in different types of innovation and the conditions for success in innovation. The function of Part One is to provide a realistic foundation for the more systematic discussion in Parts Two and Three and to demonstrate which features of the system have changed and which have been relatively continuous throughout the history of capitalist economies.

CHAPTER 2

THE INDUSTRIAL REVOLUTION

Adam Smith wrote his famous book on the *Wealth of Nations* at the beginning of the Industrial Revolution in Britain. He had travelled in many European countries and attempted to explain in his book why the British standard of living was higher than in other European countries. Unlike the French physiocrats who attributed the growth of national income to agricultural productivity, he concentrated his explanation on manufacturing industry and trade. The division of labour in manufacturing facilitated the use of new machines and the accumulation of specialized skills on the part of the operatives. The opening of markets and the reduction of barriers to trade within and between countries enabled these manufacturers to compete, to enlarge their market and to enjoy scale economies for their products, as well as permitting still further division of labour.

Smith based his theory firmly on his observations of the actual changes taking place in Britain in the 1760s and 1770s, as in his famous example of the workshop manufacturing pins. In terms of total output and employment, the physiocrats were right in thinking that agriculture was still far more important than manufacturing, both in France and in Britain, but Smith rightly perceived that the productivity of manufacturing could grow more rapidly and lead to a more opulent society because of technical change, capital accumulation and specialized skills. Almost all historians agree with Adam Smith about the importance of these factors in explaining that prodigious spurt in the growth of the British economy which is generally described as the Industrial Revolution and which took off at the time of the appearance of his book in 1776.

Historians and economists differ to some extent on the relative emphasis which they give to each one of these factors and to others in their explanations. Friedrich List (1841) believed that Smith underestimated the importance of science and technology and overstated the division of labour argument (see Chapter 12). Almost all historians agree, however, in rejecting a single factor explanation and stress the interdependence of technical, economic, political and cultural change. Supple is fairly typical in his admirably terse summary:

> Britain's economic, social and political experience before the late 18th Century explains with relatively little difficulty why she should have been an industrial pioneer. For better than any of her contemporaries Great Britain exemplified a combination of potentially growth-inducing characteristics. The development of enterprise, her access to rich sources of supply and large overseas markets within the framework of a dominant trading system, the accumulation of capital, the core of industrial techniques, her geographical position and the relative ease of transportation in an island economy with abundant rivers, a scientific and pragmatic heritage, a stable political and relatively flexible social system, an ideology

favourable to business and innovation – all bore witness to the historical trends of two hundred years and more, and provided much easier access to economic change in Britain than in any other European country.

(Supple, 1963, p. 14)

2.1 TECHNICAL INNOVATIONS AND THE INDUSTRIAL REVOLUTION

In this book we concentrate on technical innovation and its relationship with science for reasons advanced in the introductory chapter. However, this does not mean that we deny or underestimate the importance of the other factors mentioned by Supple. The embodiment of inventions in new machines through capital investment was clearly essential for the success of the Industrial Revolution to take only the most obvious example. Schumpeter included in his definition of 'innovations' not only technical innovations, but also marketing and organizational innovations. As we shall see, these were all interdependent in the Industrial Revolution, as in later waves of technical change. The revolution involved a very fundamental organizational change from a system of cottage production of textiles to a system of factory production and this could not take place without political change and conflicts as well as cultural changes, such as the work discipline of factory hours and supervision.

Precise dating of the Industrial Revolution is impossible and some authors stress the changes which had already taken place early in the eighteenth or even in the seventeenth century, such as the Scientific Revolution and the defeat of the monarchy and feudal forces in the English Civil War of the 1640s. However, economic historians appear to agree that there was a fairly sharp acceleration of British industrial output, investment and trade in the last two decades of the eighteenth century. Hoffmann calculated the rate of growth of British industrial output from 1700 to 1780 as between half and one per cent per annum, but from 1780 to 1870 at more than 3 per cent. More recent estimates (Crafts, 1994) have reduced the estimated growth rates for the later period but do not change the fundamental picture (Table 2.1). Supple sums up the consensus as follows:

Economic change did not experience a steady acceleration, rather there was a more or less precise point (which most historians place in the 1780s) after which innovation, investment, output, trade and so forth all seemed to leap forward.

(Supple, 1963, p. 35)

Although it was the surge of growth in industry in the late eighteenth century which was the principal component of the acceleration in British economic growth, Phyllis Deane (1965) estimated that the rate of growth in national income as a whole over the period from 1800 to 1860 was twice as high as the rate from 1740 to 1800. This marked a transition to a sustained rate of economic growth over a long period greater than any which had ever been previously achieved in human history. It is for this reason that the British Industrial Revolution merits intense study, even though the

Table 2.1 Sectoral growth of real industrial output (per cent per annum) in Britain

Year	Cotton	Iron	Building	Weighted average*
1700–60	1.37	0.60	0.74	0.71
1770–80	6.20	4.47	4.24	1.79
1780–90	12.76	3.79	3.22	1.60
1790–1801	6.73	6.48	2.01	2.49
1801–11	4.49	7.45	2.05	2.70
1811–21	5.59	–0.28	3.61	2.42

* 1700–90 based on 1770 weights
1790–1821 based on 1801 weights, including other industries

Source: Crafts (1994).

growth rate has since been surpassed; and it is for this reason that this book starts with a necessarily brief account of this transition.

Although we concentrate attention on the manufacturing industry and the transport infrastructure, this does not mean that technical change in agriculture was unimportant. On the contrary, productivity in British agriculture continued to increase throughout the eighteenth century, although at a slower rate than in industry. Moreover, many historians point to the social and economic changes in agriculture as one of the main factors facilitating mobility of labour and of capital, which were essential for industrialization. In those countries which are industrializing today, especially in Asia, even though manufacturing industry is increasing its output and productivity much more rapidly than the agricultural sector, in the most successful cases agriculture has also continued to raise output and rural incomes. The successful land reform in Korea and Taiwan soon after the Second World War contrasted with the lack of land reform in several Latin American countries and was a major factor in the subsequent contrasting performance of those national economies. If therefore we pass over the numerous technical, organizational and social changes in agriculture it is for reasons of space only.

The surge of growth in British industry was not simply balanced growth of all industries simultaneously but was characterized by the exceptionally rapid growth of a few leading sectors, above all the cotton industry and to a lesser extent iron (Table 2.1). The share of cotton in total value added of industry grew from 2.6 per cent in 1770 to 17 per cent in 1801. The exceptional role of the cotton textile industry has been generally acknowledged both by contemporaries and by historians ever since.

In the initial decades of the British industrial revolution it was the cotton textile industry which experienced the most spectacular expansion. Subsequently, after 1840 railroad investment and the spread of a transportation network seemed to dominate the economy and in the third quarter of the century, the steel industry and steamship construction leapt ahead.

(Supple, 1963, p. 37)

Imports of raw cotton grew from an average of 16m. lbs pa in 1783–1787 to 29m. lbs in 1787–92 and 56m. lbs in 1800 as the source changed from the

West Indies to the United States slave plantations. The rate of increase in imports was described by a nineteenth-century historian (Baines, 1835) as 'rapid and steady far beyond all precedent in any other manufacture'.

Baines attributed the extraordinary rise in the 1770s and 1780s directly to the effects of technical inventions and their diffusion: 'From 1771 to 1781, owing to the invention of the jenny and the water-frame, a rapid increase took place; in the ten years from 1781 to 1791, being those which immediately followed the invention of the mule and the expiration of Arkwright's patent, the rate of advancement was prodigiously accelerated.'

It was on the basis of a whole series of inventions and improvements (Hills, 1994; Mann, 1958; von Tunzelmann, 1995) that big increases in productivity became possible, based on their exploitation in the new system of factory (mill) based production. These improvements in process technology made possible the rapidly falling prices which in turn provided the competitive strength for British exports to undercut Indian and other Asian textiles and indeed all other producers. Exports of cotton textiles reached 60 per cent of output by 1820 and became the biggest single commodity in nineteenth-century trade, accounting still for over 30 per cent of British exports of manufactures in 1899, when Britain was still by far the biggest exporter. Whereas in the twentieth century British governments worried about the decline of the Lancashire cotton industry and the loss of employment because of Asian competition, it was the other way about in the nineteenth century. Cotton textiles thus represent the first big example in modern times of that influence of innovation on trade performance which is discussed in depth in Chapter 15.

Carlota Perez (1983) has pointed especially to the role of rapidly falling prices of key production factors in successive industrial transformations or long waves of economic development, as with steel in the late nineteenth century, or oil in the twentieth. The most obvious case is of course the contemporary orders of magnitude reduction in the price of chips for a myriad of microelectronic devices, and especially for computing. While not quite so spectacular, the fall in the price of cotton yarn was certainly remarkable, occurring as it did in the inflationary period of the Napoleonic Wars. The price of No. 100 Cotton Yarn fell from 38/- in 1786 to 6/9d in 1807.

Virtually all accounts, whether contemporary or otherwise, agree on the importance of inventions, both in the cotton industry and in other industries for the spurt in economic growth. Indeed they were often given pride of place in the older textbooks on English history. Like Adam Smith (1776) recent studies stress the continuous improvement of processes in the factory or workplace, as well as the original major inventions. They also stress the speed with which inventions became innovations and were then rapidly diffused. The number of patents sealed had been about 80 in the period 1740–49 but increased to over 100 in 1750–59 and to nearly 300 in 1770–79. Patents are an imperfect indicator but there were no changes in this period which might invalidate the series (Eversley, 1994).

There is some disagreement on the nature of the major inventions of the eighteenth century. Some authors argue that they were typically very simple; they 'leave the impression that the inventions were the work of obscure millwrights, carpenters or clock-makers, untutored in principles, who stumbled by chance on some device'. Ashton, however, argues that 'such accounts have done harm by obscuring the fact that systematic thought lay behind most of the innovations in industrial practice' and overstressed the part played by chance (Ashton, 1963, p. 154). Further: 'Many involve two or more previously independent ideas or processes, which brought together in the mind of the inventor issue in a more of less complex and efficient mechanism. In this way, for example, the principle of the jenny was united by Crompton with that of spinning by rollers to produce the mule.'

At the opposite extreme some accounts give the impression that the inventions were the result of individual genius or scientific brilliance, rather than the outcome of a continuous social process and mechanical ingenuity. In part, these differences of interpretation arise from the fact that (as still today) there is a very wide spectrum of inventions and innovations. The vast majority, then and now, are incremental improvements to existing processes and products and, as Adam Smith observed, are often made by workers who use machines in different types of workplace. They may be facilitated by specialization based on division of labour, but again, as Adam Smith observed, still other inventions result from the work of scientists whose skill is to observe dissimilar processes. Hills (1994), basing his comments on experience of actually running the machines in the North Western Museum of Science and Industry, stresses the trajectory of improvement exploited by Hargreaves and Arkwright:

> As with most inventions, their work must not be taken in isolation since it was dependent on what had been done before . . . the various methods of spinning by hand . . . are more closely linked than has hitherto been realised to the ways of spinning used by Hargreaves, Arkwright and their predecessors, Paul and Wyatt.
>
> (Hills, 1994, p. 112)

Nevertheless, the combined effect of the inventions of Hargreaves, Arkwright, Crompton and their predecessors and successors was revolutionary rather than gradual (Table 2.2). The leap in productivity at the end of the eighteenth century reduced the number of operative hours to process (OHP) 100lbs of cotton by much more than an order of magnitude. The power required to operate the later innovations meant that machinery had to be installed in purpose-built premises (factories). Even though Arkwright limited his licences to machines of a thousand spindles, human muscle and horse power were rapidly succeeded by water power and later by steam (Jenkins, 1994). Later in this chapter we shall describe his innovations in more detail. Here we present only a summary of the main features of his and other textile innovations.

In a highly original analysis of innovations in the cotton industry in the Industrial Revolution, von Tunzelmann (1995) provides strong evidence that the main inducement for innovators was time-saving and that the savings in fixed and working capital, in labour and in land were the indirect result of this time-saving objective, pursued within a general paradigm of relatively straightforward mechanization. He also brings out the role of focusing devices and co-ordination in the whole production

Table 2.2 Labour productivity in cotton:
operative hours to process 100lbs of cotton

	OHP
Indian Hand spinners (18th century)	50,000
Crompton's mule (1780)	2,000
100-spindle mule (c.1790)	1,000
Power-assisted mules (c.1795)	300
Roberts' automatic mule (c.1825)	135
Most efficient machines today (1990)	40

Source: Jenkins (1994, p. xix).

system. 'Replication of the particular components which represented the most constrictive bottlenecks was often carried out in addition to speeding them up. The cylinder for block printing could thus be replicated by up to five times' (Baines, 1835, p. 236). 'The same innovation strategy underlay the jenny, which multiplied the traditional spinning wheel initially to 8 and eventually to sometimes 120 within the one machine' (ibid., p. 15). Von Tunzelmann also quotes Baines on the productivity increases brought about by 'a series of splendid innovations and discoveries, by the combined effect of which a spinner now produces as much yarn in a day as by the old processes he [sic] could have produced in a year, and cloth which formerly required six or eight months to bleach, is now bleached in a few hours' (ibid., p. 8). The case of bleaching differs from the mechanical inventions in textile machinery since it was directly based on advances in chemistry.

The classical economists (Smith, Ricardo, Mill, Marx) put the accumulation of capital centre stage in their analysis of economic growth. While certainly not denying its importance, economic historians tend to reduce their estimates of its rate of increase in the early stages of the Industrial Revolution. Whereas Rostow (1960) had argued that 'take-off' into 'self-sustained' growth required a rise in productive investment from 5 per cent of NNP to 10 per cent and that Britain had met this condition at the end of the eighteenth century, Crafts (1994) estimates gross domestic investment as only 7.9 per cent of GDP in 1801. However, this slower rate of increase is still quite consistent with the very small size of firms at this time, the relatively low cost of the new machines and the very low rate of interest from the mid-eighteenth century onwards (the rate on consols was 3 per cent from 1757).

Although Britain already had a quite developed capital market in the eighteenth century, the cotton masters and other early industrialists usually had to acquire their capital from local sources, family, friends, country banks, and ploughed back profits. Boulton and Watt were among the many firms that obtained capital in this way when they started to manufacture steam engines at Boulton's Soho works in Birmingham.

The organized capital market and the wealthy class of landlords played a much bigger part in the finance of canals and other infrastructural investment, which grew very rapidly between 1750 and 1800 (Figure 2.1). Hobsbawm (1968) points out that the 'wide scattering' of British industry through the countryside, based on the putting out system, the coal-mining regions, the new industrial textile regions, the 'village industries' and

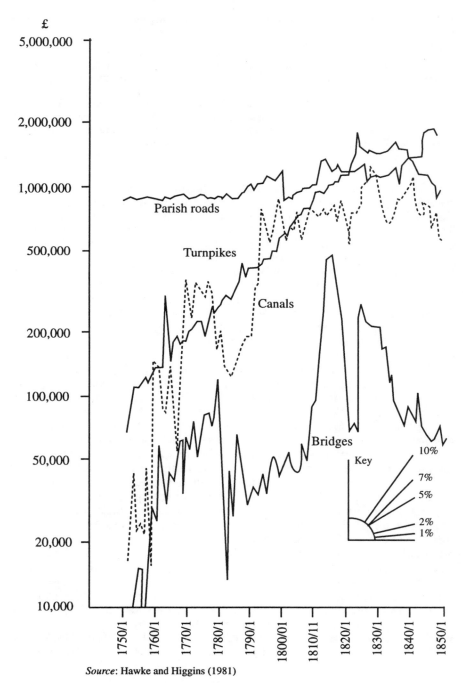

Fig. 2.1 Transport and social overhead capital
Source: Hawke and Higgins (1981).

London as a huge centre of population, trade and services (the largest in Europe) had two major consequences:

> It gave the politically decisive class of landlords a direct interest in the mines which happened to lie under their lands (and from which, unlike the Continent, they rather than the King, drew royalties) and the manufactures in their villages. The very marked interest of the local nobility and gentry in such investments as canals and turnpike roads was due not merely to the hope of opening wider markets to local agricultural produce, but to the anticipated advantages of better and cheaper transport for local mines and manufactures.
>
> (Hobsbawm, 1968, p. 16)

The second consequence was that manufacturing interests could already determine government policy, unlike other European countries and even the Netherlands where merchant and landed interests were still dominant. The oligarchy of landed aristocrats was unlike the feudal hierarchies of other European countries in several ways. They were a 'bourgeois' aristocracy. Hawke and Higgins (1981) estimate that landlords accounted for 40 per cent of total investment in British canals from 1755 to 1780 and later also for a substantial part of railway investment.

Their contribution to investment in the new transport infrastructure was remarkable but the contribution of merchants, manufacturers and professional people also showed that in the eighteenth century Britain already had a capital market capable of financing quite large investments in social overhead capital, essential for the rapid growth of industry and trade in the last quarter of the century. Foreign trade through London, Liverpool, Cardiff, Newcastle, Edinburgh, Hull, Bristol and many smaller ports kept pace with the growth of GDP from 1760 to 1820, despite the disruption of the Napoleonic Wars. Although exports did not lead the Industrial Revolution, being fairly constant at about 15 per cent of output, they were disproportionately important for the fastest growing industries.

Even though artisans or inventor–entrepreneurs and their partners or families often had to struggle to raise the capital for the numerous new firms starting up in cotton and other industries, this highly decentralized pattern imparted great flexibility to the newly emerging industries. Many historians (e.g. Ashton, 1948; Wilson, 1955) stress that social mobility was much greater in Britain than in other countries. The entrepreneurs came from very diverse backgrounds and the role of 'dissenters' (Quakers and adherents of other unorthodox religious denominations) is frequently mentioned. Ashton states that it is not easy to distinguish inventors, 'contrivers', industrialists and entrepreneurs and that they came from every social class and from all parts of the country. Aristocrats like Coke of Holkham Hall innovated in agriculture, or like the Duke of Bridgewater in canals. Clergymen and parsons, such as Cartwright and Dawson, innovated in new ways of weaving cloth and smelting iron. Doctors of medicine, such as John Roebuck and James Keir, took to chemical research and became industrialists.

> Lawyers, soldiers, public servants and men of humbler station than these found in manufacturing possibilities of advancement far greater than those offered in

their original callings. A barber, Richard Arkwright, became the wealthiest and most influential of the cotton-spinners; an inn-keeper, Peter Stubbs, built up a highly esteemed concern in the tile trade; a schoolmaster, Samuel Walker, became the leading figure in the North of England iron industry.

(Ashton, 1963, p. 156)

One reason that Dissenters were so prominent in entrepreneurship may well have been their non-conformist outlook and often their rationalism. However, Ashton also points out that the exclusion of Dissenters from the universities and from office in government forced many to make their careers in industry. Moreover, the non-conformist zeal for education led them to establish their own schools and the non-conformists constituted the better educated section of the middle classes. Presbyterian Scotland provided an unusually high proportion of the leading inventors (Watt and most of his assistants, Sinclair, Telford, Macadam, Neilson and many others) at a time when Scotland had by far the best primary education system in Europe and some of the best universities. 'It was not from Oxford or Cambridge, where the torch burnt dim, but from Glasgow and Edinburgh, that the impulse to scientific enquiry and its practical application came' (Ashton, 1963, p. 157). The Dissenters' academies, established in English towns such as Bristol, Manchester, Warrington and Northampton, did for England what the universities did for Scotland. A significant minority of the most successful entrepreneurs were well acquainted with recent science and often took the trouble to keep up these contacts. For example, Josiah Wedgwood, the leading entrepreneur and innovator in the pottery industry became a Fellow of the Royal Society (FRS) in 1783.

2.2 THE INNOVATIONS AND ENTREPRENEURSHIP OF JOSIAH WEDGWOOD

In this section we shall take the example of Wedgwood to illustrate the behaviour of innovative entrepreneurs in the fastest growing firms in the British Industrial Revolution. His success was so outstanding and his career is so well documented that he is often taken by economic historians as the archetypal example of the British Industrial Revolution entrepreneur. In the following section, we shall take the case of the cotton industry as the leading sector of the British Industrial Revolution and, in particular, the examples of Richard Arkwright, one of the most successful inventor–entrepreneurs who also made a great fortune, and of Samuel Crompton, a great inventor but an unsuccessful entrepreneur.

Josiah Wedgwood was born in 1730, the thirteenth son of a Staffordshire potter, and he was apprenticed to a local industry which scarcely sold its wares anywhere outside Staffordshire. 'To sell in London was rare, to sell abroad virtually unknown' (McKendrick, 1960, 1994). The roads were in a terrible state and the canals embryonic. The workmen 'were as likely to go drinking and wenching as to appear at work'. Wedgwood himself was born into poverty and squalor. Yet by the time he died his pottery was sold and known throughout the world and he had amassed a vast personal fortune. His was by far the most successful and fast-growing firm in the industry. A great part of this success must be attributed to the personal entrepreneurial qualities of Wedgwood himself and his partner, Thomas Bentley. He was personally responsible not only for numerous design and process innovations but also for many organizational innovations. He was motivated by ideals of political and social change as well as technical change and capital accumulation.

The network of canals and turnpike roads was vastly expanded and improved between the 1730s and 1770s (see Figure 2.1). Wedgwood was very active in promoting the construction of canals and turnpike roads in his own native county of Staffordshire. But his efforts were part of a much wider national movement, which made it possible to exploit the British and the international market. Demand for earthenware was growing in Britain as living standards were slowly rising and population grew. Tea drinking was becoming popular as cheap tea became available from India. Porcelain was too fragile and the other alternatives too expensive to compete with the potteries in satisfying the growth of demand for a variety of household wares. Many potters in Staffordshire and elsewhere took advantage of these circumstances.

Wedgwood once said of himself that he did not know whether he was an engineer, a potter, or a landed gentleman, as he was all three and indeed 'many other characters by turns'. (Letter to Thomas Bentley, 1766, cited by Jacob, 1988). However, despite this uncertainty about his profession and social status, he had a vision of the reforming, even revolutionary, role of himself and his fellow industrial entrepreneurs. He wrote to Thomas Bentley as early as 1766:

> Many of my experiments turn out to my wishes and convince me more and more of the extreme capability of our manufacture for further improvement. It is at present (comparatively) in a rude uncultivated state, and may easily be polished, and brought to much greater perfection. Such a revolution, I believe, is at hand, and you must assist in and profit by it.
>
> (Jacob, 1988, p. 136)

Among the many interesting features of this letter are his emphasis on 'experiments' and his description of his innovations as a 'revolution'.

Again in a letter to Thomas Bentley, he outlined his principle of factory organization and division of labour: 'to make such machines of the Men as cannot err'. This often quoted phrase sums up the efforts of many entrepreneurs of that age to rationalize the sequence of operations in the new factories and overcome human error, whether due to ignorance, incompetence, laziness, drunkenness, boredom or fatigue. It was a project which is by no means exhausted as the experience of Taylorism (see Chapter 3) and many current tendencies in computerization and robotics amply testify. It was an objective which was seized upon by the critics of industrial capitalism, from Marxists to romantic poets and artists to denounce the dehumanizing tendencies of industrialism, which made men and women mere appendages of machines, and where, as Werner Sombart put it, 'the soul should be left in the cloakroom on entry'. Wedgwood also introduced an elaborate system of fines and penalties to maintain discipline and correct hours of work in his factories.

However, it would be a profound mistake to portray Josiah Wedgwood as an inhuman slave-driving boss. True it was that he and other entrepreneurs were very much concerned with the pace of work and the co-ordination of the various operations of the new machines in their new factories. Wedgwood's friend Erasmus Darwin (grandfather of Charles Darwin) was the founder and leading spirit of the Derby Philosophical Society which brought together scientists, inventors and entrepreneurs to discuss such topics as the ideal factory with a central observation point from which all workshops and workers could be seen. But they also discussed town lighting, central heating, indoor toilets, and even the French Revolution and republicanism (Jacob, 1988, p. 167).

They saw themselves as idealistic but practical reformers, harnessing science, capital and machinery to usher in a new age of material improvement which would benefit everyone. Wedgwood used the same technique as Shakespeare in *The Merchant of Venice* to argue that although workmen are 'our inferiors' yet they are of the same stuff and 'are capable of feeling pain, or pleasure, nearly in the same manner as their Masters' (Letters of Josiah Wedgwood, 1762–72, quoted in Jacob, 1988, p. 168). Wedgwood consistently paid higher wages than his competitors and did show concern for improving their conditions as well as their education and skills. He was a paternalistic employer.

His imaginative vision of the future of his industry extended to almost all aspects of his enterprise. He is of course best known for his design innovations based on new techniques, such as green glaze, creamware, jasper and black basalt. He and the artists whom he hired produced vessels, plaques, cups, plates and vases moulded in clay and delicately tinted, on which minute cameo reliefs in white paste were applied while they were still soft and then fixed by firing. Often these designs were copied from Greek or Roman vases or gems (hence his choice of the name Etruria for his factory). His technical ingenuity extended also to imitating marble and seashells and other porcelain-like varieties of enamelled pottery.

Yet, although his skills and innovations as a potter, a designer, an engineer and a factory manager are often and rightly cited as a part of his success story, they still do not fully explain it. Nor yet does the general enlargement of the British market, nor the greatly improved transport system, for these benefited all Staffordshire potters and indeed those elsewhere. Moreover, the other new enterprising potters could and did rapidly imitate his designs and his process innovations, since they were not protected. (For example, his 'creamware' was being made by 25 other potters by 1784.) If the perfect competition story of mainstream economic theory were a completely true picture, then he would have been obliged to reduce his prices to the general level prevailing and indeed he was under tremendous pressure to do so. But he consistently refused to do this declaring that, 'It has always been my aim to improve the quality of the articles of my manufacture, rather than to lower their price.' As McKendrick (1960, 1994) drily observes, the testimony of his price lists is even more important evidence than his statements, interesting though these are.

His prices were always considerably more expensive than those of his fellow potters; he regularly sold his goods at double the normal prices, not infrequently at three times as high, and he reduced them only when he wished to reap the rewards of bigger sales on a product that he had already made popular and fashionable at a high price, or when he thought the margin between his prices and those of the rest of the pottery had become too great.

(McKendrick, 1994, p. 363)

It could of course be maintained that he was able to sell at far higher prices simply because of differential quality. However, this argument cannot be sustained either since other potters were able to equal or surpass his quality. He deplored the patent system, resolving 'to be released from these degrading selfish chains, these mean selfish fears of other people copying my work' (Ashton, 1963). He urged the other potters too to strive for high quality rather than low price. Indeed, he urged this course upon them in words which find an echo to this day, both in contemporary debates on quality in industry strategy, and in discussions of Adam Smith's cynical comments on the 'conspiracies' of manufacturers to raise prices. In 1771, faced with a general recession in demand for pottery, he wrote to Bentley deploring the price-cutting behaviour of some potters:

Mr Baddeley who makes the best ware of any of the potters here, & an Ovenfull of it per Diem has led the way & the rest must follow, unless he can be prevail'd upon to raise it again, which is not at all probable, though we are to see him tomorrow, about a doz^n of us, for that purpose. . . . Mr Baddeley has reduced the prices of the dishes to the prices of whitestone. . . . In short the General Trade seems to me to be going to ruin on the gallop – large stocks in hand both in London & in the country, & little demand. The potters seem sensible of their situation & are quite in a Pannick for their trade & indeed I think with great reason for *low prices* must beget a *low quality* in the manufacture, which will beget *contempt*, which will beget *neglect* and disuse and there is an end of trade.

While his high quality strategy played an important part in his success, it did so only as one element in his overall marketing strategy, which was the main source of his competitive strength. His marketing and distribution innovations included several which are often thought to have been made fifty or one hundred years later (Table 2.3).

He correctly perceived that at that time, deferential attitudes to the monarchy and aristocracy were still extremely strong, especially in matters of fashion. His strategy of elegant display of complete dinner services in attractive showrooms was highly successful in seducing this world of fashion. By pandering to the fashionable market, he succeeded in establishing an image and a reputation for his wares which enabled him to sell consistently at high prices and nevertheless steadily to widen his market. His London display of the 'Russian' service for Catherine the Great was attended by the Queen and became one of the most popular sights in London. For over a month 'the fashionable world blocked the street with their carriages'. Wedgwood had ensured its success 'for almost all of those whose country seats were represented on the service trekked from their distant homes to see the Exhibition' (McKendrick, 1994, p. 376).

Table 2.3 Marketing innovations in the pottery industry introduced by Wedgwood, 1760–90

1760s	Royal patronage obtained from Queen and other members of Royal Family. Use of trademarks such as 'Queensware'.
1765–74	London warehouse opened, followed by display rooms and warehouses in Liverpool, Bath and Dublin.
1770s	Display rooms designed as elegant, extensive and convenient centres with exhibitions as social occasions.
1760s–90s	Marketing strategy of becoming renowned as designer of high quality earthenware for royalty, gentry and ever wider markets.
1770s	Acceptance of 'uneconomic orders', e.g. from Empress Catherine of Russia for 1,282 pieces and over 1,000 separate illustrations. Used for display and advertising for repeat orders at higher prices.
1771	'Satisfaction or money back' orders.
1770s	Newspaper advertisements and 'puffs'.
1770s	Use of famous artists to advertise wares through paintings, plaques, personal commendations, etc.
1777–90s	Use of travelling salesmen to expand British market.
1769–90s	Strategy of penetrating foreign markets. Use of British ambassadors to interest royal houses and nobility abroad. Opening of warehouses in Paris and Amsterdam. Political lobbying for revision of various commercial treaties (e.g. France, Sweden). Employment of trained linguists for foreign correspondence.
1760s–90s	Use of new designs for notable events (e.g. slavery abolition campaign, peace treaties) and for famous scientists, artists, statesmen. (Only the Pope sold badly: 'Nobody nowadays troubles his head about His Holiness or his predecessors')
1792	*Travellers' Book* prescribing rules and procedures for travelling salesmen.

Source: Based on McKendrick (1960, 1994).

He pursued foreign markets with equal or even greater zeal and with the same tactics:

> He longed to serve the whole world from Etruria and constantly scanned the commercial horizon for new markets. No country – Mexico, Turkey and even China – was too distant for him to contemplate with excitement. . . . Difficulties served only as a challenge to his ambition. France – home of European porcelain, centre of rococo elegance, and safe behind a high tariff wall – was the greatest challenge of them all. Even the thought of it inspired Wedgwood. 'And do you think we may make a *complete conquest* of France? Conquer France in Burslem? My blood moves quicker. I feel my strength increase for the contest.'
>
> (McKendrick, 1994, p. 379)

He appreciated the political aspects of marketing as much as the commercial. Long before imperial Germany was accused of 'unfair competition' in the late nineteenth century, because of its use of German ambassadors for commercial objectives, Wedgwood made use of British ambassadors to open foreign markets and lobbied successfully for the revision of commercial treaties. He did not neglect national habits and tastes in any market.

All of this did not mean that he neglected the wider middle-class British market either, or even the cottages. Many of his wares (smaller ornaments, plaques, inkpots and even tableware) were specifically designed for this market. 'The servants' hall was quick to follow the mistress' lead.' It was from the huge sales of common useful objects that Wedgwood made his greatest return. His travelling salesmen were a major innovation and despite many difficulties he persevered with this technique and in 1790 issued a book of rules and travellers' procedure. The routinization of some of his numerous marketing innovations had begun and by 1850 many of them had become standard procedures for the industry.

2.3   RICHARD ARKWRIGHT, SAMUEL CROMPTON AND THE COTTON INDUSTRY

Although the pottery industry was one of the fastest growing in the late eighteenth century it was the cotton industry, as we have already seen, which led in the Industrial Revolution in rate of increase and volume of output (see Table 2.1). In this industry, among the numerous inventors and entrepreneurs who started new small firms, Richard Arkwright was certainly one of the most illustrious and became the most wealthy.

However, whereas in the case of Wedgwood there are abundant historical records, the case of Arkwright is somewhat different. Even his best-known innovation, the spinning frame, was the subject of much litigation. His precise role in the origins of the innovation is still not known and was contested in the courts in 1781 and 1785, having been the subject of patents by Arkwright much earlier, in 1769 and 1775. He employed a Lancashire watchmaker, John Kay, to assist him in the preparation of parts of his machine, being himself a barber by original trade. It was this mechanic who together with Highs, a reedmaker from Bolton, claimed priority for the invention in the court of King's Bench in 1785.

They lost the case but Arkwright met with continued hostility through-out his life, not only from those who contested priority for his inventions but even more from his fellow employers and from those who believed that their livelihood was threatened by him. From the outset of his business a combination of manufacturers refused to purchase his yarn and fought in parliament to sustain a special tax on the manufacture of calicoes by his partners. His factory near Chorley in Lancashire was destroyed by a mob, while police and military looked on.

To assess the significance of his innovations and the reasons for the litigation and the intense hostility which he encountered, it is necessary to place them in the context of the transition from cottage production to factory production in the British textile industry. The textile industry in Britain, especially wool, was already quite well developed by the eighteenth century, but it was not yet based on machines in factories. A merchant capitalist system based on 'putting out' materials to hand spinners, weavers and clothiers was inherently limited with respect to scale of machinery and division of labour, as well as co-ordination and discipline of the labour force. It was the transition to factory production which provided the opportunity not only for mechanics but for a variety of ingenious and imaginative entrepreneurs to enter the industry. The invention of the spinning jenny by Hargreaves, a Blackburn carpenter, in 1765, gave the means of spinning several threads at once with no more labour than had previously been required to spin one. But Hargreaves' invention could still be exploited in the cottage industry and had many weaknesses. It was soon overtaken by further inventions, especially those of Arkwright.

In his spinning frame, Arkwright used machine-driven rollers to impart twist and to spin a large number of threads of varying degrees of fineness or hardness. Whereas the Hargreaves' jenny could only spin soft yarn for weft and suffered frequent stoppages, the water frame was best for coarser yarns for warp and produced a more consistent higher quality yarn at greater speed. It dominated the industry for two decades until gradually replaced for most yarns by Crompton's mule in the early part of the nineteenth century.

Success in the cotton-spinning industry (weaving was not mechanized until later) did not depend primarily on product or marketing innovations. The market for cotton yarn was already well established and had the potential to grow very fast if lower prices and higher quality could under-cut and substitute for other textiles. The problem for the inventors and the entrepreneurs was to find ways of spinning a variety of coarse and fine yarns and various types of raw cotton and to do so at much lower cost and price. In the end it was Crompton's mule which best satisfied most of these requirements but in the 1770s and 1780s Arkwright's water frame led the way. Unlike most inventors, Arkwright amassed a great fortune from his inventions. This was undoubtedly due to his skill as an entrepreneur and his success in protecting his inventions in the courts.

Arkwright was a firm believer in large-scale machine production and already in the early 1770s he predicted that cotton yarn and cloth would be exported from Britain to the 'East Indies' on the basis of his factory technology. However, he lacked sufficient capital to establish his mills. Partnership with an established manufacturer of stockings, Jedediah Strutt, overcame this problem. Strutt had the experience and the mechanical knowledge to recognize the potential of Arkwright's invention and their first mill began production in 1769 in Nottingham, using horse power and based on the patent for spinning by rollers. It was followed in 1771 by a much larger mill using water power at Cromford in Derbyshire. In 1775 he took out a further patent incorporating various improvement inventions. The combination of his royalty income buttressed by successful litigation with his profitable entrepreneurship meant that he was one of the wealthiest industrialists in the country at the time of his death in 1792, by which time he had received a knighthood and had been High Sheriff of Derbyshire. Although still hated by many of his competitors and by those who lost their employment, he was the recipient of much fulsome praise:

> No man ever better deserved his good fortune, or has a stronger claim on the respect and gratitude of posterity. His inventions have opened a new and boundless field of employment; and while they have conferred infinitely more real benefit on his country than she could have desired from the absolute dominion of Mexico and Peru, they have been *universally* productive of wealth and employment.
>
> (*Encyclopaedia Britannica*, 9th edn, vol. 2, p. 543)

Arkwright's mill at Cromford employed 300 in the 1770s and by 1816 this had increased to 727, by which time there were several other spinning mills employing more than a thousand operatives. However, the dominant tech-nology had moved away from the Arkwright water frame to Crompton's mule except for the coarser yarns and the source of power was slowly shifting to the steam engine. The power loom enabled weaving to catch up with spinning in mechanization and some mills combined spinning with weaving operations.

Unlike Hargreaves and Arkwright, Samuel Crompton (1753–1827) was himself a cotton spinner and weaver. His earnings were insufficient to enable him to take out a patent on his invention of the mule in the 1770s, even though he tried to increase his income by playing the violin at the local church in Bolton (Lancashire). He therefore made known the details of the machine to a few local manufacturers for the promise of small sums, which in the end totalled only £60. The technical merits of the invention were such that it diffused rapidly in the 1780s and 1790s and in 1801 he succeeded in starting a small-scale business himself by borrowing £500 from a friend. However, his shy and unbusinesslike temperament led in the end to bankruptcy of this firm, despite the continued immense success of his invention.

The technical advantages of Crompton's mule were universally recognized so that parliament made him a grant of £5,000 in honour of his services to British industry, but this was the only substantial financial reward that he received. His mule combined some features of the jenny and the water frame, but he always claimed that he developed the roller drafting system independently of Arkwright. In his paper 'Why Three Inventors?' (1994), Hills points out that Arkwright never tried to challenge Crompton for infringement, despite the fact that his own patent was still in force. Crompton's mule separated the drafting from the twisting of the threads, thus enabling his machine to spin the finest yarns as well as the coarser ones (ibid., 1994). His ingenious arrangement of the gearing for the rollers meant that the yarn was subjected to far less strain than in Arkwright's system. Crompton's main motive initially seems to have been his disgust with the quality of the yarn he had to work with and he lacked the entrepreneurial qualities of Arkwright.

The contrast between Arkwright and Crompton is only one illustration, but an interesting one, of Schumpeter's theory of entrepreneurship in the early period of rapid growth of a new industrial sector. Arkwright had in full measure those qualities of determination to innovate (not just to invent) and persistence in overcoming obstacles which Schumpeter identified as the most important characteristics of the entrepreneur. The contrast also illustrates several other features of successful entrepreneurship in this period, notably the importance of access to capital to cover the costs of new machinery and to finance the early (loss-making) period of operation. Arkwright's partnership with Strutt solved the problem for him but Crompton never resolved it. Partnership between inventors and entrepreneurs (another famous example was the Boulton–Watt partnership) was one of the most important organizational techniques for the establishment and take-off of new innovative firms in Britain and distinguished it from other European countries.

Finally, the contrast illustrates the importance even at this early stage of the Industrial Revolution of appropriability of the profits from innovation. Arkwright's determination to protect his patent position and to exploit it to his maximum advantage was evident throughout his career and contrasted with Crompton's apparently rather feeble efforts and with Wedgwood's opposition to patent monopolies. As we have seen, Wedgwood relied on a stream of design and marketing innovations to protect his leading position in the industry. Then, as now, methods of protection and attitudes towards the various methods depended upon the sector of industry and the ease and costs of imitation, but clearly the role of appropriability and possibility of monopolistic exploitation of major innovations were already of crucial importance.

Indeed, nationalistic policies to protect inventions and innovations were among the most cherished political weapons of the old mercantilist tradition. Heavy penalties attended those who were caught exporting trade secrets and when Samuel Slater took the secrets of Arkwright's water frame to the United States in 1789 he was careful to carry no drawings and nothing in writing as he embarked on the ship. He memorized everything and made a replica of the machine soon after his arrival (Chambers, 1961). Francis Lowell did the same with the power loom twenty years later.

2.4 PROLONGED SUCCESS OF THE BRITISH COTTON INDUSTRY

The period of most rapid growth of the British cotton industry was from 1780 to 1800 and this was also the period of the most rapid increase in labour productivity. Thereafter growth continued but at a somewhat slower rate. Incremental innovations continued throughout the nineteenth century and British competitive advantage in foreign trade was sustained by these improvements as well as by scale economies affecting the entire industry. Once the innovations of Arkwright and Crompton and the power loom in weaving had been diffused, cotton was established as the main factory-based industry in Britain and settled into a relatively stable pattern of size distribution of firms. Although the industry continued to grow throughout the nineteenth century a size distribution had become established of a few large firms accounting for about a third of total output and a higher proportion of fixed capital and a long tail of medium, small and very small firms.

However, as Mansfield (1962) pointed out, Gibrat's Law on the size distribution of firms can be formulated in several different ways, depending especially on how the death of firms is treated, and all of them present some difficulties. Small firms are more likely to exit or die and the approximation in this case would be valid only if these firms were excluded. Although their ranking order changed somewhat, a dozen or so of the leading firms appear to have retained their position as leaders over quite a long period. Rather more turbulence, including bankruptcies, prevailed among the smaller firms, a group which included many firms working as subcontractors on commissions from the larger firms and merchants and a still larger number either occupying niche markets for speciality products and/or enjoying lower capital, energy or labour costs. Smaller firms absorbed the main shocks of trade recessions, but there were 33 bankruptcies in 1826, 39 in 1837, 56 in 1839 and a total of 241 from 1836 to 1842. Gatrell (1977) insists that bankruptcies also affected the larger firms but there can be little doubt that the smaller ones suffered disproportionately. This pattern of symbiosis of large and small firms will be found again in many other industrial sectors and will be discussed more fully in Chapter 9 on innovation and size of firms.

The contribution of the cotton industry alone to the aggregate growth rate of the British economy should on no account be underestimated. Floud and McCloskey (1981) have calculated that between 1780 and 1860 the productivity growth of the cotton industry accounted for 15 per cent of the entire productivity growth of the economy. By 1850 cotton yarn and cloth accounted for 40 per cent of all British exports and they still accounted for 26 per cent of a far higher total in 1900. The industry continued to grow right up to the First World War and its exports to India and China still expanded rapidly between 1890 and 1913 when two major radical innovations – ring-frame spinning and the automatic loom – were already threatening to undermine its long domination of the world market. As late as 1870 the cotton industry alone accounted for 30 per cent of all British industrial power consumption.

In their powerful explanation of the reasons for British dominance persisting throughout the nineteenth century, Mass and Lazonick (1990) attribute this 'sustained competitive advantage' to a cumulative process in which the development and utilization of several key productive factors reinforced each other. These affected labour costs, marketing costs and administrative costs. In all of these areas, industry scale economies (Marshall's external economies of scale) were important. In the case of labour:

> During the nineteenth century the development and utilisation of labour resources provided the British cotton industry with its unique sources of competitive advantage. The major machine technologies . . . required complementary applications of experienced human labour to keep them in motion. Experience gave workers not only specific cognitive skills (of which a process such as mule spinning was much more demanding . . .) but also (and more important over the long run) the general capability to work long hours at a steady pace without damaging the quality of the product, the materials or the machines.
>
> (Mass and Lazonick, 1990, p. 4)

Mass and Lazonick lay particular stress on the habituation to factory work and cumulative skills of the labour force, but they also stress that the trade union organization at that time and in that sector (surprisingly for some stereotyped ideas of British industrial relations of later periods and other industries) were particularly congruent with incentives to sustain and increase productivity. Great responsibility was given to the more skilled workers (who often had previous experience in domestic craft work) for recruitment, training and supervision of the less skilled.

> Besides the general habituation to factory work that came from growing up in factory communities and entering the mills at a young age, cotton workers developed specialised skills in spinning particular types of yarn and weaving particular types of cloth.
>
> (Mass and Lazonick, 1990, p. 5)

In common with other historians they point to the economies of agglomeration in relation to pools of specialized skilled labour in various

Lancashire towns: Bolton (fine yarns), Oldham (coarse yarns), Blackburn and Burnley (coarse cloth). Similar arguments apply to the availability of skilled mechanics adept at maintaining (and improving) the local machinery. The gains from increasing productivity were generally shared with the skilled workers, whose union power ensured this.

> By the 1870s cotton industries around the world could readily purchase British plant and equipment and even British engineering expertise. But no other cotton industry in the world could readily acquire Britain's highly productive labour force; no other industry in the world had gone through the century-long developmental process that had produced the experienced, specialised and cooperative labour force that Britain possessed.
>
> (Mass and Lazonick, 1990, p. 8)

Similar arguments apply to the machine-building industry and to mill and machine design. Whereas the early millwrights came from the earlier tradition of cornmills, windmills, etc., with the increased specialization and sophistication of machinery, special and cumulative skills became increasingly important here too. All of this led to high levels of machine utilization as well as lower initial costs of machinery.

Again in relation to material costs, the highly concentrated Liverpool cotton exchange provided Lancashire with an exceptional advantage. Foreign buyers found it cheaper to buy in Liverpool than anywhere else in much the same way that the Amsterdam flower market re-exports to the entire world today. The Manchester Ship Canal and the railway from Liverpool from 1830 onwards meant that transport costs for Lancashire were extremely low. Lancashire spinners could avoid the heavy warehousing costs of more distant competitors. It was not quite 'just-in-time' but was well in that direction. The Liverpool market gave Lancashire enormous flexibility in grades and types of cotton and spinners took advantage of price changes on a weekly basis. Furthermore, Lancashire had a unique capability to work with inferior grades of cotton for any market in the world and even to cope with the partial switch to Indian cotton at the time of the American Civil War.

The worldwide marketing structure was yet another cumulative advantage of the Lancashire industry, which like all the other factors mentioned provided external economies for the firms involved. The structure of the Lancashire industry, with very well-informed merchants, converters and finishers, meant that it had the capability to deliver rapidly to any part of the world whatever product the customer demanded. Again, inventory, transport and communication costs could be kept low through this industry-wide advantage.

We have given an account of the Mass and Lazonick theory because we believe it to be highly relevant not just to the explanation of the prolonged dominance of British cotton firms in the nineteenth century, but even more to the general explanation of the later observable patterns of 'stickiness' in world trade and industrial specialization (see Part Three). It is also helpful again in understanding some prolonged periods of productivity increase without radical technical innovation and the difficulties of 'catch-up' also discussed in Part Three.

Following this brief account of some of the major innovations and entrepreneurs in the cotton and pottery industries from the 1770s to the 1840s, we now turn in the next chapter to the experience of the United States in catching up and forging ahead of Britain in the late nineteenth century. In doing so we recall the last few pages of Chapter 1 which outlined the rapid growth of steam power and railways in Victorian Britain, thus enabling British industry to sustain its worldwide competitive advantage for a considerable time.

CHAPTER 3

THE AGE OF
ELECTRICITY AND STEEL

Chapter 2 has described, albeit very briefly, some of the major innovations of the Industrial Revolution in Britain and their subsequent effects on the growth of firms, industrial sectors and the economy as a whole. Their success was apparently due to a combination of imaginative entrepreneurship, entry into potential growth markets, access to the capital needed for investment in the new factories, and technical inventiveness, sometimes but not always protected by patents and, sometimes but not always, supported by contacts with the world of science.

Occasionally one individual, an inventor–entrepreneur, could satisfy all these requirements, but quite often success depended upon a partnership between a technical inventor and an innovative entrepreneur with business experience and acumen. Chapter 8 examines more systematically the conditions surrounding success (and failure) in industrial innovation, taking account of contemporary empirical evidence as well as the historical retrospect of Part One. This analysis shows that most of those conditions affecting industrial innovation in the Industrial Revolution are still important for success today. Indeed, by the very definition of innovation, it involves the coupling of an inventive idea with a potential market, so that what changes over time are the ways in which this matching is achieved and the national (and international) environment in which it occurs. Clearly some national environments may be more or less favourable to the advancement of science and (even more important) to its linkage with technology. They may foster or hinder innovative entrepreneurship and access to the necessary venture capital. They may or may not offer a supply of people with the skills needed to develop, produce and sell a new product or use a new process and so forth (see Chapter 12).

Recently, economists have begun to examine more systematically those features of a national economy which most affect success or failure in innovation and they use the expression 'national system of innovation' to describe this analysis (Lundvall, 1992; Nelson, 1993). We discuss some of this work in Part Three in relation to comparisons of economic growth and international trade performance, but here we limit the discussion to some special features of the United States national system. Whereas Chapter 2 concentrated on the British system and some specific innovations, especially in the cotton industry, this chapter gives more attention to the systemic aspects of innovations and the linkages between various industrial sectors in their innovative performance. This is largely because the products of the leading sectors – electricity and steel – had innumerable applications in almost all other industries. It is also because these leading sectors were

Table 3.1   Relative productivity levels (US GDP per hour = 100)

	1870	1913	1950
UK	104	78	57
France	56	48	40
Germany	50	50	30
15 countries	51	33	36

Source: Abramovitz and David (1994).

especially characteristic of that country, the United States, which moved to a position of world technological leadership during the course of the nineteenth century and whose economy and productivity grew much more rapidly than any other (Table 3.1).

Not surprisingly, the country whose 'national system of innovation' most closely resembled the British system in the eighteenth century was the former British colonies of the United States of America. However, in the first half of the nineteenth century, despite a rich endowment of natural resources and many favourable institutions, growth was still retarded by the lack of an appropriate transport infrastructure to take advantage of the natural endowment and size of the country and its market. The advent of railways and the new technologies of the late nineteenth century enabled American entrepreneurs to forge far ahead of the rest of the world. At first the United States imported much of this technology from Europe, but from the very beginning American inventors modified and reshaped these technologies to suit American circumstances. By the end of the century American engineers and scientists were developing new processes and products in most industries which were more productive than those in Britain.

3.1   THE UNITED STATES NATIONAL SYSTEM OF INNOVATION

Among those institutions most favourable to economic growth in Britain were the scientific spirit pervading the national culture and the support for technical invention. These features were readily transferred to the United States and respect for science and technology has been an enduring feature of American civilization from Benjamin Franklin onwards. As de Tocqueville observed in his classic on *Democracy in America* (1836):

> In America the purely practical part of science is admirably understood and careful attention is paid to the theoretical position which is immediately requisite to application. On this head the Americans always display a free, original and inventive power of mind.
>
> (de Tocqueville, 1836, p. 315).

The early immigrants were obliged as a matter of life and death to learn by doing about agricultural techniques in the American continent and agricultural research emerged early as an outstanding feature with strong public support. Whereas in Europe feudal institutions often retarded both agricultural and industrial development, the United States never had any feudal institutions either in agriculture or any other part of the economy. Moreover, the relative abundance of land, the westward moving frontier, the destruction of the native civilizations or their confinement to a relatively small part of the territory, all favoured a purely capitalist form of economic development with a relatively egalitarian distribution of income and wealth among the white immigrants in the early period. The big exception to these generalizations was of course the slave economy of the South. It is difficult to assess the degree to which the economic growth of the South in particular and of the Union in general was retarded by the prevalence of this slave economy, but it was in the period which followed the victory of the North in the Civil War that the United States achieved rates of growth well above any previously achieved by Britain. Even after its abolition, slavery left an enduring legacy of social and economic problems, some of which persist to this day, but the maintenance of the Union meant that the predominantly capitalist path of development in the North and West prevailed in the whole country. In these circumstances an entrepreneurial culture could flourish as nowhere else.

Historians such as Abramovitz and David (1994), who have examined American economic history after the Civil War, point to several characteristics of the United States economy which were in combination exceptionally favourable to a high rate of economic growth. These were:

(i) resource abundance in both land, minerals and forests;
(ii) an exceptionally large and homogeneous domestic market facilitating production, marketing and financial economies of scale, especially in the extractive, processing and manufacturing industries.

Abramovitz and David argue that the higher relative price of labour in North America interacted with these advantages to induce substitution of capital and natural resource inputs for skilled labour. This stimulated already in the first half of the nineteenth century the development of a specific American labour-saving, capital intensive technological trajectory of mechanization and standardized production, which at the lower end of the quality range had enabled US manufacturing to surpass British productivity levels already by 1850. As the nineteenth century advanced, 'the engineering techniques of large-scale production and high throughput rates became more fully explored and more widely diffused. American managers became experienced in the organization, finance and operation of large enterprises geared to creating and exploiting mass markets' (Abramovitz and David, 1994, p. 10).

The extent to which this specific American trajectory of tangible capital-using technology diverged from that of Europe (and Japan) can be clearly seen from Table 3.2. Until the 1880s the UK still had an overall capital/labour ratio higher than that of the United States but by 1938, like all other countries, the ratio had fallen to less than half that of the United States. The large cost reductions and productivity gains associated with this North American technological trajectory could be illustrated from numerous industrial sectors. The extraordinary productivity gains in mining and mineral processing are emphasized in particular by Abramovitz and David,

Table 3.2　Comparative levels of capital–labour ratios, 1870–1950 (USA = 100)

	Germany	Italy	UK	Average of 13 European countries	Japan
1870	73	—	117	—	—
1880	73	26	106	68	12
1913	60	24	59	48	10
1938	42	32	43	39	13
1950	46	31	46	39	13

whereas the productivity gains in agriculture are very frequently cited by other historians. In our view the examples of steel and oil are particularly noteworthy because of the key role of these commodities in all kinds of tangible capital-using investment projects, in capital goods themselves, in transport and in energy production and distribution. Cotton, iron, canals and water power were the leading sectors in the early British Industrial Revolution; steel and electricity were the leading sectors in the huge American spurt of growth from 1880 to 1913. We take up the story of oil and the automobile in Chapter 6. We concentrate our account in this chapter on steel and electric power, also because of the special characteristics of technical innovation and management innovation in these sectors.

3.2　THE UNITED STATES STEEL INDUSTRY

Steel had of course been produced in small quantities long before the Industrial Revolution, but it was expensive to make and quality control was difficult. Iron was used far more extensively (Figure 3.1). In the first half of the nineteenth century the United States still depended on imported British steel because of these quality problems and several Sheffield steel works actually had American names because of their dependence on exports to the United States. Many of the saws, axes and other tools which moved the frontier westwards were also made in Sheffield. Geoffrey Tweedale in his study of *Sheffield Steel and America* (1987) speaks of a 'century of commercial and technological interdependence', but it was in the first half of this period that the dependence on Sheffield was at its peak.

As late as 1880, US steel output was only just over a million tons and iron output was more than three times as large, but by 1913 US steel output had climbed to 31 million tons (Table 3.3). This prodigious increase (even faster than the growth of cotton in Britain from 1780 to 1810) was made possible by a number of radical process innovations in the 1850s and 1860s, notably the Bessemer process. These innovations did not originate in the United States but priority for the inventions is still disputed and certainly American inventors were not far behind their British and German counterparts in the early development of this and other process innovations.

When Carnegie built his first Bessemer plant for steel rails in 1875, it was on the basis of the technology he observed during his visit to Britain but the scaling up of production was far more rapid and led to numerous incremental innovations by US steel firms. As a result of these innovations costs and prices fell almost as dramatically as in the earlier case of cotton in

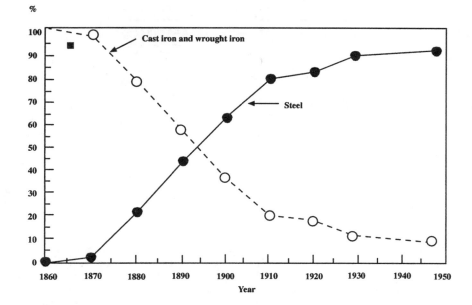

Fig. 3.1 Penetration of steel, USA (percentage of all iron and steel products)
Source: Ayres (1988).

Table 3.3 Iron and steel output (million tons)

	1880	*1913*	*1929*
Pig iron			
Great Britain	7.7	10.3	7.7
Germany	2.5	19.3	13.4*
USA	3.8	31.0	43.3
Steel			
Great Britain	1.3	7.7	9.8
Germany	0.7	18.9	16.2*
United States	1.2	31.3	57.3

* Alsace-Lorraine, annexed by Germany in 1871 but returned
to France in 1918 and therefore excluded from German
figure for 1929.

Lancashire. Landes (1969) estimated that the reduction in cost of crude steel
was between 80 and 90 per cent from the early 1860s to the mid-1890s. By
far the main application in this period was in steel rails and Table 3.4
illustrates the fall in price in this commodity in the United States. Steel rails
lasted five or six times as long as iron rails and the price decline ended
only with the formation of a cartel and the general onset of inflation.

American entrepreneurs excelled in scaling up industrial processes and
innovating in machinery and production systems designed to serve a very

Table 3.4 Price of steel rails in USA

Year	Steel rails $ per ton	Consumer price index
1870	107	38
1875	69	33
1880	68	29
1885	29	27
1890	32	27
1893	28	27
1895	24	25
1898	18	25
1910	28	28
1920	54	60
1930	43	50

Source: US Historical Statistics.

large market. As Chandler (1977) emphasizes, it was the development of the transport and communication infrastructure (rail and telegraph) in mid-century, which enabled the USA to exploit its great natural endowments and its vast size to reap scale economies unequalled elsewhere in the world. The railways also made a major contribution by the development of management techniques for large enterprises.

Andrew Carnegie, the architect of the giant American steel firm, which was later to become United Steel, and the embodiment of American entrepreneurship, brought with him a model of management derived almost entirely from his railway experience. Carnegie had himself worked on the railways from the age of 18 as a clerk and as a telegraph operator. In 1856 at the age of 24 he became a departmental head and familiar with cost accounting systems. During the American Civil War, with financial help from friends, he set up companies to make iron rails, locomotives and bridges. But following a visit to Britain soon after the Civil War, he became convinced that the Bessemer steel process would revolutionize the world iron and steel industry. On his return he opened a steelworks using that process and making steel rails; by 1880 he was already a millionaire. The son of a Chartist weaver who emigrated from Scotland in the 1840s because of the poverty, he became one of the richest men in the world by 1900. He retired from business and devoted his immense fortune to philanthropy – mainly the endowment of hundreds of public libraries in the United States and Britain, and an accident and insurance fund for his steelworkers.

The American railways had to cope with a large volume of operating statistics, long-range communication, high fixed investment and much technical development. Carnegie and his subordinates with railway experience introduced rigorous statistical cost systems, which were one of the earliest and most important achievements of the new management style:

These cost sheets were Carnegie's primary instrument of control. Costs were Carnegie's obsession. One of his favourite dicta was: 'Watch the costs and the profits will take care of themselves'.

(Chandler, 1977, p. 167)

By 1880 Carnegie's cost data were more detailed and accurate than those of almost any other enterprise in the USA. They were used to control departments and foremen and to check the quality and mix of raw materials. They were also used to make improvements in processes and products, so that technical advance and cost cutting moved together hand in hand.

Chandler describes the contribution made by Carnegie to the tempestuous growth of the US steel industry as follows:

> The history of the American steel industry illustrates effectively how technological innovation, intensified use of energy, plant design, and overall management procedures permitted a great increase in the volume and speed of throughput and with it a comparable expansion in the productivity of operation. Carnegie's pre-eminence in the industry came from his commitment to technological change and from his imaginative transferral to manufacturing of administrative methods and controls developed on the railroads. Technological and organisational innovation paid off. Carnegie's prices were lower and his profits higher than any producer in the industry. As soon as the E.T. Works was opened in 1875 it recorded profits of $9.50 a ton. In 1878 Carnegie's steel rail mill recorded a profit of $401,000 or 31 per cent on equity. It rose in the next two years to $2.0 million. As the business grew, so did its profits. At the end of the 1890s Carnegie's larger and more diversified enterprise had profits of $20 million. For the year 1900 they stood at $40million. By becoming a pioneer in the methods of high volume production in steel, Carnegie quickly accumulated, as John D Rockefeller had done in petroleum, one of the largest fortunes the world had ever seen.
>
> (Chandler, 1977, p. 169)

This admirably sums up not only Carnegie's entrepreneurial qualities but also the interdependence of scale economies, technical and organizational innovation, productivity and profitability.

3.3   APPLICATIONS OF STEEL

The United States steel industry not only led in large-scale production but, together with German and British industry, in the development of special steels and with US engineering firms in numerous new applications of steel. The advantages of steel were not simply in comparative cost, although this was a decisive incentive in the redesign of many products and processes. Steel also made possible many new products, tools and processes, especially in machinery, engineering and construction. Already in 1868 Mushet had discovered that by adding vanadium and tungsten to steel a much harder tool steel could be made. This was followed by the development of other new alloys with non-ferrous metals with much harder cutting edges in a wide range of machinery and much greater precision in the production of many engineering components. Before he became a management consultant, Fredrick Taylor worked for Midvale Steel from 1878 to 1889 when he was developing his theories of 'scientific management' and later at Bethlehem Steel where he made a major invention in high speed tool steel and a series of machinery and metallurgical inventions related to the use of tool steels.

The first electric furnaces were invented by Siemens in 1878 and Stevens in 1879 but the large-scale use of iron and steel scrap as an input in much larger types of electric furnace came only during and after the First World

War. In the early period the applications of the electric furnace were for non-ferrous metals and a new range of special steels and alloys. Originally these were produced in non-electric crucible furnaces but the electric furnace gradually took over, especially in Germany and USA. The new electric furnace permitted control of temperature between 5°C and 15,000°C which was unattainable with combustible fuel (Fabre, 1983). The new steels proved vastly superior both for high speed machinery and for abrasion resistance.

The new heat treatment technique invented by Taylor and White in 1895 at the Bethlehem Steel Co. produced alloy steel which could cut five times faster than the carbon steels previously used for machine tools and this was soon trebled again by newer alloys. The Sheffield firm of Hadfields produced manganese steel in the 1890s which had very high abrasion resistance and found worldwide applications in construction and engineering (and of course armaments). One example was in the teeth of the shovels which dug the Panama Canal. Hadfields also developed silicon steel which was at one time called 'electrical steel' and was an essential component for electrical transformers and generators because of its excellent magnetic properties and high electrical resistance (and hence low energy loss) (Pearl, 1978). The interdependence of developments in steel and electrification is evident from these examples.

The properties of stainless steel were investigated already between 1903 and 1910 by L. Guillet in France, but it was Strauss and Maurer in Germany and Harry Brearley in Sheffield who first realized the commercial possibilities. Brearley made his discovery of the extraordinary corrosion resistance of stainless steel while testing the suitability of high-chromium steels for rifle barrels. The armaments race between Britain and Germany leading up to the First World War was an important stimulus to the development of new steel alloys with exceptional properties and heavy engineering generally. Although the United States was not so heavily involved in an armaments race, both Midvale and Bethlehem worked on US naval contracts for armour plate and Taylor was also an active consultant to US arsenals and naval yards after 1906. Stainless steel found its main applications ultimately in the cutlery, food and chemical industries.

It was not only in heavy engineering, machinery and armaments that steel found a wide range of new applications. There were many consumer industries which owed either their existence or their rapid growth primarily to the availability of cheap and plentiful steel. This was the case, for example, with the canning industry, which was growing rapidly in the United States since the American Civil War, and later in Europe. The substitution of steel strip for iron in the cans and the introduction of electroplating for the tin deposition process revolutionized the industry. Steel now accounted for 98 per cent and tin for only 2 per cent of the metal used in a tin can and the industry could become a truly mass production industry. By the First World War many households were using dozens of tin cans every year for food and many other consumer goods and the armies depended on tinned food for their rations.

Another new consumer product which owed its rapid growth in the 1890s partly to cheap good quality steel was the bicycle. The techniques used for their manufacture before the 1920s were not yet Fordist mass production techniques, but they were certainly high volume production, taking advantage of the new potential of steel intensive products.

Yet another constellation of innovations which illustrates the interdependence and complementarities of technical and economic developments in the steel, heavy engineering and electrical industries was in the construction of large buildings. Before the 1880s to support the weight of a lofty building on walls of brickwork they would have to be immensely thick in the lower storeys: 'Even with the weight of the interior carried by cast iron columns, a thick very lofty external wall capable of carrying its own weight could spare only a limited area for window openings' (Hamilton, 1958).

In 1883, when William Jenny was commissioned by the US Home Insurance Company to build a 10-storey block that would be fireproof and let plenty of natural light into every room, he used cast iron columns and wrought iron beams for the six lower floors, but above that level he used steel beams. The Rand-McNally Building in Chicago in 1890 used a completely steel frame. With a steel skeleton it was possible to build walls at several levels independently and simultaneously. Soon external walls became 'weather screens' in the panels between columns and beams and the age of the skyscraper had arrived. But of course the spread of skyscrapers would have been impossible without the electric lift (1889) (and indeed the electric telephone). Hydraulic lifts were slow and not always safe to use above five storeys. Chicago and New York rivalled each other for the tallest and most spectacular building. The rolling of steel beams for large structures began at various steelworks in Britain, Germany and United States in the 1880s. In addition to large office blocks, the new types of factory using electric power and overhead cranes, as well as the power stations themselves made extensive use of new steel girders and other steel products.

From this example it is clear that it is essential to examine the complementarities between innovations. Perez (1983) was right to insist on the idea of a characteristic pattern of interrelated innovations extending over a considerable period. Steel was at the heart of a whole wave of innovations affecting every branch of industry and services. It became common sense to design big projects and new products using steel, steel alloys and electric power during the third Kondratieff wave. Perez pointed to multiple applications, falling price and universal availability as the characteristics of each 'key factor' in a Kondratieff wave. This certainly applied to steel, as to cotton and coal in previous waves and to oil and microelectronic chips in later waves of technical change (Table 1.3).

As we have seen in Chapter 1 in considering Schumpeter's long wave theory, he discussed steel mainly in terms of *steam-powered machinery*. the associated steel primarily with the second Kondratieff, which he described as a revolution 'wrought by railroads, steel and steam' (1939, p. 397). But while the major process innovations in the steel industry did indeed originate in the 1850s and 1860s, it was the iron horse, the iron rail and the iron machine which were the basis of the first big wave of steam power and 'railroadization'. The consumption of steel remained relatively small until the 1880s and 1890s (Figure 3.1).

Moreover, the complementarities between electrical machinery and processes and the uses of steel, although relatively neglected by Schumpeter, were even more important than those with steam-powered mechanical equipment. Not only did much electrical equipment use steel, alloys, and non-ferrous metals from the outset but the applications of electricity in many other industries, such as machine tools, were strongly associated with the applications of steel and its alloys. To us, therefore, it appears more justifiable to describe the third Kondratieff as an 'age of steel' and the first and second as an 'age of iron'. A purely temporal listing of innovations by date of first application obscures the economic and technical significance of steel as a key factor (in the Perez sense) in the new constellation of innovations which crystallized in the 1880s and powered the upswing of the third Kondratieff in the 1890s (Table 3.5).

Even though the innovations which led to electrification were also spread out over a long period preceding the third Kondratieff upswing, Schumpeter nevertheless states unambiguously: 'In the same sense in which it is possible to associate the Second Kondratieff with railroads and with the same qualifications, the Third can be associated with electricity' (Schumpeter, 1939, p. 397).

It is to the process of electrification of the economies of the leading industrial countries that we now turn.

3.4   ELECTRICITY AND ELECTRIFICATION

Schumpeter's assessment of the significance of electricity is shared not only by ourselves but by most historians. Leslie Hannah (1983), for example, compares it with the micro-processor: 'In the decades before the First World War, the modern equivalent of the micro-electronic industry – the sector which was transforming the efficiency of home, office, factory and indeed of urban public transport as well – was the electric power and electrical engineering industry.'

It is the perception of cheap steel, heavy engineering and electricity as a pervasive combination affecting the entire economy which is the central feature of the Perez model of change of techno-economic paradigms. In her model the new paradigm develops over a long period before it becomes dominant, embracing many radical and incremental innovations, subjected to selective economic pressures, interacting with fundamental science and responding to the limitations of the established technologies and business organizations until it crystallizes as the new common sense of engineers, designers and managers. This model of a long gestation period for a pervasive 'meta-technology', followed by very widespread diffusion over half a century or more, fits the case of steel and electricity well.

Most of the early scientific and inventive work took place in various European countries. In the 1820s, Faraday established the principle of the electric motor and in 1831 announced the discovery of electromagnetic induction in a paper to the Royal Society. In addition he had demonstrated 'transformer' action. Experimental electric motors were made already in the 1840s. Even in the eighteenth century scientists had investigated electro-chemistry and the electrical properties of many materials. With the invention of the primary battery (the voltaic file) by Volta in 1800 it became possible to extend the uses of electricity outside the scientific laboratory. The role of laboratory science in the development of electricity and its

Table 3.5 A tentative sketch of some of the main characteristics of successive long waves (modes of growth)

1	2	3	4	5	6	7	8
Number	Approx. periodization Upswing Downswing	Description	Main 'carrier branches' and induced growth sectors Infrastructure	Key factor industries offering abundant supply at descending price	Other sectors growing rapidly from small base	Limitations of previous techno-economic paradigm and ways in which new paradigm offers some solutions	Organization of firms and forms of co-operation and competition
First	1770s and 1780s to 1830s and 1840s 'Industrial revolution' 'Hard times'	Early mechanization Kondratieff	Textiles Textile chemicals Textile machinery Iron-working and iron castings Water power Potteries Trunk canals Turnpike roads	Cotton Pig iron	Steam engines Machinery	Limitations of scale, process control and mechanization in domestic 'putting out' system. Limitations of hand-operated tools and processes. Solutions offering prospects of greater productivity and profitability through mechanization and factory organization in leading industries.	Individual entrepreneurs and small firms (< 100 employees) competition. Partnership structure facilitates co-operation of technical innovators and financial managers. Local capital and individual wealth.
Second	1830s and 1840s to 1880s and 1890s Victorian prosperity 'Great Depression'	Steam power and railway Kondratieff	Steam engines Steamships Machine tools Iron Railway equipment Railways World shipping	Coal Transport	Steel Electricity Gas Synthetic dyestuffs Heavy engineering	Limitations of water power in terms of inflexibility of location, scale of production, reliability and range of applications, restricting further development of mechanization and factory production to the economy as a whole. Largely overcome by steam engine and new transport system.	High noon of small-firm competition, but larger firms now employing thousands, rather than hundreds. As firms and markets grow, limited liability and joint stock company permit new pattern of investment, risk-taking and ownership.

continued overleaf

Table 3.5 cont.

1 Number	2 Approx. periodization Upswing Downswing	3 Description	4 Main 'carrier branches' and induced growth sectors Infrastructure	5 Key factor industries offering abundant supply at descending price	6 Other sectors growing rapidly from small base	7 Limitations of previous techno-economic paradigm and ways in which new paradigm offers some solutions	8 Organization of firms and forms of co-operation and competition
Third	1880s and 1890s to 1930s and 1940s 'Belle époque' 'Great Depression'	Electrical and heavy engineering Kondratieff	Electrical engineering Electrical machinery Cable and wire Heavy engineering Heavy armaments Steel ships Heavy chemicals Synthetic dyestuffs Electricity supply and distribution	Steel	Automobiles Aircraft Telecommunications Radio Aluminium Consumer durables Oil Plastics	Limitations of iron as an engineering material in terms of strength, durability, precision, etc., partly overcome by universal availability of cheap steel and alloys. Limitations of inflexible belts, pulleys, etc., driven by one large steam engine overcome by unit and group drive for electrical machinery, overhead cranes, power tools permitting vastly improved layout and capital saving. Standardization facilitating worldwide operations	Emergence of giant firms, cartels, trusts and mergers. Monopoly and oligopoly became typical. 'Regulation' or state ownership of 'natural' monopolies and 'public utilities'. Concentration of banking and 'finance capital'. Emergence of specialized 'middle management' in large firms.
Fourth	1930s and 1940s to 1980s and 1990s Golden age of growth and Keynesian full employment	Fordist mass production Kondratieff	Automobiles Trucks Tractors Tanks Armaments for motorized warfare Aircraft Consumer durables	Energy (especially oil)	Computers Radar NC machine tools Drugs Nuclear weapons and power Missiles Microelectronics	Limitations of scale of batch production overcome by flow processes and assembly-line production techniques, full standardization of components and materials and abundant cheap	Oligopolistic competition. Multinational corporations based on direct foreign investment and multi-plant locations. Competitive

		Crisis of structural adjustment		Process plant Synthetic materials Petrochemicals Highways Airports Airlines	Software	energy. New patterns of industrial location and urban development through speed and flexibility of automobile and air transport. Further cheapening of mass consumption products.	subcontracting on 'arms length' basis or vertical integration. Increasing concentration, divisionalization and hierarchical control. 'Techno-structure' in large corporations.
Fifth*	Information and communication Kondratieff	1980s and 1990s to ?	'Chips' (micro-electronics)	Computers Electronic capital goods Software Telecommunications equipment Optical fibres Robotics FMS Ceramics Data banks Information services Digital telecommunications network Satellites	'Third generation' biotechnology products and processes Space activities Fine chemicals SDI	Diseconomies of scale and inflexibility of dedicated assembly-line and process plant partly overcome by flexible manufacturing systems, 'networking' and 'economies of scope'. Limitations of energy intensity and materials intensity partly overcome by electronic control systems and components. Limitations of hierarchical departmentalization overcome by 'systemation', 'networking' and integration of design, production and marketing.	'Networks' of large and small firms based increasingly on computer networks and close co-operation in technology, quality control, training, investment planning and production planning ('just-in-time') etc. Keiretsu and similar structures offering internal capital markets.

continued overleaf

* All columns dealing with the 'fifth Kondratieff' are necessarily speculative.

Table 3.5 *cont.*

9 Number	10 Technologic-al leaders	11 Other industrial and newly industrializing countries	12 Some features of national regimes of regulation	13 Aspects of the international regulatory regime	14 Main features of the national system of innovation	15 Some features of tertiary sector development	16 Representative innovative entrepreneurs engineers	17 Political economists and philosophers
First	Britain France Belgium	German states Netherlands	Breakdown and dissolution of feudal and medieval monopolies, guilds, tolls, privileges and restrictions on trade, industry and competition. *Laissez-faire* established as dominant principle. Repression of unions.	Emergence of British supremacy in trade and international finance with the defeat of Napoleon.	Encouragement of science through National Academies, Royal Society, etc. Engineer and inventor-entrepreneurs and partnerships. Local scientific and engineering societies. Part-time training and on-the-job training. Reform and strengthening of national patent systems. Transfer of technology by migration of skilled workers. British Institution of Civil Engineers. Learning by doing, using and interacting.	Rapid expansion of retail and wholesale trade in new urban centres. Very small state apparatus. Merchants as source of capital	Arkwright Boulton Wedgwood Owen Bramah Maudslay	Smith Say Owen
Second	Britain France Belgium Germany USA	Italy Netherlands Switzerland Austria–Hungary	High noon of *laissez-faire*. 'Nightwatchman state' with minimal regulatory functions except protection of property and legal framework for production and trade. Acceptance of craft unions. Early social legislation and pollution control.	'Pax Britannica'. British naval, financial and trade dominance. International free trade. Gold standard.	Establishment of Institution of Mechanical Engineers and development of UK Mechanics' Institutes. More rapid development of professional education and training of engineers and skilled workers elsewhere in Europe. Growing specialization. Internationalization of patent system. Learning by doing, using and interacting.	Rapid growth of domestic service for new middle class to largest service occupation. Continued rapid growth of transport and distribution. Universal postal and communication services. Growth of financial services.	Stephenson Whitworth Brunel Armstrong Whitney Singer	Ricardo List Marx

	Countries	State/political	International	R&D and education	Services	Entrepreneurs	Economists
Third	Germany, USA, Britain, France, Belgium, Switzerland, Netherlands, Italy, Austria–Hungary, Canada, Sweden, Denmark, Japan, Russia	Nationalist and imperialist state regulation or state ownership of basic infrastructure (public utilities). Arms race. Much social legislation. Rapid growth of state bureaucracy.	Imperialism and colonization. 'Pax Britannica' comes to an end with First World War. Destabilization of international financial and trade system leading to world crisis and Second World War.	'In-house' R&D departments established in German and US chemical and electrical engineering industries. Recruitment of university scientists and engineers and graduates of the new Technische Hochschulen and equivalent Institutes of Technology. National Standard Institutions and national laboratories. Universal elementary education. Learning by doing, using and interacting.	Peak of domestic service industry. Rapid growth of state and local bureaucracies. Department stores and chain stores. Education, tourism and entertainment expanding rapidly. Corresponding take-off of white-collar employment pyramid. London as centre for major world commodity markets.	Siemens, Carnegie, Nobel, Edison, Krupp, Bosch	Marshall, Pareto, Lenin, Veblen, Weber
Fourth	USA, Germany, Other EEC, Japan, Sweden, Switzerland, USSR, Other EFTA, Canada, Australia, Other Eastern European, Korea, Brazil, Mexico, Venezuela, Argentina, China, India, Taiwan	'Welfare state' and 'warfare state'. Attempted state regulation of investment, growth and employment by Keynesian techniques. High levels of state expenditure and involvement. 'Social partnership' with unions after collapse of fascism. 'Roll-back' of welfare state deregulation and privatization during crisis of adjustment.	'Pax Americana' US economic and military dominance. Decolonization. Arms race and Cold War with USSR. US-dominated international financial and trade regime (GATT, IMF, World Bank) Destabilization of Bretton Woods regime in 1970s.	Spread of specialized R&D departments to most industries. Large-scale state involvement in military R&D through contracts and national laboratories. Increasing state involvement in civil science and technology. Rapid expansion of secondary and higher education and of industrial training. Transfer of technology through extensive licensing and know-how agreements and investment by multinational corporations. Learning by doing, using and interacting.	Sharp decline of domestic service. Self-service fast food and growth of supermarkets and hypermarkets, petrol service stations. Continued growth of state bureaucracy, armed forces and social services. Rapid growth of research and professions and financial services, packaged tourism and air travel on very large scale.	Sloan, McNamara, Ford, Agnelli, Nordhoff, Matsushita	Keynes, Schumpeter, Kalecki, Polanyi, Samuelson

continued overleaf

Table 3.5 *cont.*

9 Number	10 Technological leaders	11 Other industrial and newly industrializing countries	12 Some features of national regimes of regulation	13 Aspects of the international regulatory regime	14 Main features of the national system of innovation	15 Some features of tertiary sector development	16 Representative innovative entrepreneurs engineers	17 Political economists and philosophers
Fifth*	USA Japan Germany Sweden Other EEC EFTA Russia and other Eastern European Taiwan Korea Canada Australia	Brazil Mexico Argentina Venezuela China India Indonesia Turkey Egypt Pakistan Nigeria Algeria Tunisia Other Latin American Chile	'Regulation' of strategic ICT infrastructure. 'Big Brother' or 'Big Sister' state. Deregulation and reregulation of national financial institutions and capital markets. Possible emergence of new-style participatory decentralized welfare state based on ICT and red–green alliance.	'Multi-polarity'. Regional blocs. Problems of developing appropriate international institutions capable of regulating global finance, capital, ICT and transnational companies.	Horizontal integration of R&D design, production and process engineering and marketing. Integration of process design with multi-skill training. Computer networking and collaborative research. State support for generic technologies and university–industry collaboration. New types of proprietary regime for software and biotechnology. 'Factory as laboratory'.	Rapid growth of new information services, data banks and software industries. Integration of services and manufacturing in such industries as printing and publishing. Rapid growth of professional consultancy. New forms of craft production linked to distribution.	Kobayashi Uenohara Barron Benneton Noyce Gates	Schumacher Aoki Bertalanffy Friedmann Giddens

* All columns dealing with the 'fifth Kondratieff' are necessarily speculative.

Source: Based on Freeman (1987).

applications was much more obvious and direct than in the case of mechanical technologies.

Although they were certainly important, the early low power applications were largely confined to communications (Table 3.6). The electric telegraph took off in the 1830s and was followed in the 1870s by the commercial exploitation of the telephone patented by the Canadian Graham Bell after the pioneering demonstration by Reis (1861) and the scientific work of Helmholtz. It was not until the 1850s and 1860s that the development of magnetos and dynamos reached a point where they could be used on a commercial scale for illumination. The first major application for magneto-powered arc lighting was in British lighthouses.

Following the 1858 experiments for lighthouse illumination at South Foreland, Faraday was invited to report on the trials for the lighthouse authorities (Trinity House) and was very impressed:

> I beg to state that in my opinion Professor Holmes has practically established the fitness and sufficiency of the magneto-electric light for lighthouse purposes so far as its nature and management are concerned. The light produced is powerful beyond any other that I have yet seen so applied and in principle may be accumulated to any degree; its regularity in the lantern is great, its management easy and its care can therefore be confined to attentive keepers of the ordinary degree of intellect and knowledge.
>
> (Dunsheath, 1962, p. 106)

Even though these specialized applications were spreading in the 1860s, it was only with the development of a further series of inventions and innovations (armatures, alternators, rotors, etc.) in the 1860s and 1870s that dynamo technology ('self-excited generators') reached the point where large-scale generation and transmission of electric power could be successfully achieved in the leading industrial countries. A further spurt of innovations in the 1880s included the carbon filament lamp, which meant that the new power stations could find markets in household domestic lighting as well as in public illumination (Table 3.6 and Figure 3.2). Already in 1889 the replacement of electric light bulbs was one of the many types of work studied and timed by Fredrick Taylor, thus giving rise to a long series of jokes.

Many of the innumerable applications of electric power which opened up in the 1880s and 1890s had already been explored on a small scale in the 1870s. Electricity was used for the illumination of the Gare du Nord in Paris in 1876, of *The Times* printing works and the Gaiety Theatre in London in 1878, and the Kaiser arcade in Berlin. In that year also 30,000 people in Sheffield witnessed the first ever football match played under electric lighting. Show grounds and public fairs were also illuminated by the new source of power.

One of the earliest and most important uses of electric power generation was for tramways and urban electric railways, sometimes underground. Some of the early electric power companies, especially in Japan, were set up specially for this purpose. Following earlier demonstrations of the electric railway by Siemens and Halske in Germany between 1879 and 1881, the first British electric railway was installed at Brighton by Volk,

Table 3.6 Historical evolution of electricity and its applications

	Science and invention	Electric power generation	Communications	Illumination	Industrial and transport applications
Before 1800	Frictional machines used to investigate electrical properties of materials and electro-chemistry.				
1800–30	Measurement, analysis and theory of electricity by laboratory scientists (Volta, Ampère, Orstead, Davy, Ohm, Arago, Faraday, etc.).	Cruikshank's primary battery from Voltaic pile (1800). Daniel two-fluid cell (1830) used extensively for telegraph.			
1830–50	Demonstration of electromagnetic induction by Faraday (1831) and of Magneto of Pixii (1832) and arclamps by Fourault, Dubosc, etc.	Early generators by Clarke (1834) London, Stoehrer (1840s) Leipzig, and Nollet (France) used commercially.	Rapid development and commercialization of electric telegraph service invented by Wheatstone (1837) and Morse. Multi-core cables, gutta-percha insulations.		Early 1840s: first patents for electroplating followed by rapid growth of silverplate industry in cutlery trade.
1850s	Hjorth patent for 'magneto-electric battery' 1855, Siemens armature 1856. Swan's early research on carbon filament lamps.	Compagnie d'Alliance formed to manufacture generators. Holmes demonstrates generators for lighthouses 1857–8.	Telegraph lines of 4,500 miles in Britain by 1855 owned by Electric Telegraph Co. First submarine telegraph 1851, trans-Atlantic cable 1858.	French and English lighthouses begin to use arc lighting.	First large cable factory 1858–9. Siemens Brothers established 1858.
1860s	Reis demonstrates first electric telephone (Frankfurt, 1861); 'self-exciting' generators (Wilde, 1863; Siemens, Varley 1866). 1864 Maxwells theory of radiation. Leclanche cell 1868.	1867 Wilde alternator with shuttle armature. Siemens disc armature (rotor).	First telegraph service New York to San Francisco 1861.	Experience of rubber insulation and cables in communications transferred to other applications.	1869 first electrolytic copper refinery, South Wales.

1870s	Gramme armature 1870. Bell Telephone Patents 1876. Brush invents open-coil dynamo 1878.	First continuous current reliable dynamos by Gramme early 1870s. Drum armature (1872). Siemens & Halske Plante storage batteries 1878.	First telephone exchanges in USA and Britain 1878–9.	Illumination of public buildings, naval ships, theatres, fairgrounds, factories and some streets mainly using arc lighting. 1877 formation of Edison Electric Light Co.	Electroplating used more widely for canning and other applications. 1878 Siemens high temperature electric arc furnace. 1879 first electric railway (Siemens and Halske).
1880s	Helmholtz fundamental work on telephony and radio. 1886 establishment of Physikalische und Technische Reichanstalt, Berlin. Hertz (1887) demonstrates electromagnetic radiation.	Swan lead plate 1881. Wenström generator and armature (ASEA) 1880. 1881–3 first power as a commodity (Edison). 1883 transformers (Gaulard and Gibbs). 1887 Ferranti alternator using zig-zag rotor winding. 1888 rotary converter (Bradly). 1888 Parsons 75kV turbine.	Bell manufactures 67,000 telephone sets in 1880.	Use of Swan Carbon filament lamps for lighting ships and London–Brighton train. 1881 Swan and Edison lamps go into high volume manufacture and begin to be used for domestic lighting. 1886 Westinghouse introduce AC for lighting.	Numerous tramways and urban railways in 1880s. 1887 electrolytic process for production of aluminium (Hall, USA and Herault, France). 1888 chlorine by electrolysis. 1888 Tesla develops AC motor later manufactured by Westinghouse.
1890s	1891 National Physical Laboratory. Braun (1897) cathode ray tube. Lodge (1897) demonstrates induction coil for tuning. Many radio inventions J.J. Thomson (1897) discovers electron.	1887–92 'Battle of the Systems' (AC/DC). 1893 Westinghouse high voltage transmission from Niagara Falls power scheme. Special alloy steels used for transformers and generators. 1894 Parsons 350kV turbine.	Marconi experiments with radio communications and establishes Wireless Telegraph Co. (1897). Rapid growth of telephone network.	Rapid diffusion of electricity for illumination	1891 AC power for industrial use. 1891 Oerlikon and AEG manufacture electric motors. 1891 silicon carbide. 1892 acetylene from calcium carbide in electronic furnaces. 1895 high speed tool steel alloys in USA. 1895 first power tools (Fein, Stuttgart). 1900 Héroult electric arc furnace. New generations of machine tools with unit drive from electric motors.
1900–	1901 cathode ray oscillograph. 1904 Fleming's thermonic valve. 1906 De Forest Triode. 1909 first electrocardiograph.		1901 Marconi demonstrates trans-Atlantic radio signals. Fessenden's wireless telegraphy for voice and music.	1900 mercury arc lamp. 1906 tungsten filament lamps.	

Source: Freeman (1989).

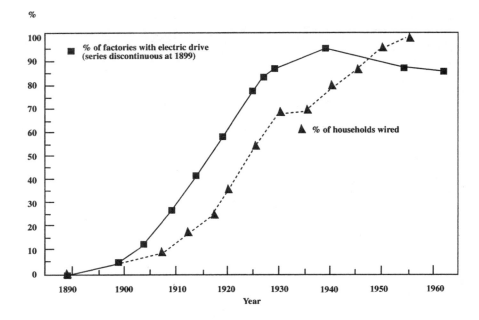

Fig. 3.2 Electrification in the USA

Sources: Historical Statistics of the United States; Ayres (1988).

mainly for recreational purposes, in 1883. Much more important were the 'tube' railways which followed in London from 1887 to 1900, again illustrating the new constellation of steel, electricity and heavy engineering technologies. It was a sign of the times that the electrical equipment was supplied by US firms.

The first electrification of part of a mainline steam railway came in 1895 with the electrification of four miles of track in a tunnel under Baltimore on the Baltimore–Ohio Railroad (Ellis, 1958). Mainline electrification came only gradually in the twentieth century, requiring as it did enormous new investment on a system already in place. But already in the 1880s and 1890s electric tramways and underground systems spread very rapidly in the big new industrial areas of Western Europe and the USA. In fact the suburban sprawl and the long journey to work often attributed to the automobile were first made possible by electric tramways and railways. It was said to be possible to travel by tram most of the way from central Germany to the Channel coast (although with numerous changes) and from the Middle West to the Atlantic. Be this as it may, it was certainly not until much later that the automobile took over as a major mode of travel to and from work and as an accelerator of urban sprawl (see Chapter 6).

More important than domestic lighting systems or transport systems were the new industrial applications of electricity. The first of these were mainly in electrometallurgy and electrochemistry. As we have seen there was an important cross-linkage between the electric furnace and later advances in the steel industry. But in the 1880s and 1890s the most important developments related to non-ferrous metals and alloys. The electrolysis of copper, first introduced on a small scale in South Wales in the 1870s but on a much larger scale in the 1890s, was particularly important for the subsequent growth of the electrical industry. Because of its high electrical conductivity and because it could be easily drawn into wire, copper was in great demand for electrical applications, which soon became its principal use. But only the best qualities of copper had the necessary properties and 'in general high-conductivity copper could be produced solely by electrolytic refining' (Chadwick, 1958). Once again the interdependence of a network of related innovations is evident linking metallurgy, chemistry and electricity.

The aluminium industry also depended on electricity almost from the outset. Hall's process in the USA and Héroult's process in France were invented almost simultaneously in 1887. The electrolysis of chlorine followed in 1888 which transformed the heavy chemical industry and led to many new applications of chlorine. In the 1880s too, the electric furnace was first used for producing acetylene from calcium carbide and this was followed in the 1890s by the production of synthetic silicon carbide (carborundum).

The convergence of heavy and electrical engineering technology with chemistry was most clearly evident in the rapidly growing German chemical industry. In his study of *Science and Industry in the 19th century*, Bernal (1953) distinguishes electricity and chemistry as the two areas where scientific research began to be directly and intimately related to industrial development. It was in the German dyestuffs industry that the inhouse R&D laboratory was invented in the 1870s. As the scaling up problems for a wide range of new and old products were encountered in the 1880s and 1890s, the co-operation of engineers and chemists in large-scale process design became the hallmark of the German industry. This will be described in greater detail in Chapter 4. Not just in chlorine production and electrolysis but in many processes, electrical technology played a central part in the extraordinary rapid growth of the German industry. Particularly notable was the German lead in acetylene chemistry which played a big part in organic chemicals (Fabre, 1983).

Most important of all in the long term, although not clearly envisaged in the early days of electricity were the applications in every industrial sector opened up by the electric motor. The *Oxford History of Technology* puts it well (vol. 5, p. 231):

The conspicuous attribute of electrical energy is mobility. It can be taken to any point along a pair of wires. Other methods of converting energy from a central plant to smaller consumers were tried, but none was as convenient or efficient as the electrical method, by which the heat energy of a boiler-furnace, or the kinetic energy of falling water, is converted into electrical energy and then again transformed into mechanical energy by the consumer's electric motors. Technologically this is by far the most important role of electricity. Electrical energy as a source of light and more recently of heat, as a means of communicating information, and as an agent in chemical processes, has clearly transformed industrial practices; but its chief significance has been to place power, great or small, in the workman's hands or at his elbow.

As early as 1894 in an address to the Canadian Society of Civil Engineers, Fred Bowman also emphasized the reliability and robust qualities of the new electric motors:

> The advantages of electrical motors for use in driving the machinery in small industries are efficiency, reduced cost of attendance, cleanliness, reduced fire risk and economy of power.
> Those who had never handled them do not realise what a well built electric motor will stand in the way of overload and general abuse . . . electric motors have passed the experimental stage and taken their place as thoroughly reliable machines.
>
> (Ball, 1987, p. 34)

Very soon electric motors were produced in tens of thousands as the advantages of a flexible, robust and reliable source of energy were recognized.

A specific example well illustrates this point. Delbeke (1982) investigated the development of the stock of engines of all types in Antwerp. The structure of industry in this region was varied but food, furniture, metals and diamonds were the major industries with many small firms. The dramatic change from steam to electricity is illustrated in Figures 3.3 and 3.4 which show that the share of electricity by number of engines rose from nil in 1885 to over 90 per cent by 1915. For a short time gasoline engines had an advantage in the late nineteenth century but in the twentieth century electricity displaced both steam and gasoline except for the largest engines. (For this reason the share of electricity by installed capacity, as shown in Figure 3.4, is a little lower than by number of engines). Most of the new electric motors were not however direct substitutes for steam or gasoline but simply new installations for small firms able to afford this cheap, new, robust and flexible source of energy to aid their mechanization. The revolution was electromechanical.

The number of electric engines rose from less than ten at the turn of the century to 1600 by the First World War, while the number of steam engines never exceeded 150. As Delbeke put it:

> A fundamental advantage of the steam engine for larger scale activities is that it provides a continual supply of a large mass of energy, which, again, is not well suited for handicraft activities, which need energy only sporadically.
> The coming of the gasoline engine changed the picture for some small scale industries, but the impracticability of energy transportation remained. Nevertheless, for many entrepreneurs, this situation was attractive in view of the lack of alternatives. The electric motor changed the opportunities radically: small motors were able to produce sufficient energy when and where it was needed and even in the tool itself. The basic innovation not only changed the organization of the factory, which now could be based on the efficient use of labor, but also and primarily it changed the situation for small scale industry, which was so widespread in Flanders.
> Thus, the introduction of electricity meant, in the first place, a very efficient way of transporting energy. Moreover, once a city-wide network became avail-

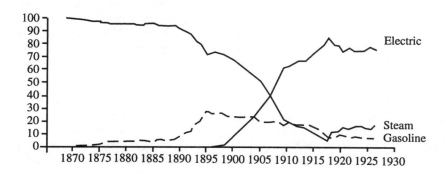

Fig. 3.3 Share of the different types of energy engines in the installed capacity (nine-yearly moving averages) in Antwerp, 1870–1930

Source: Delbeke (1982).

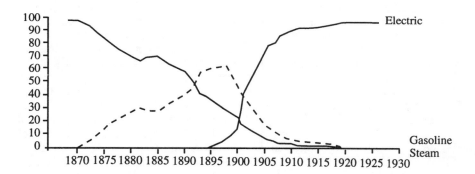

Fig. 3.4 Share of the different types of energy engines in the annual installed total number of energy engines (nine-yearly moving averages) in Antwerp, 1870–1930

Source: Delbeke (1982).

able at the end of the century, it became possible for many handicraft operations and small-scale industries to avoid the insurmountably high fixed costs of previous energy systems by buying their electricity at the door. In other words, energy became largely a variable cost.

(Delbeke, 1982, p. 16)

Ultimately, of course, electric motors were to be produced in millions and installed in every household, but the mass production and consumption of these electric consumer durables was mainly a feature of the fourth Kondratieff rather than the third Kondratieff.

To round out this picture we must add that the combination of the electric telephone and 'transmissible power' transformed the location and operation of many services as well as manufacturing industries. It greatly facilitated the administration of large organizations with dispersed branches or operations, as well as giving new flexibility to many small firms. By 1890, 228,000 telephones were in use in the USA and by 1900 nearly one and a half million. The telephone and the typewriter together began the process of office mechanization which characterized the emergence of 'bureaucracy' in large firms and central and regional governments.

Thus, after half a century of innovations in communications, illumination, generation and transmission and new industrial applications, by the late 1880s and early 1890s electricity had reached the point where myriad new investment opportunities were opening up on all sides. All the conditions were satisfied for an explosive upsurge of new investment based on cheap steel and electric power. The full exploitation of these opportunities, however, required an enormous new infrastructure. Hitherto, generators had usually been installed for each specific application, but in the 1880s Edison and others realized that electricity had to be generated and transmitted both to households and industrial consumers as a publicly available 'utility' or 'commodity'. This required a new regulatory framework, new legislation, new standards and massive private and public investment, which Weber (1922) identified as a new model of administration. The 1880s were thus a period of intense public debate which led to a range of different policies, at municipal, regional and national levels. This debate was not confined to the terms under which the new infra-structure should develop, it extended to the whole wide range of electrical applications as the new technology diffused. Engineers argued vehemently not only about the relative merits of AC and DC but even more about work organization, the implications of electric power for factory layout, machine design, location of industry, management structure and scale of enterprise. In other words the combination of cheap steel and electric power brought with it not merely a new source of energy and materials but a transformation of the whole productive system and socio-economic structure. The organizational and managerial innovations were just as important as technical innovations. The situation was comparable to the contemporary debates about robotics, teleworking, networking and other features of information technology (See Chapter 17).

Figure 3.2 shows how (in the case of the United States) the vast majority of factories and households were supplied with electricity over the half century of the third Kondratieff wave. Figure 3.5 shows that by 1920 over half the power for industry in the USA came from electricity and by 1930 over three-quarters. Figures 3.3 and 3.4 illustrated the same point for a major European city (Antwerp). Hall and Preston (1988) report estimates by German economists that the electrical industry accounted for between 30 per cent and 40 per cent of the growth of industrial production in the period leading up to the First World War.

As this vast wave of electrification got under way new constellations of interrelated innovations emerged and consolidated, linking together steel and the new industrial materials with their wide new range of properties and applications and the new flexible source of energy. First and foremost this affected factory design and layout in almost all sectors of manufacturing with new machinery and power tools replacing the old steam-

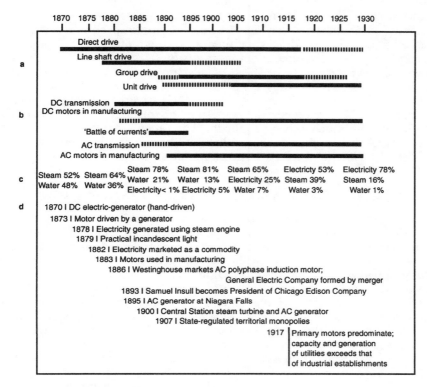

Fig. 3.5 Chronology of electrification of industry
a Methods of driving machinery
b Rise of alternating current
c Share of power for mechanical drive provided by steam, water and electricity
d Key technical and entrepreneurial developments

Source: Devine (1983).

powered machinery. Under the old system all the shafts and countershafts rotated continuously no matter how many machines were actually in use. A breakdown involved the whole factory (Devine, 1983).

Attempts were made in the 1880s and 1890s to overcome the inflexibility of the old system by using several steam engines, but once electricity became generally available it proved far superior. Although the new potential of electric power was recognized already in the 1880s, it was not until after 1900 that manufacturers generally began to realize that the indirect benefits of using the electric drives were far greater than the direct energy saving benefits. Unit drive gave far greater flexibility in factory layout since machines were no longer placed in line with shafts, making possible big capital savings in floor space. For example, the US Government Printing Office was able to add 40 presses in the same floor space. Unit drive meant that trolleys and overhead cranes could be used on a large scale unobstructed by shafts, counter-shafts, and belts. Portable power tools increased even further the flexibility and adaptability of production systems. Factories could be made much cleaner and lighter, which was very important in industries such as textiles and printing, both for working conditions and for product quality and process efficiency. Production capacity could be expanded much more easily.

The full expansionary benefits of electric power on the economy depended, therefore, not only on a few key innovations in the 1880s, but on the development of a new paradigm, style or production and design philosophy. This involved the redesign of machine tools, handling equipment and much other production equipment. It also involved the relocation of many plants and industries, based on the new freedom conferred by electric power transmission and local generating capacity.

3.5 GIANT FIRMS AND MANAGEMENT INNOVATIONS

The changes in management organization and philosophy and in the structure of firms and industries were no less profound. The managerial and organizational innovations which accompanied, followed or preceded the technical innovations mostly facilitated a larger scale of operation by the leading firms. A revolution was brought about in offices and in services, indeed in all business organizations and in the conduct of government and of war, by the electric telephone and telegraph. These new media of communications (supplemented at a later stage by radio), greatly facilitated the development of large corporations with more complex management structures controlling plants in various locations and the production and delivery of materials, components and machinery from distant locations. Much has been written in the 1990s about globalization of the economy and of large firms. For the large electrical firms this was already described by Lenin as a feature of their activity before the First World War. Already by then, the main American firms, General Electric and Westinghouse, and the dominant German firms AEG and Siemens had grown very rapidly to become giants which were among the largest in the world. Both GE and AEG grew partly as a result of mergers which involved the participation of strong financial interests. From fairly early on the electrical firms, like the railways before them, needed very large amounts of capital as they were involved in the development of a network system of generating and distributing electricity, as well as numerous electrical products. Chandler (1977) confirms one of the main points in Lenin's (1915) analysis – the tendency for a convergence of financial and manufacturing interests in the large firms emerging in the new industries in Germany and the United States. Lenin concentrated on the example of AEG but Chandler gives a much more detailed analysis of the early stages of GE's history:

> For several reasons the merger that created the General Electric Company was more important to the development of modern industrial management in the United States than were the early trusts. General Electric was the first major consolidation of machinery-making companies and so the first between already integrated enterprises. Its products and processes were as technologically advanced and complex as any of that day. And at General Electric outside financiers played as large a role as they did in any American industrial merger.

For this reason the railroad influence was particularly strong. The financiers were important because the electrical manufacturers were the first American industrialists not intimately connected with railroads who found it necessary to go to the capital markets for funds in order to build their initial enterprise.

(Chandler, 1977, p. 426)

Technological development was more complex, and expensive, and took longer than in other industries. It required specialized research, design and development departments. The new companies had to create 'an integrated system of power-generating machinery, power stations, lamps and power-using machines before they could begin to sell their products in volume' (see especially Hughes, 1982). To finance local power companies they had to go to Wall Street. From 1878 Thomas Edison was getting help from Drexel, Morgan & Company, while Elihu Thomson soon had the backing of Boston capitalists who were involved in the financing of railroads, the telegraph and the telephone. General Electric was a merger of two of the three large electrical equipment manufacturers, which were themselves the result of mergers.

Henry Villard, an eminent railroad financier who had helped to finance some of Edison's early developmental work, engineered the merger. Villard had recently returned to the United States after a three-year stay in Germany where he had become closely associated with the powerful Deutsche Bank of Berlin and with Siemens & Halske, the leading German electrical manufacturers who were already beginning to sell in the American market. He planned, according to Edison's biographer, Matthew Josephson, to create a 'world cartel'.

(Chandler, 1977, p. 427)

The idea of a world cartel was not such a distant dream. This was a period of a strong trend towards cartelization in many industries. The electrical engineering industry was one in which international patent agreements between the dominant firms played a major role at least down to the Second World War. Market sharing agreements were also far from uncommon and monopolistic purchasing policies characterized the relationships between the big utilities and their 'national' heavy equipment suppliers. In the international market GE and the German AEG systematically divided up the world. Lenin commented that this division of the world between two powerful firms did not remove the possibility of redivision as a result of uneven development or of war. Such a redivision did in fact take place when in 1922 the agreement was renegotiated following the German defeat in the First World War. The terms were more favourable to GE which had acquired 25 per cent of the newly issued stock of AEG in 1920.

The rapid growth of concentration in the electrical equipment industry was stronger than in most others but there were similar trends throughout the economic system. The establishment of a worldwide transport and communications network meant that firms could now operate on a global scale, not only in terms of exports but in terms of vertical integration with raw material suppliers, and control of manufacturing facilities and sales agencies in many countries and the finance to organize such operations. Whereas the Lancashire textile firms in the late nineteenth century enjoyed few plant economies of scale and benefited from extensive external economies (Chapter 2), the new American industries internalized many functions which were external to the small firm. Plant economies of scale which were important in steel, oil and chemicals could now be exploited in many other industries and over much larger markets. But economies of scale in finance for new investment, in procurement of materials, in establishing marketing networks and in research, design and development now became of equal or greater importance. The cumulative advantages of a highly skilled and trained labour force now extended to several echelons of professional managers in these functions.

In heavy electrical equipment they were certainly important, but although the main trend in the electrical industry was to the development of very large firms, this was offset by the consequences of electrification for many other industries and services. The new technology and the new source of energy gave opportunities for thousands of small and medium-sized enterprises (SMEs) to flourish in such diverse industries as lumber (saw mills), instruments, power tools, and other mechanical products. The flexibility of the power source as the whole country became electrified permitted the development of entirely new industrial districts, the decentralization of some industries, the rise of other entirely new industries, and the transformation of still others.

As already noted, urban transport was one such industry where electrification of tramways accelerated the trend towards concentration. Chandler (1977) emphasized the speed of this change in the United States at the turn of the century. By 1890, 15 per cent of urban transit lines were already using electric-powered street cars, and by 1904, 94 per cent. In this industry, as in so many others, the new technology brought with it a revolution in management style since it involved expensive equipment, advanced technology, complex maintenance and repair, sophisticated accounting and statistics and new forms of co-ordination and political arrangements. The new full-time salaried managers replaced the old owner–operators or municipal managers of horsedrawn systems.

The rise of a professional manager class in the United States was stimulated not only by the changes in the structure and administration of large firms and changes in technology but also by developments within the management profession and the education system. The real significance of 'Taylorism' was not that it introduced 'scientific management'. This was indeed impossible and remains impossible today. What Taylorism achieved together with other similar schemes and models was the provision of a rationale for a whole set of organizational innovations which displaced the old ways of running an enterprise and substituted a management intensive model based on the professionalization and specialization of the various functions of management – above all cost accounting, production engineering, and sales management, and (in some cases) design and development, personnel, public relations and intelligence or market research (as in the case of General Electric). The management 'bureaucracy' was the main organizational innovation of the third Kondratieff, reflected in Weber's distinctive contribution to sociology and economics (Table 3.6).

Daniel Nelson's (1980) biography of Taylor makes clear that in common with other mechanical engineers with industrial experience, he saw himself as a 'professional', as a 'scientist', who could see the need for a radical change in the old 'contractor' system to take advantage of the new technologies. The rise in fixed costs associated with rising capital intensity put pressure to sustain or improve profitability by using physical capacity more fully and to control the flow of materials and components. At the same time engineers were needed in sales to give technical service to customers. Selling too had to become an organized professional function of the firm. When the financier, Villard, reorganized General Electric, a sales force was established with seven regional offices each headed by a district manager who controlled the work of salesmen and engineers responsible for installation, service and sales. The transition from the second to the third Kondratieff was characterized by a persistent search for new forms of management organization and firm structure which were no less important than the new technologies and accompanied their diffusion, interacting with them. Weber was characteristic of German sociologists in seeing the progressive positive side of 'bureaucratic' administration, as well as some negative features.

Although Taylor's ideas were influential from the 1890s in the United States and later worldwide, no firm actually adopted his full ideal of 'power to the planning department' and extreme specialization. Most took over only some of his ideas. Chandler (1977, p. 276) observed that 'many of his basic concepts were incorporated into the modern organisation of American factories' and further:

> In the first years of the new century many factories came to be organised along the lines set out by Emerson, Taylor, Towne and other active members of the American Society of Mechanical Engineers. The contract system was eliminated, gain-sharing and incentive plans were adopted, cost-accounting based on shop orders or a voucher system of accounts was introduced: time studies were carried out; route, line, cost and inspection clerks were employed and the manager's staff was enlarged.
>
> (Chandler, 1977, p. 277)

Summing up the changes in the organization of the firm and the emergence of a new 'best practice' form of organization in the 1880s and 1890s we may identify the following features:

1. Professionalization and specialization of key management functions in response to growing complexity and scale of production, technology, markets, finance and administration. Carnegie was one of the pioneers in this change.
2. Displacement of internal subcontracting systems and replacement by professional management control using various new management and accounting systems. The establishment of the professional management bureaucracy.
3. Standardized information, accounting and administrative procedures using new office machinery and communication systems (typewriters, telephones, telegraph, etc.) and linking branch offices, plants and sales organizations with headquarters.

The diffusion of these managerial and organizational innovations interacted with the diffusion of the technical innovations and systems characteristic of the third Kondratieff: electric power, telecommunications, precision machinery, and steel intensive products, processes and structures. Firms such as General Electric were a new type of capitalist enterprise especially suited to the conditions prevailing in the United States. They enabled the US economy to continue its fast growth when the erstwhile leader, Britain, was slowing down.

The emergence of the specialized research and development department in electrical and chemical firms in the closing decades of the nineteenth century was exceptionally important from the standpoint of our analysis. Contacts between science and industry were important already in the Industrial Revolution, but they were spasmodic, indirect and unsystematic. The professionalized R&D department provided a regular point of entry into the firm for outstanding new ideas in national and international science and technology. The big new electrical firms had particularly strong R&D activities and Schumpeter (1939) pointed out that this meant that the entrepreneurial function (in his sense) could be exercised by a section head in the R&D of a giant firm. The rise of professional inhouse R&D was equally important in the chemical and oil industries and we examine this in greater depth in Chapter 4.

PROCESS INNOVATIONS
IN OIL AND CHEMICALS

In Chapter 3 we have outlined some of the main technical innovations associated with the very rapid growth of the steel and electrical industries. Firms in the United States were able to take advantage of a rich endowment in natural resources and large domestic market to develop and exploit new technologies in these industries. American firms were particularly successful in scaling up new techniques of production and marketing and became world leaders in many industries despite the fact that many of the original innovations and most of the related scientific work had been performed in Europe. The emergence of new giant firms was not confined to steel and electrical products. Similar developments took place with other materials and especially in oil and chemicals. Here too giant firms emerged both in the United States and in Europe.

Whereas in the steel industry European firms were often leaders in product and process innovations and American firms excelled mainly in the applications of steel, US firms dominated process innovations in the oil industry from the beginning down to the present time. In part this was of course due to the early exploration and discovery of oil as a domestic fuel from the 1850s onwards. But it was also due to an accumulation of technical know-how in the 'cracking' of oil to produce a variety of oil products in oil refineries. American firms still dominate in the design and construction of oil refineries worldwide based on this early lead and sustained by a remarkable series of innovations.

In this chapter we shall first of all outline the main trends in process innovation in the chemical and oil industries of the nineteenth and twentieth centuries. Then, we shall analyse the emergence of inhouse R&D in the German chemical industry from 1870 onwards and see how this affected process innovation. Following this we shall make a similar analysis of the American oil industry. All these cases illustrate the increasing costs of most process development work down to the 1970s, when some counteracting trends set in. Finally, we shall discuss nuclear power very briefly as the culmination of the trend towards increasing complexity and vast costs of process development.

4.1 FROM BATCH TO FLOW PROCESSES

For over a century the chemical industry has enjoyed a high rate of productivity advance. The most important general change in the techniques of the industry has been the move from batch to flow processes of

Table 4.1 Productivity comparison of the Burton and fluid catalytic cracking processes

Production inputs	Inputs per 100 gallons of gasoline produced		
	Burton process	Fluid process original installations	Fluid process later installations
Raw materials (gallons)	396.0	238.0	170.0
Capital ($, 1939 prices)	3.6	0.82	0.52
Process labour (man hours)	1.61	0.09	0.02
Energy (millions of BTUs)	8.4	3.2	1.1

Source: Enos (1962a, p. 224).

production. This has permitted very great economies of scale in plant construction and in labour costs for handling materials. Flow processes are also far more efficient than discontinuous batch processes in preventing heat losses, and in facilitating the monitoring and control of the chemical reactions. These advantages mean that unit costs of production have been drastically reduced for most of the major chemicals, and these reductions have affected not only labour costs but capital, energy and materials as well. At the same time, constant process improvements have led to higher quality and more uniform products. An example of the kind of economies which have been achieved is shown in Table 4.1, which compares production inputs per 100 gallons of petrol produced in the early US oil refineries before 1914 with the first fluid catalytic cracking processes in the 1940s and the improved versions available in the 1950s. The unit capacity of the early Burton process was about ninety barrels psd,[1] compared with about 13,000 for the early fluid installations and 36,000 for the later installations.

The most dramatic saving (of over 98 per cent) was in process labour costs, but almost equally impressive were the savings of over 80 per cent in capital costs and energy costs, and over 50 per cent in material inputs per unit of final output. It is this kind of technical progress which is fundamental to the growth of productivity and of the economy.

The shift to flow production techniques was facilitated by six major developments during the nineteenth and twentieth centuries. These were:

1. The enormous growth of the market for the basic chemicals such as soda, ammonia, chlorine, sulphuric acid, ethylene and propylene. These 'building blocks' are used as intermediate materials for a great variety of other chemicals as well as in many other industrial applications, outside the chemical industry.
2. The switch in base materials for organic chemicals from coal derivatives to oil and natural gas. This stimulated the development of continuous processes and chemical complexes linked to refineries.
3. The increasing availability of electricity as a source of energy and the development of electrothermal and electrolytic processes. Faraday had demonstrated the electrolysis of salt in 1833 but it was not until the end of the century that cheap power became generally available and a large-scale process was developed.

4. Improvements in materials for plant construction and in components such as pumps, compressors, filters, valves and pressure vessels. These were essential to permit the use of large-scale processes, and more severe operating conditions such as high pressures and extremes of temperature.
5. The development of new instruments for monitoring and controlling flow processes, as well as for laboratory analysis and testing.
6. The application of basic scientific knowledge to the production processes, and the development of the new discipline of chemical engineering. The design of new flow processes was linked to physical chemistry whereas the old batch processes were often based on purely empirical knowledge and mechanical engineering.

All of these trends facilitated the growth of professional inhouse industrial R&D and were stimulated by it.

This chapter first outlines the characteristic chemical processes of the nineteenth century and the role of inventor–entrepreneurs in the development and innovation of these processes. This outline is necessarily very sketchy and intended only to give the background to the new pattern of process development which emerged in the German chemical industry in the latter part of the nineteenth century, and in the American oil refining industry in the early part of the twentieth century. The new industry of process design engineering and contracting is then discussed including its relationship with the chemical and oil firms. The complexity of the design process for larger plants reached a point where hundreds of engineers and draughtsmen could be employed for months on one design. This reached its extreme limits in the design and development of nuclear reactors, discussed in the concluding section. The use of computer-aided design (see Chapter 7) has now provided some amelioration of these tendencies in design.

4.2   NINETEENTH-CENTURY PROCESS INNOVATION[2]

The characteristic plants of the early nineteenth century were the Leblanc soda and sulphuric acid plants, and the 'Alkali trades' was the name commonly given to the chemical industry as a whole. Leblanc was an ex-surgeon to the Duke of Orleans who developed his process on the basis of trial-and-error experiments on the Duke's estate from 1784–9, in response to the offer of a prize from the Academy of Sciences. Although he won the prize and was awarded a patent in 1791, he was an exceptionally unfortunate inventor–entrepreneur. His patron the Duke was guillotined and his St Denis factory was expropriated. The Committee of Public Safety ordered the revocation of his patent and full publication of all the process know-how, and although his factory was returned to him in 1801, he committed suicide. However, his process dominated the European chemical industry for a hundred years.

Some of the associated processes, such as sulphuric acid, were on a flow system but the kernel of the Leblanc system was batch production from coal furnaces, mixing one part of sodium sulphate, one part of calcium carbonate (chalk) with a half part of coal. This in turn was based on batch production of saltcake (sodium sulphate) by reacting sulphuric acid in a furnace with salt. Labour costs, material costs and fuel costs were all heavy, but big reductions in the price of alkali were achieved in the first half of the nineteenth century through improvements to the process.

The processes which replaced the Leblanc system and many of those which enlarged the whole scope of the chemical industry in the latter part of the nineteenth century were also mainly the work of inventor–entrepreneurs or individual inventors. Among the nineteenth-century inventor–entrepreneurs who established new firms to develop and exploit processes which they had invented or helped to invent were: the Swede Alfred Nobel (dynamite and other explosives), the French Brins brothers (industrial oxygen), the English chemists W.H. Perkin (aniline dyes) and John Bennett Lawes (superphosphates and other fertilizers), the German Linde (liquid air distillation), the Canadian T.L. Willson (electrothermal production of calcium carbide), the Americans Castner and Dow (industrial electrolysis), and the Belgian Ernest Solvay (ammonia-soda), as well as men like Parkes, Hyatt, Spitteler, Chardonnet and Baekeland who pioneered the early plastic materials, and to whose work ·reference will be made in the next chapter.

Almost all of these men were chemists, and although they often spent many years conducting research at their own expense and considerable personal risk (both physical[3] and financial), the scale of their experimental work was relatively small and the apparatus inexpensive. All of them attached great importance to patents (Nobel held 350 patents when he died), and most of them were successful in establishing a patent position so that production in other countries was either under a licensing arrangement or through affiliated companies. The successful enterprises which were built up became an important part of the twentieth-century chemical industry, whether they were absorbed into larger groupings (Nobel divisions of ICI, IG Farben, British Oxygen Company), or remained independent companies (Dow, Solvay, Linde). The Nobel licensee in the USA was Du Pont. The Solvay licensee in the UK was Brunner Mond and in the USA the Solvay Process Corporation which subsequently merged into Allied Chemical and Dye.

Of the immediate successors to the Leblanc system, the Solvay process was the most important and may be regarded as the beginning of the modern heavy chemical industry.

As is often the case when a product or process is threatened with extinction, many improvements were made[4] to the Leblanc process in the 1860s and 1870s and as late as 1887 the Chance process for the recovery of sulphur was highly successful. (It succeeded where Mond's process had failed largely because of improvements in carbonic acid pumps.)

Most of these process improvement inventions were made by British engineers and chemists who were involved in production problems in the Leblanc works in Cheshire and south Lancashire. Britain was the main centre of the Leblanc alkali industry from the 1830s to the 1880s, and had a thriving export business to Germany and the United States. But the improvements were not enough to affect the basic technical superiority of the Solvay process and by 1890 most of the continental soda output was based on it and the whole of the infant American industry. Only in Britain did the old Leblanc process linger on, accounting for two-thirds of total production even in 1890, despite Mond's enterprise, as one of the first foreign licensees for the Solvay process.

4.3 THE NEW PATTERN OF PROCESS DEVELOPMENT IN GERMANY[5]

The German and American chemical industries had no major investment in the Leblanc technology and did not suffer from problems of adjustment to the new techniques. The German industry in the 1870s had already established the new pattern of inhouse R&D leading to the introduction of new products and processes. Bayer, Hoechst and BASF (Badische Anilin und Soda Fabrik) were among the first firms in the world to organize their own professional R&D laboratories. Although individual inventor–entrepreneurs made the major process innovations of the nineteenth century, by the end of that century the scale of flow process experimentation was putting it beyond the reach of the individual ingenious chemist, unless he enjoyed a large private fortune or the patronage of an established chemical firm.

The new pattern first became evident in the dyestuffs industry. Whereas some of the early synthetic dyestuffs were discovered and innovated by inventor–entrepreneurs in Britain, by the end of the century the lead had passed decisively to the German industry. Synthetic 'mauve' was discovered by the 19-year-old chemistry student W.H. Perkin in 1856 and within two years, with the help of his family, he had started a factory to make aniline dyes based on coal tar. This was followed by a succession of other aniline dyes, but with the alizarin dyes Perkin was beaten by one day to the Patent Office by BASF. However, he was still able to find an alternative patent-free route and launch manufacture of alizarin dyes in 1869. Both BASF and Hoechst had started with the manufacture of aniline dyes, and the principal distributor for BASF, Rudolph Knospe, had the exclusive sales rights for Perkin's aniline mauve from 1859.

In those days a number of outstanding German chemistry graduates were working in the British industry, but with the early successes of the infant German dye industry they began to return. Caro, who had been a successful inventor in Britain, became chief chemist at BASF in 1868 and it was he who launched alizarin dyestuffs manufacture just ahead of Perkin.

Two significant features of BASF policy from its foundation in 1864 were the insistence on integration of manufacture of intermediate and final chemicals and the concentration on creating a capability for plant and apparatus manufacture and repair. Not only did Caro play a leading part in establishing the Institute of German Chemists (VDC) but also of the Institute of German Engineers (VDI). Consequently, BASF was in a position to contribute to development of new chemical processes which required something more than individual brilliance, and depended on sustained co-operation between research scientists and qualified technologists. The establishment of a Department of Chemical Technology at Karlsruhe Technische Hochschule as well as the flow of graduate chemists from German universities very much facilitated their efforts to recruit highly qualified staff capable of product and process innovation. BASF, Hoechst and Bayer were managed by chemists who considered it their business to remain in close touch with the progress of university research. The work of Kekulé on the benzole molecule provided the theoretical basis for many of the major advances in coal tar chemistry, which enabled the German dyestuffs industry to advance with extraordinary speed from 1870 to 1914. In 1880, Germany accounted for about one-third of world dyestuffs production, by 1900 about four-fifths, by which time there were 15,000 different patented dyestuff materials.

In the early period there was serious competition between the three leading German firms. All of them manufactured alizarin dyes as well as aniline dyes and all attempted the synthesis of indigo. But BASF helped to formulate the new Patent Law of 1877, which restricted price competition, and later informal understandings on markets led to the formal coordination of dyestuffs marketing through the IG (Interessengemeinschaft) from 1904.

The synthesis of indigo is a good illustration of the new importance of systematic, science-related process development. Bayer first took the lead in the 1870s in work on indigo synthesis, but without success. In 1880 Professor Baeyer, who had succeeded Liebig at Munich University, first produced synthetic indigo on a laboratory scale, followed by a whole series of indigo dyes. For this he was awarded a Nobel prize. Both Hoechst and BASF jointly took out patents on Baeyer's work, but the cost of production from his starting material far exceeded the price of the natural dye. BASF attempted to market the synthetic product but had to withdraw. In 1882 Professor Baeyer discovered a new synthesis, but again eight years' development work failed to yield an economic process. In 1890 Karl Heumann at Zurich Polytechnic discovered yet another synthesis and the patents were again acquired by BASF and Hoechst. Once more development work showed that the process, although technically feasible, was uneconomic.

After years of further effort, success was finally achieved at BASF, partly as a result of an accident during experiments on the oxidation of naphthalene, which at that time was extremely expensive. A thermometer accidentally broke and mercury flowed into the reactor vessel. It proved to be an ideal catalyst.

Hoechst had now fallen behind, but was able to retrieve its position as a result of collaboration with Degussa. At one time there were four different pilot plants each testing a different process. These trials showed that the Degussa process had decisive advantages and this became dominant at Hoechst in the early years of the twentieth century. The total cost of R&D on indigo synthesis from 1880 to 1897 was over DM20 million. Once large-scale economic processes had been successfully developed at BASF and Hoechst, there was sharp price competition with the natural indigo dyes and Indian exports fell from 187,000 tons in 1895 to 11,000 tons in 1913. The price of the natural material fell from DM11 per kilo to DM6.50 over the same period.

By the end of the century, the German and Swiss chemical firms had established their supremacy as technical and market leaders, accounting for over 80 per cent of total world production. The leading Swiss firm, CIBA, maintained close research links with BASF and with other Swiss leaders, Geigy and Sandoz, imported basic chemicals and intermediates from

Germany. The Swiss firms concentrated on research-based high quality dyes and drugs. By 1900 they were exporting 93 per cent of their output.

The new importance of an inhouse process development capacity was evident not only in the commercial introduction of new products (whether plastics, dyestuffs or drugs), but also in the production of old-established basic chemicals. In the 1880s BASF developed the new 'contact' process for sulphuric acid which is still the basic technique for a large part of present-day production. The process was based on fundamental work by Wilhelm Ostwald at Leipzig in the field of physical chemistry, but depended on intensive industrial experiments with catalysts for the conversion to sulphur trioxide. BASF attempted to keep it secret, although in 1895 it was 'betrayed' – by an employee. When patents were taken out in 1901 it was too late to prevent imitation, but the US General Chemical Company still found it worthwhile to take a licence from BASF to build the first contact plant in the USA in 1906. As so often in the chemical industry process 'know-how' was important as well as the patents themselves.

But perhaps the most spectacular example of the successful marriage between fundamental chemistry and strong process engineering capacity was the development of the Haber–Bosch process for synthetic nitrogenous fertilizers. Already, before 1900, BASF had been experimenting with various processes, but it was in 1908 that 40-year-old Fritz Haber at Karlsruhe Technische Hochschule found a way to synthesize ammonia, based on reacting nitrogen and hydrogen at very high pressure and temperature in the presence of a catalyst.

BASF made an agreement with Haber for the development of his process and a strong development group led by Carl Bosch succeeded in designing and constructing the necessary pressure vessels and compressors to launch commercial production in 1913. This also involved improving and cheapening the catalyst. Some idea of the magnitude of the achievement may be gained from the fact that it took Brunner Mond around seven years to imitate the process (from 1919 to 1926), after an inspection of BASF plant in Oppau in 1919, and after other unsuccessful attempts during the First World War. It also took other chemical firms in France and the USA about as long, but Fauser at Montecatini was able to produce an improved process with a much higher conversion rate by 1925.

The first plant had an annual capacity of only about 10,000 tons, but many more were built in Germany during the war. BASF not only designed and developed the process, they also established an agricultural experimental station at Limburgerhof in 1914. The work of this station, and the numerous advisory centres which BASF established, made it possible for Germany to survive the effect of the blockade in cutting off Chilean nitrates, and to introduce the synthetic product very rapidly in German agriculture.

The success of the Haber–Bosch process had other important consequences. It gave BASF and the German chemical industry in general a lead in the development and operation of high pressure and catalytic processes. This proved extremely important not only for methanol (BASF process developed by Matthias Pier in 1922), but even more in the production of oil from brown coal, and in other high pressure processes developed in the 1930s.

The hydrogenation processes were sold to the US oil industry in the largest single lump sum transaction ever known for process know-how in 1929. Since developments in the oil refining and chemical industry processes are so closely intertwined we now turn to consideration of some refinery process innovations.

4.4 PROCESS INNOVATIONS IN OIL REFINING

It was in 1855 that a Yale chemistry professor had first demonstrated the phenomenon of cracking, but it was not until the twentieth century that this discovery found commercial application in a succession of revolutionary new processes. These innovations have been very carefully documented by Enos (1962a). His book is one of the best which has been written on the history of innovation and this account draws heavily on his findings (see also Enos, 1962b). The background to the new processes was the rapid increase in demand for one of the light volatile products of the oil industry (petrol), a drastic fall in demand for another light product (kerosene), and a relative decline in demand for the heavier products (fuel oil). This was, of course, in turn linked to the growth of the automobile industry, and the replacement of the paraffin (kerosene) lamp by electricity.

The demand for an improved yield of refinery products was particularly strong in the Middle West, where the output of local oilfields was already declining by 1900, transport costs were high and competition from cheap coal was strong for the heavy residual fuel oil. It was generally known in the industry that intensive heating of the residual led to 'cracking' and a crude distillation process called 'coking' was in use. This amounted to cooking heavy fractions in an open vessel at atmospheric pressure, which gave a very low yield of lighter products.

The man who introduced the first really successful commercial cracking process was William Burton, who took a PhD in chemistry at Johns Hopkins University in 1889. In his boyhood he already had his own chemical laboratory and was befriended by Brush, the electrical inventor. Perhaps the most significant fact about his first industrial employment is that he was deliberately recruited and appointed by Standard Oil to run a laboratory at the Whiting refinery in the Indiana subsidiary. Although this was in an old farmhouse and he had to make many of his own instruments, he was able to make a number of improvements in the refinery methods. As a result, he was rapidly promoted to be superintendent of the refinery and two other PhDs were appointed to the laboratory.

As refinery manager Burton was able to command the resources necessary for pilot plant experimental development in 1909 and 1910. He and his colleagues were able systematically to test out the results of cracking at various temperatures and pressures on different fractions of oil.

Their experiments at higher pressures were conducted at considerable personal risk, as the equipment available was primitive, and knowledge of high pressure work only embryonic. They were limited by the size of plates and by the riveting techniques to a relatively small scale of operations and to what now appear as relatively low pressures. Nevertheless, they were successful in developing a much improved process for gas-oil. But in 1910 the parent company refused to authorize expenditure of a million dollars to build the first plant, because of fears of explosions.

However, in 1911, as a result of an antitrust decree, Standard Oil of Indiana was divorced from the parent company and the new board authorized the expenditure. Production began in 1913 and the plant was highly successful, doubling the yield of petrol (gasoline). Marketing problems due to the colour and smell of the product had to be overcome, but Enos calculated that the total profits generated by the process for Indiana Standard were 123 million dollars from 1913 to 1922; the reduction in cost initially was 28 per cent, and ultimately with various improvements (mainly inventions patented by employees) about 50 per cent.

Enos estimates that the development cost of $236,000 was paid back ten times over in the first year of operation, 1913. Subsequently, the royalty income alone amounted to over $20 million. The patent position, both on the original invention and the improvement inventions, was strong, and Indiana charged 25 per cent of the profits for using the process. Altogether 19 companies were licensed by 1921, but they were restricted to selling in certain areas under the terms of the agreements. Moreover, they acquired the patent rights only, with no accompanying technical know-how.

These circumstances provided a significant stimulus to the development of alternative processes as well as leading to the collapse of many small refineries. In 1920 four and in 1921 five more new cracking processes were introduced, which gives some idea of the intensity of the inventive effort which followed the extremely profitable Burton process. Of these the most successful were the Dubbs process and the 'Tube and Tank' process. The Dubbs process is of particular interest because it led ultimately to the formation of a unique specialist process development company – the Universal Oil Products Company (UOP) – which has played a critical part in the subsequent history of the oil refining industry. The Cross process also led to the establishment of a process company but this did not survive.

Jesse Dubbs was manager of a small and independent Californian refinery and his son Carbon Petroleum Dubbs (believe it or not) had also managed a refinery. C. P. Dubbs was working for Standard Asphalt when this company was acquired by J. Ogden Armour, who wished to find new outlets for his enormous personal fortune from meat-packing. Under his influence Standard Asphalt sought and acquired all the patents held by Dubbs Snr. These included some which appeared to offer the possibility of a flow process for cracking and a way of bypassing the Burton patents. This was not the original intention of the inventor who was concerned mainly with the peculiar features of refining and marketing Californian crude oil. As a result of rather loose specifications, several inventions in 1909 could claim originality, and this led to many conflicts.

A new company (UOP) was set up to hold the patents and to develop processes. Dubbs had a 30 per cent holding and Armour 20 per cent. Dubbs Jnr succeeded in developing and patenting a method of recirculating the residual heavy fraction back to the cracking coil. His collaboration in the UOP research laboratory with Dr Egloff,[6] an outstanding chemist, led to the successful construction of a pilot plant in 1918, which was displayed to Shell in 1919.

However, teething troubles and an explosion in 1921 on a plant being built for Shell necessitated considerable redesign and it was not until 1923 that the process became established. The total development costs had been about $6 million, which proved a heavy drain even for Armour, who went bankrupt in 1922.

The new process proved superior to the Burton process in many different ways, most of them resulting from the inherent advantages of a flow process over a batch process. Unit size and capacity were vastly increased with a corresponding reduction in labour costs. Again, many cost-reducing improvements were introduced once the process was in operation. Electric welding permitted a big improvement in the size and performance of vessels and tubing. One of the most important developments was the introduction of a pump which could perform satisfactorily under the high temperatures and pressures of the recycle stream. The original patent for the hot oil pump was taken out by Shell, but UOP's engineering department improved the design. This invention alone increased capacity by 40 per cent.

Shell was given a 25 per cent reduction in its royalty rate and this illustrates an important aspect of UOP policy. As a process development company it insisted on reciprocal exchange of know-how and process improvements and also provided technical assistance and performance guarantees. The company insisted on making the process available to all comers, without discrimination, even though it had requests for exclusive licensing. Originally the Dubbs process was licensed to Shell and many small West Coast refiners, but in 1924 Standard Oil of California also took a licence. By 1930 together with Shell it was paying over $2 million per annum in royalties to UOP. It had not expected to pay on this scale, as a lawsuit initiated in 1914 by UOP against Indiana Standard was generally expected to lead to the nullification of both sets of patents. However, the legal process dragged on for fifteen years and was only finally resolved in 1931 when Shell, California, and Indiana with two other oil majors bought UOP for $25 million. The royalty rate was gradually reduced from the original flat rate of 51 cents per barrel to 10 cents in 1934, 5 cents in 1938 and 3 cents in 1944. By this time superior processes had been developed and in particular the royalty rate on the fluid catalytic cracking process had been announced at 5 cents per barrel.

However, UOP continued to follow a relatively independent policy under the new regime. Hiram Halle, who had been an exceptionally able and enterprising president since 1916, was to stay in office for another fifteen years under the purchase agreement terms. He insisted on an offensive R&D policy and accepted Egloff's proposal to recruit some of the best scientists in the world to work on catalysis. Egloff argued that contemporary cracking processes were close to maximum efficiency and a new breakthrough would be needed. As a result UOP recruited Ipatieff, an outstanding Russian chemist, Tropsch and several other German chemists with experience of IG processes. On the basis of this far-sighted policy UOP was responsible for many other process improvements, and made a major contribution to the fluid catalytic cracking process. One of their many contributions which was adopted by almost all refineries was platinum catalytic reforming ('platforming'), introduced in 1949. The oil industry came to accept UOP's unique role and in 1944 transferred ownership to the American Chemical Society, but in 1958 it once more became an independent company, conducting R&D, selling process designs and providing technical consultancy – a firm producing and trading in new knowledge.

While the Dubbs process was largely developed by an independent process company, the other major continuous flow process of the 1920s – the 'Tube and Tank' – was the result of a deliberate decision by the largest oil company (Standard Oil of New Jersey) to set up its own specialized R&D department. This was originally named the Standard Development Company, but from 1955 became Esso (later Exxon) Research and Engineering. Established in 1919, by 1920 it already employed over fifty people and many hundreds by the 1930s. It engaged university consultants and also attached great importance to patent work, because of the tangled patent situation with many overlapping patents in the 1920s. The inventor who played the biggest part in developing the 'Tube and Tank' process was Edgar Clark. He had been one of Burton's collaborators, but the process was deliberately not given the name of an individual as it was felt to be largely the result of teamwork. Although the process was developed fairly quickly and in operation by 1921, its main advantages only became apparent later during the course of many improvements made by the development group. These made it possible to process a great variety of crude stocks, which could not be handled by the Burton process, and to raise the operating pressure from 95 to 100lbs per square inch by 1930. A big improvement in performance was made as a result of systematic analysis and redesign of heat exchange in the recycling system and the introduction of the Pacific hot oil centrifugal pump in 1929. In spite of heavy legal and patent costs (over $1 million) the process was extremely profitable, making about $300 million. The complicated patent litigation was resolved by the formation of a licence exchange agreement in 1923 and, as we have seen, the purchase of UOP by five oil companies in 1931.

It was widely realized in the 1920s that further progress in cracking was likely to come from catalytic techniques. Already in 1915 the Gulf Refining Company had attempted to introduce catalytic cracking with an aluminium chloride catalyst, but the process was abandoned because of the high costs and lack of a satisfactory method for regenerating the catalyst which became clogged with an accumulation of carbon. Several oil companies continued to experiment with catalysis in the 1920s, but the first successful process was invented outside the industry by a wealthy French engineer, Eugène Houdry. His father was a manufacturer of structural steel, and Houdry at first entered the family business, but he gave it up in 1923 to devote himself completely to experimental work on new types of motor fuel, originally from lignite. In 1925 he became interested in catalysis of oil fractions and, with the assistance of friends who included both chemists and engineers, he experimented with hundreds of catalysts. By 1927 he had succeeded in cracking experiments with a type of clay (silicone and aluminium oxides).

Houdry published his findings and the major oil companies, including Standard Oil (New Jersey), Shell and Anglo-Iranian (BP) sent technicians to visit his laboratory. However, they were sceptical of the possibilities of introducing the process commercially and, in the case of Standard Oil, had greater hopes of other process developments. Houdry continued to spend his private fortune, but reached a point where he could not make further progress without additional financial support, and even more without the participation of an oil company to help in the design, construction and test of the pilot plant equipment at a refinery. He was also disappointed over the withdrawal of official French government support for pilot plant work on another lignite process in which he was interested. These circumstances led to his emigration to the United States and the formation there of the Houdry Process Corporation, jointly with the Socony Vacuum Oil company in 1930. The development of the process continued to give difficulty and in 1932 he reached a new agreement with Sun Oil, which gave more substantial research and engineering support, as well as taking a one-third holding in the Houdry Corporation. It was only after the expenditure of about $11 million that the process was successfully introduced by Socony Vacuum and Sun in 1936–7. Of this amount, $3 million came from Houdry's own private fortune and about $4 million each from Sun and Socony.

This outline indicates clearly the difficulties confronting the independent inventor in process development, even when he had a large private fortune and almost unlimited perseverance and enthusiasm. The scale and complexities of pilot plant work had reached a point in the 1920s where collaboration with an established oil or chemical company was often essential for experimental development work. A critical factor in the ultimate success of the Houdry process, as so often in process development, was the introduction of a new piece of engineering equipment – the turbocompressor based on a Swiss design. This made the regeneration cycle for the catalyst economic for the first time, and was introduced as a result of experiments by Sun R&D staff, and a technical mission to Switzerland. The Houdry process was ultimately not only less costly than thermal cracking, but also produced much better quality products and, unexpectedly, very good aviation fuel.

The successful introduction of a catalytic cracking process by Sun Oil and Socony Vacuum was a powerful stimulus to the parallel work of Standard Oil of New Jersey. Licensing negotiations indicated that the Houdry Process Corporation expected to get about $50 million from Jersey if the Houdry process was adopted as their standard cracking process. This very tough attitude led Jersey to reject this possibility and concentrate instead on developing its own process. It was inclined in any case to do this because of the limitations of the fixed bed Houdry process, their own R&D capability, and the know-how they had acquired from IG Farben, the giant German chemical firm established in 1926 as the result of a merger between BASF, Bayer and Hoechst.

Already in the First World War, Bergius at BASF had developed a process for synthesis of oil from lignite (brown coal) by hydrogenation at high pressure. It seemed that this process might have even greater application for the heavy oil fractions than for lignite, and Standard Development reached a technical know-how exchange agreement with IG after a visit to the BASF plant in 1927. Standard agreed to supply IG with information on its heavy oil work, and in 1929, convinced of future success, it paid $35 million for the world rights (except Germany) to the IG patents

Table 4.2 Patents and royalties in the fluid catalytic-cracking process

Company	Patents contributed	Approximate share of royalties from process (%)
Standard Oil, New Jersey	296	39
Standard Oil, Indiana	96	2
Shell	38	7
Texaco	55	2
UOP	239	32
Kellogg	57	17

Source: Enos (1962a, Table 4, p. 217).

and processes. Standard had originally expected that the process would be widely applied for low-grade crude oil stocks, but although successful in the plants it built in 1930 and 1931, there were still ample world supplies of high grade crude oil. An indication of the strength of the German technical lead in high pressure process work at this time was the fact that all the valves and compressors had to be imported from Germany, because American manufacturers were not able to supply the equipment for work at pressures of 400lbs per square inch. However, what Standard did acquire was substantial experience of catalytic processes, and a very strong patent position. In the early 1930s it had already developed the 'Suspensoid' process for catalytic cracking, using a powdered catalyst, and had experimented with several other techniques.

Consequently, Standard was in a good position to challenge the Houdry process by 'leap-frogging' to a better one. From the outset it was recognized that a fully continuous flow process would be far better than the semi-continuous Houdry fixed bed catalyst system. Another major aim was to establish a process which could be used for a wide variety of crude stocks and not limited to higher grades. In pursuing these aims they made common cause with other oil companies and process companies which felt themselves threatened by the Houdry process. In 1938 a group was formed known as Catalytic Research Associates, consisting originally of Kellogg, IG Farben, Indiana Standard and Jersey Standard and soon joined by Shell, Anglo-Iranian (BP), Texaco and UOP. Interestingly, it was Kellogg, a process design and construction company, that took the initiative in convening the first meeting in London which led to the joint research programme. The group (without IG) commanded the resources of R&D facilities employing about 1,000 people (400 in Standard Development) and the work demanded the co-operation of specialists in many different fields.

The collaborative R&D, which was carried out from 1938 to 1942 to develop the fluid catalytic cracking process, was one of the largest single programmes before the atom bomb. Jersey Standard made the greatest contribution and this was recognized in the agreements which were ultimately made on patents and royalties when the process was successful in 1942 (Table 4.2). However, the two process companies, UOP and Kellogg, also made very important contributions. UOP had already developed its own catalytic processes when it joined the group and was able to contribute substantially to the patent pool. Perhaps the decisive contribution to the process was the development of the fluid bed of fine particles of catalyst propelled in the stream of oil vapours. This was achieved at Jersey Standard with the assistance of MIT Chemical Engineering Department (including two graduate theses).

Although the total costs of R&D were ultimately over $30 million, the process was extremely profitable. Jersey Standard had received over $30 million in royalties alone by 1956. By this time the process accounted for over half of total cracking capacity and had prevented the Houdry process from ever achieving more than a 10 per cent share. The original fixed bed process declined rapidly after 1943, but improved versions (TCC and Houdriflow) continued to compete. The fluid process was a triumph of the big battalions.

4.5   SCALE OF PLANT AND THE PROCESS PLANT CONTRACTOR

In both oil refineries and the chemical industry proper, new process development had become an extremely expensive business. All the main cracking processes cost more than $1 million to develop except the Burton process developed before the First World War. The two main catalytic processes cost more than $15 million (Table 4.3). The high costs arose mainly from the expense of pilot plant work, and the complexity of flow processes. Specialist groups were needed to cope with the design and engineering problems arising in each part of the plant, as well as in the overall design and heat transfer problems. At the same time economies of scale and the enormous operating advantages of flow processes put a premium on scaling up existing processes.

The first Dubbs process unit had a capacity of 500 barrels per day; by 1931 this had increased to 4,000 barrels per day. The biggest Houdry units had a capacity of 20,000 barrels per day; by 1956 fluid catalytic crackers had a capacity of 100,000 barrels per day. At the same time the optimal size of plant for many other chemicals was rapidly increasing, for example, ethylene and ammonia plant, which typically had a capacity of 30,000 tons per annum just after the war, were being installed with a capacity ten times as great by 1965, and with much lower costs of production (Table 4.4).

The technical economies of scale in large process plants arise to a considerable extent from reduced capital cost, due to the simple fact that capacity is a function of volume while capital cost is a function of surface area. While the volume of a cylinder is $\pi r^2 h$, the surface area (of the metal required to construct it) is $2\pi r h$. This means that as the volume is increased the amount of metal increases by only about half as much. Most parts of a chemical plant are columns, cylinders, pipes, spheres, etc., so that the 'plant factor' is usually about six-tenths. This relationship is sometimes known as Chilton's Law. Where repetition of batch processes is involved then the plant factor will be much closer to one (i.e. no technical economies of scale). But as we have seen, large scale of operations not only yields great economies in capital costs, but also savings in supervisory labour, maintenance labour, operating labour, energy, feedstock, etc. (Similar considerations apply to oil tankers.)

Table 4.3 Estimated expenditure of time and money in developing new cracking processes[a]

Process	Development of new process		Major improvements to new process		Total	
	Time	Estimated cost $000	Time	Estimated cost $000	Time	Estimated cost $000
Burton	1909–13	92	1914–17	144	1909–17	236
Dubbs	1917–22	6,000	1923–31	>1,000	1909–31	>7,000
'Tube and tank'	1918–23	600[b]	1924–31	2,612	1913–31	3,487
Houdry	1925–36	11,000[c]	1937–42	n.a.	1923–42	>11,000
Fluid	1938–41	15,000	1942–52	>15,000	1928–52	>30,000
TCC and Houdriflow	1935–43	1,150	1944–50	3,850	1935–50	5,000

[a] Excluding preliminary activities (background research and patent acquisitions).
[b] Including $100,000 legal expenses.
[c] Including some lignite process costs.

Source: Enos (1962a, p. 238).

Table 4.4 Influence of capacity on production cost (in £ thousand) of ethylene, 1963

Production costs	Capacity and output, 000 tons		
	50	100	300
Capital investment			
Battery limits plant	2,200	3,100	5,400
Off-site facilities	650	900	1,600
Total (excluding working capital)	2,850	4,000	7,000
Current costs per year			
Net feedstock	250	500	1,500
Chemicals	50	100	300
Utilities	300	600	1,800
Operating labour and supervision	50	50	50
Maintenance at 4 per cent of battery limits plant cost	90	125	215
Overheads at 4 per cent of battery limits plant cost	90	125	215
Depreciation	250	355	620
Total	1,080	1,855	4,700
Current costs (£) per ton of ethylene	21.6	18.6	15.7

Source: Wynn and Rutherford (1964).

However, there were signs in the 1970s that diseconomies of scale were beginning to affect the construction and operation of some of the larger plants. During a period of very rapid economic growth, such as the 1950s and 1960s, new and larger plants can quickly be brought into full capacity working provided the technical problems of larger scale design are satisfactorily resolved. This means that during such periods the operation of Verdoorn's Law (which associates growth of productivity with growth of total output) can be explained in terms of technical progress associated with increasing scale of plant and similar scale economies. But the slowdown in economic growth in the 1970s, together with the bunching of major investment decisions affecting very large plants, led to the emergence of serious over-capacity in several branches of organic chemical production, especially in synthetic fibres and plastics. Prolonged below capacity working has an extremely adverse effect on the economics of large plant operation because of the very high fixed capital costs. Such below capacity working may be induced either by a failure of the market to grow sufficiently rapidly or by technical problems in the construction, commissioning and operation of new large types of plant. In either case, losses can be on a very large scale. Similar problems of diseconomy of very large plants set in during the 1970s in electric power generation and in steel after a period of rapid scaling up in size of power stations. In the steel industry and several others, as well as some types of chemical plant, new processes and new designs restored some advantages to smaller plants.

But despite these diseconomies of scale affecting several highly capital intensive industries, the combined effect of the shift to flow process, the use of catalytic processes, the complexity of plant design and the use of petrochemicals as the basic material has been to give the large chemical and oil companies a predominant position in new process development since the First World War. This must, however, be qualified by the observation that a new group of companies has emerged, which is also making a significant contribution: the specialist design, development and construction company (plant contractors and process companies).

The NIESR survey of 6,000 chemical plants erected in the 1960s (Freeman et al., 1968) showed that while some of the largest chemical firms still preferred to design, engineer, procure and construct their own new plants with their own 'captive' design engineering organizations, the majority of plants were engineered and built by process plant contractors. These results were confirmed by two later surveys in the 1970s and 1980s. All of these surveys pointed to the fact that the design and construction of new process plants all over the world is now a huge industry carrying with it a large volume of associated exports of mechanical equipment and instruments. This worldwide business is now dominated by the specialized plant contractors, rather than the chemical firms. However, the role of the chemical firms is still a crucial one, since they originate most of the major technical innovations in process design (Freeman et al., 1968; Mansfield et al., 1977, Chapter 3) which are subsequently exploited by the contractors under appropriate licensing and know-how arrangements. The oil industry in the United States (through firms like Kellogg, Foster-Wheeler and UOP) pioneered this change. Although chemical firms try to safeguard their process secrets by keeping contractors out of sensitive areas, they are increasingly using their services.

Only a few of the contractors have strong research and development of their own and are capable of designing and developing their own new processes. The disparity in technical strength is clearly shown by the patent statistics (Table 4.5). Mostly they depend on licensing processes from the big oil and chemical companies. But as a result of their experience in detailed design, engineering and construction of many process plants in different parts of the world for a variety of clients, they are often able to suggest and implement minor process improvements, and sometimes major ones.

Table 4.5 Patents taken out in London by chemical companies and by contractors, July 1959 to November 1966[a]

USA	UK	West Germany	France	Italy
Part A: Chemical companies				
Du Pont 1,731	ICI 2,998	Bayer 2,161	Rhône–Poulenc 307	Montecatini 709
Esso Research & Engineering 1,191	BP 641	Farbwerke Hoechst 1,431	St. Gobain 231	
Union Carbide 1,136	Distillers 491	BASF 1,136	Institut Français Petrole 92	
Monsanto 723	British Oxygen 379	Wackerchemie GmbH 195	Solvay 90	
Dow Chemical 632	Courtaulds 277	Chemische Werke Hüls 166	Soc. Nationale des Pétroles Acquitaine 22	
Rohm & Haas 399	Shell Research 224	Ruhrchemie AG 134		
Olin Mathieson Chemical 358	Albright & Wilson 100	Deutsche Erdöl AG 21		
Allied chemical 350	Laporte 96	Scholvenchemie AG 18		
W. R. Grace 350	Fisons 87			
Hercules Powder 191				
Ethyl 145				
Part B: Contractors				
Foster–Wheeler 108	Simon–Carves 106	Lurgi (Metallgesell.) 246 (377)	Heurtey 22	SNAM Progetti 24
Scientific Design 85	APV 67	Linde 141 (236)	L'Air Liquide 21	Oronzio de Nora Impianti
M. W. Kellogg & Pullman (chemical patents only) 59	Power Gas 38	Klockner–Humboldt–Deutz 131 (42)	Krebs et Cie 2	Elettrochemical 7
Air Products & Chemicals 47	Woodhall Duckham 35	Heinrich Koppers 117 (65)		
Chemico 46	Humphreys & Glasgow 25	Didier Werke 33 (15)		
	Whessoe 21			
	Matthew Hall Eng. 17			
	Petrocarbon Devt 14			

[a] Patents taken out in London and therefore with a national bias to British-based firms. German patent statistics are shown for German contractors for comparative purposes. The German statistics exclude coke ovens and coal installations. Patent protection for merger innovations is normally sought.

Sources: *Name Index to Complete Specifications*, Patent Office, London. Figures in parentheses from Patent Office, Munich.

Moreover, the NIESR plant index showed that although over 80 per cent of new plants used processes originally developed by the major chemical and oil companies, a significant minority was either originated by specialist process companies or embodied modifications to basic processes developed by contractors, and marketed as proprietary processes.

Excluding construction site labour, plant contractors typically employ over half their staff on process design and engineering (i.e. detailed flow sheets and drawings) and the remainder on procurement, sales and administration, with very small numbers on research and development. But a few contractors, such as Scientific Design and Lurgi, and process companies, such as UOP and Houdry, had a significant R&D activity. Although their total size is relatively small compared with the chemical companies they are sometimes able to afford the expenditure required to develop a new process. Scientific Design developed new catalysts for producing maleic anhydride and ethylene oxide in the early 1950s, and as a result their processes accounted for over a third of world capacity by the mid-1960s. But this is exceptional, and in each case development costs were over $1 million. Much more typical is the situation where a process company or a contractor collaborates with a large chemical or oil company in process development or improvements. The examples of Houdry and UOP have already been discussed and American companies have generally been more ready for this kind of collaboration than European.

Mansfield et al. (1977) has shown that in the United States the contribution of the four largest chemical firms to major innovations was greater for product innovations than for process innovations in the period from 1930 to 1971 (1966 for product innovations). But in any case these firms accounted for over 55 per cent of all innovations, except in the case of process innovations from 1950 to 1971, when their contribution fell to just over 40 per cent. This fall can be attributed to the increased contribution of the process plant contractors, such as UOP and Scientific Design, and of the oil companies, such as Shell. The chemical industry in the United States is less concentrated than in Europe and Mansfield's study provides valuable independent confirmation of some of the main conclusions of the NIESR/SPRU work.

Further examples of design and development collaboration were the Sohio–Badger acrylonitrile process and the ICI–Kellogg ammonia process. It is likely that the large oil and chemical companies will remain the major source of new processes, because of the scale of their R&D, and their experience in plant operations. Advances in science might conceivably reverse this trend to some extent. Pilot plant work in the development of new processes is still often necessary because scientific knowledge of the probable behaviour of liquids and gases is still not sufficiently precise to be able to predict with complete assurance the effects of scaling up. Insofar as the growing precision of scientific knowledge and the use of computers for design calculations make it possible to eliminate the pilot plant stage in process development, this could redress the balance in favour of smaller firms and R&D groups. However, the complexity of the design problems would still militate in favour of large R&D laboratories. The relevance of these R&D scale problems to size of firm are discussed more fully in Chapter 9.

While corporate R&D has come to dominate the major revolutionary leaps in technology, it is important not to overlook the steady growth in productivity brought about by relatively minor technical improvements to existing processes. Enos's account of technical innovation in the oil industry confirms the results of Hollander's (1965) very detailed study of technical change in Du Pont rayon plants on this point. Hollander shows that the greater part of productivity increase was attributable to minor technical advances introduced largely as a result of the activities of the engineering department and technical assistance groups. The process of technical change in industry thus takes two main forms: radical innovations in products and processes which have increasingly originated in professional R&D laboratories in universities, industry and government; second, incremental improvement of products and processes associated with increasing scale of investment and learning from experience of production and use. However, it would be dangerous to conclude that scale economies have reached their limits in these or other industries. In his interesting discussion of this problem, Gold (1979) points to the fact that the Japanese doubled the size of blast furnaces beyond the established optimum level.

The brief account of process innovation given here has stressed the shift from the inventor–entrepreneur of the nineteenth century towards large-scale corporate R&D. This differs sharply from the interpretation given by Jewkes and his colleagues in their classic study of *The Sources of Invention* (Jewkes *et al.*, 1958). They minimize the differences between the nineteenth and twentieth centuries and generally belittle the contribution of corporate professional R&D. They argue that most important twentieth-century inventions were the result of the work of individual inventors as in the nineteenth century, either freelancing or working in universities. Part of this difference in emphasis is due to the fact that Jewkes was concerned primarily with invention, whereas this account is concerned primarily with innovation. Jewkes and his colleagues concede that the costs of development are often so high that large-scale corporate R&D may be necessary to bring an invention to the point of commercial application. Of the 40 major twentieth-century inventions which they attribute to individual inventors (compared with 24 attributed to corporate R&D) at least half, according to their own account, owed their successful commercial introduction to the development work and innovative efforts of large firms.

Thus Jewkes *et al.* emphasize most strongly the contribution of Houdry as an individual inventor, whereas this account has emphasized his collaboration with the oil companies in the later stages of his work, and the even heavier costs of the fluid catalytic cracking process, developed almost entirely through corporate R&D. However, it is important to note that Enos supports Jewkes' view of the origin of inventions in the oil refining industry.

It can reasonably be maintained that, from the standpoint of economics, it is innovation that is of central interest, rather than invention. This is not to deny the importance of invention, nor the vital contribution of creative individuals to both invention and innovation. There is no inconsistency between Jewkes' emphasis on the importance of university research and invention and the interpretation which is given here. Nor is it denied that the lone wolf and the inventor–entrepreneur still play an important part.

(This is discussed at greater length in Chapters 8 and 9.) But here too, even on Jewkes' own account of major inventions, there has been a shift over time since the early part of the twentieth century towards a larger contribution from inventors associated with corporate R&D. Moreover, he conceded that this contribution has been particularly important in the chemical industry as compared, say, with mechanical engineering. If, as is argued here, the chemical pattern is gradually becoming more typical of industry as a whole, then the differences between nineteenth- and twentieth-century invention and innovation cannot be so lightly dismissed.

Not surprisingly, in view of the difference in approach, Jewkes and his colleagues dismissed nuclear power somewhat contemptuously and did not even count it in their list of important twentieth-century inventions (although they included rockets, cinerama, the self-winding wrist watch, and the zip fastener). The omission was deliberate and was explained on the grounds that the subject was too large and still surrounded by too much secrecy. They also justifiably poured much cold water on the over-optimistic estimates of the economic advantages of nuclear power.

It is perfectly true that some of the early hopes associated with nuclear power have not been fulfilled and in particular that there has been a major problem of cost escalation and disappointingly low load factors with most types of reactor, except the Canadian CANDU (Surrey and Thomas, 1980). There have also been other legitimate public anxieties over the possibility that the rapid worldwide diffusion of civil nuclear power may increase the dangers of nuclear weapons proliferation and the ultimate possibility of very serious accidents arising from human errors. These fears and anxieties have greatly increased since the very serious accident in the Ukraine at Chernobyl. The issue of future energy supplies and technologies is dealt with more fully in Part Four.

It would be a reasonable objection to the very high priority accorded to nuclear power in public allocations for R&D that it pre-empted almost all the available resources for new energy technologies. The strength of the political lobbies associated with the centralized powerful atomic energy authorities was an important factor in this outcome and is further discussed in Part Four.

Public involvement in the civil development of nuclear power arose from the government's role in weapons development and Jewkes et al. may be right in believing that military security slowed down the early civil work on reactors for power systems, increased its cost and inhibited the participation of some groups in industry and universities which might have been able to make a fruitful and critical contribution. They may also be right in believing that some public nuclear R&D programmes were unnecessarily lavish because of the lack of normal commercial constraints. Duncan Burn (1967) has provided strong supporting evidence. However, it is difficult to see how reactors could have been developed except at very high cost and without a considerable degree of government participation in financing the R&D. In fact, nuclear reactors have not been developed anywhere in the world without such public sector involvement. The new fast breeder reactors under development in the 1980s and 1990s have proved even more dauntingly expensive. A very disquieting feature of the R&D on fast breeder reactors has been the large-scale systematic and persistent underestimation of real development costs which has character-ized all the main projects (Keck, 1977, 1980, 1982).

The need for such massive public investment arose because nuclear engineering processes carry to an extreme degree all the tendencies which have been discussed in relation to chemical and oil refinery processes. The very heavy costs and long gestation period arose from the extraordinary complexity of the design problems, involving new materials, instruments, components and equipment of all kinds to satisfy the exacting require-ments and safety standards of the new technology.

NOTES

1. Per stream-day; approximately 7.5 barrels per metric ton.
2. On the history of the chemical industry see Haber (1958, 1971); Hardie and Pratt (1966); Achilladelis (1973), Achilladelis *et al.* (1987, 1990).
3. Nobel's brother was killed and explosions or fires affected Solvay seriously.
4. A New Zealand scientist has christened this phenomenon the 'sailing ship effect' (Ward, 1967, p. 169).
5. On the early history of BASF and Hoechst see BASF (1965); Baumler (1968); Beer (1959).
6. Ultimately Egloff became Director of Research and held 300 patents.

CHAPTER 5

SYNTHETIC MATERIALS

In Chapters 3 and 4 we have illustrated the changes in the pattern of innovation which took place in the late nineteenth and early twentieth century. The inventor–entrepreneur or the inventor collaborating with an entrepreneur in a business partnership increasingly gave way to the professional R&D department. In the fastest growing industries, such as oil, steel and electrical products, some very large firms emerged, profiting from scale economies in production, marketing and finance, as well as in R&D. A very similar pattern also emerged in the chemical industry which was to become one of the most R&D intensive industries. The new R&D departments often collaborated with independent inventors, universities and government institutes, but the inhouse R&D laboratories became characteristic institutions of the industry. Small firm innovation certainly did not disappear but its relative importance declined. In this chapter, we concentrate on one of the fastest growing sectors of the chemical industry which was also one of the most research intensive – synthetic materials.

Plastics, although almost entirely a twentieth-century industry, are already one of the world's main groups of industrial materials. World plastics consumption by weight already exceeded that of non-ferrous metals by 1970 and in volume terms it is far greater. Synthetic rubber overtook natural rubber consumption in the 1960s, and synthetic fibres already accounted for nearly half of total fibre consumption by 1990 (Table 5.1). In the 1980s and 1990s, the growth rate in the production of both metals and synthetic materials slowed down considerably. In part this was the result of generally slower growth in the world economy, in part because of higher energy costs following the oil price increases in 1973 and 1979 and in part because of reduction in materials intensity in some branches of production. However, synthetic materials continued to increase their share of total materials consumption.

Their growth rate has been extremely high, largely because they have outstanding technical and cost advantages in a wide range of applications, and because of actual and anticipated shortages of naturally occurring materials. From the 1940s to the 1970s, competitive substitution for older materials played a big part in the very high rates of growth of production and consumption. Inevitably, as they came to account for a high proportion of the combined total consumption, their rate of growth slowed down in the 1970s and began to asymptote towards the slower rate of the natural materials. Plastics are light, easy to fabricate and install, frequently have good electrical insulation and excellent resistance to corrosion and pests. With synthetic rubbers they can increasingly be tailormade or blended to meet the requirements of any particular application, but they have the disadvantage of the rather limited temperature range within which most of

Table 5.1 World production of various materials

Material	Volume of world production[a] (million metric tons)						
	1913	1938	1950	1960	1970	1980	1990[b]
Plastics	0.04	0.3	1.5	5.7	27.0	40.0	52.2
Aluminium[e]	0.7	0.5	1.3	3.6	8.1	11.2	16.7
Zinc[e]	0.8[d]	1.4	1.8	2.4	4.0	4.8	5.2
Copper[e]	1.0	1.8	2.3	3.7	6.1	7.1	8.5
Steel	53.0[e]	88.0	153.0	241.0	448.0	480.0	491.0
Synthetic rubber	—	0.01	0.5	1.9	4.5	7.7	9.8
Natural rubber	0.12	0.92	1.9	2.0	2.9	3.7	4.4
Synthetic fibres[f]	—	—	0.12	0.65	4.5	8.4	11.2
Cotton, lint	..	5.2	6.0[g]	7.1	7.7	9.1	10.1
Wool, raw	..	1.6	1.7	2.1	2.2	2.2	2.3

— Nil.   .. Not available.
[a] excluding USSR, China, Eastern Europe.
[b] provisional estimates for 1990.
[c] primary refined production.
[d] zinc 1909.
[e] steel 1910.
[f] pure synthetics, i.e. excluding rayon, including China, E. Europe
[g] cotton 1951.

Sources: UN Yearbook of Statistics, New York: Saechtling (1961); *UN Monthly Bulletin of Statistics*; author's estimates.

them can be used. Some of the newer plastic materials can be used at very high temperatures and are extremely strong, but are still relatively expensive. Synthetic fibres also have properties of strength, durability and resistance to pests, which in many cases surpass or supplement those of natural fibres, greatly enlarging the range of possibilities for the textile industry. Paradoxically their very virtues lead to problems of waste disposal and pollution.

As they are usually defined, synthetic materials differ from similar older manmade materials, such as glass and ceramics, in their organic origin. They are composed of giant molecules of organic substances based on long chains of carbon atoms. For casein and cellulosics, such as rayon, these chains or polymers are of natural origin, but the vast majority of the newer materials are synthesized from simple chemical units or monomers, such as ethylene (polyethylene) or styrene (polystyrene). These polymers can be made to flow, on the application of adequate heat and pressure, to any desired shape, which is maintained when the heat or pressure is withdrawn. Thus the same basic material, such as nylon, or polypropylene, may be used as a fibre, or as a sheet, or film, or moulded to form a component or product of a specific shape. The fundamental chemical knowledge required to manufacture and blend the true synthetics is much greater than for the cellulosics, which were the nineteenth-century plastics innovations. Advances in plastics technology have in turn strongly influenced metallurgy, so that the new materials science has developed embracing both.

The major synthetic materials are manufactured by the chemical industry, using raw materials such as petroleum products, natural gas and coal, or intermediates derived from these materials such as ethylene, propylene or acetylene. The basic chemical producers may deliver the materials in the form of solid or liquid resins, moulding or extrusion compounds or emulsions to the fabricating industry, just as the metal producers deliver metal to the engineering industry. Alternatively, the chemical producers may turn out film, sheet, fibre, rods, tubes and other mouldings and extrusions, or fabricate more complex products themselves. The extent of vertical integration varies but because of their know-how and the economics of integrated production, chemical firms have become increasingly involved in the textile industry, and to a lesser extent in the manufacture of building and packaging materials and engineering components. Another important factor which impelled the chemical firms to enter these industries was the need to develop new markets for their materials in areas which were resistant to technical change or unwilling to experiment with new materials. Although their advantages are now widely recognized, in the early days of synthetic materials they were treated with almost universal scepticism. Frequently it was the shortage and high price of natural materials or the prospect of war which led to experiments with new materials and ultimately to innovations.

5.1   THE EARLY SYNTHETIC MATERIALS
AND THE INVENTOR–ENTREPRENEURS

Of the sixty or so major synthetic materials (plastics, rubbers and fibres), all except about a dozen were innovated by large chemical firms (see Freeman *et al.*, 1963; Hufbauer, 1966). These exceptions were mainly the earlier innovations which were usually brought to the point of commercial application by inventor–entrepreneurs, such as Baekeland (bakelite), Chardonnet (rayon), Parkes and Hyatt (celluloid). The professionalization of R&D and the role of the industrial R&D laboratory is clearly apparent as we move into the twentieth century.

None of the synthetic materials was invented, developed or innovated by the suppliers or fabricators of the natural materials and metals for which they were largely substituted. But after the new materials had been introduced a few of these firms took an interest in them, made improvements and developed new applications. In the case of fibres and rubbers, some of the principal users of natural materials, such as tyre-makers, played a similar role.

Perhaps the man who had the strongest claim to be the original inventor of the first true synthetic was the British chemist Alexander Parkes, who called his material Parkesine. Patented in 1865, it was made from cellulose nitrate and oils and was the forerunner of celluloid. But the Parkesine Company which Parkes established to exploit his invention went bankrupt. The combs and other products which he sold were defective, because he had failed to solve the problem of suitable plasticizers and solvents satisfactorily. Although he was not formally trained as a chemist, Hyatt successfully used camphor as a plasticizer so that his company in the United States established celluloid as the first commercially viable plastic material. This illustrates at the very outset the international character of the inventive process in this industry and its uncertainty.[1]

In a similar way in the rayon industry, a series of British inventions solved some of the critical problems of inflammability which caused the

French inventor–entrepreneur, Count Chardonnet, to suspend production of his new nitrocellulose rayon in Besançon. Like Parkes, Chardonnet could claim to be a pioneer, but he too launched production (in 1884) before his product was in a satisfactory state. The fibre was weak and brittle and the available textile machinery was not adapted to it. However, Chardonnet had the resources to persist and was able to resume production in France and Germany in the 1890s. But viscose rayon and cellulose acetate later proved far more successful than the original nitrocellulose rayon. The viscose process was the invention of a British consulting chemist, C. G. Cross, in 1892, but it was some time before a spinning technique was developed.

Another early plastic material to be pioneered by a new firm was Galalith made from casein and formaldehyde and widely used for buttons. The German chemist, Spitteler, collaborated with Krishe, a businessman, in setting up the successful International Galalith Company (Vereinigte Gummiwaren) in 1899.

Although the first innovations in the rayon industry were made by inventor–entrepreneurs, it was not long before much larger firms with greater financial resources became involved. The Viscose Spinning Syndicate which had been formed in 1900 to exploit Cross and Bevan's viscose rayon process, sold out to Courtauld's, the dominant firm in the British silk industry, in 1904. From then onwards Courtauld's held a monopolistic position in the British rayon industry and later a very powerful position in the world industry.

Similarly, Hyatt's dental plate company in Albany, which first made a commercial success of celluloid, was soon taken over by a much stronger financial grouping. Brandenberger, the Swiss-born French chemist who invented cellophane film and took out world patents in 1912, came to an arrangement with the French rayon cartel to form a new subsidiary, known as 'La Cellophane', to manufacture and sell his product. It came on the market in 1917. Thus a new pattern began to emerge in twentieth-century industry, in which the role of the inventor–entrepreneurs became less significant, and the large rayon firms dominated product and process improvement.

However, the most successful of the early inventor–entrepreneurs was undoubtedly Leo Baekeland, the Belgian chemistry professor, who set up his own firm in the United States in 1910 to manufacture the condensation plastic which bore his name. He had previously worked for several years as a private inventor and in American industrial research. Not only did he take out the original patents for Bakelite, a phenol-formaldehyde resin, he also personally pioneered most of the early applications. He built up a very successful commercial enterprise with subsidiaries and licensees in many countries. Although it accounts today for less than 5 per cent of total plastics consumption, Bakelite was the most important synthetic resin in the inter-war period.

5.2 THE MAIN SYNTHETIC MATERIALS

The synthetic rubbers and almost all of the other major synthetic materials and fibres introduced after the First World War were innovated by established large chemical firms, with extensive research and experimental development facilities. Very often, of course, fundamental chemical discoveries and inventions were made in university laboratories, and in particular Staudinger's work at Freiburg on long chain molecules provided the theoretical basis for many of the industrial advances of the 1930s. But years of intensive applied research and pilot plant work have usually been necessary to take a material from the stage of a laboratory curiosity to that of a commercially viable process for a reasonably homogeneous and stable product. In addition, a great deal of applied research and experimental development work has been necessary to explore the vast number of potential applications, and to modify the materials to create the variety of grades or blends to suit each particular end-use.

Noreen Cooray (1980), in her study of the substitution of synthetic rubbers for natural rubber, found that one of the main comparative advantages of the synthetics was the breadth and depth of the applications research, which meant that there were tailormade modifications, blends and specialities for a great variety of specific requirements. She came to the conclusion that natural rubber could only compete effectively if the producers of natural rubber organized a comparable applications research programme and did not confine themselves simply to improving the techniques of cultivation, even though that type of research had been very successful in producing high yielding varieties of tree and reducing costs and prices of natural rubber.

To some extent the advantages of polymer R&D and applications research were cumulative, reinforcing the technical leadership of the strongest industrial laboratories. The early inventor–entrepreneurs usually financed their own experiments and took out their own patents before establishing an innovating firm. The pattern changed completely with the newer synthetics. An example which indicates this change very clearly is that of Terylene, the ICI polyester fibre (known under the trade name of Dacron in the USA and Trevira in Germany). This fibre was invented in 1940 in the small R&D laboratory of a moderate sized textile firm, the Calico Printers' Association. But Calico Printers were unable to develop the process and innovate the fibre. It was licensed to Du Pont in the United States and to ICI in Britain, which were able to bear the extremely heavy expense of pilot plant work, applications research, trial production and trial marketing of the new fibre. It was estimated that the total costs were of the order of £10 million for ICI and a similar sum for Du Pont, before they were able to market the product in 1950.

In many other cases both the inventions and the development took place in the R&D laboratories of large chemical firms, as with nylon, polyethylene, PVC, and Corfam, most of which are briefly described at the end of this chapter. But even where the original discovery or invention was made in a university (as with neoprene and methyl methacrylate) or by a smaller firm (as with Terylene), the costs of development work, the problems of marketing and the scale of investment in new plant led to the actual innovation being launched by the larger firms.

Among the firms which had an outstanding record in technical innovation in synthetic materials were IG Farben, Du Pont and ICI, the largest chemical firms in Germany, the USA and Britain respectively. Not only did these companies account for a number of the most important synthetic innovations themselves (PVC, nylon, polyethylene, polyesters, acrylics, polystyrene, buna, neoprene, etc.), they were usually among the first 'imitators' or 'adopters' of innovations made by others and played a considerable part in developing new machinery and a wide variety of new applications. One of the major innovative achievements of each of them is discussed, but first the overall contribution of the German chemical firms is reviewed.

5.3 RESEARCH AND EXPERIMENTAL DEVELOPMENT AT IG FARBEN

As we have seen in Chapter 4, German chemical firms had already established a strong tradition of generating their own new products and processes during the nineteenth century, building up the synthetic dyestuffs industry to become world leaders on this basis. Thus they were already accustomed to heavy long-term investment in R&D programmes long before they formed the IG Farben Trust in 1925. Regular statistics of expenditures on R&D were not kept in any country until the 1950s, but some firms who pioneered professional R&D kept records for earlier periods, and fortunately we have such figures for IG Farben (Ter Meer, 1953). It is true that these estimates are not completely consistent with modern definitions[2] but discussion with German chemists suggests that these differences were small.

From 1925 to 1939 it is fairly clear that, in absolute terms, the R&D activity in IG Farben in synthetics was far greater than in any other firm, and indeed its total R&D programme was the biggest in the world. IG Farben's total R&D expenditure averaged just over 7 per cent of turnover from 1925 to 1939. This is a higher ratio than in most large chemical firms since the war, which are mainly in the range of 3 to 5 per cent. From 1925 to 1931, IG's expenditure was between 7 and 10 per cent, but in the world recession it was cut back fairly drastically to 4.9 per cent in 1933. It is notable that the main economies were in development expenditure, and that the research staff were maintained throughout at more than 1,000 qualified scientists and engineers. From 1934 to 1939 research and experimental development expenditure rose again to between 5 and 6 per cent of turnover. These figures exclude technical services and extra-mural grants and donations to universities (which are sometimes included with R&D to give a misleadingly inflated figure) and exclude capital expenditure on new laboratories and instruments etc. This would normally add about one percentage point to the figures, making them even more impressive. Throughout this period the firm spent more on research than it distributed in dividends. According to one account, it was the need to concentrate research resources and make large investments in high polymer chemistry that finally persuaded the constituent firms to come together in 1925 to form the trust. It is an indication of the high priority given to research that there was no pruning before the world depression, but a substantial increase from an already high level of expenditure.

Another notable feature of IG's pre-war R&D was the importance attached to close contact and co-operation with fundamental researchers in the universities and other academic institutes. This continued the approach of the constituent firms, Bayer, BASF and Hoechst, all of which had a tradition of employing outstanding academic consultants, including a number of Nobel Prize winners. They funded a great deal of research in German universities and tried to create conditions in their own laboratories which would attract the best chemists. Both the management and the R&D departments were dominated by graduate chemists. In the case of synthetic materials, the outstanding world authority on macromolecular chemistry, Professor Staudinger of Freiburg, was an active consultant with IG throughout the inter-war period. It is notable that Ziegler, who made the major theoretical contribution to the post-war plastics industry, was already a consultant to Hoechst when he made the critical discoveries at Mülheim which led to the innovation of low-pressure polyethylene in the 1950s. A new consultancy agreement was then signed which covered licensing arrangements for his process.

Finally, we may note that IG Farben sometimes followed a strategy of parallel teams in experimental development work. No doubt this was partly due to the local pressures of the constituent firms, but it was also a deliberate strategy of the co-ordinating research apparatus, which adopted a philosophy of experimenting with several alternative routes in developing processes for the new synthetic materials.

5.4   PATENTS AS A MEASURE OF INVENTIVE OUTPUT

We may now turn to the output of IG Farben's research and experimental development. The scale of their R&D was undoubtedly much larger than that of any other privately owned firm in the world in the inter-war period. But how effective was it? In order to answer this question we must first spend some time discussing patents and other output measures.

The measurement of efficiency in R&D is one of the most complex problems in management economics, and there is no simple answer to this question. Inputs into R&D can be measured and reduced to a common financial denominator, but even here there are serious complications, such as the attribution of information inputs from outside the formal R&D structure. When it comes to the measurement of output the difficulties are even greater. They are discussed in some detail in UNESCO (1970) and in a special issue of *Research Policy* on output measurement (1987, vol. 16, nos. 2–4). Here it is only possible to indicate two complementary ways of approaching the measurement of effectiveness of IG Farben's R&D – in terms of numbers of patents and numbers of innovations. It is not suggested that either of these is satisfactory or would be employed if better information were available. But taken together they do enable us to give some kind of answer on the basis of cost effectiveness, although not on the basis of profitability of innovations. To assess the profitability of IG Farben's R&D we would need far more detailed information about the costs and profit margins on their new products than is ever likely to be published. There is also the enormous complicating factor of the German war economy.

Patents are a measure of inventive output rather than innovative success, and therefore should be used together with some measure of innovation. But provided their limitations are kept in mind they are probably much more useful than is commonly believed. Schmookler did more than any other economist to demonstrate their value in economic history and he concluded from his extraordinarily thorough studies of inventions in the railroad industry, petroleum refining, paper-making and construction that patent statistics provided a more satisfactory indicator of inventive output in the United States from 1850 to 1950 than lists of important inventions. In his view, they reflected all the minor and improvement inventions and avoided the bias inherent in any subjective assessment of importance (Schmookler, 1966). While accepting Schmookler's point about the value of aggregate patent statistics as a measure of incremental invention, we cannot follow him on the question of key patents as a measure of radical inventions.

One of the limitations of patent statistics is the variation between industries and firms in propensity to patent. For example, in defence-oriented industries there is usually a lower propensity to patent, and this might account, to some extent, for the much lower ratio of patents per unit of R&D expenditure in aircraft and electronics (Tables 5.2a and 5.2b).[3] However, in the chemical industry there is a high propensity to patent. It is quite exceptional to find a chemical firm which does not attach importance to securing patent protection for its inventions, and it is difficult to identify any major technical advances in the plastics industry which were not the subject of patenting activity.

It is generally agreed by those who have attempted to use these statistics that annual fluctuations and variations in quality are such that it makes more sense to analyze the figures for groups of five or ten years. Another complicating factor arises from the variations in national patent legislation.[4] To overcome these difficulties, Pavitt and Soete (1980) have pioneered the use of statistics of foreign patents taken out in the United States, which puts all countries (except the USA) on a similar basis. As the USA is the most important single market, it is reasonable to assume that firms will register their important patents in that country and they have demonstrated extremely interesting results on the relationship between foreign trade performance and patents taken out in the United States (Soete, 1981; Pavitt, 1982). More recently it has become possible to do similar work with European Patents since the European Patent Office began to make available its statistics. The EU publication on *European Science and Technology Indicators* (1994) provides a good comparison.

After this brief discussion of some of the problems of patent statistics, we may now return to consideration of the role of IG Farben in the development of plastics. Fortunately, in the case of synthetic materials a three-volume classified anthology of all international patents was painstakingly assembled by Delorme (1962) for the entire period, 1791–1955. This avoids double counting and any bias which exists in the statistics derived from this source is probably towards French and American rather than German firms. The outstanding feature of an analysis of these figures is the dominant position of IG Farben in the period from 1931 to 1945, and of its predecessors before 1930. The firm accounted for over a third of all patents taken out by the thirty largest firms (Table 5.3), and since large firms accounted for a high proportion of all patents this was equivalent to 17 per cent of the world total from all sources. No other firm except Du Pont has

Table 5.2a Patents delivered in various branches of British and French industry compared with research expenditure, 1961

| Industry | Percentage of total number of patents delivered | | Percentage of total research expenditure, manufacturing industry | | | |
| | UK | France | UK | | France[a] | |
			Excluding aircraft	Including aircraft	Excluding aircraft	Including aircraft
Aircraft	1.7	1.8	—	38.4	—	32.7
Electrical engineering and electronics	22.2	17.3	38.3	23.5	35.0	23.6
Instruments	6.3	10.6	4.1	2.5	0.7	0.5
Chemicals and oil products	24.0	20.6	20.3	12.5	27.5	18.4
Vehicles	5.0	6.6	4.3	2.7	11.1	7.5
Engineering	18.0	16.3	13.1	8.0	9.8	6.6
Metals and metal products	9.5	7.5	6.1	3.7	5.1	3.4
Building materials, wood and furniture, building	6.4	11.0	6.0	3.7	5.4	3.6
Textiles and clothing	5.4	6.3	3.8	2.4	3.6	2.4
Food, drink and tobacco	1.5	1.9	3.2	2.0	1.8	1.2
Total	100.0	100.0	100.0	100.0	100.0	100.0

[a] France, 1966.

Sources: Fabian (1963); Report of the Comptroller General of Patents (1961); Bulletin de la Propriété Industrielle-Statistiques (1961); Report of Advisory Council on Scientific Policy, Cmnd. 1920 (1963); 'Les moyens consacrés à la recherche et au développement dans l'industrie française en 1966' (1968), Le Progrès Scientifique, numéro spécial.

Table 5.2b Sectoral shares of US industrial patenting compared to other activities

Sector	Patents granted (1973) (%)	Total R&D expenditures (1973) (%)	Industry financed R&D (1974) (%)	Qualified scientists and engineers (1975) (%)	Manufacturing employment (1974) (%)	Manufacturing sales (1974) (%)
Food and kindred products	1.0	1.2	2.1	3.5	6.9	10.2
Textiles	0.9	0.3	n.a.	1.2	3.3	1.8
Chemicals (except drugs)	11.3	6.7	15.6	13.1	7.9	9.6
Drugs	1.3	2.9	4.4	3.1	3.4	(14.9)
Petroleum and related products	1.3	2.3	n.a.	3.0	2.7	2.3
Rubber	4.1	1.3	1.3	2.3	2.5	1.6
Stone, glass, clay, concrete	1.9	0.8	2.2	4.9	7.8	7.8
Ferrous metals	0.7	0.7				
Non-ferrous metals	0.6	0.6	2.0	7.3	5.1	3.4
Fabricated metals products	11.2	1.3	15.3	12.5	11.2	8.5
Non-electrical machinery	25.9	10.2	21.2	21.5	16.6	10.6
Electrical and electronics	20.0	30.6	23.8	18.5	16.5	14.7
Motor vehicles	2.8	11.3				
Aerospace	1.8	23.6	5.8	4.9	3.6	2.5
Scientific instruments	9.8	4.2	6.4	4.2	12.5	14.1
Other manufacturing	5.5	2.0				
Total	100.0	100.0	100.0	100.0	100.0	100.0

Source: Pavitt (1982, pp. 33–51).

Table 5.3 Patents for plastic materials taken out by leading firms

30 leading firms' patents taken out in UK, USA, France, Germany				30 leading firms' patents taken out in UK only					
1791–1930	No.	1931–45	No.	1946–55	No.	1954–8	No.	1959–62	No.
1 IG Farben	346	1 IG Farben	889	1 Du Pont	637	1 ICI	299	1 ICI	485
2 Eastman Kodak	169	2 Du Pont	321	2 Monsanto	283	2 Du Pont	288	2 Du Pont	428
3 Du Pont	78	3 Röhm and Haas[c]	145	3 American Cyanamid	266	3 Standard Oil/Esso	243	3 F. Bayer[a]	346
4 Celluloid	66	4 Hercules Powder	132	4 Shell/N.V. de Bataaf	263	4 F. Bayer	199	4 Union Carbide	327
5 Bakelite Corp	59	5 GE	120	5 ICI	253	5 US Rubber	170	5 Standard Oil/Esso	246
6 Bayer[a]	55	6 Eastman Kodak	120	6 Röhm and Haas[c]	210	6 Midland Silicones	168	6 Hoechst[a]	207
7 Meister, Lucius and Brüning[a]	55	7 Dow	115	7 Dow	187	7 Monsanto	143	7 Monsanto	205
8 CIBA	42	8 Kodak–Pathe and Kodak	115	8 B. F. Goodrich	160	8 GE	128	8 CIBA	168
9 Bakelite GmbH	40	9 ICI	90	9 US Rubber	156	9 Celanese	113	9 Dow Chemical	162
10 BASF[a]	38	10 Carbide and Carbon	88	10 Eastman Kodak	140	10 Courtaulds	109	10 Phillips Petroleum	153
GE	38	11 Phrix Arbeits-gemeinschaft	73	11 Standard Oil/Esso	131	11 Shell/NV de Bataaf	108	11 BASF[a]	145
12 British Thomson-Houston	35	12 Celanese	67	12 BASF[a]	115	12 Union Carbide	107	12 Röhm and Haas	132
13 Consortium für Elek[b]	33	13 A. Wacker Ges.	65	13 F. Bayer[a]	111	13 BASF[a]	106	13 Shell/NV de Bataaf	118
14 British Celanese	28	14 American Cyanamid	60	14 CIBA	101	14 Dow Chemical	100	14 American Cyanamid	112
15 Chem. Fab. Albert	26	15 CIBA	56	15 St. Gobain	77	15 Röhm and Haas	87	15 US Rubber	110
16 Barrett	25	16 Ellis-Foster	51	16 Distillers	74	16 American Cyanamid	76	16 Midland Silicones	109
Ellis-Foster	25	17 Bakelite	48	17 Gen. Aniline and Film	73	Dow Corning	76	17 Monsanto	100
ICI	25	18 Deutsche Hydrier-werke	46	Celanese	73	18 B. F. Goodrich	75	18 GE	92
19 Cie. Fr. Thomson-Houston	24	19 Cie. Fr. Thomson-Houston	44	19 Wingfoot	72	19 Hoechst[a]	68	19 B. F.Goodrich	83
Naugatuck	24	20 British Celanese	42	20 Cie. Fr. Thomson-Houston	69	20 Hercules Powder	64	20 Dunlop Rubber	79
21 Kroll	23	21 Standard Oil	41	Carbide and Carbon	69	Distillers	64	21 Minnesota Mining	77
Canadian Electric	23	22 Bakelite Ges.	40	22 Hercules Powder	65	22 Dunlop Rubber	60	22 Courtaulds	73
Pathe	23	Cie. Fr. Thomson-Houston	40	23 Phillips Petroleum	57	23 Phillips Petroleum	57	23 Celanese	63
24 AG für AF	21	24 Monsanto	37	24 Hoechst[a]	55	24 CIBA	56	24 Hercules Powder	60
Kunstharz Pollak	21	25 Deutsche Kelluloid	35	25 Kodak–Pathe	47	25 Wingfoot	38	25 Distillers	53
26 Chem. Fab. Griesheim	16	26 B. F. Goodrich	32	American Viscose	45	26 Minnesota Mining	28	26 Chemstrand	51
27 Carbide and Carbon	15	27 Thuringische Zellwolle	30	27 Chemstrand	44	27 Chemstrand	23	27 Rhône–Poulenc	47
PF Instruments	15	28 Rhône–Poulenc	26	28 Rhône–Poulenc	43	Rhône–Poulenc	23	28 Chem. Werke Hüls[a]	42
29 E. Schering	14	Harvel Research Consortium für Elek[b]	26	GE	43	Wacker–Chemie	23	29 Wacker–Chemie	41
30 Hercules Powder	13		24	30 Chem. Werke Hüls[a]	42	30 Gen. Aniline and Film	20	30 Dow Corning	34
Soc. Chem. des Usines du Rhône	13			Koppers	42				

a Part of IG Farben.

b Undertaking research in association with A. Wacker.

c The German and American parts of Röhm and Haas are listed together here.

Sources: Delorme (1962) and Patent Office, London.

Table 5.4 Patents issued for the principal groups of plastics, 1791–1955

	1791–1930		1931–45		1946–55	
Patents taken out by	Number	Per cent of total	Number	Per cent of total	Number	Per cent of total
Individuals	1,803	43	791	15	489	8
Firms	2,436	57	4,341	85	5,749	92
Total	4,239	100	5,132	100	6,238	100

Source: Delorme (1962).

as many as one-sixth of the number of patents taken out by IG in this period. In the field of vinyl patents it accounted for over a quarter of all world patents. Between 1925 and 1930 IG registered twice as many plastics patents as any other firm in the world for the whole period from 1791 to 1930.

Altogether more patents were taken out in plastics during the fourteen years 1931–45 than in the previous 140 years, and a notable feature of the long-term trend was the decline in patents taken out by individuals by comparison with corporate patents (Table 5.4). In the post-war period patents awarded to individuals had declined to less than 10 per cent of the total. This reflects the increased contribution of professional R&D organizations within firms by comparison with the private inventor, and the increasingly science-based character of the inventive process in this industry.

In the period up to the Second World War, German and American firms were responsible for over 80 per cent of all patents taken out by firms but, with the exception of Du Pont, the leading US chemical firms came relatively late into this field. The early plastics manufacturers, Bakelite and Celluloid, were patenting on a significant scale, as were Eastman Kodak from the film side, and General Electric, one of the most research intensive American firms with very wide interests in new materials. Both played an important part in the earlier period and are still among the leaders, for example in polyacetals and polycarbonates. Rôhm and Haas was at this time based mainly in Germany and was principally concerned with the development of acrylic materials (methyl methacrylate). By the 1930s, ICI was among the leaders although still behind IG Farben in the range of its plastics research and production. Over a long period, the highly research intensive Swiss chemical firm, CIBA, was consistently among the leading firms in numbers of patents taken out and had several major innovations to its credit, notably in epoxy resins. But with the exception of ICI and CIBA, the remainder of the European chemical industry was far behind the German and American firms.

From 1946 to 1952, IG Farben was being reorganized by the Allied military governments and was not in a position to take out any patents. Moreover, many of its secrets were compulsorily made available to British, French and American firms in 1945–6 by Allied investigation teams. It was not until 1952 that the successor firms to the dissolved combine were able to resume normal production and research activity. Consequently, the patent statistics for 1946–55 showed American firms in a dominant position with eight of the ten leading firms, and Du Pont as the established

world leader. However, the combined total of the successor firms to IG
Farben, even in this period, was greater than that of any other firm except
Du Pont.

Unfortunately, Delorme's anthology does not extend beyond 1955, but
using British Patent Office statistics it is clearly apparent that, in the 1950s
and 1960s, the leading German chemical firms continued their recovery
(Table 5.3) although not achieving the pre-war dominance of IG Farben.
British national patent statistics are biased in favour of British domiciled
firms but probably not biased in favour of German as against American
firms or vice versa, except possibly in the case of Monsanto and Standard
Oil, which had major subsidiaries operating in the UK.

Because of the importance of distinguishing the most important radical
inventions from the much larger number of incremental improvement
inventions, it is also desirable to use a separate measure of key inventions
(Baker, 1976; Clark et al., 1981). This was done at the NIESR with the
assistance of a specialist consultant, C. A. Redfarn. By this means 117 major
technical advances were identified over the period 1790–1955. An advance
might be embodied in one key patent or in several related patents. Of these
117, IG Farben were responsible for 30 out of the German total of 51, Du
Pont for 12 out of the American total of 43, and ICI for 7 out of the British
total of 15. All other countries accounted for only 8.

5.5   PATENTS AND INNOVATIONS

The evidence of the patent statistics and key technical advances can now be
compared with the achievements in innovation. The simplest method of
measuring innovations is to list all the new synthetic materials and to
identify that firm in each country which was responsible for the first
commercial production. This gives a list of innovations but is, of course,
subject to the criticism that it omits most innovations in new applications
of a material, and in new processes of manufacture. Nevertheless, it
provides a rough guide to innovative achievement, and can be adjusted to
allow for the relative importance of each material.

Hufbauer prepared such a list of 56 plastics, synthetic rubbers and fibres
and identified the first producer in a large number of countries. One
interesting result of his study was that it enabled comparisons between
countries and firms not only in terms of numbers of innovations (world's
first producer) but also in terms of imitation (first producer in a particular
country after the innovation). Hufbauer used the data to measure imitation
lags and demonstrated the extremely important result that countries with
the highest innovation rate also had the shortest imitation lag. On average
it was about three years before Germany or the USA imitated an innovation made abroad, but for Britain and France it was several years longer,
and for all other countries more than ten years. For most it was more than
twenty. This result has major implications for the theory of foreign trade,
which were explored by Hufbauer and other economists (see Posner, 1961;
Hufbauer, 1966; Vernon, 1966). Size of national market is, of course, also an
important factor affecting the imitation and diffusion process. Here we are
concerned primarily with imitation as an additional indicator of research

Table 5.5 Patents and innovations in synthetic materials (percentage world total)

Patents and innovations	Total	Percentage world total		
		IG Farben[a]	Du Pont	ICI
All plastics patents taken out by firms 1791–1945	6,777	20	6	2
All plastics patents taken out by firms 1931–45	4,341	20	8	2
'Major technical advances' in patent literature 1791–1945	117	26	10	6
Innovations in synthetic materials 1870–1945	56	32	9	2
'Major innovations' 1870–1945	20	45	10	5
Innovations 1925–45	36	44	11	3
First 'imitations' 1870–1945		14	4	8

[a] Including predecessors and successors.

Sources: Author's estimates from Delorme (1962), Hufbauer's analysis of innovations (1966) and Redfarn survey (see p. 60).

and innovative capacity. The links between innovation and foreign trade are further explored in Chapter 13.

A firm with a strong research capacity may be able to assimilate and imitate more quickly. It may also innovate almost simultaneously. Thus, during the Second World War IG Farben was able to launch independently the production of polyethylene (ICI innovation) and of nylon (Du Pont innovation), while ICI was able to launch production of PVC in 1940 and the US chemical and rubber industries were able to launch a range of synthetic rubbers (IG Farben innovations).

Comparing the patent statistics with those for innovation and imitation, it is clear that IG Farben scores well on almost all counts (Table 5.5), and so to a lesser extent do Du Pont and ICI. The IG performance is significantly better in numbers of innovations than in patents and much better in innovation than in imitation. The more successful a firm has been as an innovator in new products the less need for it to be an imitator, even where it has the capacity to imitate. The profit margin on original innovations may be much better than on imitation,[5] and the lead time over competitors encourages the firm to concentrate on its own new products. However, this must be balanced against the high risks and heavy losses associated with unsuccessful original innovations, as in the case of Corfam discussed later in this chapter. Alternative strategies relating to imitation are discussed in Chapter 11.

Where a major chemical firm does imitate the innovation of a foreign competitor, whether under licensing arrangements or independently, it will often attempt to develop a better process or a major improvement in the product. In some cases the modification or improvement may be just as important as the original innovation. When cellophane was licensed to Du

Pont in 1924, two of their research chemists found a way to make it moisture proof, which greatly increased its range of applications as a packaging material. The IG Nylon 6 differed in important respects from Du Pont's Nylon 66 and was the result of an independent R&D programme.

Furthermore, it must always be remembered that after the introduction of a new process or the construction of a new plant, many minor technical improvements will be made. Japanese firms, which were importing many technologies from Europe and North America, were apparently exceptionally good at the improvement of both processes and products but they could only do this because of their own strength in R&D (Freeman, 1987; Fukasaku, 1987). These will not necessarily be recorded in patent statistics, both for reasons of secrecy and of patentability. Hollander found that in the case of Du Pont's rayon plants, many of the minor technical improvements were not patented but were together more important in their contribution to productivity than the major changes. Moreover, they were mainly initiated by the engineering department or technical assistance groups, rather than central research. In the absence of any detailed data on the minor technical improvements in IG Farben plastics plants, it is not possible to assess IG's performance in minor innovations, except indirectly from their general competitive performance by comparison with the world industry (including exports) and by that part of minor innovation which is reflected in patent statistics.

The fact that IG scores much better in major innovations and in major technical advances than in patents, and probably better in innovations than in proportion to its R&D expenditure, may reflect in part its efficiency as an innovating organization, but it also probably reflects its powerful monopolistic position in the German war economy. Not only did IG dominate the various trade associations and consultative bodies for the chemical industry, it also controlled some important marketing outlets for its new products. This enabled it to exploit some of the cumulative advantages of scale in R&D programmes. With its immense resources and government backing for the synthetic substitute programme, IG was assured of a market for some of those synthetic materials which it successfully developed, especially synthetic rubbers.

This was by no means true in the USA or UK, except during the Second World War. Even then the greater availability and lower price of many natural materials lent less urgency to the development of synthetics. However, the influence of a strong government-backed demand was also seen in the case of polyethylene in the UK and to a far greater extent in the synthetic rubber programme in the USA, when Far Eastern natural rubber supplies were temporarily interrupted (Solo, 1980; Morris, 1989).

At one time in the 1920s IG had offered to sell its synthetic rubber patents to the world natural rubber cartel in the belief that the natural product would retain its major price advantage, but with large-scale rearmament, development work was pursued with extreme urgency and a major new industry was established, which was essential to the German war economy. However, it is notable that several of the most important plastics were developed by IG before Hitler's accession to power, including PVC, polystyrene, polyvinyl acetate, urea-formaldehyde glues and melamine formaldehyde. Consequently, IG's outstanding performance cannot be explained purely in terms of exceptional wartime demand, or rearmament, unless it is postulated that covert rearmament dominated IG research policy even in the 1920s.

Throughout the post-war period West Germany continued to enjoy the world's highest level of per capita production and consumption of plastic materials, although its synthetic rubber production was temporarily interrupted after the war. The IG Farben Trust was dissolved but the successor firms continued to make important innovations and develop new applications. Thus the overall picture which emerges is one of an industry in which the innovative process was dominated since the First World War by the largest chemical firms with strong professional R&D facilities. However, it would be misleading to think of the process in terms of massive planned research programmes leading almost automatically to the introduction of new materials. A high degree of technical and commercial uncertainty was characteristic of all these innovations, as is quite evident when we consider three of them in a little more detail.

5.6 PVC

A good example of IG Farben's research and innovation is the story of PVC, later one of the three highest tonnage plastics in world consumption. It has been relatively neglected in the Anglo-American literature, but a study by a British chemist gave a valuable detailed account which is summarized here (Kaufman, 1969).

Vinyl chloride was first prepared and described by a young French chemist, Regnault, at Liebig's laboratory in 1835. Subsequently, the polymer was also described by other academic chemists but without any inkling of its potential industrial preparation or applications. In 1912 and 1913 an industrial research chemist, Fritz Klatte, working at Griesheim (Hoechst), took out a series of patents which anticipated the industrial processes used twenty years later and some of the future applications of PVC. However, PVC is an unstable material which deteriorates on exposure to light, is extremely hard to work and may liberate hydrochloric acid when heated. The monomer is difficult to prepare and Klatte's polymerization process was unsatisfactory. It was not until these problems were solved and suitable plasticizers, stabilizers and compounding elements were developed that PVC could find extensive commercial applications. It was mainly in the laboratories of IG Farben that solutions were found to these intractable problems in the 1920s and 1930s, and then only when fundamental knowledge of macromolecular structures permitted a higher degree of understanding and control of the process.

Laboratory work continued at Hoechst during the First World War, in the hope of developing substitutes for natural materials during the blockade. About four tons of polyvinyl chloracetate were produced at Griesheim. All three of the leading German chemical firms had research programmes on synthetic rubber well before this, and Bayer was successful in producing the first synthetic rubber, methyl butadiene, during the war. Production was discontinued in 1919, as natural rubber was still superior in quality and much cheaper. In the case of PVC, Klatte's patents were allowed to lapse in 1926, although research work continued. At this time other vinyl polymers appeared to be more promising and polyvinyl acetate was produced on a commercial basis by both Hoechst (IG) and Wacker in 1928–9. This success, together with the deeper understanding provided by Staudinger's papers on macromolecules, led to a renewed interest in PVC and co-polymers, and to the first commercial production early in the 1930s.

Early efforts to polymerize vinyl chloride had been based on thermal methods, on photopolymerization (as in Klatte's original patents), or on solution polymerization (as in the Russian chemist Ostromislensky's patent of 1912). Although further patents were taken out on all these methods during the 1920s, notably by Du Pont, the breakthrough to a commercial production process came with the development by IG Farben and Wacker of dispersion polymerization processes. The emulsion polymerization process developed by Fikentscher at the IG Ludwigshafen laboratory from 1929 to 1931, although now superseded, was used in the full-scale plant at Bitterfeld throughout the Second World War.

The successful development of this process owed a good deal to the earlier work on the rubber Buna-S, a butadiene-styrene co-polymer, at Bayer, as well as to the close co-operation between Staudinger and IG Farben's research workers. This was a case where the cumulative benefits of research experience on a variety of polymers and co-polymers (acrylonitrile, styrene, methyl methacrylate, vinyl acetate, vinyl chloride, etc.) greatly facilitated IG's progress. In a few areas other leading chemical firms, particularly Union Carbide, B. F. Goodrich and Du Pont in the United States, and ICI in Britain, were not far behind IG. Indeed, some of the early advances in PVC plasticizers were made at B. F. Goodrich rather than at IG Farben. But German research on plasticizers overtook Goodrich and was well ahead in the applications of PVC and co-polymers, not only for sheathing and insulation in the cable industry, but also in soles for footwear, in floor tiling, chemical plant, packaging and many other uses. An extremely important early application was in the development of magnetic tape as a result of a joint research programme with the German electrical firm AEG. All of this work, of course, enjoyed a powerful stimulus after 1933 from war preparation and fears of shortage of natural materials.

From this outline it is apparent that it is hardly possible to speak of the invention of PVC in the same sense as the invention of the safety razor. Its successful introduction into the German economy was the result of a long series of experiments, inventions and discoveries extending over a period of thirty years with many set-backs and disappointments. It was really not one but a family of materials and even during the Second World War its use was still on a relatively small scale, reaching only about 10,000 tons early in the war. Even after the war, the possibilities of PVC were often greatly underestimated, partly because of the very poor quality of some early products in such fields as rainwear. It was the end of the 1950s before world production reached the level of a million tons. In the post-war period many further improvements were made in the production process, notably as a result of the substitution of ethylene for acetylene as the basic intermediate and the use of suspension polymerization.

US, German and British firms shared in these process developments and in the applications research which led to the establishment of PVC as a major material in construction (drainpipes, gutters, flooring, etc.), clothing and footwear, packaging and engineering, as well as wire and cable. It is now well established as a bulk tonnage plastic material, although some new environmental concerns have affected its growth.

5.7   POLYETHYLENE

The most important single plastic material was not deliberately sought but was the indirect result of other research.

The links between chemical research and ICI's polyethylene innovation have been thoroughly discussed by Allen (1967). Like nylon, polyethylene also owed its discovery to a programme of fundamental research. But, unlike the Du Pont programme and Staudinger's research, this was not oriented to work on the structure and synthesis of long-chain molecules, but to the study of high pressure reactions. The Alkali Division of ICI had a strong research tradition from its origins as Brunner Mond. Mond himself had made a number of improvements in the Solvay ammonia-soda process and established the research laboratory at Winnington. The Research Director in the 1920s, Freeth, encouraged a long-term approach to research and in 1925 recruited a young British research chemist R. O. Gibson, who had collaborated closely with Professor Michels of Amsterdam. When Gibson began to work at ICI this association continued and Michels became a consultant for ICI, because of his outstanding work on the effects of high pressure. Special equipment designed by Michels was installed at the ICI laboratory at Winnington in 1931 and was used in a series of studies of chemical reactions.

In the course of these experiments, polyethylene was discovered and its properties recognized. The discovery owed a good deal to chance in that on one occasion in 1933 a defect in the apparatus led to the polymerization of ethylene. The research engineer, W. R. D. Manning, made important improvements in the apparatus, making it relatively safe to continue the high pressure experiments. It would be impossible to identify any one man as the inventor of polyethylene, as Swallow, Fawcett and Perrin also made important contributions to the work. Patents were taken out in 1935, and the difficult and expensive development of a safe high-pressure process took several years longer. Michels assisted in the compressor design that was critically important for the pilot plant which started production in March 1938. Although ICI was quick to identify some important properties of the new material, it was a long time before most of the present-day applications of polyethylene were realized or even conceived. Although its possibilities as an insulator were recognized almost immediately, it was expensive to make and before the war it was assumed that its main application would be in submarine cables. Joint applications development with BICC had established its importance in this field already by 1937. It proved extremely important in war-time radar applications, but with the post-war drop in defence demand, the closing down of the main ICI plant was apparently seriously considered.

However, the versatility of the material was gradually recognized and many new applications were established in the 1950s as process development work lowered the costs of production, and applied research led to the

Table 5.6 High pressure LDPE processes by type of reactor and initiator (number of licensees in brackets), 1987

Type of initiator	Type of reactor		
	Autoclave	Tubular	No details available
Peroxide type	Gulf Oil (2)	Arco (1)	
	ICI (22)	BASF (6)	
		Dow (2)	
		Imhausen (3)	
Oxygen type	Sumitomo (2)	ANIC (1)	
		Sumitomo (0)	
		Atochem (7)	
No details available	CdF (7)	El Paso (3)	Du Pont (2)
	USI (7)	DSM (0)	Koppers (1)
	Esso (0)	Union Carbide (4)	
		Scientific Design (1)	

Source: Bejar (1994).

necessary modifications and new grades of the basic polymer. A powerful impulse to the growth of the world market also came from the decision of the United States courts in 1952 to compel ICI to license several other US chemical firms, in addition to Du Pont, the original licensees. Although bitterly contested at the time, on the grounds that the court had no jurisdiction over ICI, the decision may well have been a blessing in disguise even for ICI, in that it almost certainly led to a more rapid growth of new applications, particularly in the domestic field and in packaging, as well as to a substantial increase in licensing fees and know-how payments. Among those that took licences were Monsanto, Dow, Koppers and Spencer.

A further stimulus to the growth of the total world market for polyethylene came from the introduction of many process improvements. These affected both the cost of production and the range and quality of new applications. Union Carbide and Du Pont, after originally obtaining licences from ICI, changed the technique to such an extent that they could then themselves grant licences on the basis of their own technology. As early as 1938, BASF had an independently developed pilot plant using a somewhat different type of reactor (Schott and Müller, 1975). Many new types of polyethylene were introduced to suit the special requirements of particular end uses and these depended on the ability of the resin producer to control the melt index and density very closely. Product and process improvements were thus intimately related. Schott and Müller (1975) distinguished as many as ten distinct high-pressure processes and even more were distinguished by Bejar (1994). (See Table 5.6)

In the 1950s came the more radical innovation of low-pressure (high-density) polyethylene. Although there were some fears at the time that the original ICI high-pressure process might be adversely affected by these new developments, it turned out that they were partly complementary rather than competitive. The new types of polyethylene had a higher density than the original ICI low-density polythene and rather different properties. Thus at first they tended to enlarge the total range of applications rather than to diminish the importance of the high pressure process. Ziegler's low-pressure process was discovered in 1950 in the course of a programme of research on catalysts at the Max Planck Research Institute in Mulheim. He was at the time already a research consultant with Hoechst and arrangements were soon made for large-scale development of the new process based on the use of his aluminium catalyst and his patents. Production began in both Germany and the USA in 1956.

As a result of a continuing programme of intensive R&D by ICI and its many licensees, numerous improvements in both products and processes continued to be made. Although the original patents had long since expired (including the special extension which was granted because of the war), ICI continued to enjoy a considerable income from the sale of know-how and the use of more recent patents throughout the 1960s. The technical leadership of the company and the continued importance of minor improvements were such that new producers still found it desirable to make substantial payments for this know-how. Although the total amount of these payments has not been revealed, together with those for Terylene they must have accounted for a substantial proportion of the £10 million or so which ICI was receiving for licence and know-how payments annually in the 1960s. (In 1971 total ICI receipts from licences amounted to £13 million and expenditures for licences were £3 million (ICI, 1971). However, during the 1970s the situation in high-pressure polyethylene underwent a drastic change. The pressure of rising costs and the entry of many new producers led to the erosion of profit margins and substantial over-capacity. Newer processes proved more profitable.

In 1977 Union Carbide introduced a new low pressure process to make low density polyethylene LLDPE (hitherto produced by high pressure processes). It offered major economies in capital costs (over 50 per cent) and energy consumption (75 per cent) and as Bejar (1994) points out, it also produced some new grades of LLDPE with superior properties for some applications, especially packaging: 'LLDPE emerged as a third product group in the polyethylene markets. This product has on the one hand properties and applications which are close to those of LDPE. On the other hand, it is produced by a very similar process to that used by HDPE' (ibid., p. 224) 'Swing plants' can produce either LLDPE or HDPE and have blurred the old boundaries. It was these developments which led ICI to exit from the polyethylene business in the 1980s, after making an unsuccessful joint effort to develop a proprietary LLDPE technology with an American firm. The ICI chairman commented at the time that ICI had made 'a misjudgement by not exploring the technology' (ibid., p. 250). BASF did succeed in developing its own process but had to make a drastic reduction in its HPLDPE plant capacity. Other American oil and chemical companies became the main licensors of new plants built in the 1980s and 1990s (Tables 5.7 and 5.8). The combined use of all types of polyethylene now makes it the largest tonnage plastic in the world.

Further work on catalytic polymerization resulting from Ziegler's discoveries led to the development of a process for the synthesis of poly-propylene by Dr Natta of Milan Polytechnic, working closely with the leading Italian chemical company, Montecatini. This process, too, was widely licensed throughout the world, but some US companies contested

Table 5.7 Major polyethylene technologies number and location of plants, 1992

	Total operating	North America	Western Europe	Eastern Europe	Far East	Latin America	Middle East	Africa
North American	83	23	20	6	21	7	5	1
Dow	13	4	4	—	4	1	—	—
Du Pont	6	1	1	1	2	1	—	—
Phillips	19	5	7	1	5	—	1	—
Quantum	12	2	3	1	3	2	1	—
UC	33	11	5	3	7	3	3	1
West European	105	10	40	15	29	9	1	1
Solvay	6	3	2	—	—	1	—	—
Atochem	11	—	5	3	2	1	—	—
Enichem	17	—	9	1	5	1	1	—
BASF	11	2	5	1	3	—	—	—
Hoechst	13	1	5	—	5	2	—	—
Stamicarbon	3	—	1	—	2	—	—	—
BP	13	2	3	—	8	—	—	—
ICI	31	2	10	10	4	4	—	1
Japanese	20	1	3	4	10	2	—	—
Mitsui	15	1	3	4	6	1	—	—
Sumitomo	5	—	—	—	4	1	—	—
Total	208	34	63	25	60	18	6	2

Source: Bejar (1994).

Table 5.8 Number of polyethylene technologies available by country of origin in 1987

	LDPE	HDPE	LL/HDPE
Total	19	14	15
USA	10	4	4
Japan	2	5	3
Germany	2	2	1
France	2	0	2
Italy	1	1	1
UK	1	1	1
Netherlands	1	1	1
Belgium	0	1	0
Canada	0	0	1
Spain	0	0	1

LDPE: Low Density Polyethylene (high pressure)
HDPE: High Density Polyethylene (low pressure)
LL/HDPE: Linear Low Density Polyethylene/High Density Polyethylene

Source: Bejar (1994).

the patents and claimed that they had developed a polypropylene process independently. Its rate of growth was extremely rapid, as it is cheap and tough and provides worthwhile applications for a refinery byproduct (propylene) which was often wasted before. It soon became one of the major bulk tonnage plastic materials along with PVC, polyethylene and polystyrene.

5.8   CORFAM (SYNTHETIC LEATHER)

This last example differs from the others in that it turned out to be a commercial failure, and resulted in a loss estimated at $100 million. It is all the more instructive for this reason. Encouraged by its outstanding success with nylon, by the late 1930s the Du Pont Central Research Department had developed several techniques to make porous poromeric films, and work continued on these techniques throughout the 1940s (Lawson *et al.*, 1965). Permeable films were wanted not only for leather substitutes for shoe uppers but also for other textile, packaging and coating applications. Several of Du Pont's product groups (industrial departments) collaborated with the Central Research Department in development work in the early 1950s, but from 1956 the work was concentrated in the Fabrics and Finishes Department. Field trials in 1956 and 1957 for applications in shoes and garments led to a decision to go ahead with full-scale development and pilot plant construction in 1959. Big engineering problems had to be resolved in scaling up production and it took longer than anticipated to achieve a satisfactory process.

Leather prices were rising and market studies suggested that the potential market for grain leather-substitute shoes was large and stable. A sales manager was appointed in 1960 and regular quarterly sales research product planning sessions were held to iron out the difficulties involved in preparing product launch. Initially it was decided to concentrate on the higher priced women's fashion shoes. 'We felt this marketing concentration would create a much more desirable effect than just having shoes shipped helter-skelter into all markets.' An intensive study was made of the economics of the shoe industry and of the leather market, and another special study on the attitude of consumers to the retail price of shoes.

More than 16,000 pairs of shoes were made in 200 different shoe factories and extensive field trials carried out to ensure that the product was acceptable, hygienic and comfortable. A computer 'venture analysis' model was used to test out various assumptions about future market growth, the effects of advertising and other factors. Finally, after a tremendous advertising campaign full-scale production was launched at a plant in Tennessee, and sales built up rapidly both in North America and Europe in the period 1966 to 1968.

However, although several million pairs of shoes were sold using Corfam, the product did not bring the rewards anticipated by Du Pont and in 1970 it was announced that the plant would be closed down and the company would withdraw from the market. No full explanation has ever been given but it may be conjectured that the availability of cheaper synthetic substitutes (PVC and Porvair) for ladies' shoe uppers was one of the factors leading to this decision. Porvair itself, although temporarily rather more successful than Corfam, continued to face great uncertainties and risks throughout the 1970s (Gibbons and Littler, 1979).

The example is instructive mainly because it demonstrates that strong R&D, experience in innovation, thorough market research and trials, or careful new product planning cannot in themselves ensure success in innovation. The implications of this degree of uncertainty confronting even the largest and most successful innovating firms are discussed in Chapter 10.

5.9 PRODUCT DEVELOPMENT AND PROCESS DEVELOPMENT

This summary account of the background to the innovation of several of the major synthetic materials has concentrated on the materials and their applications. But it has become evident that their successful development, and perhaps even more the shift to very large tonnage plants, depended upon the kind of innovative process engineering described in Chapter 4. New processes had to be developed not only for the new materials, but also for the intermediate products used in their manufacture. To a large extent the chemical companies also undertook this engineering work, although they collaborated with other firms for particular items of process equipment such as compressors, pressure vessels and valves. It involved the specialization of the plant design and engineering functions and the consequent emergence of the new profession of chemical engineering.

The new materials were often first isolated or discovered by relatively inexpensive methods, or indeed even accidentally. Frequently this occurred in university or other laboratories concerned with fundamental research, often long before commercial application. The reason for the proportionately large contribution of the giant chemical firms to the innovation of these materials on an industrial scale lay in the expense and difficulty of developing a satisfactory process, both for the material and for the intermediates and byproducts. In every one of the cases which we have considered this proved to be very costly and took a long time. It required a combination of skills in fundamental physical chemistry, process design, mechanical engineering and other types of engineering. It was difficult to bring together such a combination of skills except within the framework of a professionalized R&D system in industry, or within the framework of a public corporation such as the Atomic Energy Authority or similar nuclear power institutes.

5.10 INSTRUMENTS

The increasingly intimate relationship between new materials, new process development and fundamental research is nowhere more apparent than in the field of instrumentation. It would have been impossible to develop nuclear power or many new chemical processes and materials in the last fifty years without new scientific instruments. The use of on-line chromatography, analogue controllers, and specialized sensors and transducers has revolutionized chemical plant design and made possible the precise monitoring and remote control of complex flow processes. It has also led to enormous improvements in standards of purity and quality control.

The design of new laboratory instruments, particularly spectroscopes, was an integral part of the advance of physical chemistry. The development of mass spectrometers, other types of spectrometer and the electron microscope made it possible to ascertain molecular structure and the arrangement of specific groups within a molecule. The innovation of these and other new types of instrument strengthened the links between chemical technology and fundamental scientific knowledge, because most of the instruments originated in basic research laboratories. As Shimshoni pointed out:

Physical chemical analysis began with the observation of the visual part of the spectrum. It was found very early that substances could be identified by the characteristic special patterns of the wavelengths emitted when a sample of a substance is excited, or of the wavelengths absorbed from an external source by a sample. The subsequent story of the applications of physical methods to the study of matter is that of the extension of the spectral range until the whole of the electromagnetic spectrum could be used, of obtaining increased efficiency in discrimination or detection, and of the design and marketing of lower-cost instruments.

(Shimshoni, 1970, p. 64)

In the course of their fundamental chemical work, the universities frequently designed and built new instruments for their own use, and these were often commercialized by scientist–entrepreneurs setting up as instrument manufacturers. Many of the important instrument companies in the United States and Europe started in this way.[6] Sometimes scientists from chemical companies also assisted instrument companies in the development of new products. The particularly intimate nature of this collaboration has been documented in detail by von Hippel (1976, 1978). Together with university laboratories and government laboratories, the R&D departments of chemical and oil companies were the main users of laboratory analytical instruments. The infra-red spectrophotometer, which was innovated by Perkin–Elmer in 1943, was largely designed by Barnes and Williams of Cyanamid Laboratories, who needed the instrument for their analytical work. The role of scientists who set up their own instrument companies was decisive and Daniel Shimshoni has documented in detail the critical role of these inventor–entrepreneurs in establishing the main products.

Most of the recent advances in laboratory nuclear and process instrumentation have been in the development of electronic instruments, and the electronic computer is now of critical importance in laboratory and design calculations, as well as in process control. This is the link between the industries which we have so far discussed and the electronics industry to which we turn in Chapter 7.

5.11   NEW TRENDS IN MATERIALS TECHNOLOGY: CONCLUSIONS

All the large chemical firms confronted great uncertainties in the course of their innovative activities, both with respect to technological development and to markets. In the case of IG Farben and its constituent firms, these were compounded by huge political uncertainty. Synthetic rubber was almost abandoned for a time, PVC had big setbacks, oil from coal was saved only by very large tariff protection first from the Weimar Republic and later from the Hitler government. After the war the very existence of the firms was in doubt and their property and technology were sequestrated because of their role in the Hitler regime.

The problems and uncertainties confronting large chemical firms outside Germany were not quite so severe, but they too have experienced some great failures and the necessity to withdraw completely from some erstwhile promising markets, for example, ICI with linear low density polyethylene (LLDPE) or Du Pont with Corfam.

The large chemical and oil firms have proved extraordinarily resilient. No doubt this is derived in part from financial resources and lobbying power but it is also related strongly to their innovative capability and their knowledge accumulation. Frustrated or outcompeted in some areas they have been able to respond either with new products or new processes or both. In the 1980s and 1990s they confronted an entirely new situation, not only because of the slowdown in the world economy, but even more because of changes in science and technology. The economy was experiencing a change of techno-economic paradigm from a system based primarily on energy intensive mass or flow production of standardized commodities to one based on information intensive computerized technologies capable of producing a wider range of more specialized and often customized products and services. This transition is analysed more fully in Chapters 7 and 17. Nowhere was it more evident than in materials technology.

Advances in fundamental science, facilitated by electronic instrumentation innovation and far more powerful computers (see Chapter 7) have made possible a proliferation of advanced materials with greatly enhanced properties.[7] This proliferation has been so widespread that both industry and academics have had difficulty in defining and classifying the scope and limits of the new materials. However, most analysts agree that it includes both polymers and metals, organics and inorganics (such as ceramics) and innumerable composites of these. Tables 5.9a and 5.9b illustrate the properties and applications of some of these new (and older) materials in the case of polymers and ceramics respectively.

The increase in production and consumption of advanced materials has been difficult to track because of the definition and measurement problems, but OECD (1990) estimated world production in the late 1980s at about 2 million tons. However, it is not so much the growth in weight or volume of these materials which is remarkable as the rapidly expanding range of their properties and functions and the amazing improvements in their quality and performance. The strength to density ratio (Figure 5.1) is one example of this that has been especially important in structural materials, such as carbon fibres, which are basically acrylic fibres bound together with epoxy resins to make a material which is very light but also extremely tough.

In their brief account Cook and Sharp (1991) showed that carbon fibres are yet another example of the extreme uncertainty which continues to confront innovators with the newer materials, as with the old. They were originally developed in the 1960s in a British governmental research institute, the Royal Aircraft Establishment, in collaboration with Rolls Royce, which intended to use them for its RB211 engine. When this attempt failed (and Rolls Royce was temporarily bankrupted) the material was offered to ICI for development which, however, rejected the offer. Another firm, Courtauld's took it up and began to supply what was then a very small market; they also licensed the technology to Hercules, an American chemical company, which developed many new uses and went to the big Japanese chemical firm, Sumitomo, for supplies of the intermediate (acrylic fibre). This gave Japanese firms access to the technology and partly because of their strong interest in advanced materials research, the Japanese industry soon became the biggest suppliers to the now rapidly growing world market.

Table 5.9a Functions and applications of new polymers

Functions	Examples of materials	Examples of applications
Mechanical Functions		
High strength and durability	Polyester, polyamide	Various structural materials
Elasticity	Synthetic rubber, foamed plastics	Various structural materials
Shock and sound absorbing	Foamed plastics	Various structural materials
Surface protection	Coating films, electron beam hardened plastics	Coating materials, various paints
Adhesiveness	Polychloroprene	Various adhesives
Thermal functions		
Heat resistance	Polyimide, silicone resin	Heat resistant structural mats
Low temperature resistance	Silicone rubber, fluoro-rubber	Low temperature resistant rubber
Thermal insulation	Foamed plastics	Heat insulation materials
Electrical functions		
Electric conductivity	Polyacetylene	Battery, electric wire
Insulation characteristics	Polyimide, polyethylene, terephthalate	Printed circuit board, condenser conductor
Energy convertibility	Polyvinylidene fluoride, doped polyacetylene	Sensor, electro-acoustic transducer device
Optical functions		
Light transmitting	Polymethyl methacrylate acid polycarbonate	Optical fibre, plastic lens
Photo-active property	Photo-setting plastics	Copying materials, photo-mask
Double refraction property	Liquid crystal	Display device
Biological functions		
Compatibility to blood	Polyethylene terephthalate	Artificial blood vessel, artificial heart
Histocompatibility	Silicone polymer	Artificial skin, artificial organ, artificial bone
Separating functions		
Ion exchangeability	Styrene group, acryl group	Ion exchange resins
Separation of mixtures	Cellulose acetate group, aromatic polyamide group polypropylene	Reverse osmosis membranes, air/gas separation, biological separation membranes and drug delivery systems
Chemical functions		
Corrosion resistance	Polybutane-1, polyamide, neopreme	Roofing materials, offshore, structural materials
Chemical resistance	Polychloroprene, butadien acrylonitrile	Flexible structure storage tank, fertilizer tank

Source: Lastres (1992).

Table 5.9b Functions and applications of advanced ceramics

Functions	Examples of materials	Examples of applications
Mechanical functions		
High temp. strength	Silicon nitride, silicone carbide	Gas turbine, diesel engine,
Cuttability	Titanium carbide, tit. nitride, tungsten carbide, boron carbide	Cutting tools
Lubricity	Boron nitride, molyb. disulphide	Solid lubricant
Wearproof property	Alumina, boron carbide	Bearing, mechanical seal, drills
Thermal functions		
Heat resistance	Alumina, silicon nitride, silicon carbide, magnesium oxide	Electrode for MND generator heat resistant bearing
Thermal insulation	Potassium oxide-titanium oxide, aluminium nitride, zirconia	Heat insulators for high temp. furnace, nuclear reactor
Heat transfer characteristics	Boron oxide, silicon nitride, aluminium nitride, alumina	Electrical and electronics parts, radiator
Optical functions		
Light transmitting	Alumina, yttrium oxide, barium oxide	Sodium vapour lamp, high temperature optical lens
Light inducing	Silicon oxide	Optical communication fibre, gastro-camera, photo-sensor
Light deflecting	(Zirconium, titanium) acid (lead, lanthanum)	Photo-memory device (reversible)
Fluorescence	GaÅs-rare earth ceramics, neodymium-yttrium series glass	Semiconductor laser, light emitting diode
Photo-sensitivity	Silver halide containing glass	Sunglasses, image memory mats
Electrical functions		
Superconductivity	Yttrium-barium-copper oxide, bismuth-strontium-calcium-copper oxide	Power generator, magnet, supercomputer, maglev train, linear motor car
Semiconductivity	Zinc oxide, barium titanate	Varistor, heater, solar cell, sensor
Piezoelectricity	Quartz crystal, lead zirconate titanate, lithium niobate	Ignition device, piezoelectric oscillator
Insulation characteristics	Alumina, silicon carbide, beryllium oxide	Multilayer wiring board, IC package, IC printed board
Inducivity	Barium titanate, strontium titanate	IC microcondenser, high voltage service condenser
Ion/ionic conductivity	Zirconia, β alumina	Enzyme sensor, solid electrolyte
Electron radiation	Lanthanum bromate	Cathode mat. for electron gun
Magnetic function		
Magnetism	Iron oxide-manganese, iron oxide-barium oxide	Ferrite magnet, magnetic tape, memory device
Biological function		
Histocompatibility	Alumina, apatite	Artificial teeth, artificial bone
Chemical functions		
Absorbing property	Porous silica, alumina	Absorbent, catalyst carrier, bioreactor
Catalysing property	Zeolite	Catalyst for environm. protection
Corrosion resistance	Zirconia, silicon oxide, alumina	Electrode for MHD generator

Source: Lastres (1992).

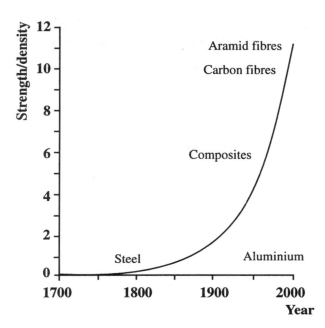

Fig. 5.1 Progress in materials strength-to-density ratio as a function of time (in., $\times 10^6$)

Source: Adapted from US National Research Council (1989); Lastres (1992).

Helena Lastres (1992) in her study of Japanese policies for advanced materials shows that many of the new materials there have been developed and innovated by electronic companies. Chemical, metals and engineering firms have also played a part but Japanese capability in microelectronics has been a major source of competitive strength. For this reason too, Japanese interest in superconductivity has been intense since the discovery of high temperature superconductivity in the research laboratory of IBM in Switzerland in 1986 (for which two physicists were awarded the Nobel Prize). Much of the new materials development and applications research is done by collaboration between materials suppliers and users, a pattern identified by Lundvall (1985, 1988b) as of central importance in machinery, and still earlier by Adam Smith.

The large European and American chemical firms not only faced the challenge of declining profitability in their bulk tonnage standard commodities, but also the challenge of entirely new technologies based on biology and biochemistry, in a wide range of products and processes, especially in pharmaceuticals and agrochemicals. These enormous changes in technologies and markets have led to a process of structural change in the industry in the 1980s and 1990s. The larger firms have tried to develop new specialized products themselves, but as they did not have the necessary scientific research experience they were often obliged to collaborate with universities (Monsanto, Hoechst and ICI all made major agreements with selected university departments and hospitals in the field of biotechnology),

Table 5.10   Acquisitions and mergers in world chemical industry, 1988

Acquisitions and mergers by region		
	No. of transactions	% of total
USA	723	56
W. Europe	404	31
Japan	24	2
Other	132	10
Total	1,283	100

Acquisitions and mergers by location of acquirer		
	No. of transactions	% of total
USA	613	48
W. Europe	520	41
Japan	56	4
Other	94	7
Total	1,283	100

Acquisitions and mergers by type of business		
	No. of transactions	% of total
Speciality chemicals	217	17
Pharmaceuticals and healthcare	211	16
Fabricated plastics and rubber	148	12
Tonnage chemicals	136	11
Consumer products	80	6
Other	491	38
Total	1,283	100

Source: Cook and Sharp (1991).

or to make acquisitions of small specialized firms which had the expertise in narrow fields. These small firms were often themselves spin-offs from universities or research institutes whose scientists had identified a big potential application but lacked the financial resources and business skills to undertake development and commercialization over a long period. On the other hand, some of the larger firms wished to divest themselves of some of their low-profit or loss-making plants in the other bulk chemicals. This meant that diversification went hand-in-hand with concentration of some older products in the strongest firms, sometimes based on 'swaps' of products and plants between firms. Table 5.10 illustrates the scale and nature of the acquisition process in the 1980s.

In the face of all this turbulence the large chemical firms generally succeeded in maintaining a very strong position in the industry. The German and Swiss chemical firms, world leaders in R&D in the 1930s remained highly R&D intensive in the 1990s. In absolute terms they still had both a larger sales volume and larger R&D activity than most of their competitors (Figure 5.2). This quality of survival of the large firms and their symbiosis with the new small innovative firms is discussed in Chapter 9. The role of research tradition and of knowledge accumulation in relation to the observed long-term 'stickiness' in patterns of international trade is discussed in Part Three.

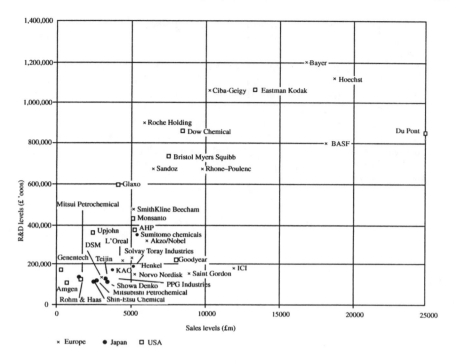

Fig. 5.2 Sales vs R&D levels for the chemical sector, 1992

Source: MERIT.

NOTES

1. On the early history of the plastics industry see Kaufman (1963); Hufbauer (1966); Yarsley and Couzens (1956); see also De Bell (1946).
2. For definition of R&D see OECD (1963a). Research intensity of a firm or an industry may best be measured by relating R&D expenditures to net output. This enables satisfactory inter-firm and inter-industry comparison because it adjusts for the differences in value of bought in materials and components. However, since figures for net output are seldom available, frequently the less satisfactory measure of R&D expenditure as a ratio of turnover has to be used, or of R&D personnel as a ratio of total employment. In the case of IG Farben the figures available are for R&D as a ratio of turnover. The net output ratio would probably be nearly twice as high, but cannot be exactly estimated.
3. There are, of course, other reasons for these differences. The large number of patents in instruments and mechanical engineering relative to R&D is partly due to the fact that invention in these industries is still conducted to a significant extent outside the formal industrial R&D system. There are also classification problems for the two series of statistics.
4. In some countries patents are granted without examination for originality, whereas in others there is an examination procedure. This means that in countries such as France, Belgium and Italy about 90 per cent of patent applications lead to the grant of a patent, but the proportion falls to about 60

per cent in the USA and UK, and still lower in Germany, the Netherlands and Scandinavia. An international comparison would thus tend to be biased against German firms in comparison with British and American firms, and still more with French. The most important patents will be taken out in all the major manufacturing countries, but the statistics for any one country will normally be biased towards the firms domiciled in that country (including foreign owned subsidiaries operating there).

5. An ICI study of the productivity of its research found that the highest yield was from new products and processes based on inhouse research. Although it is true, as Mueller has demonstrated, that many of Du Pont's innovations were not the result of its own R&D, nylon made by far the biggest contribution to corporate profits (see Holroyd, 1964; Mueller, 1962).

6. For example, Perkin–Elmer and Hewlett–Packard.

7. For an account of these new developments see Lastres (1992); US National Research Council (1989); Forester (1988); Barker (1990).

CHAPTER 6

MASS PRODUCTION AND THE AUTOMOBILE

In Chapters 4 and 5 we have demonstrated the huge productivity gains which were made in the oil and chemical industries through the introduction of flow processes and scaling up in large plants (see, for example, Table 4.1). In parallel with this trend in flow production processes comparable gains were made in some other industries through the introduction of interchangeable parts and the assembly line, above all in the automobile industry. In Chapter 3, we pointed to the capital intensive trajectory of technical change in United States industry in the period from 1870 to 1940 and the special circumstances which led to that country forging ahead of others during that period. One of the reasons for that trend was the shortage of skilled labour which led American entrepreneurs and engineers to seek for production techniques which would both substitute machinery for labour and facilitate the use of fewer skilled craft workers relative to the unskilled and semi-skilled.

The use of interchangeable parts was essential for this purpose as some skilled machinists and fitters could then be replaced by less skilled assembly workers. Charles Babbage (1834) had already clearly indicated this as a desirable strategy for manufacturers. He is of course best known for his work as the inventor of the computer and as Professor of Mathematics at Cambridge University. Rosenberg (1994) deserves the credit for rediscovering his merits as an economist, especially his 'Economy of Machines and Manufactures' in which he not only extols the virtues of what later came to be called 'time and motion study' but also ingeniously extended Adam Smith's principle of the division of labour in new directions:

> The master manufacturer by dividing the work to be executed into different processes, each requiring different degrees of skill or of force, can purchase exactly that precise quantity of both which is necessary for each process; whereas if the whole work were executed by one workman, that person must possess sufficient skill to perform the most difficult, and sufficient strength to execute the most laborious, of the operations into which the art is divided.
>
> (Rosenberg, 1994, pp. 175–6)

However, while this theoretical principle was clearly enunciated and commended by Babbage as early as the 1830s, its practical realization later in the nineteenth century took place primarily in the United States and was closely related to the introduction of interchangeable components of engineering products, whether made of metals or other materials. As long as parts were irregular in shape and not machined to a precise specification, skilled craftsmen were required to fit the parts and assemble each product. As we have seen in Chapter 3, greater precision and speed of machine tools was facilitated by the increasing availability of steel alloys in the late nineteenth century and F.W. Taylor himself made an important contribution as an inventor before he became a management consultant.

Already in the late nineteenth century the use of interchangeable parts had spread from its first highly innovative application in the US Ordnance Department's Springfield Armoury to other American industries. In the early twentieth century many new applications were being found but none of these, not even the Armoury, had been able to dispense with skilled craft workers to file and fit at least some of the components, and usually quite a high proportion. What existed in the so-called 'American system of production' before Ford was in fact a hybrid system or 'late craft system' in which some interchangeable parts co-existed with many craft shaped parts in the final product. This was the situation in such industries as sewing machines, small arms, agricultural machinery and bicycles.

The first true application of mass production techniques was by Henry Ford in his Highland Park Plant at Detroit. Between 1908 and 1914 he gradually eliminated craft-made components in the manufacture of the Model T and this process culminated with the introduction of the moving assembly line itself in 1913. As the MIT book on Ford's International Motor Vehicle Project (IMVP) emphasizes (Womack et al., 1990), the assembly line itself was only made possible by the introduction of machines and presses, to cut, shape or stamp out each one of the components.

In this chapter, we shall discuss mainly the organizational innovations in the automobile and related industries, rather than the technical innovations, although some consideration is given to both. We concentrate on organizational and social change because the 'mass production paradigm' or, more simply, 'Fordist paradigm' dominated management philosophy for more than half a century and is only late in the twentieth century giving way to a new style of management thinking and organization, which we discuss in later chapters. Fordist production technology illustrates better than any other the phenomenon of 'lock-in' to a dominant range of design of products and processes which has been identified and explored by Brian Arthur (1988a, 1989). Finally, Fordism strongly influenced consumer behaviour as well as production technology in such industries as refrigerators, washing machines, and other 'white goods'. It made possible the 'consumer durable' revolution, as it brought many appliances into the range of relatively cheap goods affordable by the average household. While it is true that there were forerunners of mass production technology in the nineteenth century, such as the use of some interchangeable components in ordnance and the manufacture of sewing machines, or the 'disassembly' line of subdivision and specialization of tasks in the Chicago meat-packing industry, it was the Ford assembly line which established both the philosophy and practice of mass production as an archetypal twentieth-century American technology. During the Second World War the huge scale of American production of trucks, tanks, aircraft and landing craft, together with the petroleum products which fuelled them, were in the end decisive factors in the Allied landings in Europe and their advance into Italy and Germany. Following that war the mass consumption life style and the mass production of automobiles and other durable goods was not only consolidated in the United States but spread rapidly to Europe and Japan. The quarter century after the Second World War was the period of most rapid economic growth the world has ever known and it was largely based on oil, automobiles, aircraft, petrochemicals, plastics and consumer durables.

This triumph of mass and flow production was, however, achieved only after a very painful period of worldwide structural adjustment in the 1920s and 1930s. Mass production capacity for automobiles and other goods had outstripped the absorptive capacity of the (then) very limited market for the new goods. Only with social innovations such as consumer credit arrangements, new wage structures, a new highway infrastructure and Keynesian-type management of the economy was it possible to harmonize the new technological potential and the socio-institutional frameworks. As we have seen in Chapters 4 and 5, cheap oil not only provided an extremely flexible, universally available source of energy, for aeroplanes as well as cars and trucks, it also provided a very cheap feedstock for the huge new family of chemical products so well suited to the needs of mass consumption: plastic products and packaging, throw-away containers and above all components for the mass production of consumer electronics. All of these interdependent technologies were part of the Fordist techno-economic paradigm, which dominated the world economy in the quarter century after the Second World War.

6.1   INTERNAL COMBUSTION, STEAM OR ELECTRICITY?

Authors from MIT (Womack *et al.*, 1990) took the lead in a major study of the automobile industry, known as the International Motor Vehicle Project (IMVP). The chapter titled 'The Rise and Fall of Mass Production' recounts how when a wealthy English MP, Ellis, went to buy a car in 1894, he did not go to a dealer as none existed. He went to the French machine tool manufacturer, Panhard et Levassor, and 'commissioned' an automobile, (the word itself is from the French). Since 1887 Panhard et Levassor had a licence to manufacture Daimler's internal combustion engine and by the early 1890s they were building a few hundred per annum, each one customized and made in various craft workshops in the Paris region. Ellis wanted a special body constructed by a coachbuilder and the transmission brake and engine controls moved from their usual place. The IMVP authors point out that 'for today's mass producer this would require years – and hundreds of millions of dollars to engineer' but for the craft producers of that time, such a request was quite normal. After testing his automobile in the streets of Paris, Ellis came back to England in June 1895 and drove the 56 miles to his country house in 5 hours 32 minutes, at an average speed of 9.8 miles an hour. The official speed limit for non-horsedrawn vehicles was 4 miles per hour but in 1896 Ellis led the way in promoting a new law in Parliament with a 12 mph limit.

The United States did not lead Europe in the early days of the automobile industry; in fact, almost all the early inventions and innovations were made in Germany and France. By 1905, however, hundreds of small companies were producing automobiles in the United States as well as the main European countries. They all used craft techniques and general purpose machine tools scattered in small machine shops. The craft workers were highly skilled and co-ordinated by the assembly entrepreneur. A few firms today still make a very small number of cars by craft techniques but they are very expensive and account for only a tiny proportion of total output. Most of the small firms have long since gone bankrupt (including Panhard et Levassor) or were taken over as the industry made the transition to mass production. It was Henry Ford more than anyone else who was responsible for this transition.

It was actually by no means clear at the turn of the century whether the internal combustion engine would be preferred to the steam engine or the electric engine. The basic innovations for all three had been made and Klein (1977, p. 91) pointed out that in 1900 steam and electric vehicles accounted for about three-quarters 'of the four thousand automobiles estimated to have been produced by 57 American firms'. However, by 1917 about three and a half million automobiles had been registered in the United States, of which less than 50,000 were electric and steam vehicles were disappearing. The last major steam manufacturer, the Stanley Motor Carriage Company, produced 730 steam vehicles in 1917 – fewer than Ford produced in one day before lunch (Volti, 1990, p. 43).

The simple explanation of the demise of steam and electric vehicles seems to be, with the benefit of hindsight, that the internal combustion (gasoline) engine was 'better' or even 'optimal'. However, in his fascinating article 'Why Internal Combustion?' Rudi Volti (1990) shows that things were by no means so simple. In the very early days both steam and electric cars had many technical advantages and the IC automobile had some severe disadvantages, notably the sliding gear transmission invented by Emile Levassor in 1891. His own description of his invention became famous: 'C'est brutal mais ça marche!'. Another big problem was the starting handle which could break the thumb or the wrist or injure the chest of the unskilled or the unwary. This problem was not resolved until long after the invention of the first electric starter by Kettering for the 1912 Cadillac because of the relatively slow diffusion of his invention.

Steam cars and electric cars were both more smooth running in these early days but they suffered from the major (and increasing) disadvantage of their short operating range, due to the weight of their boiler or batteries and consequent refuelling problems. The dashboard of a 'Stanley Steamer' was festooned with gauges that required regular attention: boiler water level, steam pressure, etc., and 'just to start the engine required the manipulation of thirteen valves, levers, handles and pumps' (Volti, 1990, p. 44). The electric car was simpler to start and to drive, having no clutch or transmission; moreover it was quiet, reliable and odourless. Yet by the 1920s, the internal combustion engine completely dominated the car market, leaving the steamers and electrics to very specialized niche markets or museums.

Longer operating range was undoubtedly one of the decisive advantages of the IC engine but this was not purely a technical matter. The chain of refuelling stations, repair and maintenance facilities, could conceivably have been organized on a different basis, given different strategies and policies of the utilities, manufacturers and regulators. Indeed, in the 1990s policies are now being developed to cope with battery recharging services, etc. in California and elsewhere, because of the pollution problems caused by millions of IC engines. However, the 'lock-in' to the IC engine makes any such change to an alternative system a truly massive undertaking (see Part Four). There were over 500 million automobiles in use in the world by the mid-1990s.

6.2 HENRY FORD AND MASS PRODUCTION

The main reason for this vast lock-in to the internal combustion engine was, of course, the success of Ford's assembly line which reduced the cost and price of the Model T dramatically. The price of electric cars was rising at the time because of the introduction of better batteries, but the price of a Model T fell from $850 in 1908 to $600 in 1913 and $360 in 1916 because of a combination of organizational, technical and social innovations. The price of electric cars in 1913 was $2,800. Not surprisingly, the sales of the Model T increased fifty times over and market share increased from 10 per cent in 1909 to 60 per cent in 1921. Profits on net worth were sometimes as high as 300 per cent per annum and the USA attained a dominant position in world export markets. This was indeed 'fast history' analogous to the tempestuous growth of the semiconductor industry half a century later, with its similar drastic price reductions, rapid changes in market shares, sudden profits for innovating firms and world export hegemony for the leading country until imitators caught up (see Chapter 7).

The 'basic innovation' which enabled Ford to achieve these dramatic results was, of course, assembly line production. In one sense it was purely an organizational innovation, but it entailed and stimulated a great deal of technical innovation:

> Once the organisational change was made, the automobile firms found many opportunities for developing more efficient machines by making them more automatic. For example, replacement of the vertical turret lathe by a more automatic horizontal lathe doubled output per worker. Or to take a more spectacular example, an automatic machine for making cam-shafts increased output per worker by a factor of ten, and literally dozens of cases can be found in which better machines permitted output per worker to increase by a factor of between two and ten.
>
> (Klein, 1977, p. 97)

Incidentally, Klein casts some doubt on the 'heroic' entrepreneur idea so far as it applied to Ford, quoting the comment of Nevins and Hill (1954):

> It is clear that the impression given in *My Life and Work* that the key ideas of mass production percolated from the top of the factory downwards is erroneous: rather seminal ideas moved from the bottom upwards. To be sure, Ford took a special interest in the magneto assembly. But elsewhere able employees like Gregory, Klann and Purdy made important suggestions, Sorensen and others helped them work out, while Ford gave encouragement and counsel. *The largest single role in developing the new system, however, was played by the university-trained thinker [Avery] so recently brought in from his school-room* [our italics].
>
> (Klein, 1977, p. 98 quoting Nevins and Hill, 1954, p. 474)

Table 6.1 Craft production versus mass production in the assembly hall, 1913 versus 1914

Minutes of effort to assemble:	Late craft production, fall 1913	Mass production, spring 1914	Percent reduction in effort
Engine	594	226	62
Magneto	20	5	75
Axle	150	26.5	83
Major components into a complete vehicle	750	93	88

Note: 'Late craft production' already contained many of the elements of mass production, in particular consistently interchangeable parts and a minute division of labour. The big change from 1913 to 1914 was the transition from stationary to moving assembly.

Calculated by the authors from data given in Hounshell (1984, pp. 248, 254–6). Hounshell's data are based on the observations of the journalists Horace Arnold and Fay Faurote as reported in their volume *Ford Methods and the Ford Shops*, New York: Engineering Magazine, 1915.

Source: Womack et al. (1990, p. 29).

At first, in 1908, when Ford had achieved a high degree of interchangeability of parts, it was the workers who moved from car to car around the assembly hall and already the division of labour into specialized tasks had achieved big increases in productivity. Womack *et al.* believe that these increases were greater than those achieved by the moving assembly line, when it was introduced in 1913. But the moving line was a highly visible change and therefore attracted more attention. Table 6.1 shows a contemporary calculation of the productivity gains made between 1913 and 1914 by the introduction of the moving assembly line at the Highland Park plant in Detroit. The designation of the 1913 system as 'late craft' is questionable as by then the filing and fitting of each inaccurate part had been virtually eliminated. In fact, the Ford production system differed already from all the earlier efforts to develop interchangeability of parts as Hounshell (1984) showed in his classic work: *From the American System to Mass Production*.

The US Ordnance Department did indeed originate a radical change in production technology with its introduction of interchangeable parts at the Springfield Armoury. The epochal significance was clearly recognized at the time by British visitors to the Armoury, as well as by many American observers. Moreover, the Ordnance Department made considerable efforts to extend the use of interchangeable parts to other suppliers, such as the Colt Firearms Company, which in its turn further disseminated the ideas. However, American historians have shown that neither the Springfield Armoury, nor any other nineteenth-century producer actually achieved full interchangeability. In fact, the success of such firms as Singer (sewing machines), McCormick (harvesters) and Pope (bicycles) was not, as at one time suggested, due to mass production using interchangeable parts. Hounshell (1984) has shown that other manufacturers of sewing machines tried to use interchangeable parts earlier and to a greater extent than Singer, who overcame their competition, not by new production technology, but like Wedgwood long before (see Chapter 2) by higher quality and by marketing and advertising innovations. Thanks mainly to these, Singer was able to take over several of the competing firms which had gone much further in attempting to displace craft methods with inter-changeable parts. Both Singer and McCormick did make moves in this direction but they were beset with production problems and still had to rely on craft methods to a great extent.

Their efforts and those of many other American firms to achieve the interchangeability of parts, although never completely successful before Ford, were certainly not fruitless. Rosenberg (1976) showed that their successive attempts played a very important role in the development of the American machine tool industry, which grew from the small arms indus-try, especially the Colt Firearms Manufacturing Company. Indeed, the young machine tool firms worked closely together with their customers to solve the production problems in each industry. By this means they greatly improved their own machinery for boring, milling, cutting, planing and otherwise shaping both metals and wood. This was yet another example of what Lundvall (1985) later designated as user–producer interaction. In the end it was the developments in new materials, cutting precision and speed of machining (see Chapter 3) at the turn of the century which brought the industry to the point where full interchangeability became possible.

Ford himself in his own accounts of the triumph of mass production, emphasized, on the one hand, the resemblance to flow production in flour milling and to the disassembly line of the Chicago meat packers and, on the other hand, the mass consumption due to the cheapness of large-scale production. Singer, McCormick and Pope, who have sometimes been described as pioneers of mass production, all sold their products at high prices. Like Wedgwood (see Chapter 2), their prices were the highest in their respective industries (Hounshell, 1984, p. 9). Ford, on the other hand, insisted that the Model T was truly designed as a car for the masses. The reduction of prices was an essential part of his doctrine and Peter Drucker (1946) maintained that his demonstration of falling costs and prices with the increasing scale constituted a revolution in economics as well as a management technique. Moreover, unlike many other early IC engine cars, the Ford T was deliberately designed for ease of maintenance and ease of operation. The 64-page owner's manual was intended to help the owner to solve many of the common problems which might arise. Ford assumed, however, that many of his early customers would be farmers with experi-ence of farm machinery and some tools (Womack et al., 1990). Geographical and social characteristics of the United States market and Ford's attention to them also contributed to his success. Personal mobility at low cost was the decisive advantage of the internal combustion engine to the consumer.

The radical change in the production system was not an easy one for the workforce:

> The assembler on Ford's mass production line had only one task – to put two nuts on two bolts or perhaps to attach one wheel to each car. He didn't order parts, procure his tools, repair his equipment, inspect for quality, or even understand what the workers on either side of him were doing. Rather he kept his head down and thought about other things. The fact that he might not even speak the same language as his fellow assemblers or the foreman was irrelevant . . . the assembler required only a few minutes training.
>
> (Womack et al., p. 31)

The machining of parts was also reduced to a few simple operations by the redesign of the tools and presses to carry out one simple repetitive task. In this way, the need for skilled workers was reduced to a minimum and the plant was controlled and co-ordinated by the new profession of industrial (production) engineers and an army of foremen and indirect workers responding to their orders. Discipline was strict and unions were banned. Ford tried hard to implement Wedgwood's principle (Chapter 2): 'to make such machines of men as cannot err'. As we shall see, he failed in this objective. Poor industrial relations were a major problem for Ford after the Model T and GM took advantage of this in overtaking Ford in the 1920s (Lazonick, 1990).

With respect to wages, the new system did not work out quite as Babbage had imagined. The speed of the line, the boring nature of the work and other unpleasant features of employment, such as the work discipline, led to an extremely high turnover of labour during the first year of operation. In 1913 labour turnover reached nearly 400 per cent so that Ford was obliged on 5 January 1914 to introduce the '5-dollar day' which meant in effect that the wages of Ford workers were doubled: 'With highly mechanised production, moving line assembly, high wages and low prices on products, "Fordism" was born' (Hounshell, 1984, p. 11). The much higher productivity of the new system of production meant that these high wages could easily be paid and Ford could still remain by far the most profitable automobile company in the United States and in the world.

At an early stage Ford and his collaborators had to accept the obvious fact that some defective parts were an inevitable feature of his system. The solution adopted was not to improve the skills or responsibilities of the production line workers but to have an inspection and 'rework' department at the end of the line. Even so, of course, some defective cars got through this procedure too, so that customer complaints always accompanied mass production. One of the main objectives of the Japanese producers who challenged the Fordist system after the Second World War was to make a drastic reduction in the number of defective parts or subsystems. As we shall see, they had some success in achieving this objective, but for half a century American and European producers simply accepted defects and rejects as a cost of mass production which was greatly outweighed by the benefits of a vast output of cheap and fairly efficient machines.

6.3   THE DIFFUSION OF FORDISM AND ITS MODIFICATION

Ford's immense success obliged other American firms to introduce the assembly line, to become small niche producers or to go to the wall. A few of them were combined in a company, General Motors, which offered a successful challenge to Ford in the 1920s and 1930s by modifying some of his more eccentric and autocratic ideas. Alfred Sloan, the chief architect of this giant corporation, introduced a more sophisticated strategy based on a greater range of models, more frequent model changes and steady incremental improvements coming partly from production engineers but also from a large R&D activity. Sloan also introduced a divisionalized corporate structure with the various divisions taking responsibility for specific segments of the market. Many of Ford's key personnel left him and joined GM (Lazonick, 1990, p. 240).

Sloan (1963) also supported the introduction of a basic research activity in GM and in his book justified the decision on the grounds that company strategy should be based on long-term profitability, not on short-sighted profit maximization. GM has in fact had a highly successful R&D activity, although paradoxically some of its most important inventions and innovations have been made outside the automobile industry, in diesel-electric locomotives and in chemicals. Two outstanding inventors, Charles Kettering and Thomas Midgley, were brought into GM's research laboratories, through the acquisition of the Dayton Engineering Laboratories Co. in 1919 and they played a leading part in the success of these and several other important innovations. Kettering persuaded GM in 1930 to purchase two companies already quite advanced in the development of diesel-electric technology:

> The combination of the diesel-engine and electric traction equipment was not new but General Motors by acquiring two small pioneering firms, by placing the large resources of their research organisation to the task, and by taking advantage of a large potential domestic market, finally established the great commercial advantage of the system.
>
> (Jewkes et al., 1969)

Womack et al. (1990, p. 129) point out that GM's research laboratory 'proved vital to the welfare of the entire world auto industry when its scientists and engineers – at very short notice – perfected the exhaust catalyst technology now used by every car company in the world to produce automobiles that meet emission standards'. However, they also offer some interesting comments on why GM failed to introduce radical innovations in the manufacture of automobiles: 'Unfortunately, in the absence of a crisis – a situation in which the future of the company was at stake and normal organizational barriers to the flow of information were suspended – new ideas percolated from the research center to the market very slowly.' They list the Corvair Project (1950s), the Vega project (1960s), the X-Car Project (1970s) and the high-tech factories for the GM-10 products in the late 1980s among GM's failures: 'In each case innovative ideas for new products and factories foundered when implementation could not live up to the original technical targets.'

Womack et al. offer another particularly interesting explanation of why GM was not more innovative in automobiles in the inter-war and post-war years. They point out that although Sloan was an MIT graduate in electrical engineering he maintained that:

> It was not necessary to lead in technical design or run the risk of untried experiments [provided that] our cars were at least equal in design to the best of our competitors in a grade.
>
> (Sloan, 1963, p. 180)

Womack et al. appear to suggest that it was the existence of a tight oligopoly of only three large producers with GM as the dominant firm, which restrained radical innovation:

When GM had sewn up half the North American auto market, any truly epochal innovation – say a turbine-powered truck or a car with a plastic body – could have bankrupted Ford and Chrysler. The auto makers' plight would certainly have attracted the attention of a US government intent on preventing monopoly in its largest industry. So caution made sense. GM hardly wanted to innovate its way to corporate dismemberment.

(Womack et al., 1990, p. 128)

Utterback (1993) has shown that the pattern of evolution displayed in the first half of the twentieth century by the US automobile industry was characteristic of several other industries, both before and since, such as typewriters, bicycles, sewing machines, TV and semiconductors. An early radical product innovation leads to many new entrants and to several competing designs. Process innovations and scaling up of production then lead to the emergence of a dominant robust design, the erosion of profit margins and a process of mergers and bankruptcies, ending with an oligopolistic structure of a few firms. Incremental innovations then tend to prevail in both product and process (Figure 6.1).

The account of mass production which has been given makes it clear how far removed this technology was from the scurrilous schoolboy limerick of the 1920s:

There was a man named Henry Ford;
He took some tin, he took some board.
The board he nailed,
The tin he bent,
He gave it a kick and the damn thing went.

Nevertheless, among both European consumers and European managers and engineers, there was still a certain arrogant contempt for mass-produced goods, which contrasted with American cultural traditions and probably also was a retarding factor in diffusion. However, this resistance fairly rapidly diminished once per capita incomes began to rise in the post-war world and the masses actually had the opportunity to purchase cars and other consumer durables. Most of Western Europe caught up in per capita incomes between 1950 and 1975 and the pattern of ownership of durables came to resemble that of the United States fairly closely. West European car output surpassed that of the United States in the 1960s (Table 6.2).

This catch-up process was mainly based on the now successful diffusion of American industrial technology and management techniques. The Marshall Plan, the European Recovery Agency and its successors the OEEC (later OECD) all laid great stress on this transfer and many industrial missions went to study US productivity in American firms in the early post-war years. United States companies and especially Ford and General Motors had long established production plants in several European countries which also facilitated the absorption of American techniques. However, European manufacturers were neither purely passive recipients of American technology, nor simply imitators. They were active innovators themselves and were especially innovative in design (see Womack et al., 1990, pp. 46–7). They had a big export success with small cars, sports cars and some luxury cars.

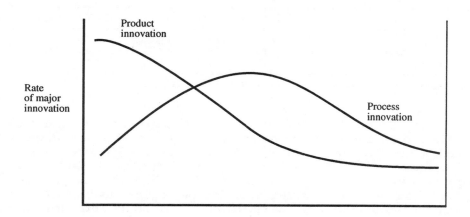

Predominant type of innovation	Frequent major changes in products	Major process changes required by rising volume	Incremental for product and process, with cumulative improvement in productivity and quality
Competitive emphasis on	Functional product performance	Product variation	Cost reduction
Innovation stimulated by	Information on users' needs and users' technical inputs	Opportunities created by expanding internal technical capability	Pressure to reduce cost and improve quality
Product line	Diverse, often including custom designs	Includes at least one product design stable enough to have significant production volume	Mostly undifferentiated standard products
Production processes	Flexible and inefficient; major changes easily accommodated	Becoming more rigid, with changes occurring in major steps	Efficient, capital-intensive, and rigid; cost of change is high.
Equipment	General purpose, requiring highly skilled labour	Some subprocesses automated, creating 'islands of automation'	Special purpose, mostly automatic with labour tasks mainly monitoring and control
Materials	Inputs are limited to generally available materials	Specialized materials may be demanded from some suppliers	Specialized materials will be demanded; if not available, vertical integration will be extensive
Plant	Small scale, located near user or source of technology	General purpose with specialized sections	Large scale, highly specific to particular products

Fig. 6.1 A model for the dynamics of process innovation in industry

Source: Abernathy and Utterback (1978, pp. 2–9).

Table 6.2   World automobile production, 1950–95 (000s)

	1950	1960	1970	1978	1989	1995
USA	6,950	7,001	7,491	10,315	7,835	8,363
Western Europe	1,110	5,120	10.379	11,321	13,749	12,636
Japan	2	165	3,179	5,748	9,052	7,611

Source: Graves (1991). Automotive News Market Data Book (1996).

Among the important innovations of West European producers in the 1960s and 1970s were front-wheel drive, disc brakes, fuel injection, unitized bodies, five-speed transmissions and high power-to-weight ratios. American firms had led with 'comfort' innovations such as power steering, air conditioning, stereos and automatic transmission. The rise in oil prices in the 1970s gave some advantages to the Europeans, especially in small fuel-efficient cars and they took this opportunity to expand their exports to the United States. From being the dominant producer and exporter of cars for half a century, the United States became a net importer. However, the main reason for the loss of American dominance in the world market was not so much European competition as the meteoric rise of Japan. As with Ford in 1913 this rapid rise was based on a radical redesign of the entire production system, to which we now turn.

6.4   POST-WAR INNOVATION IN JAPAN

Since the Meiji Restoration of 1868 in Japan, there was great emphasis on the improvement of imported technology by process innovation. The method of assimilating and improving upon imported technology was mainly some form of 'reverse engineering' (Freeman, 1987; Tamura, 1986; Pavitt, 1985). The widespread use of reverse engineering in the 1950s and 1960s had several major consequences for the Japanese system of innovation, affecting especially the characteristic R&D strategy of the major Japanese companies. Japanese management, engineers and workers grew accustomed to considering the entire production process as a system and to thinking in an integrated way about product design and process design. This capability to redesign an entire production system has been identified as one of the major sources of Japanese competitive success in industries as diverse as shipbuilding, automobiles and colour television (Jones, 1985; Peck and Goto, 1981). Whereas Japanese firms made few original radical product innovations they did redesign many processes and make many incremental innovations (Figure 6.2) in such a way as to improve productivity and raise quality. The automobile industry is probably the most spectacular example (Altshuler et al., 1985; Jones, 1985; Womack et al., 1990; Graves, 1991).

Japanese engineers and managers grew accustomed to the idea of 'using the factory as a laboratory' (Baba, 1985). The work of the R&D department was very closely related to the work of production engineers and process control and was often almost indistinguishable. The whole enterprise was involved in a learning and development process and many ideas for improving the system came from the shop floor. Since almost all case studies of the management of innovation in Western Europe and the

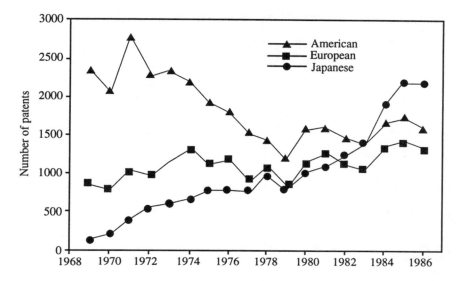

Note: Figures are for patents granted by the US Patent Office to assembler and supplier firms located in each main region. In case of subsidiaries whose parent is headquartered in one region but which operate in another region, the patents were counted in the region of operation. For example, Alfred Teves is a German subsidiary of the US-headquartered ITT. Teves' patents have been counted in the European region.

Patenting by supplier firms was estimated by developing a list of major automotive suppliers headquartered in the three principal regions, using the following sources:
Japan: Dodwell Consultants, *The Structure of the Japanese Autoparts Industry*, Tokyo: Dodwell, 1986.

North America: Elm International, *The Elm Guide to Automotive Sourcing, 1987–88*, East Lansing, Michigan: Elm International, 1987.

Europe: PRS, *The European Automotive Components Industry 1986*, London: PRS, 1986.

This list was then compared with data on patents by company, provided by the US Office of Technology Assessment. Adjustments were made to exclude nonautomotive patenting by large multi-product firms, such as Allied Signal in the United States and Hitachi in Japan.

Fig. 6.2 Motor vehicle industry patenting, 1969–86

Source: Estimated by the Science Policy Research Unit of the University of Sussex from data supplied by the US Office of Technology Assessment, Washington, DC. Womack *et al.* (1990, p. 134)

United States point to the lack of integration between R&D, production management and marketing as a major source of failure (see Chapter 8), the integrative effect of learning by creative reverse engineering conferred a major competitive advantage on many Japanese firms. It also gave production engineering a much higher status than is usually the case in Europe or even the USA. Table 6.3 illustrates the big reductions in lead times for development which the Japanese firms achieved by this kind of technique.

Horizontal information flows increasingly characterized Japanese management organization rather than the vertical flows so characteristic of the hierarchical US corporation (Aoki, 1986). It is interesting that this feature of Japanese management was singled out not only by an academic

Table 6.3 Product development performance by regional auto industries, mid-1980s

	Japanese producers	American producers	European volume producers	European specialist producers
Average engineering hours per new car (millions)	1.7	3.1	2.9	3.1
Average development time per new car (in months)	46.2	60.4	57.3	59.9
Number of employees in project team	485.0	903.0	904.0	
Number of body types per new car	2.3	1.7	2.7	1.3
Average ratio of shared parts	18%	38%	28%	30%
Supplier share of engineering	51%	14%	37%	32%
Engineering change costs as share of total die cost	10–20%	30–50%	10–30%	
Ratio of delayed products	1 in 6	1 in 2	1 in 3	
Die development time (months)	13.8	25.0	28.0	
Prototype lead time (months)	6.2	12.4	10.9	
Time from production start to first sale (months)	1.0	4.0	2.0	
Return to normal productivity after new model (months)	4.0	5.0	12.0	
Return to normal quality after new model (months)	1.4	11.0	12.0	

Sources: Kim B. Clark, Takahiro Fujimoto and W. Bruce Chew (1987) 'Product Development in the World Auto Industry', *Brookings Papers on Economic Activity*, no. 3; Takahiro Fujimoto (1989), 'Organizations for Effective Product Development: The Case of the Global Motor Industry', PhD thesis, Harvard Business School, Tables 7.1, 7.4, and 7.8; as reproduced in Womack *et al.* (1990).

economist (Aoki) but also by the head of research of one of the most successful worldwide competitors of the Japanese electronic companies – Northern Telecom – in his centenary address to the Canadian Engineering Institutes (Sakus, 1987).

Reverse engineering in such industries as automobiles and machine tools also involved an intimate dialogue between the firm responsible for assembling and marketing the final product and numerous suppliers of components, subassemblies, castings, materials and so forth. The habits, attitudes and relationships engendered during this prolonged joint learning process did much to facilitate the high degree of co-operation with subcontractors which finds expression in the 'just-in-time' system. Another factor fostering such intimate relationships was the conglomerate structure of much Japanese industry.

The emphasis on high quality of products which is characteristic of Japanese technology also owed much to the experience of reverse engineering. In the 1950s the first production models, whether in automobiles, TV sets or machine tools, were often of relatively poor quality (Jones, 1985; Baba, 1985). A determined effort to overcome these defects included a systematic review of all the possible sources of weakness. This led in turn to the ready and widespread acceptance of such social innovations as 'quality circles' (originally an American innovation) and to the development of greatly improved techniques of quality control, not simply at the end of the production run but at every stage. Some of the most important (and most closely guarded) Japanese innovations have been on-line inspection, test and quality control equipment and instrumentation arising in this process. Where the quality of components from subcontractors was particularly bad, as in the case of castings, this led to intense pressure from MITI to restructure the entire industry.

The most spectacular example of the redesign and reorganization of a production process and the associated network of subcontractors was of course the 'lean production system' of the Japanese automobile industry. The worldwide exports of cars and consumer electronics were probably the most vivid evidence of Japanese technological strength for large numbers of people all over the world. In the United States especially the reversal of their prolonged dominance in the archetypal mass production industry was bound to make a very deep impression.

This was demonstrated in two major research projects in which MIT engineers were heavily involved. The first of these was based on co-operation between MIT engineers and social scientists and led to the publication *Made in America* (Dertouzos *et al.*, 1989). The book was based on MIT's own exceptionally thorough knowledge of technology in American industry and on interviews with firms in many countries; using this evidence it endeavoured to interpret the relative decline in American competitiveness in several different industries. The main conclusion was that in too many cases US management still clung to outdated conceptions of a 'mass production' philosophy which was no longer appropriate.

The second project was the in-depth study (Womack *et al.*, 1990) of the world motor vehicle industry, to which we have already made extensive reference. This study was based on close co-operation between most of the leading automobile firms in the United States, Japan and Europe with a team of academic researchers, mainly based at MIT, but also including a group from the Science Policy Research Unit at the University of Sussex (Daniel Jones and Andrew Graves). Members of the team undertook extensive interviews at about one hundred plants in fifteen countries and had access to very detailed plant level and company level data over a five-year period. The participating companies discussed the findings in depth during the progress of the project between 1985 and 1990. This was undoubtedly one of the influences which led to the determined attempts by most American and European firms to imitate some features of the Japanese lean production system in their endeavours to catch up with competitors.

6.5 OHNO AND THE LEAN PRODUCTION SYSTEM IN TOYOTA

The origins of the lean production system can be traced back to developments in Toyota in the late 1940s and 1950s. The firm started in the textile machinery industry late in the nineteenth century and already had a reputation for technical innovation when it was urged by the Japanese government in the 1930s to enter the motor vehicle industry and to specialize in trucks for the army. After the war, although it had only made a few prototype cars, Toyota decided to go into automobiles and commercial truck manufacturing. This was not at all easy in early post-war Japan because of shortages of both materials and capital. It was not feasible to launch into large-scale production, taking into account the small market as well as these circumstances. However, the Toyota Chief Engineer, Taiichi Ohno, was an extremely creative and innovative man. He visited Detroit several times and became convinced that the Fordist system was extremely wasteful in materials, labour and capital. He began to experiment with new methods, at first in his own plant and later with the suppliers.

One of Taiichi Ohno's early experiments, described in the MIT study, was in the system of changing the dies for stamping sheet metal. A slight misalignment of the dies would cause defects in the parts stamped out and in the Fordist system die change was assigned to specialists. They took a long time to make the change but this was accepted because the large scale of production made it possible to change dies rather infrequently. To Ohno, however, this was unacceptable, because Toyota at that time was making only a few thousand cars a year and had to change dies frequently since they could not afford to buy large numbers of presses, each for a specialized stamping operation. He therefore innovated with new methods of changing dies and after numerous experiments reduced the time for changing dies from a whole day to three minutes. Moreover, he realized that the production workers could learn to change the dies as well.

These changes had several very important consequences. In the first place the rapid change of dies for small batches made any faults immediately obvious and since the production workers were themselves changing the dies they could eliminate the faults at an early stage and save a good deal of waste. In the second place, and in the long run far more important, the system required more responsible, more skilled and more highly motivated production line workers than the Fordist system. Through these and other similar experiments and changes the Toyota–Ohno system evolved further and further in this direction. Ohno devolved more and more responsibility to production teams in which the leader himself (they were all men) worked with the rest of the team and any worker had the right (and the means) to stop the line and with the help of the team to remedy a fault. Ultimately, this led to the notion that almost all defects could be eliminated before a car reached the end of the line. As we have seen, in the Fordist system it was accepted that quality inspection of assembled cars would reveal many faults and a special group of employees (rework men) dealt with the rectification of these faults. Even so, many cars reached the dealer chain with various faults, so that customer complaints were quite frequent. While the Japanese ideal of 'zero defects' has not been realized anywhere, the Toyota–Ohno system (and analogous systems in the electronics industry) pressed strongly in that direction. Workers discussed problems with engineers in 'quality circles' or similar organized groups and proposed ways to improve design of product or process. All of these experiments and changes led to a system which produced a higher quality car at lower cost.

It took about twenty years for the Toyota–Ohno system, including the supply chain (see below) to evolve to a point where it could out-perform the world competition. During that time, the Japanese government maintained a ban on foreign investment in automobiles and a protective tariff. The end result can be judged both by the world trade and production

Table 6.4 General Motors Framingham assembly plant versus Toyota Takaoka assembly plant, 1986

	GM Framingham	Toyota Takaoka
Gross assembly hours per car	40.7	18.0
Adjusted assembly hours per car	31.0	16.0
Assembly defects per 100 cars	130.0	45.0
Assembly space per car	8.1	4.8
Inventories of parts (average)	2 weeks	2 hours

Note: Gross assembly hours per car are calculated by dividing total hours of effort in the plant by the total number of cars produced.

Adjusted assembly hours per car incorporates the adjustments in standard activities and product attributes described in the text.

Defects per car were estimated from the J. D. Power Initial Quality Survey for 1987.

Assembly space per car is square feet per vehicle per year, corrected for vehicle size.

Inventories are a rough average for major parts.

Source: IMVP World Assembly Plant Survey, Womack et al. (1990).

statistics and by specific comparisons of performance at company and plant level. The IMVP made many inter-plant comparisons and Table 6.4 illustrates one of these between a typical GM plant and the Toyota–Takaoka plant in 1986. Such comparisons are difficult to make because of variations in design of car, degree of vertical integration of various operations within the plant, lifetime of model and so forth. The MIT team endeavoured to adjust for these factors to a 'standardized' model (line 2, Table 6.4). By the 1970s and 1980s the lean production system had opened up a considerable productivity gap in comparison with old Fordist plants whether in the United States or in Europe. The IMVP team was anxious to emphasize that lean production is not necessarily the same as the 'Japanese' production system. Not all of the six large Japanese companies developed the full Toyota system and some non-Japanese plants and companies have been quicker to adopt it than some Japanese ones.

Nevertheless, just as specific national circumstances strongly influenced the evolution of industrial technology in eighteenth and nineteenth century Britain (Chapter 2) or capital intensive, materials intensive and energy intensive production systems in the United States (Chapters 3 and 4) so undoubtedly the lean production system was powerfully influenced by various Japanese national institutions. This does not mean that the technology could not be adopted in other countries, just as British and American technologies were successfully transferred. But it does mean that such transfer involved institutional as well as technical change and could take a long time. Nor does it mean that the technology could not be improved during the process of transfer and assimilation, just as the Japanese improved upon the technologies which they imported. As we have seen, the experience of reverse engineering practised in many industries stimulated the close co-operation between production, design and development which proved characteristic of many Japanese industries, including automobiles. Whereas a somewhat profligate use of materials and energy was characteristic of some American industries, economy in their use was essential in an island economy with few natural resources and capital saving was also more important.

What is perhaps more surprising is the intense effort at *labour* saving in the Toyota–Ohno system. Neoclassical economists, both in Japan and elsewhere, had urged post-war Japan to specialize in labour intensive industries and to eschew capital intensive industries like automobiles. However, G. C. Allen, one of the first European economists to learn Japanese and study Japan intensively, pointed out that the advice of MITI prevailed with the Japanese government, rather than that given by the Bank of Japan:

> Some of these advisors were engineers who had been drawn by the war into the management of public affairs. They were the last people to allow themselves to be guided by the half-light of economic theory. Their instinct was to find a solution for Japan's post-war difficulties on the supply side, in enhanced technical efficiency and innovations in production. They thought in dynamic terms. Their policies were designed to furnish the drive and to raise the finance for an economy that might be created rather than simply to make the best use of the resources it then possessed.
>
> (Allen, 1981, p. 74)

Some commentators (including the IMVP Report) lay great stress on the rejection of MITI's advice on restructuring by the vehicle firms, when MITI favoured concentrating the industry into only three firms. However, the strong support of the Japanese government in ensuring priority for the industry in the days of post-war shortages and difficulties was probably of greater importance in the long run, as Allen perceived. In the evolution of the lean production system, two other factors were of still greater importance: in the first place, the conglomerate system characteristic of relationships between groups of Japanese firms facilitated technical collaboration between the leading firms and their subcontractors. Secondly, the close involvement of engineers with production workers facilitated numerous incremental innovations.

6.6 SUBCONTRACTORS IN THE LEAN PRODUCTION SYSTEM

The Toyota innovations in the supply chain were probably just as important as those in the assembly factory. Obviously, a regular supply of high quality components was essential to the success of the whole production system. In order to achieve this, Toyota established very close relationships with its 'first-tier' suppliers. Engineers moved freely between plants and exchanged detailed information; in many cases Toyota engineers were loaned to supplier firms. Relationships of long-term trust were cultivated by a variety of methods; cross-holdings of shares between the companies were common. Toyota also acts as a bank for its suppliers. Considerable responsibility for design of subsystems was assigned to first-tier suppliers, subject to meeting certain performance specifications. Each first-tier supplier organized its own second-tier suppliers for simpler components. Some of Toyota's former inhouse component supply was spun off by forming new independent companies in which Toyota retained some part

Table 6.5 Sub-contracting networks

	(British) ACR*	(Japanese) OCR*
Dependence	Lo	Hi
Procedure	Bids ⇨ Price ⇨ Contract	Order before price
Disputes	Strict adherence	Case by case
Risk sharing	Lo	Hi
Communication	Narrow Minimal	Multiple Frequent
Period	Short-term contract life	Long-term commitment
Set-up costs	Lo	Hi
Inspection	Thorough	None
Contract	Detailed clauses	Oral communication
Technology	Not negotiated	Strong interchange

* ACR = Arms length contractual relationship
* OCR = Obligated contractual relationship

Source: Sako (1992).

of the equity. They were encouraged to take outside business but had defined responsibilities for 'just-in-time' supply to Toyota. These were not contractual as in most American and European supplier networks but 'obligated' relationships. The IMVP describe the system as follows:

> Ohno developed a new way to coordinate the flow of parts within the supply system on a day-to-day basis: the famous just-in-time system, called *kanban* at Toyota. Ohno's idea was simply to convert a vast group of suppliers and parts plants into one large machine, like Henry Ford's Highland Park Plant, by dictating that parts would only be produced at each previous step to supply the immediate demand of the next step. The mechanism was the containers carrying parts to the next step. As each container was used up, it was sent back to the previous step, and this became the automatic signal to make more parts.
>
> (Womack *et al.*, 1990, p. 62)

This made the whole system extremely fragile by eliminating almost all buffer inventories but increased the sense of responsibility and commitment throughout.

The characteristic features of the Japanese lean supply chain were compared with British suppliers in the same industry by a Japanese researcher at the London School of Economics, Mari Sako (1992). Her summary of the main differences is shown in Table 6.5. This brings out the degree to which national and firm behaviour can vary in the development of inter-firm relationships. Such differences underlie the concept of 'national systems of innovation' which is discussed in Part Three of the book. This does not mean that international diffusion of organizational innovations is impossible, only that it may take a long time and faces problems which are non-technical. Both the example of the assembly line and that of the Toyota–Ohno system provide ample illustration of this point.

6.7 CONCLUSIONS

Not all commentators agree about the future system of automobile production. Volvo in Sweden attempted to introduce an alternative system at their plants in Udevalla and Kalmar. This was based on a team of ten workers assembling four cars a day on a stationary platform. The system did not last long enough to provide a full test and opinions still differ on the reasons for closure. It was introduced in an effort to overcome the problem of negative worker attitudes and high labour turnover, which has dogged the industry from the beginning of mass production.

As we have seen, when Ford introduced his assembly line, he immediately faced a crisis of labour turnover, which he only overcame by doubling wages. Ever since then Fordism has faced a continual problem of recruitment and wages, whether in Europe or North America. It remained a very unattractive type of work, boring and exhausting. The automobile companies have tackled this problem in two main ways: first, like Ford, by paying above the average wages. Research by sociologists (e.g. Platt et al., 1979) shows that, as in Detroit, many European workers have been ready to trade adverse working conditions against higher pay. Second, by recruiting many of their workers from recent immigrants, some of whom may have little occupational choice (Turks in Germany, Mexicans and other Hispanos in USA, Algerians in France, etc.).

There were no large immigrant groups in Japan and soon after the war, Toyota in common with most large companies was faced with a very serious crisis in industrial relations. The American occupation authorities insisted on recognition of trade unions and strikes were widespread. One result of this wave of industrial unrest was that the distinction between blue-collar and white-collar workers almost disappeared. Another was that many large companies offered security of employment to a large part of their workforce (so-called lifetime employment) in exchange for the acceptance of wage increases based on seniority and annual bonus payments linked to company profitability. Toyota workers also agreed to flexibility in type of work. This system gave a very powerful incentive for workers to stay with the company even if they disliked the work. Older workers who quit would face a massive drop in earnings. It also gave a strong incentive to the company to invest in training and to motivate the workforce.

These changes were not universally welcomed and some critics maintain that the Toyota system is still essentially a variant of Fordism ('neo-Fordism') with some special Japanese features. One trend in the automobile industry in the 1990s has been to relocate assembly in low-wage countries (from Western to Eastern Europe, from North America to Mexico); the first plant closure has occurred in Japan and there has even been German relocation in the USA. In addition to the pressure to pay high wages in the richer countries, the automobile industry is faced with the necessity to redesign the product to deal with environmental problems and to incorporate more microelectronics in components and subsystems (see Chapter 7). Despite all these problems, indeed partly because of them, the IMVP authors remained extremely optimistic about the future of lean production and very sceptical about the Volvo experiments:

> We believe that once lean production principles are fully instituted, companies will be able to move rapidly in the 1990s to automate most of the remaining tasks in auto assembly – and more. Thus by the end of the century we expect that team-assembly plants will be populated almost entirely by highly skilled problem-solvers whose task will be to think continually of ways and means to make the system run more smoothly and productively.
>
> The great flaw of neocraftmanship is that it will never reach this goal, since it aspires to go in the other direction, back toward an era of handcrafting as an end in itself.
>
> We are very skeptical that this form of organisation [Udevalla or Kalmar] can ever be as challenging or fulfilling as lean production.
>
> (Womack *et al.*, 1990, p. 102)

It remains to be seen in the twenty-first century how far this optimism might be justified. Paradoxically, this lean production dream of the IMVP project resembles the old Marxist dream of abolishing the distinction between mental and manual labour and turning all workers into skilled knowledge workers who would enjoy work as a necessity rather than as a burden.

Mass production ideas have affected not only manufacturing but also many service industries, although of course the delivery of services cannot be quite the same as the output of goods. The ideas of standardization and flow delivery have certainly affected the fast food industry to a remarkable extent, as well as such processes as packing of eggs or vegetables. Perhaps the best illustration was analyzed by Auliana Poon (1993) in her work on tourism. She showed how the combination of paid holidays, cheap bus and air travel and standardized hotels and tourism produced the 'packaged holiday' in the 1950s and 1960s, enabling millions of Europeans, Americans and Japanese to enjoy cheap holidays in the sun which would have been out of the question for their parents.

Whereas the proportion of unskilled workers in manufacturing has been declining and the ratio of white-collar to blue-collar employees has been rising, this does not appear to be so true, or at any rate, not to the same extent, of the services. There has been a proliferation of low-skill, low-paid, often part-time jobs in the service industries. Information and computer technology are opening up entirely new possibilities for future employment in every industry and service and it is to this technology that we turn in Chapter 7 and later chapters of the book.

CHAPTER 7

ELECTRONICS AND COMPUTERS

7.1 INTRODUCTION

During the 1950s and 1960s electronic communications and computing systems made it possible to perform in minutes or even in a fraction of a second calculations and operations which previously took weeks, months or years, or could not be performed at all, and to perform them with a higher degree of reliability and at a lower cost than by older methods (Table 7.1). Beginning with radio communications in the 1890s and television in the 1930s, the applications of electronics have spread first to systems of detection and navigation (radar), and since the war to computers for data processing and to the control of a great variety of industrial processes.

The introduction of reliable low-cost electronic computers into the economy was the most revolutionary technical innovation of the twentieth century. While it is true that the older mechanical and electromechanical calculators and other devices could already perform some of the functions of modern computers before and during the Second World War, it was the electronic computer which totally transformed both the range of potential applications and their cost. Table 7.1 illustrates the dramatic increase in capacity and reduction in costs of computing which occurred in the 1950s and early 1960s, from the early valve (tube) computers to those using semiconductor technology and integrated circuits. Since that time the microprocessor revolution of the 1970s and 1980s has further increased the number of components per cubic cm and reduced their cost by at least two more orders of magnitude, so that the computers of the 1950s now appear incredibly expensive and cumbersome.

Already in the 1960s electronic computers had greatly increased the efficiency with which enormous quantities of data could be stored and processed, such as payroll calculations, invoicing, insurance premiums, design calculations and so forth. In addition, electronic equipment is so fast and reliable that automatic feedback systems can be used to control operations in real time, even where this involves fairly complex calculations with several variables, as in chemical processes or aircraft navigation. A computer of some sort is at the heart of such systems, and it is electronics which made it possible to automate a much greater variety of operations and processes than was hitherto possible.

In a sense, there is nothing really new in automation, as the thermostat, invented in 1625, already represented an automatic feedback control system. It is no accident that some of the pioneering firms in electronic equipment for automation also produced thermostats and regulators (for example, Honeywell and Elliott-Automation). The difference is one of degree. However, electronics has increased the applications of the automatic feedback

Table 7.1 Technical progress in computers

Measure	Vacuum-tube computers (early 1950s)	Hybrid integrated circuits-360 system (late 1960s)
Components per cubic foot	2,000	30,000
Multiplication per second[a]	2,500	375,000
Cost ($) of 100,000 computations	1.38	0.04

[a] A single multiplication on mechanical or the first electromechanical computers took more than one second.

Source: Fortune (1966), September.

control principle so rapidly that it is not unreasonable to look upon automation primarily as a post-war change, associated with the electronic computer, electronic sensing and detecting devices, and process instrumentation.

Experience in war-time radar, gun-control systems and missile guidance devices formed the basis for revolutionary advances in industrial process control systems and numerical control of machine tools. Together with new types of electronic instruments and robots these are gradually transforming engineering processes so that many of them will increasingly come to resemble automated flow processes, as we have seen in the case of automobiles in Chapter 6. Thus, computers will increasingly dominate the design and manufacturing system in the traditional heartland of capital goods manufacture. The Molins 'System 24' designed by Williamson in the 1960s was an early forerunner of the automation of machine shops which is now being realized by a growing number of firms, based on a combination of computers, NC and robotics.

These new manufacturing systems are often described as 'flexible manufacturing systems' (FMS) but in Fleck's thorough studies (1988, 1993) of the spread of robotics and computerization of machining systems, he has shown that each system has unique characteristics and craft skills that cannot yet (if ever) be completely eliminated. He has coined the expressions 'innofusion' or 'diffusation' to describe the complex combination of innovation and adaptation which has accompanied the diffusion of CNC and robotics. He attributes the early lead of Japanese firms in the diffusion of robotics to their recognition of the need for this customization of each system in collaboration with the users, and in hybridization of equipment. Kodama (1995) confirms Fleck's account and gives some detail showing the higher degree of customization in Japanese FMS. Some firms in North America, Japan and Europe were approaching close to 'CIM' (computer-integrated manufacturing) in the 1990s but it has proved a long and difficult road.

Since the Second World War, military applications of radar have been extended to complex early warning systems, missile guidance and so forth. Civil applications have also grown rapidly – in air traffic control, airborne and marine radar and navigation systems, space exploration and aids for fishing vessels (sonar). Both in civil and military applications there is also a close link between the growth of these installations and of computer networks. The American SAGE (Semi-Automatic Ground Environment Control) system linked a big chain of radar stations to very powerful computers, and the same principle was used in NADGE and in the civil air traffic control systems in Western Europe. On a smaller scale, airborne navigation systems such as the Decca 'Navigator' and Marconi 'Doppler' equipment made use of very small specialized computers.

Originally, the radio industry formed the basis of the electronics industry, followed in the 1930s by television. Communication equipment and entertainment and information systems remain an extremely important part of the industry, but they too are being transformed by the electronic computer. The convergence of computer and telecommunication technology has led to the growing use of the expression 'information and communication technology' (ICT) to describe this transformation. The applications are so widespread that it has become common parlance to talk about the 'information society' or, more recently with the spread of the Internet linking millions of computers worldwide, the 'on-line society'. The telephone system increasingly uses electronic exchanges and switching computers and is used for data transmission and for communication between computers. Services such as 'Prestel', despite teething troubles, already demonstrated the way in which domestic television and VCRs could be linked up through a greatly expanded telephone network to provide a wide range of information services both for households and for business. Teleshopping and telebanking are now diffusing fairly rapidly and the potential is enormous. Word processors are already in use on a large scale and electronic mail services are increasingly important for business communication as well as research. Consequently, electronic innovations in the twentieth century are revolutionizing service activities and households, as well as all branches of manufacturing. In Chapter 17 we discuss in more depth the contemporary problems of the newly emerging information society and the policy implications of this computerization of every sector of the economy. Here in this chapter we concentrate on the background history which led to these revolutionary developments. Developments in telephone technology, such as optical fibres and mobile telephony, were as important as those in electronics and computer technology but for reasons of space could not be included here.

It is seldom possible to talk meaningfully about the 'inventor' of any of the major electronic products. Their successful realization, first in a laboratory and later on a commercial scale, depended on contributions from many scientists and engineers in several countries over a long period. They represent systems involving a large number of components, all of which are subject to change and improvement. The interplay of component innovations, materials innovations, software innovations and new capital goods and consumer products is one of the most important features of the industry's development. In particular the introduction of transistors in the 1950s, and later of integrated circuits and micro-processors, facilitated major new advances in the design of electronic products and systems, and reductions in their cost and size.

Any analysis of this industry must study innovation problems particularly carefully. For none of the products existed before the beginning of the century, and most of them did not exist in their present form even fifteen

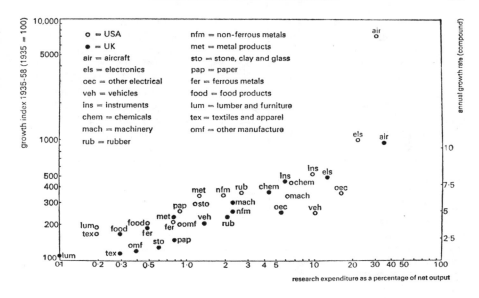

Fig. 7.1 Research expenditure as a percentage of net output in 1958, and growth of industries, 1935–58

years ago. The entire industry is based on research and in the 1950s and 1960s was one of the most research intensive of all industries (Figure 7.1). Like plastics, it illustrates the transition from the inventor–entrepreneur to the corporate R&D department. In the historical account which follows compare, for example, innovation in valves with semiconductors or radio with television. Along with the growth of corporate professionalized R&D, governments too have played a very big part both as sponsors and directly in the conduct of R&D. This culminated in the space research programmes, but was already very important in radar. However, the inventor–entrepreneurs continued to play a significant role in industry, especially in electronic instruments. Even in computers their role has been more important than is commonly assumed. The account below briefly summarizes the major developments in each of the five main sectors: radio; television; radar; computers; components.

7.2 RADIO[1]

For radio, Maxwell's theory of electromagnetism in the 1860s was the foundation for Hertz's first practical laboratory demonstrations of the production and detection of wireless waves in Germany in the 1880s, for Branly's coherer demonstrated in 1890, and for Lodge's demonstration of wireless reception at the British Association in 1894. At about the same time Popoff demonstrated an improved system of reception at the University of Kronstadt. All these men were academic scientists engaged in fundamental research. It was not until Marconi formed his Wireless Telegraph Company in London in 1897 that systematic applied R&D work began. It was Marconi who gave the first practical demonstrations of wireless communication between ships and shore, between shore stations, between countries and finally across the Atlantic in 1901. It was his new company which established the first regular wireless telegraph services, both between countries and from ships, followed closely by Telefunken in Germany. His role in the electronics industry was analogous to that of Baekeland in plastics, as a highly successful inventor–entrepreneur.

Up to and during the First World War there were innumerable improvements in the components, circuits and techniques used in radio communication, made by inventors in many countries, but the most important of them originated in Britain, Germany and the United States. It would be difficult, if not impossible, to evaluate precisely the relative contribution from scientists and engineers of the three principal countries. Some of the developments, such as the feedback circuit, were almost simultaneous in all three countries, and there was bitter patent litigation which went on for twenty years. While Professor Fleming of University College, London (who was employed as a consultant to the Marconi Company), invented and patented the first thermionic value in 1904, it was the American, de Forest, who invented the triode valve in 1907, which later proved far more effective for reception and transmission. Other American inventors, notably Langmuir, Armstrong, Alexanderson and Fessenden made important contributions to the improvements of valves, circuits, alternators, and aerials and in the introduction of radio telephony. The American Telephone and Telegraph Company (AT&T) pioneered the use of valves for relays in the telephone system, having purchased rights to de Forest's triode patents. Both the research laboratories (Bell Labs) and the manufacturing arm (Western Electric) of AT&T were to play a key role in components invention and innovations, as well as many spinoff firms from AT&T.

But in this early period, up to the end of the First World War, the leading companies in the industry were not American but British and German. The largest manufacturer of radio in the USA was a Marconi subsidiary, and the world market was dominated by Marconi and Telefunken, which between them controlled most of the ship and shore installations all over the world, including those in the United States. In 1915 the Marconi companies controlled 225 out of 706 coast installations, and 1,894 out of 4,846 marine installations. Until the outbreak of the First World War the share of Telefunken was probably somewhat larger. Although starting a year or two later than Marconi, the largest German electrical companies, Siemens and AEG, rapidly developed effective radio communication systems – Professor Braun's system at Siemens and the Slaby–Arco system at AEG. They received strong backing from the German government, particularly the Navy, and were persuaded by the Kaiser to merge their interests in 1903 and form the jointly owned Telefunken Company. The company was concerned primarily with R&D and the sale, installation and maintenance of radio stations. AEG and Siemens continued to manufacture parts and equipment. Already by 1906 Telefunken had more installations than Marconi. While the Marconi patents were very strong and upheld in most countries, their priority was not accepted in Germany until

1912 when Telefunken and Marconi reached a worldwide agreement on patents, licences and know-how. Similar cross-licensing and know-how agreements were renewed after the First World War in 1919, and again more recently.

Thus both Marconi and Telefunken had well-organized industrial R&D programmes and strong patent positions. They were able to assimilate and imitate the technical advances made in other countries. Both of them were able to provide worldwide technical service, had their own schools for training radio operators, and could repair and maintain leased equipment. (The first Marconi school was opened at Frinton in 1901.) But Telefunken had the advantage over Marconi in consistency of government support and of its financial resources. Although the Post Office had originally encouraged Marconi in his first experiments in Britain, and the Admiralty had also been sympathetic, later relationships with the Post Office became difficult and the attitude of the government was sometimes unhelpful and even obstructive. There were, of course, difficult issues involving the problem of private monopoly in communications and the Marconi company's determination to uphold and exploit its strong patent position.

Like many major innovations, radio did not prove profitable for a long time and the Marconi Company paid no dividends from its inception in 1897 until 1910.

In view of the common tendency to ascribe almost superhuman attributes to US management, it is important to note that, by contrast with Marconi and Telefunken, the early American radio companies were poorly managed, and some of the pioneering inventors, such as de Forest and Fessenden, were failures as innovators and entrepreneurs. De Forest produced a stream of inventions and patents, but although his Wireless Telegraph Company had orders from the Navy and War Department it failed to produce reliable communications equipment. It was not until the formation of the Radio Corporation of America in 1919 that a really successful specialist electronics enterprise was started. The big electrical companies, General Electric and Westinghouse, and the major telephone company, American Telephone and Telegraph, all had an interest in radio and relatively strong R&D organizations which had made major contributions to the development of radio telephony. They were blocked, however, by the fact that in international communications the British Marconi Company dominated the field, and in the United States the control of key patents by opposing interests contributed to a stalemate that retarded the best utilization of radio. An imaginative solution to this deadlock was found as a result of government initiative. With strong encouragement from the United States Navy, Owen D. Young of General Electric set about buying out Marconi's American subsidiary and setting up a powerful unified American-owned radio company. The Navy's motivation was partly commercial and partly strategic. Both the Secretary of the Navy (who favoured a publicly owned communications network) and the Assistant Secretary, Franklin Roosevelt, wanted an American company because they could not accept a position where a vital communications network was controlled by a foreign, even though friendly, power. With other officials of the Navy they realized the great commercial potential of radio.[2] Owen Young became chairman of the new Radio Corporation of

America (RCA) with General Electric supplying sets and valves (Westinghouse came into the arrangements for manufacturing in 1921). David Sarnoff, of the old Marconi Company, became commercial manager (and later president), and RCA immediately concluded patent and cross-licensing agreements with British Marconi, Telefunken and the French Compagnie Générale de Télégraphie sans Fil (CSF), which was also based on a former Marconi subsidiary. After some bargaining, agreements were reached within the United States which ensured that RCA had the right to use over two thousand American patents, including all the important ones; RCA also made a cross-licensing arrangement with AT&T.

Shortly after the formation of RCA, the radio industry was transformed by the growth of public broadcasting. The United States industry took the lead in the scale of manufacture and improvement of design of home radio receivers, and an American company, Motorola, took the lead in the mass production of radios for automobiles. However, American companies never achieved that supremacy in the world export market for consumer goods which they later reached in capital goods. Many of the important inventions concerning radio receivers between the wars were made by European companies, especially Philips and its subsidiaries. In 1937 Dutch radio exports were as big as those of the USA. After the Second World War, Japanese companies were quick to appreciate the possibilities of using transistors in radio sets, taking American licences for the semiconductor devices, and achieving a very large share of world exports of electronic consumer goods, first in radio and later in television and VCR.

The big post-war developments in capital goods were the introduction of worldwide short-wave communication networks in the 1920s and the introduction of frequency modulation (FM) in the 1930s. Short-wave communications were developed primarily by amateurs and by the Marconi Company. The development of directional aerials by Franklin at Marconi's made it possible for them to propose a Commonwealth radio chain using a beamed short-wave system. His proposals were accepted by the first Labour government in 1924 in spite of opposition from the Post Office, which was wedded to cable communications, and in 1926 the first stations were opened. One of the immediate effects was a sharp reduction in cable rates; and eventually the Post Office took over the entire 'Imperial chain'.

Frequency modulation was pioneered by an independent inventor, Edwin Armstrong, who was a professor at Columbia University. He did not meet with much sympathy from the broadcasting networks or from RCA, and consequently had to build his own station to demonstrate his system. Partly because of these delays the first experimental operational FM network was set up not in the United States but by Telefunken for the German army in 1936. Its success led to the establishment of a large-scale FM network during the war which covered the whole of German occupied Europe and Africa. The introduction of FM in Britain came very much later in 1955.

New developments since the Second World War have vastly improved the quality and range of radio communication. They have transformed the global telephone network through the use of communication satellites, mobile telephony and radio links and made possible the provision of broadband networks for a vast array of services.

7.3 TELEVISION

The possibilities of television were foreseen as early as 1884, when Paul Nipkow took out a patent in Berlin for his invention of the scanning disc. The invention and improvement of the photoelectric cell, and of the cathode-ray oscilloscope, also took place mainly in Germany, but it was Professor Boris Rosing of St Petersburg who first suggested using Braun's cathode-ray tube to receive images from a remote source, in 1907. Similar suggestions were made by Campbell Swinton in England. Zworykin was a pupil of Rosing's, who had already worked on a cathode-ray receiver before going to America in 1919. He had already conceived a complete electronic system for transmission and reception and in the United States patented the iconoscope which made it possible to transmit television pictures successfully. But it was not until he began a much more ambitious R&D programme at RCA that the numerous development problems were overcome, in the years from 1924 to 1939. RCA launched television commercially in 1939, but pictures had been successfully transmitted on a laboratory basis many years earlier.

Although EMI in England started later, they moved more quickly and a team led by Blumlein independently developed an iconoscope known as 'Emitron'. Marconi co-operated in the development of transmitters and the two firms were so successful in developing their system that the BBC were able to begin regular television broadcasts in 1936.[3] In the same year EMI and RCA made a licensing and know-how agreement. There were already financial links, and Sarnoff from RCA was a director of EMI. Telefunken, too, was able to develop an electronic television system before the Second World War and had a licensing and know-how agreement with EMI. Experimental transmissions were made of the Olympic Games in 1936 and regular broadcasts were made to some German troops in 1939–40. The war prevented the introduction of a regular service, although Telefunken had made advanced preparations for the mass sale of a popular model. Thus, although Zworykin's contribution to television was outstanding and other American inventors, such as Farnsworth, also made important advances, European countries were close behind the United States, and a public service was launched earlier in Britain. It was in the 1939–41 period and the early post-war years that the American industry went ahead.

The part played by RCA in the development of television is particularly notable. Between 1930 and 1939 RCA spent $2.7 million on television research and development, and a further $2 million on patents and patenting. Another $1.5 million was spent on testing the system. Telefunken and EMI probably spent similar sums in the pre-war period. Such an investment was impossible for most smaller firms or independent inventors, but Farnsworth is said to have spent $1 million privately. The teams at RCA and EMI were quite small in the early days; Zworykin had only four or five assistants before 1930, and Shonberg, the research director at EMI, had only a few when he started in 1931. But the scale of effort built up as the introduction of a commercial system approached; the problems could not have been solved without resources of a fairly large organization.

This applied even more to colour television. RCA spent $130 million on launching colour television before it became profitable in 1960. After about four years' development work it demonstrated its first colour tube in 1950, but the Federal Communication Commission (FCC), after considering two competing systems, mechanical non-compatible and all-electronic compatible, gave its approval to the first, thus in effect banning the RCA system from the market. This decision was only reversed after several years, through court action by RCA. During this period, development work continued intensively and the first sales were made in 1954. Growth was very slow at first, because of the attitude of the broadcasting networks and the high price of sets, but sales had reached about five million sets per annum by 1970 in America.

Europe and Japan were both originally a good many years behind the United States with colour television, but Japan not only closed the gap in the 1960s and 1970s but went ahead to undisputed world leadership in this industry. By 1977 Japan accounted for over half of world production in colour television and three-quarters of world exports. They were exporting about 5 million sets, compared with about 1 million from Germany and 250,000 from the UK, despite the fact that Japan was still limited in many markets in Europe by the PAL patents and other restrictions. Later in the 1970s, Japanese direct exports declined especially to the USA, because of Japanese investment overseas and because of agreements between Japanese and American producers.

Both in television and in video cassette recorders (VCRs) the standards adopted by firms, by countries and by international organizations played an extremely important role. The PAL (German) and SECAM (French) colour TV transmission standards afforded some degree of protection against American imports in the 1960s, and in national markets within Europe. However, in the case of VCR, although American and European firms had led the way for many years, in the later stages of development Philips was forced to give way to Japanese competitors and license the Matsushita VHS technology in 1985. The VHS system was said by many to be technically inferior to another Japanese standard – the BETAMAX system of Sony. However, because of superior sales and distribution within and outside Japan, based partly on an alliance of Matsushita with other companies, it ultimately became impossible for Sony to sustain the competitive struggle. This became the classic case of 'lock-in' by standards, not necessarily to the optimal system. In the much earlier case of the QWERTY system in typewriters and also in the electrical industry, Paul David (1985, 1993) had demonstrated the economic importance of 'lock-in' and B. Arthur (1988, 1989) further elaborated the consequences for the evolution of technologies in electronics. One reason for the success of the VHS system and of Japanese firms in its manufacture and export was the complexity of the precision engineering and machinery which was needed. Even after Japanese firms transferred the greater part of their production of VCRs and TV to cheaper manufacturing locations in the 1980s and 1990s, the precision machinery was still supplied from Japan.

As in the case of the introduction of transistors into the radio industry, this extraordinary Japanese success was not based on simple carbon copy imitation, but involved a whole series of product improvements and process innovations. After comparing the performance of the American, European and Japanese industries, Sciberras concluded:

> Japanese firms have been the most successful innovators [in the 1970s]. By applying advanced automation in assembly, testing and handling to large production volumes, the Japanese have achieved drastically superior performance in terms both of productivity and of quality.
>
> (Sciberras, 1981, p. 590)

Although he found that the main advantage of the new automated techniques was in improving product reliability, Sciberras calculated that Japanese man hours per set were 1.9 compared with 3.9 in Germany and 6.1 in the UK. He attributed the opening up of this remarkable productivity gap mainly to the integrated approach to automation technology and to the intensive training of personnel at all levels in Japanese firms. Peck and Wilson (1982) also pointed out that the Japanese manufacturers were the first to introduce integrated circuit technology into the colour television industry (with the important economies in assembly labour that this involved). The success of this innovation was based on a joint research effort starting in 1966 and involving five television manufacturers, seven semiconductor manufacturers, four universities and two research institutes and the overall backing of MITI (Ministry for Trade and Industry). This example illustrated the capacity of the Japanese social system to achieve a flexible mobilization of resources to make and diffuse decisive innovations quickly (Allen, 1981; Fransman, 1990).

An American engineer resident in Japan produced the interesting diagram shown in Figure 7.2. It demonstrates the increasing contribution which Japanese firms made to successive new products during the post-war years. In the most recent opto-electronic technologies Kumiko Miyazaki (1995) has shown in detail that Japanese research as well as development work in the large firms was an essential feature of their success, for example, in liquid crystal displays and in consumer electronics generally. Her analysis is amply confirmed by Kodama's (1995) account of Japan's innovation system.

Another Japanese researcher, Yasunori Baba (1985), in his comparison of Japanese and US management techniques in colour television innovations, showed that Japanese firms in this industry, as in automobiles, shortened lead times by close integration of research, development, production and marketing. In his own phrase, they 'used the factory as a laboratory' and made product and process innovations together.

7.4 RADAR[4]

As with television, so with radar: the possibilities were conceived long before they were realized in practice. As early as 1904 a Düsseldorf engineer, Christian Hülsmeyer, took out a patent for a process for 'detecting distant metallic objects by electrical waves'. But he failed to produce a working prototype. In a lecture in June 1922, Marconi also foresaw the use of such a system for detection in darkness or in fog, but the Marconi Company did not

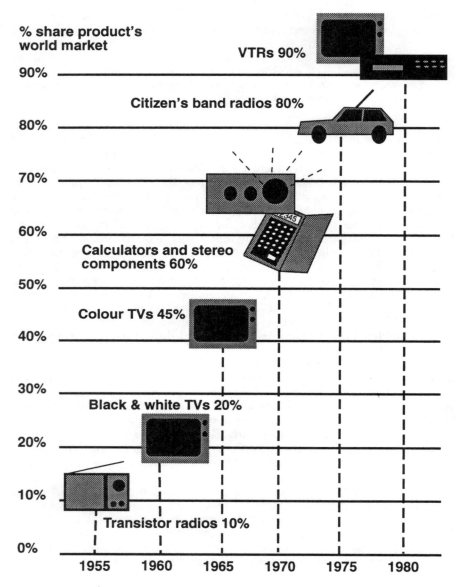

Fig. 7.2 How increasing original technology affects Japan's market share
Source: Gregory (1986).

do any development work in this field before the Second World War. In 1923 an American engineer, Loewy, also patented a radar device in America.

It was only when a government-sponsored R&D programme was started in Britain and in Germany in the 1930s that practical results were achieved. Radar (radio detection and ranging) became an invaluable military aid in the Second World War. Sir Robert Watson Watt demonstrated that the reflection from electromagnetic waves could be projected on a fluorescent screen (an oscilloscope), and received high level support for a crash programme to set up a chain of radar stations for air defence. British firms with experience in the development and production of high-powered transmitters (Metro-Vickers) and of cathode-ray tubes (EMI and Cossor) were associated with the programme before the war, whereas in Germany almost the entire R&D effort was concentrated in Telefunken. By the end of the war, eight to ten thousand people were engaged on R&D at Telefunken, but this included work on radio communications and control systems as well as on radar. Total numbers engaged were smaller in Britain and were mainly in government establishments, especially the Telecommunications Research Establishment (TRE), later the Royal Radar Establishment (RRE). Perhaps three thousand people were engaged in government establishments and another one thousand on radar development work in industry at the peak of the effort. Marconi's R&D staff were engaged entirely on communications work during the war. Because industry's development facilities were inadequate, the government teams often carried projects right through to production drawings, and sometimes undertook the first stages of manufacture as well. An extremely important feature of the whole programme was the intimacy of the direct contact between users of the equipment, manufacturers, and government (TRE) research teams, as embodied in the 'Sunday Soviets'.

The high priority assigned to radar work, the direct involvement of the best scientists in the country from universities and especially of Cockcroft and his colleagues from Cambridge, the relative freedom of the development teams and the large resources committed, resulted in an extraordinary rapid and successful flow of new devices and equipment, without previous parallel in the history of the industry. Among the new equipment successfully put into active service within a few years were the home chain of radar stations for intercepting enemy aircraft (CH), air interception equipment for fighter aircraft (AI), air-to-surface equipment for locating ships and surfaced submarines (ASV), equipment in ships and aircraft for identification as friend or foe (IFF), gun-laying equipment for control of anti-aircraft fire (GL), navigational and positional aids for aircraft and ships ('Gee' and 'Oboe') and display systems such as the Plan Position Indicator (PPI) which, combined with airborne radar, provided an accurate and detailed map of the ground below (H_2S, Home Sweet Home). Before his tragic death in an air crash, Blumlein led the development work at EMI on H_2S which was the most sophisticated type of radar.

Perhaps the most outstanding British achievement was the invention of the resonant cavity magnetron by a team at Birmingham University led by Professor Oliphant in 1940. This made possible the 'centimetric revolution', the use of very short wave equipment as low as 3cm which permitted far more accurate performance for various types of radar using smaller aerials. It involved an increase in the peak pulse power of the radar equipment of several thousand times and was far more difficult to jam. This invention gave British radar a decisive lead over German equipment for a year of two, until captured equipment at Rotterdam permitted them to catch up. This invention also proved decisive in persuading the United States to pool all radar know-how during the war.

Although British radar developments were, in most cases, a little ahead of American and German work, the lag was short. All the main countries had been doing government sponsored work in the 1930s. France had developed the first civil application of marine radar on the 'Normandie', and some radar know-how was brought to England in 1940. Germany had a number of operational radar devices at the outbreak of war, but they were relatively unsophisticated.[5] Telefunken took out a patent for PPI already in 1936, but the work was not pursued energetically because U-boat chasers and night bombers were given a low priority in the German weapons programme. Work on centimetric equipment was also slowed down in 1940, as it was thought that it would be unnecessary. Once the centimetric revolution was taken seriously, with the aid of the Rotterdam equipment Telefunken soon caught up. The official British history records:

> At the end of the war the Germans were developing a centimetric ground equipment of very high discrimination, against which no economic method of jamming could be foreseen. With this development the radar defence had caught up with the radar attack and the tactics of night bombing would, if the war had continued, have required drastic revision.
>
> (Postan *et al.*, 1964, p. 427)

The principal developments since the war have been in three directions. First, the know-how acquired during the war was applied to new civil uses. On the whole, British firms led in this field. However, the firms which were prominent in radar during the war (Metro-Vickers and Cossor) were not the leaders in the post-war civil applications. Marconi and Decca, firms only marginally involved in wartime work on radar, took the lead in the development of new types of equipment, such as the Decca marine radar and the Doppler airborne equipment developed by Marconi after feasibility studies by the RRE. Their technical leadership was strong enough for them to achieve and maintain a relatively high share of world exports of radar equipment for a considerable time.

Second, new military equipment has been developed – early warning systems with long-range capability, low flying interceptor systems and very short wave airborne reconnaissance radar. Third, much more complex control systems have been devised, based on computer links with radar chains. These have civil applications as well as air traffic control. On the whole, the United States has been ahead in these developments and American firms dominate world production and exports.

Finally, there have been new developments both in space and defence arising from the wartime work on radar servo-control systems and guidance systems. For reasons of brevity these are not dealt with here, but the most spectacular applications have been in the United States space programme and in communication satellites. With NASA expenditure in the 1960s equivalent to the total R&D expenditure of a large European country, the United States had an unchallenged technical lead in this whole area, which has been maintained despite increased European and Japanese efforts. The former Soviet Union was the only serious competitor in the military and space fields.

7.5 COMPUTERS[6]

Following the pioneer work on calculating machines by Leibnitz, Pascal, Schickard, Jacquard and others, Babbage began work in the 1830s on an 'analytical engine' which already embodied all the main features of the modern computer. He received one of the first large government development grants, amounting to £17,000 over twenty years for the development of his 'difference engine', but neither this (a special purpose calculating machine) nor the analytical engine (a general purpose machine) was ever completed, mainly because the available components and techniques were inadequate for the purpose.

It is widely believed in Britain and the United States that the first successful computers were built in these countries; but in fact one of the earliest, the Z3, was developed by Zuse in Berlin between 1936 and 1941. Zuse began his work while still an engineering student at the Technical High School and much of his early development (the Z1 and Z2) was done with only a few colleagues assisting. Already in 1942 a second Zuse model, the Z4, was used for aircraft design calculations at the Henschel works. It would probably have been in operation still earlier if Zuse had not been called up in 1939 and only released in 1940. The development then had some official support from the German Experimental Aeronautical Institute with some assistance from the Darmstadt and Charlottenburg Technical High Schools.

The Zuse Z3 and Z4 were both electromechanical and slow by electronic standards; so also was the first Harvard-IBM computer, the Automatic Sequence Controlled Calculator (ASCC), which was under construction from 1937 to 1944. Multiplication of two numbers took about five seconds on the ASCC and on the Z3, but addition and subtraction took only about 0.07 seconds on the Z3 compared with 0.3 seconds on the ASCC. The total development cost of the German machines was far less than the $400,000 spent on the ASCC at Harvard, or the $500,000 on the later American ENIAC (Electronic Numerical Integrator and Calculator).

Perhaps the most remarkable computing machine developed during the Second World War was at Bletchley Park and was designed to crack the German 'Enigma' machine code by British intelligence. It was developed with the advice of Alan Türing, an outstanding mathematician who played a leading role in the development of computer technology through his work on the fundamental algebra and logic of computers. The Bletchley machine was one of the greatest British technical achievements of the Second World War and could perform some tasks only surpassed by the much later development of parallel processing machines (Jones, 1978).

Its importance has been underestimated because of the secrecy which surrounded it but the machine has now been reconstructed for museum purposes.

Even the first electronic machines were over a thousand times faster than the electromechanical ones for both multiplication and addition (Table 7.1). Zuse was also the first to begin work on an electronic computer, in collaboration with Dr Schreyer at Charlottenburg. Special valves were ordered from Telefunken and a breadboard model in 1942 was very promising. But the work was stopped when Schreyer was called up and official support for the project was refused. This and the disruption at the end of the war meant that the lead in computer development passed to the United States and, to a lesser extent, Britain.

In the United States work on the first electronic computer, ENIAC, began at the University of Pennsylvania in 1942 and was completed in 1946. It received financial support from the Army and was mainly designed for calculating trajectories of shells and bombs. It used 18,000 valves, whereas the first Zuse electronic machine would have used only 1,500.

Katz and Phillips, in their fascinating account of the early history of the US computer industry, make particularly interesting comments on the reasons why private funds were not committed to the commercialization of the electronic computer in the early post-war period:

> The general view prior to 1950 was that there was no commercial demand for computers. Thomas J. Watson Senior, with experience dating from at least 1928, was perhaps as acquainted with both business needs and the capabilities of advanced computation devices as any business leader. He felt that the one SSEC machine which was on display at IBM's New York offices 'could solve all the scientific problems in the world involving scientific calculations'. He saw no commercial possibilities. This view, moreover, persisted even though some private firms that were potential users of computers – the major life insurance companies, telecommunications providers, aircraft manufacturers and others, were reasonably informed about the emerging technology. A broad business need was not apparent.
>
> (Katz and Phillips, 1982, p. 425)

From 1945 to 1955, rapid progress was made in solving some of the problems of logic design, memory storage systems and programming techniques. The US Army, Navy and Air Force, the Atomic Energy Commission and the National Bureau of Standards all placed major contracts for the development of improved computers with universities and with the first firms which began design and manufacture – especially Remington Rand (Univac) and, later, IBM. The major technical advances were made in these large computers and most of the medium and small computers incorporated the developments of their larger brethren scaled down for less complex EDP and scientific applications. Almost all the early demand in the United States was from the military market. Few people then envisaged the large-scale use of computers for data processing, and both government and industry thought mainly in terms of military and scientific applications.

Even after they produced the 650 under the pressures of the Korean War, IBM was still underestimating completely the potential future market.

Their Product Planning and Sales Department forecast that there would be no normal commercial sales of the 650, whereas the Applied Science Group forecast a sale of 200 machines. In the eventual outcome, over 1,800 machines were sold and the 650 became known as the 'Model T' of the computer industry. This indicates very strongly the limitations of theories of market demand leadership for radical innovations and the key role of patient government sponsorship in the early period of radically new technology.

It was perhaps particularly surprising that IBM of all firms did not then appreciate the potential commercial EDP market. The dominant personality in the company was Thomas J. Watson Senior, who was notable for his insistence on product innovation rather than price cuts as the way to enlarge market share, and for his flow of suggestions to the development and engineering departments for improvements in equipment. He seldom admitted any distinction between sales or market research and technical research. Confronted with financial difficulties at IBM when he took over, he borrowed $40,000 and used $25,000 to develop a new tabulator. One of his associates wrote: 'It required a great deal of courage to authorize the tremendous expense for this development and there were many of us who seriously doubted that the customers would stand for the increased rental necessary for the increased complication of the machines; but it is now evident that Mr Watson had correctly estimated the final result.' It took four years to develop the machine. He continued to insist on the import-ance of new product development, particularly during the depression of the 1930s. 'We had some new machines and ideas to give our salesmen. . . . If we had to depend on the line we had five years ago, it would have been a different story.' It was also very characteristic of him to insist on the importance of education. Few companies can have given so much attention to the selection and training of their own employees and to the training of their customers. The IBM Department of Education in 1956 (that is, before their big expansion in electronic computers) had a budget of $14 million – nearly 3 per cent of turnover.

Powers/Remington Rand was the technical leader and had a stronger R&D effort than IBM (Hoffmann, 1976), but by the 1930s IBM had by far the largest market share based on the success of its world-wide service and sales organization and field-force of engineers.[7] It also had manufacturing subsidiaries in France, Germany and several other countries. It had over three-quarters of the punched card market, whose users were the obvious outlet for EDP. It had fairly important R&D facilities before the Second World War, and was taking out more patents in London in the field of calculating machinery than any other firm. (Until 1949 the British Tabu-lating Machinery Company, BTM, had an exclusive franchise for all IBM products in Britain, and took out patents on IBM inventions. It did not conduct its own R&D.) Not only IBM, but also other American companies, such as NCR, had a very strong position in patents in this area for over thirty years. US companies consistently accounted for about half of all London patents for calculating machinery for about fifty years and a much higher proportion of US patents.

The existing data processing equipment in the early 1950s (mainly punchcard calculators and tabulators) was profitable both for IBM and for

British companies such as BTM and Power-Samas. Consequently, there was some reluctance to make this equipment obsolete by introducing electronic computers. Many firms in this kind of situation have failed to innovate, or left it too late. What was remarkable about IBM at that time was the speed with which it recovered from a situation in which it was in danger of falling behind and embarked on the large-scale development and manufacture of the new products once their advantage had been demonstrated. This change was also associated with a change in management and with the settlement of an anti-trust suit brought by the Department of Justice against IBM over its dominant position in the punched card market. The story has been told by Watson's son (Belden and Belden, 1962, p. 100):

One of the most exciting chapters in IBM's post-war history has to do with large-scale electronic computers and data processing. Many very large engineering computational jobs and a fair number of accounting applications were being hampered by the slowness of the calculating machines available in the later 1940s. However, Drs Eckert and Mauchly of the Moore School at the University of Pennsylvania had built a large electric computer – the Eniac – for the Army to make ballistic curve calculations. Many of us in our industry, including me, had seen the machine, but none of us could foresee its capabilities. Even after Eckert and Mauchly left the Moore School and began privately to manufacture a civilian counterpart to the Eniac – the Univac – few saw the potential.

The company was finally absorbed by Remington Rand in 1950, and soon had installed several machines in the US Government, including one in the Census Bureau which replaced a number of IBM machines.

During these really earth-shaking developments in the accounting machine industry, IBM slept soundly. We had put the first electronically-operated punched card calculator on the market in 1947. We clearly knew that electronic computing even in those days was so fast that the machine waited 9/10 of every card cycle for the mechanical portions of the machine to feed the next card. In spite of this, we didn't jump to the obvious conclusion that if we could feed data more rapidly, we could increase speeds by 900 per cent. Remington Rand and Univac drew this conclusion and were off to the races.

Finally we awoke and began to act. We took one of our most competent operating executives with a reputation for fearlessness and competence and put him in charge of all phases of the development of an IBM large-scale electronic computer. He and we were successful.

How did we come from behind? First, we had enough cash to carry loads of engineering, research and production, which were heavy. Second, we had a sales force which enabled us to tailor our machine very closely to the market. Finally, and most important – we had good company morale. All concerned realized that this was a mutual challenge to us as an industry leader. We had to respond with all that we had to win, and we did.

By 1956, it became clear that to respond rapidly to challenge, we needed a new organization concept. Prior to the mid-1950s the company was run essentially by one man – T. J. Watson, Sr. He had a terrific team around him, but he made the decisions. If we had organization charts, there would have been a fascinating number of lines – perhaps thirty – running into T. J. Watson.

In the early 1950s the demands of an increasing economic pace and the Korean War were calling for more rapid action by IBM at all levels than our monolithic structure was able to respond to adequately. Increasing customer pressures – plus a few more missed boats of lesser consequence than the Univac situation – forced us to decide on a new and vastly decentralized organization.

Here, we hope we responded a little more rapidly than we did in the case of Univac. The new organization was 180 degrees opposed to the old in fundamental concept, but we made the move.

In late 1956, after months of planning, we called the top one hundred or so people in the business to a three-day meeting in Williamsburg, Virginia. We came away from that meeting decentralized.

(Belden and Belden, 1962, p. 100)

The major innovations in electronic digital computers after ENIAC and EDVAC (University of Pennsylvania) were made by Univac (Remington Rand) and MIT rather than IBM (magnetic core memory stores, machine translation of instructions, magnetic drums and discs). Hoffmann (1976) has shown in his study of R&D strategies in the computer industry that IBM tended to follow a strategy of rapid imitation, picking up the most important scientific and technical advances from universities and from competitors. In the early days IBM hired von Neumann, one of the leading mathematicians from the EDVAC team, and also received important help from Sperry Rand for their 650 series, first sold in 1954. In the 1950s, government contract R&D accounted for about 60 per cent of total expenditure at IBM; now private venture expenditure accounts for the greater part of the (much larger) total. By the 1960s, about 15,000 people were engaged on research and development, about 90 per cent of them in eighteen development laboratories (thirteen in America and five in Europe). Some very large computers embodying more advanced features were built on contract for government agencies, but the highly successful transistorized 1401 series (1960), the 360 series (1965) and the 370 series (1971) were private venture projects.

The launching of the 360 series is a particularly interesting illustration of the change in scale of innovative work in this industry compared with the early wartime and post-war computers. The total development costs including software have been estimated at $500 million. It was preceded by the failure of IBM in the larger computer systems with the STRETCH machine, on which it lost $20 million. STRETCH did not meet the promised specification and the price had to be reduced to $8 million from $13.5 million, at which level it sold at a loss and very few were sold. The loss would have been much greater but for government support with the development costs.

There was a strong temptation in the early 1960s to stay with the highly successful 1401 series (General Products Division of IBM) and the 7000 series (Data Systems Division). However, these were overlapping increasingly in range and each division had its own development programme for extending the range. After sharp internal power struggles and after several new development projects had been killed off, including one based at IBM's Hursley Laboratories in England, Learson and Evans, now in charge of development at Data Systems, decided to concentrate on a completely new compatible series and not to proliferate the existing lines. The new series should cater for both business and scientific users. They also decided to use the new hybrid integrated circuits which were just becoming available. 'It was roughly as though General Motors had decided to scrap its existing makes and models and offer in their place one new line of cars, covering the entire spectrum of demand, with a radically redesigned engine and an exotic fuel' (Wise, 1966).

This not only involved overruling competing development projects, and concentrating on the NPL (later 360) series, it also involved a huge programme of plant expansion (five new plants) and IBM's entry into circuit manufacture. Between 1964 and 1967 IBM budgeted $4.5 billion in addition to R&D costs for investment in rental machines, plant and equipment.

The risks involved in launching the series were very great and were recognized, but the critical factor in the decision seems to have been the view that if they did not launch a new series they risked a serious decline in growth rate and erosion of their position as market leaders, particularly in larger computers. Evans' justification of this course of action was echoed almost word for word in relation to a similar crucial decision by Rolls-Royce to launch a new generation of aero-engines (the Spey). He said that the 360 series was a lot less risk than it would have been to do anything else or to do nothing at all (Wise, 1966).

Altogether, over a dozen other American firms entered the EDP market in addition to IBM, but none of them made any profits from EDP computers before the 1960s. As with colour television, RCA had to lay out over $100 million and together with General Electric withdrew from the market in the early 1970s after a heavy investment. In the case of General Electric this included an attempt to establish a strong foothold in Europe by the purchase of the main French manufacturer, Bull. While IBM succeeded in maintaining its overall market dominance throughout the 1970s, it was successfully challenged in special sectors of the market, such as minicomputers and the largest systems. The advent of the micro-computer presented a more serious threat, but once again IBM was able to react through the introduction of its own machine, although belatedly.

According to most accounts IBM was too fixated on mainframe large computers to recognize the change in the world computer market arising from the PC. By 1980 when IBM got round to launching a crash development programme for its own, there were already several firms well established in the new market, such as Atari, Apple, Commodore and Radio Shack and sales had already reached $1 billion. A special team was assembled outside the main inhouse R&D establishment, first under Jim Lowe and later under Don Estridge. They were given one year to develop a saleable product. They succeeded but only by the unusual procedure (for IBM) of buying in most of the parts, both hardware and software. In Robert Cringeley's (1994) iconoclastic but very entertaining account ('All IBM stories are true'), he suggests that IBM made some serious strategic errors in bringing in Microsoft for the Operating System, which led ultimately to the dominance of Microsoft in the PC software market. Other accounts, however, argue that IBM could not have foreseen the fateful long-term consequences of its deal with Bill Gates, head of what was then a very small software company (50 employees).

Although it succeeded in keeping a position in the world market for personal computers IBM encountered far more serious competition in the 1980s and 1990s than ever before. It was obliged to reduce its labour force worldwide by more than 50 per cent and suffered a drastic decline in profitability. The relative decline of large mainframe computers compared with the proliferation of PCs and work stations led to a large number of new entrants, taking advantage of the drastic change in scale economics. Competition came from many sides, including Japanese firms, such as NEC, Fujitsu (large machines) and Toshiba (portables), and from numerous small firms cloning the main features of the PCs. It also came from software firms as the relative importance of software grew in relation to hardware. The story is beautifully if somewhat cynically told in Cringeley's (1994) aptly entitled book, *Accidental Empires*. No recent book has shown more vividly the degree of turbulence and uncertainty in both software and hardware innovations in the United States, or the role of very young computer enthusiasts ('the triumph of the nerds').

New developments in the 1980s for the most powerful computer systems were based on the 'transputer' and parallel processing, enabling a number of computers to combine together. The interaction between industrial R&D and university groups has again been important in developing new architecture, as well as the interaction between semiconductor firms and computer firms (Molina, 1989). IBM's capability in semiconductor technology continues to be a source of strength although it has also had to collaborate with INTEL and other component firms. The rise of network computing has provided new opportunities for IBM in the 1990s.

7.6 ELECTRONIC COMPONENTS[8]

The development process in the electronic capital goods industry consists largely of devising methods of assembling components in new ways, incorporating new components to make a new design, or developing new components to meet new design requirements. This is not quite so simple as it may sound. There are more than a hundred thousand different components in a large computer, more than a million in a big electronic telephone exchange and ten million in the Apollo rocket system. There must be close collaboration in design work between end-product makers and component makers, and in the more complex products there must be sophisticated programming of component supply and subassemblies, and of testing arrangements. Increasingly in complex systems the software development costs may outstrip those of the hardware and become the crucial factor in the entire system (Hobday, in press).

If component makers succeed in developing new products with outstanding improvements in performance, this will most quickly benefit those who are in close touch with them. Before the Second World War European firms were probably not at any disadvantage here: some of the most important advances in valves and tubes were made in Europe by Philips, Telefunken,[9] GEC, AEI (BTH) and Marconi, as well as by European universities. Since the war the position has changed completely. Almost all the important inventions and innovations in components have been made in the United States and more recently in Japan, and there has been a lag of one to four years before manufacture began in Europe, often by American or Japanese subsidiaries. The major breakthrough was the invention of the transistor at Bell Laboratories (AT&T), but there have also been revolutionary improvements in the design and performance of most other components, including the so-called passive components (such as resistors and capacitors).

These innovations and improvements depended on advances in fundamental research as well as in development. The discovery of the transistor was made in the laboratories of a firm which had always spent heavily on both research and development. Bell Laboratories estimated total R&D expenditure on transistors and transistorized equipment at £2.7 million up to 1953, at £28 million up to the end of 1960, and £57 million up to September 1964. In the 1960s, probably only Philips or Siemens among European firms could match this scale of expenditure. The leading Japanese firms (in terms of R&D) included half-a-dozen electronic firms which by the 1970s were in the American league although lacking the heavy US contribution from government for military electronics.

The Bell basic patents in transistors were made available to all-comers on payment of a $25,000 advance royalty, partly because of an anti-trust suit (filed by the Department of Justice in 1949 and finally settled by a consent decree in 1956). But licensing arrangements subsequent to the decree were negotiated individually with each company. Those who could offer know-how in return obtained more favourable terms, often paying only the $25,000 lump sum.

Between 1952 and 1963 the Western Electric Co. (Bell) received over £9 million in total income for royalties from companies all over the world, excluding cross-licence benefits, and of this the company estimated that over £3 million was attributable to transistors. By far the greater part of this income came from companies outside the United States (£556,000 from licensees having their principal office in the United Kingdom). United States concerns were able to use Bell patents prior to the consent decree of 1956 royalty free, but pay royalties (varying in scale with the individual agreement) on patents issued since the decree.

Total expenditure on the negotiation and administration of licence agreements was £6.3 million from 1952 to September 1964, of which £1 million was for transistors and transistorized equipment. Cross-licence benefits in respect of agreements involving the basic transistor patent were estimated at £2.6 million. Only £4,000 of this was attributable to UK companies. Thus licensing income did not lead to the direct recovery of more than a small part of Bell's R&D costs.

Technological change continued at a rapid rate throughout the 1960s and 1970s and, as a result of their technical lead, other American component firms, in addition to Bell, enjoy a considerable royalty and know-how income from the licensing agreements which they have concluded with European firms. The planar technology patents of Fairchild were particularly important, and in addition to European firms, leading American firms, such as Texas Instruments and ITT, made licence agreements with Fairchild to obtain this technology. There were, it is true, a few component developments in Europe which were licensed to the USA: for example, the Lucas development work in industrial semiconductors, which began in 1954, resulted in the successful development of high voltage devices for ignition systems, which were licensed to Delco in the USA. Production of the Gunn diode was first launched in the UK and Siemens licensed

Table 7.2 Major product innovations in the semiconductor industry since the integrated circuit (1960s–1970s)

Innovations	Firm	First commercial production
MOS transistor	Fairchild	1962
DTL integrated circuit	Signetics	1962
Gunn diodes	IBM	1963
Light-emitting diodes	Texas Instruments	1964
TTL integrated circuit	TRW	1964
MOS integrated circuit	General Microelectronics General Instruments	1965
Magnetic bubble memory	Western Electric	
MOSFET (MOS field-effect transistor)	Western Electric Philips	1968
Schottky TTL	Texas Instruments	1969
CCD (charge coupled device)	Fairchild	1969
Complementary MOS	RCA	1969
Static RAM	Intel	1969
Silicon-on-sapphire (SOS)	RCA	1970
P-MOS		1971
3-transistor cell dynamic RAM (1K bits)	Intel	1971
CMOS		1971
Microprocessor	Intel	1972
I^2L integrated circuit	Philips	1973
1-Translator cell dynamic RAM (4K bits)	Intel	1974
VMOS integrated circuit	AMI	1975
C^2L integrated circuit		1976
MNOS		1976
Micro-computer (8048)	Intel	1977
V-MOS	Mitsubishi	1978
64-K bits memory	Fujitsu	1978

Source: Dosi (1981).

Westinghouse for their ultra-pure silicon process. But, in total, American companies have a substantial positive balance in 'technical payments' in the component field; this reflects their lead in most areas of new component development since the war. Developments in the microprocessor field in the 1970s generally enhanced this lead with US firms accounting for almost all the major innovations (Tables 7.2 and 7.3).

Diffusion of technical know-how does not simply depend on ability to pay. It owes a great deal to personal contacts and discussion, or to the movement of people, and here American firms enjoyed a major advantage. The research director of Texas Instruments came from Bell and so did other key personnel in the American semiconductor industry. As Tables 7.2 and 7.3 suggest, Western Electric (then the manufacturing arm of AT&T), Texas Instruments, Fairchild and Intel made an especially important contribution to the advance of component technology since the original Bell innovations. Golding documented in detail the great importance of the movement of key R&D personnel in the development of other firms. Dozens of new small firms were started in this way and, although many did not survive, some became important. Moreover, individuals and groups also joined some large established companies enabling them to start semiconductor operations. American firms in Silicon Valley and elsewhere not only

Table 7.3   Major process innovations in semiconductor industry (1950s–1970s)

Innovation	Firm	Date of development
Single crystal growing	Western Electric	1950
Zone refining	Western Electric	1950
Alloy process	General Electric	1952
3-5 Compounds	Siemens	1952
Jet etching	Philco	1953
Oxide masking and diffusion	Western Electric	1955
Planar process	Fairchild	1960
Epitaxial process	Western Electric	1960
Plastic encapsulation	General Electric	1963
Beam lead	Western Electric	1964
Dielectric isolation	Motorola	1965
Collector diffusion isolation	Western Electric	1969
Ion implantation	Mostek	1970
Self-aligned silicon gate	Intel	1972
Integrated injection logic	Philips	1973
Vertically oriented transistor	AMI	1975
Double polysilicon process	Mostek	1976
E-beam mask projection		1976
Plasma nitride processing		1976
Automatic bonding on 'exotic' (35 mm film) substrate	Sharp (Japan)	1977
Vertical injection logic	Mitsubishi	1978

Source: Dosi (1981).

benefited from this but also from the close contacts with the equipment makers and universities, such as Stanford. These advantages of recruitment and close contact and communication have led many of the leading European and Japanese firms to set up subsidiaries in the American electronics industry, often through the purchase of small or medium-sized US firms. The enormous success of Japanese, and later South Korean companies, in catching up with American semiconductor companies in the design and manufacture of memory chips, was partly due to this strategy and partly to the recruitment of returning graduate students from American universities and former employees of American companies.

After the major discoveries at Bell, the US government took a hand in advancing component technology. Since 1950 it has financed R&D on a large scale, helped the principal firms to enlarge their capacity, and has placed large orders for devices. A study by Arthur D. Little found that:

> Due to its considerable interest in semiconductors and particularly in transistors, the government has throughout the 1950s tried to stimulate the development of improved types. Around the middle of 1950 they were convinced transistors were needed for future military equipment so they accelerated the production investment to provide developmental and production facilities for making certain types which were considered to be desirable for future military electronic equipment.
>
> The contracts for a total of thirty different types of germanium and silicon transistors were placed with about one dozen of the major semiconductor companies, and this helped some of these to gain a foothold in the industry. In many cases, this investment was matched by similar amounts of capital equipment or

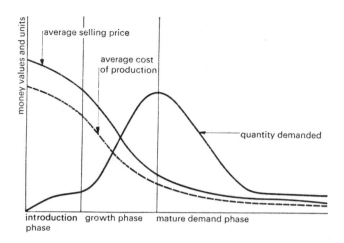

Fig. 7.3 The pattern of a typical semiconductor product cycle

Source: Golding (1972).

plant space supplied by the contracting companies. Thus a total potential capacity of over a million transistors a year was created.

(Little, 1963, p. 20)

These conclusions were confirmed by the results of Golding's research (1972) on economies of scale in the semiconductor industry. He found that dynamic scale economies associated with the learning curve were particularly important for new devices because of importance of the yield phenomenon in semiconductor manufacture. The difficulties of maintaining very high purity and cleanliness in the manufacture of microscopically small components mean that yields initially may be as low as 5 to 10 per cent. Obviously, very great economies can be secured by improving this yield, and to a considerable extent this proved to be a function of production volume. Similar phenomena have been found in other industrial processes subject to rapid technical change, and were first documented in connection with the aircraft industry. Golding (1972) estimated that a unit cost reduction of around 20 to 30 per cent was associated with cumulative doubled quantities (Andress, 1954; Sturmey, 1958; Beloff, 1966). These are twice as great as those estimated for aircraft production in the Plowden Report (1965) (see Figure 7.3).

These 'dynamic' economies of scale arise from adaptive learning of the labour force and management engaged in the production process. They are additional to, but intimately related to improvements arising from R&D, and from the normal 'static' economies of scale – reduction in unit costs arising from the spread of fixed costs over a larger production and sales volume.

Evidently, the increases in production volume for new devices arising from US government orders gave an enormous competitive advantage to

Table 7.4   Leading semiconductor manufacturers worldwide, 1988–9

Rank 1989	1988	Company	1989 SC revenues ($m)	Market share 1989 (%)
1	1	NEC	4,964	8.9
2	2	Toshiba	4,889	8.8
3	3	Hitachi	3,930	7.0
4	4	Motorola	3,322	5.9
5	6	Fujitsu	2,941	5.3
6	5	Texas Instruments	2,787	5.0
7	8	Mitsubishi	2,629	4.7
8	7	Intel	2,440	4.4
9	9	Matsushita	1,871	3.4
10	10	Philips	1,690	3.0
11	11	National	1,618	2.9
12	12	SGS-Thomson	1,301	2.3
13	18	Samsung	1,284	2.3
14	15	Sharp	1,230	2.2
15	20	Siemens	1,194	2.1
16	14	Sanyo	1,132	2.0
17	17	Oki	1,125	2.0
18	13	AMD	1,082	1.9
19	16	Sony	1,077	1.9
20	19	AT&T	873	1.6

Source: Dataquest, cited in Electronic Business, 16 April 1990, Hobday (1991).

US manufacturers, compared with those in Europe, which were faced with much smaller demand and much greater uncertainty. This was reflected in the dominance of the world market by the US manufacturers, the high proportion of semiconductor imports into Europe from the USA, the establishment of US subsidiaries in several European countries, and the great difficulties experienced by all but the largest European firms in survival in this industry.

However, Japanese firms were able to use the consumer electronics market to build up production volume for the products of the semiconductor industry in rather the same way that American firms originally used the defence markets. They were so successful in this strategy that by the late 1980s four of the five largest producers of semiconductors in the world were all Japanese firms (Table 7.4).

In Dosi's view, the industry became a mature international oligopoly during the 1970s. This oligopoly was temporarily reinforced by the inter-governmental and inter-industry agreements limiting Japanese exports to the United States. However, these agreements were relatively unstable because of the speed of technical change. Aided by co-operative R&D and by government support during the 1990s, some American firms fought back and once more led Japanese firms in world market share (Table 7.5). Furthermore, South Korean firms succeeded in breaking into the leading group in the 1980s and 1990s. Their surprising success in a 'catching up' country is further discussed in Chapter 12.

The manufacture of semiconductors is still an extremely complicated and difficult process, requiring more than a hundred different steps of coating,

Table 7.5 Worldwide top ten merchant semiconductor suppliers

Rank			Company	1994 total semi sales ($M)	1994 IC sales ($M)	1994 discrete sales ($M)	1994/1993 percent change
1994	1993	1991					
1	1	5	Intel	9,850	9,850	—	30
2	2	1	NEC	8,830	7,855	975	25
3	3	2	Toshiba	8,250	6,614	1,636	32
4	4	4	Motorola	7,011	5,870	1,141	21
5	5	3	Hitachi	6,100	5,300	800	20
6	6	8	TI	5,550	5,500	50	35
7	7	12	Samsung	5,005	4,365	640	61
8	8	7	Mitsubishi	3,959	3,286	673	34
9	9	6	Fujitsu	3,335	2,975	360	14
10	10	9	Matsushita	2,925	2,145	780	26
			Total	60,815	53,760	7,055	30

Source: Integrated Circuit Engineering Corporation (1995).

baking, etching, etc. Appleyard *et al.* (1996), after studying a number of firms, concluded that many of these steps involved art and tacit knowledge, rather than science:

> They are not well-understood and easily replicated on different equipment or in different facilities, and they impose demanding requirements for a particle-free manufacturing environment. Product innovation depends on process innovation to a much greater extent than is true of automobiles. The introduction of a new automobile requires substantial time and investment to manufacture dies and tooling for stamping and forming body parts and components, but a new model rarely demands significant change in the overall manufacturing process. Semiconductor product innovations, on the other hand often require major changes in manufacturing processes, because of the tighter link between process and product characteristics that typifies semiconductors, Moreover, imperfect scientific understanding of semiconductor manufacturing means that changes in process technologies demand a great deal of experimentation. New equipment, with operating characteristics that are not well understood, often must be introduced along with a new 'recipe', also not well understood, in order to manufacture a new product. The complexity of the manufacturing process also means that isolating and identifying the causes of yield failures requires considerable time and effort.

> (Appleyard *et al.*, 1996, p. 5)

This high degree of uncertainty, together with the huge costs of investment in new plant and new R&D for each generation of chips create formidable entry barriers. Similar considerations, although of course on a much smaller scale, apply to the costs of experimentation for 'application specific integrated circuits' (ASICs). For this reason, microelectronics has remained a very R&D-intensive industry, with R&D often accounting for about 10 per cent of sales. The large multi-product firms, such as Siemens, Philips, Hitachi or Matsushita, devote quite a high proportion of their total R&D budgets to microelectronics and patents in that area can account for as much as 20 per cent of their total patents (Figure 7.4). Turbulence and

Total EPO patents by company

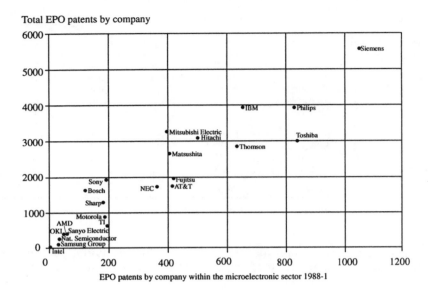

EPO patents by company within the microelectronic sector 1988-1

Fig. 7.4 Patenting within the microelectronics sectors vs total company patenting activity, 1988–92

Source: European Commission (1994).

uncertainty continue to rule in this still very fast-growing industry. The implications of costs of innovation in relation to size of firm are discussed further in Chapters 8 and 9, but this chapter concludes by considering the effects of microelectronic technology on other sectors of the economy.

7.7   THE MICROELECTRONIC REVOLUTION

The fact that a new technology has many potential applications does not of course mean that all of these will occur simultaneously, or even over a short period. On the contrary, the assimilation of a major new technology into the economic and social system is a matter of decades, not of years, and is related to the phenomenon of long cycles in the economy, as Schumpeter (1939) originally suggested (see Chapter 1). This prolonged diffusion process is more or less inevitable, because a major new technology, such as steam power, electric power or electronics, involves a great many enabling educational, social and managerial changes, as well as clusters of technical innovations in a variety of applications, and many scaling-up processes.

Carlota Perez (1983, 1989) has argued that inertia in various social institutions had a retarding effect on the diffusion of each major wave of new technology, and that a social process of trial and error was necessary before new institutions developed which 'matched' the new technology. However, once the necessary institutional changes were made they could facilitate and accelerate technical change.

In the case of electronics, visionary scientists and engineers, such as Wiener (1949) and Diebold (1952), were essentially right fifty years ago when they foresaw the many potential applications of electronic computers throughout the economic system in both factories and offices. Where some forecasts went badly wrong was in their estimation of the time scale. Wiener failed to take account of the long time lags in building up a capital goods supply industry and a component industry on a sufficient scale to provide all the computer power, peripherals and instrumentation for this vast transformation. Even more, perhaps, he underestimated the time scale needed to educate and train millions of people in the design, redesign, operation and maintenance of a huge variety of processes incorporating the new technology. Finally, he took insufficient account of the relative costs of the new technology which was still unattractive in purely economic terms for many potential applications.

As we have seen, during the 1950s, 1960s and 1970s there was indeed a massive expansion of the supply potential of the electronic capital goods and component industries and an enormous improvement in their relative costs and reliability compared with older electromechanical systems. But this still did not mean that there could be a sudden introduction of the new technologies throughout the economy, for several very important reasons. First, there was still a skill shortage especially in those sectors which never had any wide experience of electronics technology. Second, the software costs in designing entirely new applications could be extremely high, even though the hardware costs have been drastically reduced. Third, full-scale automation may only be possible in association with other heavy re-equipment investment. Finally, in some important service sectors it was only possible as a result of legislative, organizational, managerial and social changes which took a long time to bring about.

A pervasive new technology is likely to find application first in the rapidly growing new sectors of the economy, where a lot of new investment is taking place in any case and where there is greater acceptance of innovation. Thus the main applications of microelectronics were initially in the electronics industry itself.

The second group of firms to make early and extensive use of the new technology was those where electronic subsystems represented a large part of total product cost and where the necessary skills were either already available (as sometimes in the case of scientific instruments and cash registers) or could be injected by aggressive strategies on the part of component suppliers (as in the case of calculators).

A third group of firms to use electronic technology rather quickly were already operating large-scale flow production systems, especially those with innovative management facing expanding markets, as in the case of the chemical industry and the electric power industry in the 1950s and 1960s. Flow processes were already partly automated using older techniques. All of these three types of firm were able to achieve very big and sustained increases in labour productivity in the post-war period through the combined advantages of the new technology and the exploitation of scale economies.

When, however, we come to consider the older and slower growing (or declining) sectors of the economy, especially those with a low endowment of qualified engineers, then the problems of diffusion are very different. Empirical diffusion studies suggest that thirty years is by no means uncommon for the time period over which revolutionary innovations are diffused through the majority of a potential adopter population and it can be much longer.

De Bresson (1991) has suggested that the most pervasive and important technologies start in the public domain in government laboratories and universities, then move to instrument makers and capital goods manufacturers with strong R&D and only in the final stage to large consumer goods industries and services.

Signs of a more rapid diffusion of computers in the service sectors have multiplied during the 1980s and 1990s including a growth of R&D in some service industries, heavy investment in telecommunications and computers, and growth in the staff of qualified engineers and scientists. We discuss these changes more fully in Chapter 17.

NOTES

1. For the early history of radio and television see Sturmey (1958); Maclaurin (1949); Briggs (1961); Telefunken (1928, 1953); Radio Corporation of America (1963).
2. Owen Young's account states: 'When Admiral Bullard arrived in my office, he said that the President, whom he had just seen in Paris, was concerned about the post-war international position of the United States and had concluded that three of the key areas on which international influence would be based were shipping, petroleum and radio. But in radio the British were now dominant and the United States, with her technical proficiency, had an opportunity to achieve at least a position of equality' (Maclaurin, 1949, p. 101).
3. Broadcasts using the less satisfactory Baird mechanical system had begun as early as 1929, and there was a ridiculous xenophobic campaign in the 1930s to try and persuade the BBC to discriminate in favour of the 'British' mechanical system against the allegedly 'American' EMI System.
4. For the early history of radar see Postan et al. (1964); Telefunken (1928, 1953); Gartmann (1959).
5. For example, the Lorenz blind approach beam system used to guide bombers and its successor, X-Gerät, both needed continuous wave systems rather than pulse generation. These were less accurate and more easily jammed. Pulse techniques were not used until 1944. For similar reasons the German ground chain, Freya, using Würzburg sets (53cm) was virtually immobilized by British bombers in 1943.
6. For the early history of computers see Zuse (1961); Watson (1963); Hollingdale and Toothill (1965); Katz and Phillips (1982).
7. IBM's gross income grew from $116 million in 1946 to $696 million in 1955, and employment from 22,000 to 41,000. By 1964 gross income reached $3,239 million and employment was 149,000. In 1971 gross income was $8.273 million and there were 265,000 employees. Employment reached over 400,000 at its peak but

declined again very sharply to 200,000 in the far more competitive market of the 1980s and 1990s.

8. For a thorough treatment of component innovation see Golding (1972); OECD (1968); Tilton (1971); Sciberras (1977); Braun and MacDonald (1978); Dosi (1982).

9. For example, the pentode value was developed at Philips and the hexode by Steimel at Telefunken.

PART ONE   REVIEW ARTICLES, LITERATURE SURVEYS AND KEY REFERENCES

Abramovitz, M. and David, P. A. (1994) *Convergence and Deferred Catch-up: Productivity Leadership and the Waning of American Exceptionalism*, CEPR Publication no. 401, Stanford, Stanford University Press.

Ayres, R. U. (1991) *Computer-integrated Manufacturing*, vol. 1, London, Chapman and Hall.

Bruland, K. (1989) *British Technology and European Industrialisation: the Norwegian Textile Industry in the Mid-nineteenth Century*, Cambridge, Cambridge University Press.

Chandler, A. D. (1977) *The Visible Hand: The Managerial Revolution in American Business*, Cambridge, MA, Harvard University Press.

Church, R. A. and Wrigley, E. D. (eds) (1994) *The Industrial Revolution* (11 vols, but see especially vols, 2, 3, 8, 9) The Economic History Society, Oxford, Blackwell.

Clark, N. (1985) *The Political Economy of Science and Technology*, Oxford, Blackwell.

Cringeley, R. (1994) *Accidental Empires: How the Boys of Silicon Valley Make their Millions, Battle Foreign Competition and still cannot get a Date*, London, Penguin.

Dosi, G. (1984) *Technical Change and Industrial Transformation*, London, Macmillan.

Dyer, J. H. (1996) 'How Chrysler created an American Keiretsu', *Harvard Business Review*, July–August, pp. 42–61.

Enos, J. L. (1962) *Petroleum Progress and Profits: A History of Process Innovation*, Cambridge, MA, MIT Press.

Gerschenkron, A. (1962) *Economic Backwardness in Historical Perspective*, Cambridge, MA, Harvard University Press.

Hobday, M. (1995) *Innovation in East Asia: The Challenge to Japan*, Aldershot, Elgar.

Hounshell, D. (1982) *From the American System to Mass Production, 1800–1932*, Baltimore, Johns Hopkins University Press.

Hounshell, D. and Smith, J. K. (1988) *Science and Corporate Strategy, Du Pont R&D, 1902–1988*, Cambridge, Cambridge University Press.

Hughes, T. (1989) *American Genesis*, New York, Viking.

Landes, M. (1969) *The Unbound Prometheus: Technological and Industrial Development in Western Europe from 1750 to the Present*, Cambridge, Cambridge University Press.

Lazonick, W. (1990) *Competitive Advantage on the Shop Floor*, Cambridge, MA, Harvard University Press.

Metcalfe, J.S. (1997) *Evolutionary Economics and Creative Destruction*, London, Routledge.

Mokyr, J. (1990) *The Lever of Riches, Technological Creativity and Economic Progress*, New York, Oxford University Press.

Morris, P. J. T. (1982) The development of acetylene chemistry and synthetic rubber by IG Farben, 1926–1945, DPhil thesis, Oxford University.

Mowery, D. (1983) 'The relationship between intra-firm and contractual forms of industrial research in American manufacturing, 1900–1940', *Explorations in Economic History*, vol. 20, pp. 351-74.

Nye, D. E. (1990) *Electrifying America: Social Meanings of a New Technology*, Cambridge, MA, MIT Press.

Pursell, C. W. (ed.) (1991) *Technology in America: A History of Individuals and Ideas*, Cambridge, MA, MIT Press.

Rosegger, G. (1996), *The Economics of Production and Innovation: An Industrial Perspective*, Third Edition, Oxford, Buttersworth.

Rosenberg, N. (1976) *Perspectives on Technology*, Cambridge, Cambridge University Press.

Rosenberg, N., Landau, R. and Mowery, D. C. (eds) (1992) *Technology and the Wealth of Nations*, Stanford, Stanford University Press.

Sloan, A. (1963) *My Years with General Motors*, New York, Doubleday.

von Tunzelmann, G. N. (1995) *Technology and Industrial Progress: The Foundations of Economic Growth*, Cheltenham, Elgar.

Walsh, V. (1984) 'Invention and innovation in the chemical industry: demand-pull or discovery push?', *Research Policy*, vol. 13, pp. 211-34.

Womack, J. T., Jones, D. T. and Roos, D. (1990) *The Machine that Changed the World*, New York, Rawson Associates.

PART TWO

THE MICRO-ECONOMICS
OF INNOVATION: THE THEORY
OF THE FIRM

INTRODUCTORY NOTE

Part One has shown many instances of dramatic increases in productivity achieved by a combination of technical and organizational innovations as, for example, in cotton spinning, catalytic cracking of oil, scaling up of steel and chemical plants, assembly line production of automobiles, or miniaturization of integrated electronic circuits. It has also illustrated the widening and cheapening of the range of products available to consumers through product innovations in many industries, for example, pottery, consumer durables, new materials, radio and television.

This book now attempts to examine more systematically the conditions which have promoted such successful innovations, first at the level of the individual firm and innovative project (Part Two) and then at the level of the individual nation or region (Part Three). Chapter 8 first discusses those empirical research projects which have sought to identify the pattern of successful innovation in firms. Many innovative attempts end in failure, so that a systematic comparison of success and failure yields some interesting results, as is shown in the case of Project SAPPHO, described in some detail. This leads on to a discussion of the characteristics of those firms which have repeated success with innovations.

The discussion in Chapter 9 shows that size of firm certainly influences what kind of projects can be attempted in terms of technology, complexity and cost but does not in itself determine the outcome. In some areas and in some industries small firms play a very important role in innovation, as indeed the historical account has also shown. They have advantages of speed and flexibility in decision-making and often of lower costs in development work. The historical dimension is again shown to be crucial as the stage of development of a technology and/or an industry is one of the principal determinants of the relative contribution of large and small firms to innovations, and the types of innovation which they are able to make.

Although, as Part One has shown, technologies have certainly changed in rapid succession, and although firms have grown much larger in many industries and introduced entirely new management techniques, such as industrial R&D, nevertheless there are some things which have changed little if at all in their fundamentals. One of these is the prevalence of uncertainty with respect both to future technological change and to future market change. Chapter 10 shows that despite the introduction of numerous sophisticated mathematical techniques into project evaluation and decision-making, generally it has not proved possible for firms to make accurate forecasts of the future costs of development or the time such development will take. Even greater errors are typical of forecasts of future market size and rate of return on investment. Typically firms underestimate cost and overestimate speed of development, sometimes by very wide margins, as in such well-known cases as the aircraft Concorde or the fast breeder nuclear reactor. Errors of estimation in relation to future markets can go in either direction and are often wildly inaccurate. As Part One has shown, the future market for computers, for polyethylene and for synthetic rubber was grossly underestimated. In the case of nuclear power it has been vastly overestimated.

Forecasting errors are greatest in the case of the more radical innovations. With regard to small incremental improvements and new applications of existing products, much greater accuracy is possible and project evaluation techniques can be very useful management tools. The use of such techniques is in any case of some value as a means of mobilizing the necessary combined efforts within the firm and of disclosing potential difficulties at various stages of development and product launch. True uncertainty, where the future simply cannot be known is of course most characteristic of fundamental research and the most radical inventions.

This should not, however, necessarily lead to the conclusion that the most risky and uncertain projects should never be undertaken at all. On the contrary, again as the historical account in Part One has shown, the benefits from such R&D may be very great indeed over the long term. Many of today's most useful technologies owe their very existence to programmes of fundamental research in physics, chemistry and biology conducted over very long periods mainly in university laboratories. Many of today's most valuable products would not exist if determined entrepreneurs and inventors had not been prepared to devote their fortunes, their careers and even their lives to their development.

However, the uncertainty and the risks are such that most firms will not be able to contemplate basic research or the more radical types of innovation. This means that typically in all countries public expenditure has accounted for by far the greater part of basic research and has made a substantial contribution to generic technologies, such as biotechnology, and to information technology and various radical innovations. This type of public expenditure is discussed in Part Four, which deals with public policies for science and technology. In Part Two, the final chapter deals with the strategies of firms, confronted as they are with all the hazards and uncertainties attending technical innovation, whether by themselves or their competitors.

Chapter 11 attempts to classify the strategies which firms adopt as either offensive, defensive, imitative, dependent, traditional or opportunist. Firms which follow an offensive strategy are that very small minority which attempt to make radical innovations, sometimes but not always based on the conduct of fundamental research. A larger number of firms follow defensive strategies, responding fairly quickly to the innovative efforts of others with new products and processes of their own. This is sometimes described as a 'fast second' strategy. Much larger numbers of firms follow a simpler imitative strategy, sometimes on the basis of licensing, franchising or subcontracting from more innovative firms. Imitators may become completely dependent or may start out in a dependent role, as is often the case with firms in developing countries importing technology.

There are some industries where there is little technical change or where there is actually a competitive advantage in making a traditional product with long-established techniques. Fashion and design innovations may nevertheless be important in such cases but not necessarily technical innovations. Finally, the variety of changing circumstances is so great, both in markets and technology, that there will always be possibilities of identifying product niches and moving into them on a purely opportunist, entrepreneurial basis.

Any attempt to classify firm strategies in this way is necessarily an oversimplification. For example, multi-product firms may follow different strategies in different sectors of their business and these may, in fact almost certainly will, change over time. The effort at classification is nevertheless valuable in bringing out the variety of ways in which firms make use of R&D and STS or, of course in many cases, do not do this. It is particularly valuable to conceptualize the efforts of firms in 'catching up' countries as attempts to upgrade their strategies as they learn to modify the imported technology and make increasing use of their own R&D and STS. This discussion at the end of Part Two thus leads on directly to the further discussion of national systems of innovation and catching up in Part Three.

CHAPTER 8

SUCCESS AND FAILURE IN INDUSTRIAL INNOVATION

8.1 INNOVATION AS COUPLING OF NEW TECHNOLOGY WITH A MARKET

We now consider some tentative generalizations about the technical innovation process in the firms and industries described, and discuss how far it is possible to test the validity of such generalizations, and to relate them to other industries and the economy as a whole.

Jewkes and his colleagues (1958) have argued that the nineteenth-century links between science and invention were much greater than is commonly assumed. Certainly, the classical economists were well aware of the connection between scientific advances and technical progress in industry, in the eighteenth and early nineteenth centuries. The quotation from Adam Smith (1776), with which this book begins, illustrates this point. Nevertheless, the evidence of the previous seven chapters suggests that there were profound changes in the degree of intimacy and the nature of the relationship between science and industry.

As already explained in Chapter 1, this does not imply the acceptance of a linear model of R&D with a simple one-way flow of ideas from basic science through applied research to development and commercial innovation. On the contrary, there has always been and there remains in the modern science-related industries a strong reciprocal interaction between all these activities (Soete and Arundel, 1993) and in particular a powerful influence of technology upon science. Gazis (1979) has given some examples of this interaction in the case of IBM's research laboratories. However, the effectiveness of this two-way movement of ideas depends on the ability of both communities to communicate with each other.

The new style of innovation in the industries which we have considered was characterized by professional R&D departments within the firm, employment of qualified scientists as well as engineers with scientific training, both in research and in other technical functions in the firm, contact with universities and other centres of fundamental research, and acceptance of science-based technical change as a way of life for the firm. Some of the firms we have considered had very strong scientific and technical resources, such as ICI, BASF, Du Pont, IBM, NEC, GM, Toyota, Siemens, GE, Hoechst, RCA, Marconi, Telefunken and Bell. An extreme case was the development of nuclear weapons and atomic energy.

During the twentieth century the main locus of inventive activity shifted away from the individual inventor to the professional research and development (R&D) laboratory, whether in industry, government or academia.

The nineteenth century was the heroic period of both invention and entrepreneurship. Names like Eli Whitney, 'Blacksmith, Nail-maker, Textile and Machine Tool inventor and innovator' ('he can make anything') spring to mind. Henry Thoreau, remembered now as a solitary philosopher, when asked to describe his profession ten years after graduation, replied that he was a Carpenter, a Mason, a Glass-pipe maker, a House Painter, a Farmer, a Surveyor and of course, a Writer and a Pencil-maker. He was in fact responsible for numerous inventions in pencil-making and there was a time when he could think of little else but improving the processes in his little pencil factory (Petroski, 1989). These men were not untypical of American and European nineteenth-century inventors. The British nineteenth-century Industrial Revolution owed much of its success to such men as these.

With an inventive career extending into the twentieth century, Thomas Edison embodied the transition from the 'great individualists', of which he was certainly one, to the large-scale R&D laboratories, that he helped to establish. He made a host of inventions and took out more patents (1,093) than any other single individual, but this was possible partly because he set up large contract research laboratories, first in Newark and later in Menlo Park. Among Edison's staff at these laboratories were some of the outstanding engineers and scientists who later helped to build up corporate inhouse R&D in Germany and Britain, as well as in the United States.

By the first decade of the twentieth century, although Edison was still making inventions, the focus of inventive effort was shifting from the contract laboratory typified by Menlo Park, Tesla's laboratories or that of Edward Weston to the inhouse industrial laboratories established by such firms as Kodak (1895), General Electric (1900), or Du Pont (1902). As Thomas Hughes (1989) shows in his classic study of the 'torrent' of American inventions from 1870 to 1940, by the time of the First World War, corporate R&D had displaced the contract laboratory as the centre of American inventive activities. Even as embryonic military–industrial complex had come into existence with the sponsorship of industrial research by the US Navy and especially the strong links established with Sperry Gyroscope.

Most of the major innovations we have considered were the result of professional R&D activity, often over long periods (PVC, nylon, polyethylene, hydrogenation, catalytic cracking, nuclear power, computers, television, radar, semiconductors). Even where inventor–entrepreneurs played the key role in the innovative process (at least in the early stages) such individuals were usually scientists or engineers who had the facilities and resources to conduct sustained research and development work (Baekeland, Fessenden, Eckert, Houdry, Dubbs, Marconi, Armstrong, Zuse). Some of them used university or government laboratories to do their work, while others had private means.

Frequently university scientists or inventors worked closely as consultants with the corporate R&D departments of the innovating firms (Ziegler, Natta, Haber, Fleming, Michels, Staudinger, Von Neumann). In other cases special wartime programmes led to the recruitment of outstanding university scientists to work on government sponsored innovations (the atomic bomb and radar). Intimate links with basic research through one means or another were normal for R&D in these industries and their technology is science based in the sense that it could not have been developed at all without a foundation in theoretical principles. This corpus of knowledge (macromolecular chemistry, physical chemistry, nuclear physics and electronics) could never have emerged from casual observation, from craft skills or from trial and error in existing production systems, as was the case with many earlier technologies. The same is true of recent biotechnology.

The rise of these new science-related technologies has had major economic and social repercussions over and above the growth of professional industrial R&D. It changed not only the development procedures, but also the production engineering, the sales methods, the industrial training and the management techniques. Quite often the majority of employees in firms in the new industries were not employed in production or handling of goods at all, but in generating, processing and distributing information and knowledge. In the extreme cases the computer software or process plant design and consultancy firm may employ hundreds of people but have no physical output other than paper or computer printout. But even in quite 'normal' electronic or chemical firms, the combined employment in research, development, design, training, technical services, patents, marketing, market research and management may be greater than in production. The complexity of the technical information involved and of the data processing means that specialized information storage, handling and retrieval systems are increasingly necessary. One of the most successful firms in the global telecommunications industry, Ericsson, employed fewer than ten per cent of its workforce in production by the mid-1990s. This proliferation of 'non-production' occupations is often treated as a form of Parkinson's Law or conspicuous waste. Even a scientist–inventor such as Gabor (1964) treated it as unnecessary in economic terms (although perhaps desirable on social grounds). No doubt Parkinson's Law does operate and labour savings can be made in some of these occupations (as they can in production). But it is essential to this analysis that the major part of this growth is due to the changes in technology, and to the new forms of competition which this has brought about.

So far we have discussed the new industries mainly in terms of the scientific basis of their new products and manufacturing technologies, but it is impossible to disregard the pull of the market as an essential complementary force in their origins and growth. In many cases the demand from the market side was urgent and specific.

The strength of the German demand for 'ersatz' materials to substitute for natural materials in two world wars spurred on the intense R&D efforts of IG Farben and other chemical firms. The strength of the military–space demand in the American post-war economy stimulated the flow of innovations based on Bell's scientific breakthrough in semiconductors and the early generations of computers. The urgency of British wartime needs spurred the successful development of radar of all kinds while the German government sponsored the development of FM networks, as well as radar. The Japanese government persuaded Toyota to enter the truck industry for military aims.

Conversely, the absence of a strong market demand for some time retarded the development of synthetic rubber in the USA, the growth of the European semiconductor industry, the development of radar in the USA before 1940, or of colour television in Europe after the war.

This does not mean that only wartime needs and government markets can provide sufficient stimulus for innovations, although they were obviously important historically. A strong demand from firms for cost-reducing innovations in the chemical and other process industries is virtually assured, because of their strong interest in lower costs of producing standard products and their technical competence. The demand for process innovations is related to the size of the relevant industry and here again the American oil industry provided a key element of market pull for the innovative efforts of the process-design organizations. The market demand may come from private firms, from government or from domestic consumers, but in its absence, however good the flow of inventions, they cannot be converted into innovations.

Innovation is essentially a two-sided or coupling activity. It has been compared by Schmookler (1966) to the blades of a pair of scissors, although he himself concentrated almost entirely on one blade. On the one hand, it involves the recognition of a need or more precisely, in economic terms, a potential market for a new product or process. On the other hand, it involves technical knowledge, which may be generally available, but may also often include new scientific and technological knowledge, the result of original research activity. Experimental development and design, trial production and marketing involve a process of matching the technical possibilities and the market. The professionalization of industrial R&D represents an institutional response to the complex problem of organizing this matching, but it remains a groping, searching, uncertain process.

In the literature of innovation, there are attempts to build a theory predominantly on one or other of these two aspects. Some scientists have stressed very strongly the element of original research and invention and have tended to neglect or belittle the market. Economists have often stressed most strongly the demand side: 'necessity is the mother of invention'. These one-sided approaches may be designated briefly as 'science-push' theories of innovation and 'demand-pull' theories of innovation (Langrish et al., 1972). Like the analogous theories of inflation, they may be complementary and not mutually exclusive.

In a powerful critique of demand-pull theories of innovation, Mowery and Rosenberg (1979) pointed to the inconsistent use of the concept of demand in this literature and insist that the results of empirical surveys of innovation cannot legitimately be used (although they often have been) to support one-sided market-pull theories. The example of the electronic computer cited in Chapter 7 is a good example of their point that the market cannot evaluate a revolutionary new product of which it has no knowledge.

It is not difficult to cite instances which appear to give support to either theory. There are many examples of technical innovation, such as the atomic absorption spectrometer, where it was the scientists who envisaged the applications without any very clear-cut demand from customers in the early stages. Going even further, advocates of 'science-push' tend to cite examples such as the laser or nuclear energy, where neither the potential customers nor even the scientists doing the original work ever envisaged the ultimate applications or even denied the possibility, as in the case of Rutherford. Advocates of 'demand-pull' on the other hand tend to cite examples such as synthetic rubber, cracking processes or Whitney's cotton gin where a clearly recognized need supposedly led to the necessary inventions and innovations.

While there are instances in which one or the other may appear to predominate, the evidence of the innovations considered here points to the conclusion that any satisfactory theory must simultaneously take into account both elements. Since technical innovation is defined by economists as the first commercial application or production of a new process or product, it follows that the crucial contribution of the entrepreneur is to link the novel ideas and the market. At one extreme there may be cases where the only novelty lies in the idea for a new market for an existing product.[1] At the other extreme, there may be cases where a new scientific discovery automatically commands a market without any further adaptation or development. The vast majority of innovations lies somewhere in between these two extremes, and involves some imaginative combination of new technical possibilities and market possibilities. Necessity may be the mother of invention, but procreation still requires a partner.

Almost any of the innovations which have been discussed could be cited in support of this proposition. Marconi succeeded as an innovator in wireless communication because he combined the necessary technical knowledge with an appreciation of some of the potential commercial applications of radio. The Haber–Bosch process for synthetic ammonia involved both difficult and dangerous experimental work on a high pressure process and the development of a major artificial fertilizer market, stimulated by fears of war and shortage of natural materials. Despite their early complete underestimation of the market, IBM was for some time the most successful firm in the world computer industry because it combined the capacity to design and develop new models of computers with a deep knowledge of the market and a strong selling organization. Firms such as General Electric and RCA with similar or greater scientific and technical strength, but much less market knowledge and market power in this field, in the end had to withdraw.

We may indeed advance the proposition that 'one-sided' innovations are much less likely to succeed. The enthusiastic scientist–inventors or engineers who neglect the specific requirements of the potential market or the costs of their products in relation to the market are likely to fail as innovators. This occurred with EMI and AEI in computers and with several British firms in radar, despite their technical accomplishments and strong R&D organizations. Professionalization of industrial R&D means that there is now often an internal pressure group which may push 'technologically sweet' ideas without sufficient regard to the potential market, sales organization or costs.

On the other hand, the entrepreneurs or inventor–entrepreneurs who lack the necessary scientific competence to develop a satisfactory product or process will fail as innovators however good their appreciation of the potential market or their selling. This was the fate of Parkes with his plastic comb and of Baird with television. The failures may nevertheless contribute to the ultimate success of an innovation, even though the individual efforts fail. The social mechanism of innovation is one of survival in the market. The possibility of failure for the individual firms which attempt to innovate arises both from the technical uncertainty inherent in innovation and from the possibility of misjudging the future market and the competition. The notion of perfect knowledge of the technology or of the market is remote from the reality of innovation.

The fascination of innovation lies in the fact that both the market and the technology are continually changing. Consequently, there is a kaleidoscopic succession of new possible combinations emerging. What is technically impossible today may be possible next year because of scientific advances in apparently unrelated fields. Although Usher developed the concept mainly in relation to invention rather than technical innovation, this 'Gestalt' theory probably comes close to representing the imaginative process of matching ideas. What cannot be sold now may be urgently needed by future generations. An unexpected turn of events may give new life to long-forgotten speculations or make today's successful chemical process as dead as the dodo. Patents for a float glass process and for radar were taken out before 1914. The stone that the builders rejected is the cornerstone of the arch. The production of polyethylene was nearly suspended after the Second World War because the peacetime markets were thought to be too small. IG Farben offered to sell their synthetic rubber patents to the natural rubber cartel because they thought the synthetic product would not be able to compete in peacetime in price or quality. The early computer manufacturers expected that the market would be confined to government and scientific users. A century after early experiments electric road vehicles were again being seriously investigated by major automobile manufacturers. The apparently random, accidental and arbitrary character of the innovative process arises from the extreme complexity of the interfaces between advancing science, technology and a changing market. The firms which attempt to operate at these interfaces are as much the victims of the process as its conscious manipulators. Innovation works as a social process but often at the expense of the innovators. The implications of this high degree of uncertainty are discussed in Chapter 10.

These considerations lead to three conclusions of fundamental importance. First, since the advance of scientific research is constantly throwing up new discoveries and opening up new technical possibilities, a firm which is able to monitor this advancing frontier by one means or another may be one of the first to realize a new possibility. Strong inhouse R&D may enable it to convert this knowledge into a competitive advantage. Second, a firm which is closely in touch with the requirements of its customers may recognize potential markets for such novel ideas or identify sources of consumer dissatisfaction, which lead to the design of new or improved products or processes. In either case, of course, they may be overtaken by faster moving or more efficient competitors or by an unexpected twist of events, whether in the technology or in the market. Third, the test of successful entrepreneurship and good management is the capacity to link together these technical and market possibilities, by combining the two flows of information and new ideas.

Innovation is a coupling process and the coupling first takes place in the minds of imaginative people. An idea gels or clicks somewhere at the ever-changing interfaces between science, technology and the market. For the moment this begs the question of creativity in generating the inventive idea, except to note that almost all theories of discovery and creativity stress the concept of imaginative association or combination of ideas previously regarded as separate. But once the idea has clicked in the mind of an inventor or entrepreneur, there is still a long way to go before it becomes a successful innovation, in our sense of the term. Rayon was 'invented' 200 years before it was innovated, the computer at least a century before, and aeroplanes even earlier.

The coupling process is not merely one of matching or associating ideas in the original first flash; it is far more a continuous creative dialogue during the whole of the experimental development work and introduction of the new product or process. The one-man inventor–entrepreneur like Marconi or Baekeland may very much simplify this process in the early stages of a new innovating firm, but in the later stages and in any established firm the coupling process involves linking and co-ordinating different sections, departments and individuals. The communications within the firm and between the firm and its prospective customers are a critical element in its success or failure. As we have seen, in many cases the original idea may take years of even decades to develop, and during this time it continually takes on new forms as the technology develops and the market changes or competitors react. Consequently, the quality of entrepreneurship and good communications are fundamental to the success of technical innovations.

Summing up this discussion and the evidence in Part One, we might conclude that among the characteristics of successful innovating firms in the twentieth century in the industries considered were:

1. Strong inhouse professional R&D.
2. Performance of basic research or close connections with those conducting such research.
3. The use of patents to gain protection and to bargain with competitors.
4. Large enough size to finance fairly heavy R&D expenditure over long periods.
5. Shorter lead times than competitors.
6. Readiness to take high risks.
7. Early and imaginative identification of a potential market.
8. Careful attention to the potential market and substantial efforts to involve, educate and assist users.
9. Entrepreneurship strong enough effectively to co-ordinate R&D, production and marketing.
10. Good communications with the outside scientific world as well as with customers.

We might hypothesize that these are the essential conditions for successful technical innovations.

Up to a point, such tentative generalizations about the characteristics of innovation may be tested by analysing and comparing case studies of a large number of innovations. One difficulty about such case studies (many of which have been cited in Part One) is that we do not know how far they are representative of the innovative process. Indeed, much of the literature on industrial innovation falls into two categories: scattered case histories lacking comparability of coverage or theoretical analysis lacking systematic empirical foundations.

As a result, there are many plausible, half-tested hypotheses and many interesting conjectures in innovation theory, but insufficient firm evidence to refute or support them. The historical account in Part One suggests interesting conclusions, but it is difficult to find ways to substantiate them, or to assess their relative importance. Yet such systematic testing of generalizations and hypotheses is essential to advance our understanding.

The remainder of this chapter is therefore devoted to the description of a project which was deliberately designed to test such generalizations about innovation. The project was called SAPPHO and it was carried out at the Science Policy Research Unit during the 1970s. The original project was designed in 1968 by R. C. Curnow, but later stages of the work were led by R. Rothwell.

8.2   PROJECT SAPPHO

The basic idea of the project was to attempt to substantiate (or refute) generalizations about technical innovation, by the systematic comparison of pairs of successful and unsuccessful attempts to innovate in each branch of industry in turn. This method of course rests on the observation that competitive technical innovation is a fairly general characteristic of many branches of industry in industrialized capitalist societies.

Since the introduction of a new product or a new process in any branch of industry may render older products and processes obsolete or uneconomic, firms which wish to survive and grow must be capable of adapting their technologically based strategy to this competition. This does not necessarily mean that every firm has to be research minded or to innovate itself. Various alternative strategies are possible for the firm even in an industry subject to rapid technical change. Some companies may even prefer to disappear rather than to innovate. These alternatives are considered more systematically in Chapter 11. For the time being we are concerned with those firms which do attempt to innovate, whether in products or processes. Some of these firms may attempt to be the first to introduce a new product or process hoping thereby to gain a technological lead and temporary monopoly profits. This strategy is sometimes designated as offensive innovation. Others may act only defensively in response to innovations introduced by competitors. In either case the firm will need some capacity to develop and launch new products or processes (even if under a licensing agreement or by simple imitation). Frequently a firm which attempts to be first may not succeed, and multi-product firms may be offensive in some fields and defensive in others. But in the long run their survival and growth will depend on whether they succeed or fail in their innovations, whether offensive or defensive.

The first stage of project SAPPHO, which is summarized here, was a study of 58 attempted innovations in chemicals and scientific instruments (listed in Table 8.1). Those in the chemical industry were mainly process innovations, whereas the instrument innovations were all product innovations. The instruments were mainly electronic and the chemical processes

Table 8.1   List of SAPPHO pairs

Scientific instruments	Chemicals (process innovations)
Amlec eddy-current crack detector	Accelerated freeze-drying of food (solid)
Atomic absorption spectrometer	Acetic acid
Digital voltmeters	Acetylene from natural gas
Electromagnetic blood-flow meter	Acrylonitrile I
Electronic checkweighing I	Acrylonitrile II
Electronic checkweighing II	Ammonia synthesis
Foreign-bodies-in-bottles detector	Caprolactam I
Milk analysers	Caprolactam II
Optical character recognition	Ductile titanium
Roundness measurement	Extraction of aromatics
Scanning electron microscope	Extraction of n paraffins
X-ray microanalyser	Hydrogenation of benzene to cyclohexane
	Methanol
	Oxidation of cyclohexane
	Phenol
	Steam naphtha reforming
	Urea manufacture

related mainly to intermediates derived from petroleum (SPRU, 1972). In later work, additional pairs of innovations were studied in these same industries and then, using a somewhat different methodology, paired comparisons were made in various sectors of the mechanical engineering industry (Rothwell *et al.*, 1974; Rothwell, 1976, 1992, 1994).

By pairing attempted innovations it was hoped to discriminate between the respective characteristics of failure and success. The technique had of course been widely used in the natural sciences, especially in biology (McKay and Bernal, 1966). When the two halves of a pair differ with respect to a particular characteristic or set of characteristics, this indicates a possible explanation of innovative success or failure. Where there is a significant and repeated variation between the pattern of success and failure, across a large number of pairs, this provides systematic evidence for the validity of particular hypotheses or groups of hypotheses. Such explanations as appear to have a significant statistical foundation may then be tested again on a new sample of innovations. In this way a structured and tested foundation for theoretical work may be built up.

It was expected that the success and failure halves of a pair would resemble each other fairly closely in many respects, and this proved to be the case. It could be assumed from previous experience that firms attempting to develop a particular new process or product would often have many characteristics in common. The analysis of similarity is complementary to the analysis of divergence for two reasons. First, it enables the identification of some characteristics which are shared by all firms attempting innovation in particular industries. These may be necessary conditions for entry into the race, and may be regarded as such unless other success cases can be found which disprove such tentative generalizations. But second, and more important, they enable us to focus attention on those characteristics in which the pattern does diverge between success and failure. In future research it will be possible to concentrate in greater depth on these significant differences through a process of elimination of unnecessary hypotheses.

The pairs were not 'identical twins'. Their similarity was defined in terms of their market, not necessarily in terms of their technology. For example, two firms might both be seeking a new, cheaper and better way to produce phenol or urea. They might adopt somewhat different technical solutions. It was an assumption of this project that this very choice constituted part of the success, and the wrong choice part of the failure. In a few cases the resemblance was very close, as where several licensees shared access to the same basic technical knowledge. But even here the design varied when two manufacturers attempted to satisfy the same demand. Success depended partly on developing the 'right' design, having regard to the available scientific and technical knowledge and to the potential uses of the innovation.

Concentration on innovation rather than invention has many consequences in terms of method. The most important of these is that the marketing aspects of the process assume much greater importance, whereas the role of that individual, usually described as the inventor, recedes into a wider social context. Our comparisons did not include those numerous experiments by inventors and would-be innovators which are discarded or shelved long before they reach the point of commercial introduction. Such studies are undoubtedly of interest in the management of R&D, but the focus here was on the wider problem of the management of innovation. The failures were products or processes which were brought to the point of commercial introduction, and usually were in fact on the market for some years. Attention was therefore concentrated both on the various stages of development work and also on the preparations for production and sale and the experience of marketing the innovation.

Since the project was concerned with technical innovation in industry, the criterion of success was a commercial one. A failure is an attempted innovation which failed to establish a worthwhile market and/or make any profit, even if it worked in a technical sense. A success is an innovation which attained significant market penetration and/or made a profit. This chapter analyses 29 successes and 29 failures. Often a failure was clear-cut, e.g. a firm went bankrupt, or closed a plant down, or withdrew a product or failed to sell it, whereas success was not always so self-evident. A product might achieve a worldwide market, but take a long time to show a profit. One case (Corfam) which was originally expected to be a success was withdrawn from the market on these grounds (Chapter 5). Even with failures it was not always simple to make an assessment. There were varying shades of grey between the extremes of success and failure. The project deliberately tried to investigate the fairly clear-cut 'black and white' cases of failure and success. In two cases in the chemical industry, and one in instruments, it proved to be feasible to complete two pairs, as there were several commercial successes and several less successful attempts in each case.

Earlier work on the literature survey had shown that there were many possible explanations of success and failure. The project was therefore designed to test a large number of single hypotheses and simultaneously to test a large number of possible combinations of factors. The aim was to identify a characteristic pattern of failure or success. Altogether about two hundred measurements of each case of success or failure were attempted. Some of the measures were comparative, some absolute. Thus, for example, it was possible to test the hypothesis that large size is generally advantageous for innovation, both by testing in how many pairs the smaller of the two firms failed, and by checking what proportion of firms with fewer than 500 employees succeeded (or 100, 1,000 or 10,000 employees).

But most of the measures were comparisons between the success and failure halves of the pair, enabling statements to be made such as 'successful attempts were characterized by greater . . . or less . . . or smaller . . . or shorter . . . or more . . .' than attempts which failed, but the aim was to link all those comparisons together to derive a pattern of success. The main hypotheses which the project attempted to test related to various measures of size (employment, R&D department and project team); measures of market research, publicity, education of users, involvement of users; modification of the innovation and checks on its progress at various stages; the role of engineers and scientists and of various key individuals, their previous experience, education and background[2], the management, control and planning system in the firm; the communication network with the outside world; degree of dependence on outside technology and familiarity of the firm with the innovation; effectiveness and methods of organizing R&D work, patent policy, competitive pressures, speed in development work and date of commercial launching.

The results of the analysis may be classified under three headings:

1. Factors which were common to almost all attempts to innovate, whether successful or not.
2. Factors which varied between innovative attempts, but in which the variation was not systematically related to success or failure.
3. Measures which discriminated between success and failure.

8.3   RESEMBLANCE BETWEEN SAPPHO PAIRS

Taking the first 29 pairs, involving 58 attempts to innovate, there were many resemblances between both halves of the pairs (Table 8.2). Almost all attempts in these two industries took place within a formal R&D structure which was used to develop the innovation. This confirms the professionalization of R&D described in Part One. Only in the instrument industry were there cases of attempted innovation without such a structure. Most of these were designs for a new product brought from an outside environment.

Since almost all of the firms involved in attempts to innovate had this formal R&D structure, it might be expected that critical differences would exist in the way in which such departments were organized, R&D was planned, projects evaluated, or incentives provided for engineers and scientists. A great deal of the management and sociological literature has concentrated on these aspects of the efficiency of industrial research.

The inquiry did not uncover systematic differences of this kind with respect to R&D organization or incentives. As with previous empirical studies it was found that many supposedly best practice techniques in long-term planning and project assessment were honoured more in the breach than in the observance. But the successful innovators differed only a little from the failures in this respect. In the chemical industry, although not in instruments, there was some evidence that better management and planning techniques were associated more frequently with success. Some of the difficulties inherent in R&D forecasting and project evaluation are discussed in Chapter 10. There was no evidence that successful innovators expected rewards or penalties differing from the less successful nor was there any evidence of unusual incentive schemes for R&D personnel, or greater freedom in successful cases.

One possible explanation of more successful attempts to innovate might lie in patent priority, but again it was not possible to identify differences here. Almost all innovators, both successful and unsuccessful, took out patents and regarded them as important. But the failures did not attribute their lack of success to the patent position of their rivals, except in one case. The results confirm the evidence of the historical account that innovating firms usually take trouble over patents because of their importance as bargaining counters, and to ensure rights of entry into a field, but that patents do not necessarily prevent any competitive developments.

Nor was it found that successful innovators differed from unsuccessful ones in the way in which they organized their project teams. One hypothesis had suggested that the less successful attempts might be characterized by departmental organization on disciplinary lines. But this was not the case. Where firms had a large R&D organization, they sometimes had laboratories working on conventional subdisciplinary lines, but this did not really affect the project development team which was set up in a similar way in both successes and failures.

Another hypothesis for which no supporting evidence was found was the view that business or technical innovators might be less well qualified academically in unsuccessful attempts (or better qualified). There were important differences between business innovators which did distinguish between success and failure, but this was not one of them. Most of the business innovators, and almost all the technical innovators, were qualified scientists or engineers in both halves of the pair. There was a slight tendency for the PhDs to be the more successful in chemicals. Obviously, in these two industries amateurs are rarely chosen to manage innovations. In the two cases when accountants were the business innovators, both failed, but this would be too small a number on which to construct any general theory. It would probably not be possible to find a sufficient number of cases in these industries where innovators were not technically qualified, to test any hypotheses relating to the supposed merits or demerits of amateurism. The difficulty of finding such cases is, however, further evidence of the professionalization of the innovative process.

8.4   VARIATION UNRELATED TO SUCCESS OR FAILURE

Many other measures did show considerable variation between attempts to innovate, but the variance was not closely related to success. Among them were measures relating to size of firm, size of R&D department and numbers of qualified engineers and scientists in R&D. These results need considerable care in interpretation. There was no strong systematic evidence that larger or smaller firms or R&D departments were more or less successful. For example, of the cases involving firms employing more than 10,000 people, six were successes and seven were failures. Where large firms were in competition with smaller firms there was a tendency for them to be more successful, but it was by no means clearcut. At first sight this finding is perhaps at variance with some of the evidence from the historical descriptive account in Part One, which suggested that the heavy costs and long time scale of many innovations would give an advantage to large firms.

However, this result should not be interpreted as implying that size of firm is completely irrelevant in relation to innovation in these two industries. Comparative size measured within a pair did not differentiate between success and failure clearly but in chemicals only 4 out of 34 attempts were made by firms employing fewer than 1,000. Clearly size is relevant to the type of innovations which are attempted at all, and inter-industry differences are very important. The next chapter is devoted entirely to a critical discussion of this problem of size in relation to innovation.

No relationship was found between success and the number of scientists and engineers on the main board of the innovating company, although this proportion varied considerably. However, in almost all cases there were some engineers or scientists on the main board, and it may be that this is the critical threshold factor since the innovation process requires a combination of technical, financial, marketing and management skills.

Perhaps surprisingly, for those who believe in the amenability of innovation to planning techniques, no relationship was found between success and the capacity to set and fulfil target dates for particular stages of the project plan, or in the general approach to planning of the innovators. This finding too needs considerable care in interpretation and is discussed more fully in Chapter 10.

Contrary to some theories, there was no association between failure and the attempt to innovate in areas unfamiliar to the firm. Where firms differed significantly in their familiarity with the field, the outcome was evenly distributed between success and failure.

Another set of measures which did not discriminate between success and failure related to the growth rate of the firm and its competitive environment. There were of course variations between firms in the growth which they had experienced before the innovation, and in the competitive pressures to which they were subject. But these differences apparently did not affect their degree of success in attempting to innovate. Again, it is important not to overstate this finding. This does not mean that competitive pressures or declining growth may not be important in stimulating attempts to innovate, only that they do not ensure success.

A rather surprising finding was that development lead time was not strongly correlated with success. It had been expected that the more successful innovators would be those who found ways of shortening the development phase and telescoping the stages from prototype or pilot plant to commercial launch. But support for this hypothesis came only at the earlier stage of applied research. In the chemical industry successful firms were quicker to get through this early stage. The absence of any evidence of a shorter development stage associated with success provides support for those who have argued that hardware development is a gestation process akin to animal reproduction in that it cannot easily be artificially shortened (Burke, 1970). It may also indicate that successful firms take more trouble at the development stage to get rid of all the bugs, so that later stages are trouble free. There was considerable indirect support for this interpretation from those measures which did discriminate between success and failure.

More recently, there has been some evidence of systematic differences in development lead times between Japanese firms on the one hand and American firms on the other (Mansfield, 1988; Womack *et al.*, 1990; Graves, 1991). Shorter Japanese lead times appear to be related to specific Japanese organizational techniques in the management of innovation, at least in some industries (Baba, 1985; Nonaka and Takeuchi, 1986). Some of these techniques were described in Part One but project SAPPHO did not include any Japanese firms and did not reveal systematic differences between American and European innovations.

8.5   THE PATTERN OF SUCCESS

Of the two hundred measures attempted, only a small number differentiated clearly between success and failure, and these varied a little between the two industries. The principal measures are shown in Table 8.2.

Those which came through most strongly were directly related to marketing. In some cases they might be regarded as obvious, but the case studies showed that even the most obvious requirements were sometimes ignored. Successful attempts were distinguished frequently from failure by greater attention to the education of users, to publicity, to market forecasting and selling (particularly in the case of instruments where it was most relevant) and to the understanding of user requirements.

The single measure which discriminated most clearly between success and failure was 'user needs understood'. This should not be interpreted as simply, or even mainly, an indicator of efficient market research. It reflects just as much on R&D and design as well as on the management of the innovation. The product or process had to be designed, developed and freed of bugs to meet the specific requirements of the future users, so that understanding of the market had to be present at a very early stage. The work of von Hippel (1976, 1978) on 'customer-active' paradigms in new product development, and of Teubal *et al.* (1976) on 'market determinateness' in the Israeli medical electronics industry, both point to the same conclusion. It has been further explored by the interesting work of Mansfield *et al.* (1977) on the integration between marketing and R&D in project selection systems and the ways in which this influences probability of success, and in the work of Lundvall (1985, 1988b, 1993).

This interpretation was confirmed by the strong evidence on the occurrence of unexpected adjustments and bugs after development in the failure cases, and of the need for user adaptations in nearly half the failures. About three-quarters of the cases of failure showed greater after-sales problems. Thus, 'user needs understood' is just as much a discriminating

Table 8.2 Part 1 Measures which did not differentiate between success and failure

Question	Chemicals			Instruments			Both industries			
	S > F	S = F	S < F[a]	S > F	S = F	S < F[a]	S > F	S = F	S < F[a]	Binomial test
Was the innovation more or less radical for the firms concerned?	5	7	5	4	3	5	9	10	10	0.5
At what level was the decision to proceed with the innovation made?	2	11	4	1	10	1	3	21	5	0.363
Was a time limit set?	4	12	1	1	8	3	5	20	4	0.5
Were patents taken out for this innovation by the organization?	—	17	—	3	8	1	3	25	1	0.313
Did one organization accept the innovation as being more in its natural business than the other?	5	8	4	6	1	5	11	9	9	0.412
Did one organization have a more serious approach to planning than the other?	6	7	4	2	7	3	8	14	7	0.5
Was there a systematic and periodically reconsidered R&D programme?	6	7	4	—	10	2	6	17	6	0.613
What was the company's publishing policy?	5	7	5	1	11	—	6	18	5	0.623
Were there any incentive schemes to encourage innovation effort?	2	15	—	—	12	—	2	27	—	0.25
What outcome was the project expected to have on the careers of members of the project team in the event of success?	3	11	3	1	11	—	4	22	3	0.5
Was the innovation part of a general marketing policy?	5	8	4	2	9	1	7	17	5	0.387
What was the degree of coupling with the outside scientific and technological community in general?	2	12	3	2	8	2	4	20	5	0.5
Would the firm have recruited more QSEs if it could have done so at the time of the innovation?	—	17	—	1	10	1	1	27	1	0.75
In each case, when was the decision to innovate formalized on paper?	5	5	7	1	10	1	6	15	8	0.395

continued overleaf

Table 8.2 Part 1 *cont.*

Question	Chemicals			Instruments			Both industries			
	S > F	S = F	S < F[a]	S > F	S = F	S < F[a]	S > F	S = F	S < F[a]	Binomial test
How many months elapsed from prototype or pilot plant to first commercial sale?	7	3	7	5	3	4	12	6	11	0.5
Was there a formal R&D department in the organization	1	16	—	2	8	2	3	24	2	0.5
What was the scale of growth of the organization up to the time of marketing (measured by annual growth of turnover in the five years prior to the marketing of the innovation)?	1	13	3	4	4	4	5	17	7	0.387
How many years did the business innovator spend in the educational system?	5	7	5	2	5	5	7	12	10	0.315
Was the R&D department regarded as a profit centre?	6	8	3	1	10	1	7	18	4	0.274
Was there any need to find or use new materials?	—	17[b]	—	—	10	1	—	27	1	0.5

[a] S > F Success more than failure, greater than failure, etc., or in success but not in failure.
S = F No measurable difference between success and failure.
S < F Success less than failure, smaller than failure, etc., or in failure but not in success.
[b] Data not available in one case.
— nil.

Table 8.2 Part 2 Measures which differentiate between success and failure

Question	Chemicals			Instruments			Both industries			Binomial test
	S > F	S = F	S < F[a]	S > F	S = F	S < F[a]	S > F	S = F	S < F[a]	
Was the innovation more or less radical for world technology?	10	6	1	2	9	1	12	15	2	0.0065
How deliberately was the innovation sought, comparatively?	7	8	2	6	6	—	13	14	2	0.0037
Was there opposition to the project within the total organization on commercial grounds?	1	9	7	1	7	4	2	16	11	0.0112
Was more use made of development engineers in planning and costing for production in one case than in the other?	5	9	3	4	8	—	9	17	3	0.073
Did one organization have a more satisfactory communication network than the other externally?	5	10	2	5	7	—	10	17	2	0.0193
Was the R&D chief more senior by accepted status in one case than the other?	9	5	3	2	8	2	11	13	5	0.105
Was the sales effort a major factor in the success or failure of the innovation?	7	10	—	9	3	—	16	13	—	0.000015
Were any modifications introduced after commercial sales as a result of user experience?	1	8	8	2	6	4	3	14	12	0.0176
Were there any after-sales problems?	—	4	13	1	2	9	1	6	22	0.000005
Were any steps taken to educate users?	8	9	—	6	5	1	14	14	1	0.00049
If new tools or equipment were needed for commercial production, were any ordered before the decision to launch full-scale production?	8	7	2	2	10	—	10	17	2	0.227

continued overleaf

Table 8.2 Part 2 *cont.*

Question	Chemicals			Instruments			Both industries			Binomial test
	S > F	S = F	S < F[a]	S > F	S = F	S < F[a]	S > F	S = F	S < F[a]	
What was the degree of coupling with the outside scientific and technological community in the specialized field involved?	8	9	—	5	6	1	13	15	1	0.00092
How much attention was given to publicity and advertising?	6	10	1	4	7	1	10	17	2	0.0193
Did the innovation have to be adapted by users?	—	10	7	—	7	5	—	17	12	0.00024
Were there unexpected production adjustments?	1	7	9	1	7	4	2	14	13	0.00636
Did any 'bugs' have to be dealt with in the early production stage?	1	6	10	1	5	6	2	11	16	0.0049
Was any systematic forecasting by the marketing (or sales) department involved in the decision to add the innovation to production lines or to existing processes?	5	7	5	6	5	1	11	12	6	0.166
Were user needs more fully understood by the innovators in one case than in the other?	15	2	—	9	3	—	24	5	—	0.0000001
Did the business innovator have a more diverse experience in one case than in the other?	8	8	1	8	2	2	16	10	3	0.00377
Did the business innovator have a higher status in one case than in the other?	8	8	1	5	4	3	13	12	4	0.0245

Question										
Did the business innovator have more or less authority (power) in one case than in the other?	9	7	1	6	4	2	15	11	3	0.000656
To what extent was dependence on outside technology a help or a hindrance in production?	10	6	1	6	4	2	16	10	3	0.00221
How large a team was put to work on the innovation at the beginning of the project?	12	2	3	4	4	4	16	6	7	0.0466
How large a team was put to work on the innovation at the peak of the project?	9	4	4	7	4	1	16	8	5	0.0133
How many years had the business innovator spent in industry?	9	7	1	3	4	5	12	11	6	0.119
Had the business innovator had any overseas experience?	3	14	—	5	6	1	8	20	1	0.0352
Did the business innovator have a greater degree of management responsibility in one case than in the other?	10	7	—	4	5	3	14	12	3	0.00636

[a] S > F Success more than failure, greater than failure, etc., or in success but not in failure.
S = F No measurable difference between success and failure.
S < F Success less than failure, smaller than failure, etc., or in failure but not in success.
— nil.

Source: Science Policy Research Unit (1971).

measure of efficiency in R&D performance as of marketing and overall management.

Size of project team emerged as a clear-cut difference, whereas other size measures did not differentiate. Since in a number of cases the smaller firms deployed a larger team, this implies a greater concentration on the specific project. This consideration is important in considering the relative advantages of the small firm in innovation in Chapter 9. Another measure which strongly suggests the advantages of specialization in R&D is that related to coupling with the outside scientific community.

Carter and Williams (1959a) already emphasized good communications with the outside world as one of the most important characteristics of the technically progressive firm. The most backward firms would not of course be found among those attempting to innovate. But among those who were making such attempts there were significant differences in their general pattern of communications. Better external communications were associated with success, but the strongest difference emerged with respect to communication with that specialized part of the outside scientific community which had knowledge of the work closely related to the innovation. General contact with the outside scientific world did not discriminate between success and failure.

All of these differences may of course be related to the quality and type of management, so that measures relating to the business innovator are perhaps the most interesting. First, it should be noted that the business innovator was hardly ever the same person as the chief executive in the chemical industry, but was frequently so in the instrument industry. The most interesting difference between successful and unsuccessful business innovators, and one which was unexpected, was that greater seniority was associated with success. The successful man (they were all men) had greater power, higher formal status, and more responsibility than the unsuccessful. He was also older and had more diverse experience. Some of these differences were not so clear cut in the instrument industry which may reflect the greater mobility and smaller size of firm, together with more hierarchical structure of management in the chemical industry. Usually the successful chemical innovator had been longer with the innovating firm and in the industry, whereas this was not true in the instrument pairs.

The higher status and greater power of the more successful innovators may be associated with their readiness to take greater risks and to recruit larger teams for their projects. In the chemical industry there was a strong association between success and a more radical technical solution. But taking a variety of measures relating to risk acceptance, there was only very slight evidence that successful innovators assumed greater risks. In the chemical innovations the successful cases were usually the first to market but in the instrument pairs those who came later were usually more successful.

The fact that the measures which discriminated between success and failure included some which reflected mainly on the competence of R&D, others which reflected mainly on efficient marketing, and some which measured characteristics of the business innovator with good communications, confirms that view of industrial innovation as essentially a coupling process, which was suggested at the outset. One-sided emphasis on either

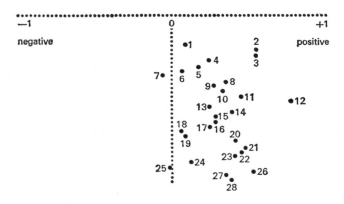

Fig. 8.1 Values of index variable for success points

Source: Science Policy Research Unit (1972).

R&D or sales does violence to the real complexity of the process. This was strongly confirmed by the multivariate statistical analysis illustrated in Figure 8.1. Composite index variables were formed consisting of several measures relating to one factor. The percentage of points correctly classified was greatest by the combination of the following composite measures: R&D strength, marketing and user needs. In the case of chemicals, composite variables relating to management techniques and management strength were also important. Management strength relates mainly to the status and responsibilities of the business innovator.

The critical role of the entrepreneur (whatever individual or combination of individuals fulfil this role) is to match the technology with the market, i.e. to understand the user requirements better than competitive attempts, and to ensure that adequate resources are available for development and launch. This interpretation of the key role of the quality of entrepreneurship is in line with the findings of Barna (1962), Penrose (1959) and the earlier work of Schumpeter on the theory of the firm (1912, 1942).

In the large firm the business innovators must be high enough in the hierarchy to command resources and get things done. They must have enough knowledge of the way the firm works to know *how* to get things done. In the small firm it frequently means that it will be the chief executive, or a person sufficiently close to ensure the necessary concentration of effort. In either case they must be sufficiently powerful and clear about marketing objectives to ensure that the various screening and testing procedures during the course of development and trial production prevent an unsatisfactory product or process coming onto the market. Premature launch may be more dangerous than slowness.

These conclusions will not necessarily be valid for consumer goods innovations where some different mechanisms are at work. In capital goods it is essential to satisfy certain minimal technical performance criteria. The extent to which these generalizations may apply to other industries is discussed in

Chapter 16. Here it is necessary to consider some other limitations of the analysis, before going on to consider in greater depth the question of size of firm in Chapter 9 and the problem of uncertainty, risk and planning in Chapter 10.

Thus, the results of SAPPHO confirm points 1, 3, 8, 9 and 10 among the tentative generalizations advanced on p. 203. Point 2 is discussed more fully in Chapter 11. Point 4 requires great care in interpretation and the whole of Chapter 9 is devoted to this. Points 5 and 7 were not supported by the evidence of SAPPHO. The approach of the innovator to risk again requires much more detailed consideration, and this is attempted in Chapter 10, followed by discussion of the implications for theory of the firm in Chapter 11.

The generalizations so far made about innovation based on the historical descriptive material in Part One thus find some confirmation from this test. Although they may provide a fairly plausible interpretation of some aspects of industrial innovation since the rise of professionalized R&D, it is certainly not claimed that they are securely based statistically or empirically. The sample was not random and the 'universe' is not known. However, a further sample of fourteen pairs confirmed the original results in all essentials. Moreover, the interpretation of innovation as a coupling process is strongly supported by much additional empirical evidence, as well as by logic and common sense. Earlier studies by Carter and Williams (1957, 1959a, 1959b) had led them to formulate the concept of the 'technically progressive firm' embodying many of the combined characteristics of the SAPPHO success cases. Another major series of case studies of industrial innovation in Britain was conducted at Manchester quite independently of SAPPHO at about the same time, and the authors of these concluded:

> Perhaps the highest level generalization that it is safe to make about technological innovation is that it must involve synthesis of some kind of need with some kind of technical possibility. The ways in which this synthesis is effected and exploited take widely differing forms and depend not only on systematic planning and the 'state of the art' but also on individual motivations, organizational pressures and outside influence of political, social and economic kinds. Because the innovation process extends over time, it is important to retain continuous sensitivity to changes in these factors and the flexibility to perceive and respond to new opportunities.
>
> (Langrish et al., 1972, p. 200).

8.6   FURTHER STUDIES OF INNOVATION

Additional important empirical evidence for some of the main SAPPHO conclusions came from a Canadian survey of 47 new small firms, started by technologically oriented entrepreneurs. Like SAPPHO this included the study of failures as well as successes. Litvak and Maule (1972) concluded from their survey that:

> The marketing performance of the entrepreneurs was weak, and was a major factor for the apparent high mortality rate of the projects. Most of the entrepreneurs were unable to see the linkage between product innovation and

marketing innovation. . . . Most of the new product development was carried out
and implemented before any attempt was made to assess the market potential
and the costs of penetrating the market. . . . The point to be made is that the love
that the entrepreneur has for his product innovation often blinds him from
perceiving his real opportunities and the state of market competition.

(Litvak and Maule, 1972, p. 47)

The point about underestimation of user needs and understanding of the
market must be heavily underlined, since the SAPPHO inquiry constantly
found in discussion with R&D managers and entrepreneurs that they tend
to dismiss the point as obvious, but nevertheless continue to ignore it in
practice.

Further confirmation of some of the SAPPHO conclusions came rather
unexpectedly from the Hungarian electronics industry (Szakasits, 1974) and
from the OECD's international studies of industrial innovation, particularly
Pavitt's (1971) cross-country comparisons of the relative innovation success
of firms in various member countries, and from the studies of innovation
sponsored by the National Science Foundation (NSF) in the United States.
An interesting example of the way in which some industrial firms accepted
and used the SAPPHO findings in their own management of innovations is
given by Leonard-Barton (1995) in her account of the highly successful
R&D-intensive American company, Hewlett Packard (pp. 132–3).

However, even when the statistics are relatively good, as in relation to
size of firm (Chapter 9), generalizations still need to be heavily qualified.
One reason for this is that the 'universe' of innovations or inventions is not
known and therefore no strictly random sample can be drawn.[3] Consequently, although attempts may be made to study a representative group of
inventions and innovations, as in the Jewkes study or in the research project
described in this chapter, we cannot be sure that such a sample is truly
representative. This reservation is particularly important when we come to
consider so-called secondary or improvement inventions and innovations.
There is a tendency in case study and historical work to concentrate on the
more spectacular inventions and innovations. But it can be argued as, for
example, by Gilfillan (1935) and Hollander (1965), that the myriad of minor
improvements and new models are as important for technical progress as
the more radical breakthrough innovation. Moreover, it can also be
plausibly maintained that non-specialists and non-professionals may make a
much bigger contribution to the secondary type of innovation than to the
breakthrough. It is also probable that knowledge of the market plays a
bigger part in this secondary type of invention and innovation than contact
with scientific research or, in the case of process innovations, direct experience of operating the process.

Schmookler (1966) showed that in several American industries, over a
long period of more than a century, invention (as measured by statistics of
relevant patent numbers) tended to follow behind demand (measured by
statistics of investment), with a time lag of a few years. However, another
project on the chemical industry at the Science Policy Research Unit
showed that in some instances in the chemical industry, there was evidence
of 'counter-Schmookler' patterns of growth in the early stages of the
emergence of radical new technologies, such as synthetic materials and

drugs in the 1920s and 1930s (Walsh *et al.*, 1979). In these periods, surges of inventive activity and of science discovery tended to precede the take-off of sales and investment, as in the analogous case of the electronic computer discussed in Chapter 7. This apparent contradiction may be explained in part if some distinction is made between the more radical inventions and innovations, which are relatively few in number, and the very large numbers of secondary and improvement inventions and innovations, multiplying rapidly as an industry grows and responding directly to market signals and investment behaviour. Thus Schumpeter's theory which puts the emphasis on autonomous innovative activity by entrepreneurs as the mainspring of economic development rather than market demand, can be reconciled with the Schmookler statistics, which measure something rather different.

Insofar as patent statistics capture both minor and major inventions, then Schmookler (1966) showed that professionalized corporate R&D in the 1950s accounted for about half of industrial inventions in the United States and probably for a higher proportion of those which were exploited, i.e. translated into innovations. Many of the other inventions also originated from professional R&D in government and universities. The development and exploitation of inventions emanating from university, government or private inventors probably also involved some professional R&D work in industry in the great majority of cases. But this would still leave a significant number of inventions and innovations which could not be attributed to specialized professional R&D. It is also likely that an even higher proportion of non-patented technical advances are attributable to those outside the professional R&D system.

However, the extent of our ignorance should not be exaggerated. There is firm empirical evidence that most professional industrial R&D is concentrated on product and process improvement and on new generations of established products. What is not known is the relative contribution to technical progress of the R&D work by comparison with the inventions and improvements generated entirely outside the formal R&D system. It is a plausible hypothesis that the proportionate contribution of the formal R&D system is much higher in the research intensive industries, but it also seems likely that technical progress will be most rapid where there is a very strong interaction between the professional R&D groups and all other personnel associated with the process or product who may themselves contribute to the solution of many problems as well as to their identification. This was confirmed by a detailed study of a major technical innovation in the coal-mining industry – the Anderton shearer-loader. Townsend (1976) demonstrated that the highly successful introduction and diffusion of this machine was based on an interplay between a series of more radical inventions and innovations introduced by the machinery makers (co-operating closely with the research establishments of the National Coal Board) and numerous improvement inventions made as a result of operating experience and encouraged by an awards scheme. Both British and German manufacturers contributed to major improvements in the design of this machine, derived in part from their own R&D, and in part from the incorporation of improvements specified by the National Coal Board in Britain. Hollander's work emphasized especially the contribution of the engineering department and technical assistance groups to technical change, but sometimes in association with R&D.[4]

The later stages of project SAPPHO shifted the emphasis away from the individual cases of success or failure with particular innovations to the study of success or failure of firms over a fairly long period. This enabled the project to take account both of individual major innovations and of incremental innovations. Work was concentrated in sectors of the engineering industries, such as textile machinery (Rothwell, 1976), mining machinery (Townsend, 1976) and agricultural machinery (Rothwell, 1979). Whereas firms could and did succeed for short periods by concentrating on incremental improvements, sometimes even without any formal R&D organization, they were often trapped in the long run by an inability to cope with the more radical types of technological competition (such as the Sulzer weaving machine). The results showed that long-run success depended on an ability to combine occasional more radical innovations with a flow of minor improvements in design, responding to customers' wishes and experience. Strong R&D was increasingly necessary to sustain this combination of technical change in the 1960s and 1970s. The earlier SAPPHO case studies, although oriented towards individual projects, had also pointed in this direction. Especially in the scientific instruments industry, one of the hallmarks of the successful cases was almost always a capacity to incorporate successive design improvements in a series of new models, as for example in the case of the milk analyser (Robertson and Frost, 1978).

One important piece of empirical work lent support to the view advanced here that specialized R&D and other technical services have been increasingly important both for the major radical inventions and innovations and for the minor improvement inventions and innovations. This was the work of Katz (1971) in Argentina. He set out to measure the contribution of technical progress to the growth of a large number of enterprises in several branches of the Argentine economy. He was able to collect very comprehensive time series for a large number of firms (250) and to relate his results to measures of the scale of adaptive R&D and other technical activities carried out by the enterprises. From his preliminary interviews he had ascertained that many Argentine firms, while not making original radical innovations themselves, nevertheless made many adaptations and improvements to the processes and products which they had acquired either from foreign parent companies or by imitation or licensing. He hypothesized that such adaptive R&D would confer important competitive advantages by enabling the firms to meet the peculiar requirements of the Argentine market more satisfactorily, or to adapt to the specialized operating conditions. The conclusion from this work and the research of Martin Bell (1984, 1991) in other developing countries is that 'learning' in whatever country is not simply a function of time but depends on deliberate organized activities, whether preformed in what is nominally an R&D department or elsewhere in the firm.

Katz's results showed conclusively that: (1) the growth of enterprises was closely related to their technical progress; (2) their technical progress was strongly associated with the performance of adaptive R&D, and of specialized technical services, although the professional group responsible for this work might be called process development or technical department rather than research department. His results also suggested the important conclusion that imitative or adaptive R&D is more certain in its outcome than offensive or defensive R&D, since studies of firm growth and R&D intensity in the USA and UK have not shown such a strong association. Hollander goes so far as to claim that many minor technical improvements are virtually risk free.

The Federation of British Industries' comparisons of UK firm growth rates and R&D intensity (1947, 1961) did show positive but weak correlations, and fairly strong association at the extremes. These results, taken together with those of the SAPPHO project, suggest that:

1. Firms performing little or no R&D in industries of rapid technical change are likely to stagnate or disappear.
2. Firms performing a great deal of R&D may sometimes enjoy exceptionally high growth rates through offensive success.
3. In the defensive middle zone, variations in R&D intensity show no statistical association with growth, and uncertainty predominates.

Although the statistical association between R&D intensity and subsequent growth by firms is not very strong, the association between successful innovation and subsequent growth of the firm *is* strong. Both Mansfield (1968a, b) and other economists have provided convincing confirmatory empirical evidence of the conclusions which common sense suggests – that successful technical innovation leads to the rapid growth of the firm. On the other hand, as we have seen, unsuccessful innovation may lead to bankruptcy, however large the scale of R&D. The implications of the high degree of uncertainty associated with radical innovation are discussed further in the succeeding chapters.

8.7 CONCLUSIONS

Following Rothwell's work on project success and firm success in the 1970s and 1980s, he attempted to synthesize all the results of this and other empirical research on innovation in the 1980s (1992, 1994). While maintaining that this work had generally confirmed the findings of the SAPPHO project in other firms and other industries, he indicated a greater emphasis on management planning and control procedures (Table 8.3) without, however, giving strong evidence of the effectiveness of such procedures. At the corporate level he stressed top management commitment and long-term strategy. This is discussed further in Chapter 11. He cited the work of Maidique and Zirger (1985), Dodgson (1991) and Prahalad and Hamel (1990) in support of his contention that repeated success with innovation depended on a process of know-how accumulation over fairly long periods.

Basil Achilladelis, who did the research for many of the SAPPHO project innovations in the chemical industry, has subsequently made some extremely thorough studies of the innovation performance of firms in various sectors of that industry, notably pesticides, petrochemicals and pharmaceuticals (Achilladelis *et al.*, 1987, 1990). Among his many interesting findings, one of the most significant has been his demonstration that

Table 8.3 Success factors

Project execution factors
- Good internal and external communication: accessing external knowhow.
- Treating innovation as a corporate-wide task: effective inter-functional coordination: good balance of functions.
- Implementing careful planning and project control procedures: high quality upfront analysis.
- Efficiency in development work and high quality production.
- Strong marketing orientation: emphasis on satisfying user needs: development emphasis on creating user value.
- Providing a good technical service to customers: effective user education.
- Effective product champions and technological gatekeepers.
- High quality, open-minded management: commitment to the development of human capital.
- Attaining cross-project synergies and inter-project learning.

Corporate level factors
- Top management commitment and visible support for innovation.
- Long-term corporate strategy with associated technology strategy.
- Long-term commitment to major projects (patient money).
- Corporate flexibility and responsiveness to change.
- Top management acceptance of risk.
- Innovation accepting, entrepreneurship accommodating culture.

Source: Rothwell (1992, 1994).

chemical firms which had an original success with a radical innovation were frequently able to follow this with an accumulative series of further successful innovations in the same field (Table 8.4). His work showed further that the success of the large chemical firms in synthetic materials extended also to other sectors of the industry (Table 8.5). These results confirm Rothwell's conclusions on the role of knowledge accumulation in successful firms.

Finally, Rothwell (1992, 1994) studied the influence of ICT on innovation management and innovative success. This led him to an increasing emphasis on the importance of various forms of 'networking' in what he designated as the 'fifth generation' innovation process (Table 8.6). Systemic factors have always been important for successful innovation, as has been clearly demonstrated in the historical evidence on textiles, chemicals, electrical engineering and automobiles, but it is becoming increasingly clear that ICT has redoubled their significance. In the first place, ICT has provided vastly more efficient means for the accumulation and speedy transmission of data within and between individuals and organizations. Second, many innovations now incorporate some electronic devices or elements of computerization which often necessitate some type of collaboration with electronic hardware or software firms. Studies of the rapidly increasing scale of collaborative agreements between firms in the 1980s and 1990s, showed that a high proportion involved firms in ICT, in biotechnology and in advanced materials (Hagedoorn and Schakenraad, 1990, 1992). The complexity of technological development in these and other technologies now often rules out going it alone in R&D and impels firms into collaborative arrangements of one kind or another. A special issue of *Research Policy* on 'Networks of Innovators' (DeBresson and Amesse, 1991)

Table 8.4 Some examples of corporate technological traditions

No.	Company	Technological tradition	Radical innovation	Year
1	American Cyanamid	Aminoplasts	Urea melamine resins	1935
2	American Cyanamid	Organophosphorus insecticides	Thimet	1956
3	BASF	Organic chemical intermediates	Ammonia synthesis	1913
4	BASF	Polystyrene plastics	Polystyrene	1928
5	BASF	Magnetic recording tapes	First magnetic tape	1935
6	Bayer	Organophosphorus insecticides	Parathion	1942
7	Bayer	Synthetic rubber	First synthetic rubber	1910
8	Bayer	Polyurethane plastics, foams	Polyurethane	1942
9	B. F. Goodrich	PVC	PVC	1930
10	Celanese	Synthetic fibres	Cellulose acetate	1924
11	Celanese	Organic chemical intermediates	Acetic acid	1933
12	Ciba-Geigy	Insecticides	DDT	1939
13	Ciba-Geigy	Herbicides	Triazines	1957
14	Ciba-Geigy	Vat dyestuffs	Ciba violet	1905
15	Dow	Halogenated hydrocarbons	Chloroform	1903
16	Dow	Polystyrene	Polystyrene	1932
17	Dow	Pesticides	Pentachlorophenol	1930
18	Du Pont	Synthetic fibres	Nylon	1936
19	Du Pont	Fungicides	Nabam	1936
20	ICI	Herbicides	MCPA	1942
21	ICI	Reactive dyes	Procion dyes	1956
22	Monsanto	Herbicides	Randox	1955
23	Montedison	Organic chemical intermediates	Ammonia	1924
24	Montedison	Polypropylene plastics, fibres	Polypropylene	1954
25	Rohm & Hass	PMMA-acrylics	Polymethylmethacrylate	1932

Source: Achilladelis et al. (1990).

Table 8.5 Concentration and technological accumulation in chemical innovations, 1930–80*

1 Type of innovations and companies	2 Pesticides	3 Pesticides	4 Synthetic materials
	Top 5 companies	Top 10 companies	Top 5 companies
	Bayer Geigy ICI Dow Du Pont	Col. (2) plus: BASF Hoechst Shell Cyamid Sumitomo	Bayer BASF Hoechst Du Pont ICI
% of all innovating companies	6	12	5
% of all product and process patents	19	27	30
% of all new products	31	44	58
% of all radical innovations	38	54	60
% of major market successes	35	55	66

* Synthetic materials 1930–55

Sources: Achilladelis et al. (1987); Freeman et al. (1963).

Table 8.6 The fifth generation innovation process: Systems Integration and Networking (SIN)

Underlying strategy elements
- Time-based strategy (faster, more efficient product development).
- Development focus on quality and other non-price factors.
- Emphasis on corporate flexibility and responsiveness.
- Customer focus at the forefront of strategy.
- Strategic integration with primary suppliers.
- Strategies for horizontal technological collaboration.
- Electronic data processing strategies.
- Policy of total quality control.

Primary enabling features
- Greater overall organizational and systems integration:
  - parallel and integrated (cross-functional) development process
  - early supplier involvement in product development
  - involvement of leading-edge users in product development
  - establishing horizontal technological collaboration where appropriate.
- Flatter, more flexible organizational structures for rapid and effective decision-making:
  - greater empowerment of managers at lower levels
  - empowered product champions/project leaders.
- Fully developed internal databases:
  - effective data sharing systems
  - product development metrics, computer-based heuristics, expert systems
  - electronically assisted product development using 3D-CAD systems and simulation modelling
  - linked CAD/CAE systems to enhance product development flexibility and product manufacturability.
- Effective external data links:
  - co-development with suppliers using linked CAD systems
  - use of CAD at the customer interface
  - effective data links with R&D collaborators.

Source: Rothwell (1992).

and the Conference on 'Technological Collaboration' at Manchester in 1993 (Coombs *et al.*, 1996) were two of many instances of the rapid growth of research interest in this field. Still lacking were studies of the evolution of networks with a few exceptions such as the brilliant study of imaging networks in Sweden by Anders Lundgren (1991).

Nevertheless, there is now sufficient evidence on the role of networking in innovation to postulate that the typical pattern of nineteenth-century innovation (the inventor–entrepreneur) and of twentieth-century innovation (the inhouse corporate R&D department with good external communications) is now increasingly giving way to a pattern of networking collaborative systems innovation in the twenty-first century. Among the driving forces of this change, two of the most important factors are the increasing complexity of technical change and the systemic nature of many ICT innovations. The example of IBM illustrates this change very well: in the 1950s and 1960s IBM had hardly any collaborative R&D arrangements and came very close to autarchy in its own immense R&D facilities; in the 1980s and 1990s IBM has made dozens of collaborative arrangements with other firms, large or small, in a variety of industries. Parts Three and Four will further explore the growth of networking in both national and international systems of innovation.

NOTES

1. While this may be described as innovation, it cannot be legitimately described as technical innovation. Non-technical organizational innovations are extremely important and often associated with technical innovations as in the case of mass production and lean production or the marketing innovations of Wedgwood.
2. Four key roles in the conduct of innovation were defined as follows:

   1. Technical innovator: the individual who made the major contributions on the technical side to the development and/or design of the innovation. He would normally, but not necessarily, be a member of the innovating organization. He would sometimes, but not always, be the inventor of the new product or process. (They were all male.)
   2. Business innovator: that individual who was actually responsible within the management structure for the overall progress of this project. He might sometimes be the technical director or the research director. He might be the same man as the technical innovator. He could be the sales director, or chief engineer. Occasionally, especially in smaller firms, he could be the chief executive for the organization as a whole. (They were all male.)
   3. Chief executive: the individual who was formally the head of the executive structure of the innovating organization, usually but not necessarily with the job title of managing director. In every case there was an identifiable chief executive, and almost always an identifiable business innovator, but quite often there was no identifiable technical innovator. No attempt was made to force individuals to assume these roles if they were not readily identifiable, since one of the objects of the inquiry was to assess the contribution of outstanding individuals.
   4. Product champion: any individual who made a decisive contribution to the innovation by actively and enthusiastically promoting its progress through critical stages. He might sometimes be the same individual as the technical innovator, or chief executive. Although these roles have been recognized in much of the earlier innovation literature, they are not always identifiable from formal titles used in firms. The job title might vary a good deal, but it was the role which was important. (They were all male.)

3. The second Manchester study of innovation (Gibbons and Johnston, 1972) was based on an ingenious attempt to develop a random sample (see p. 258).
4. This may sometimes be just a question of nomenclature. What is called 'engineering', 'OR' or 'technical department' in one firm may be called 'process development' or 'R&D' in another.

CHAPTER 9

INNOVATION AND SIZE OF FIRM

The historical account in Part One showed that in synthetic materials, in chemical processes, in nuclear reactors and in some electronic systems large firms predominated in launching the innovations. But a blanket hypothesis of 'bigness wins' could not be sustained, either from Part One or from the SAPPHO project. In scientific instruments in particular, new small firms made outstanding contributions. Inventor–entrepreneurs establishing new firms had apparently also been important in the early days of the chemical industry, the automobile industry, the semiconductor and radio industries. They continued to flourish in the microcomputer industry and in computer software. How far is it possible to test generalizations about the relative contribution of large and small firms to industrial innovation?

The evidence from project SAPPHO, so far as it goes, suggests that as between competitive attempts to innovate, size in itself does not affect the outcome very much. However, it is apparent that there is a range of innovations which is not attempted at all by really small firms so that, for example, the competition in the chemical industry or turbine generators is mainly between various large or giant firms. The relative contribution of large and small firms varies a great deal from industry to industry, and investigations such as SAPPHO cannot answer the question of the aggregate contribution of large or of small firms to research and innovation in the economy as a whole.

The size structure of industry and its relationship to problems of monopoly and competition is a problem which has preoccupied economists for a long time (Turner and Williamson, 1969; Cohen, 1995; Scherer, 1992c) and there is now a considerable amount of statistical information. Unfortunately, in our field of interest most of this relates to R&D, or patents rather than innovation, so that there are big problems of interpretation. This chapter aims to provide such an interpretation and concludes by reviewing some attempts at the direct measurement of the numbers of innovations by large and small firms in the manufacturing industry (Freeman, 1971; Kleinman, 1975; Townsend et al., 1982; Pavitt et al., 1987; Acs and Andretsch, 1988; Kleinknecht and Reijnen, 1992b).

9.1 SIZE OF FIRM AND EXPENDITURE ON R&D

Whereas in the 1950s there was very little reliable empirical evidence outside the USA on the degree of concentration in the performance of industrial research and experimental development, such evidence became available in the 1960s. As a result of the efforts of the OECD in standardizing definitions and methods,[1] reasonably comparable data for a dozen countries (OECD, 1967, p. 46) became available. The picture that emerged

Table 9.1 Percentage of total industrial R&D performed in firms ranked by size of R&D programmes

Country	Number of firms ranked by size						
	4	8	20	40	100	200	300
USA	22.0	35.0	57.0	70.0	82.0	89.0	92.0
UK	25.6	34.0	47.2	57.9	69.5	75.0	77.0
France	20.9	30.5	47.7	63.4	81.0	91.2	95.6
Japan	—	—	—	47.7[a]	52.1[b]	63.1[c]	71.4[d]
Italy	46.4	56.3	70.4	81.6	92.5	—	—
Canada[f]	30.3	40.8	58.4	71.5	86.2	93.2	—
Netherlands	64.4[e]	—	—	—	—	—	—
Sweden	33.2	43.0	54.0	71.0	85.4	90.0	—
Belgium	38.5	51.8	72.6	82.7	92.8	97.5	99.4
Norway	29.5	38.8	55.7	70.6	88.2	97.9	100.0
Spain	25.2	47.0	73.9	91.5	—	—	—

[a] The first 54 firms.
[b] The first 85 firms.
[c] The first 180 firms.
[d] The first 289 firms.
[e] The first 5 firms.
[f] Current intramural expenditure.

Source: OECD (1967).

was consistent and confirmed the hypothesis of those economists who had postulated a high degree of concentration. The hundred largest R&D programmes accounted for more than two-thirds of all industrial R&D in all countries except one, and for more than three-quarters in most cases. The forty largest programmes accounted for more than half of all industrial R&D in all cases except one, and the eight largest for more than 30 per cent in all countries for which figures were available (Table 9.1). In the Netherlands the five largest programmes accounted for two-thirds of all expenditures (Philips, Shell, Unilever, AKU, DSM).

Since the 1960s, there has been a slight reduction in concentration. This has been associated with the very rapid growth of new, small NTBFs (new technology based firms) in ICT, advanced materials and biotechnology. There is less complete but fairly conclusive evidence that the vast majority of small firms in OECD countries still do not perform any organized research and development. For France, Britain and the United States, and probably most other countries, the proportion of small firms performing R&D is almost certainly less than 5 per cent (if small is defined as fewer than 200 employees).

It is true that the official statistics of research and experimental development expenditures may not capture research or inventive work which is performed by managers, engineers or other staff incidentally to their main work. It may be that this part-time amateur inventive work is very productive, and the evidence will be discussed for the view that small firms account for an exceptionally high proportion of significant inventions and innovations. But so far as specialized professional R&D activity is concerned, there is pretty firm evidence that this is highly concentrated in large firms in all countries for which statistics are available.

Table 9.2 Percentage of R&D, net sales and total employment by companies with largest R&D programmes, USA, 1970

Programme size	Total R&D	Federal R&D	Net sales	Total employment
First 4	18	20	6	8
First 8	32	40	9	11
First 20	55	71	16	19
First 40	66	85	23	27
First 100	79	93	38	39
First 200	87	96	50	50
First 300	91	97	63	62

Source: National Science Foundation (1972, pp. 46–7).

The OECD statistics which have been cited (Table 9.1) measured the degree of concentration by size of R&D programme, and not by size of firm in terms of total employment, turnover or assets. However, for the major countries some statistics are available on concentration by size of firm, although not as a consistent classification. Firms with more than 5,000 employees accounted for 89 per cent of all industrial R&D expenditures in the United States in 1970, and for 90 per cent in 1978. They accounted for about 75 per cent in the German Federal Republic in 1979 and probably about the same proportion in the UK. Firms employing more than 3,000 accounted for about two-thirds of Japanese industrial R&D in 1978–9.

However, the degree of concentration was much less marked by size of firm (classified by total employment) than by size of R&D programme. In the United States there were 466 firms with more than 5,000 employees performing R&D in 1970. But many of them had relatively small R&D programmes, whereas some medium-sized firms (1,000–4,999 employees) had rather large ones. Thus the 300 largest programmes were approximately equivalent to the outlays of the 470 largest firms, each accounting for about 90 per cent of the total (Table 9.2). R&D programmes were far more concentrated than sales or employment (Table 9.2). In France the 200 largest programmes accounted for about 91 per cent of total expenditures, but the 200 largest firms (measured by employment) accounted for about 72 per cent. There are some industries in which even the largest firms perform little or no research, and others in which even small firms perform a good deal.

The major source of variations in research intensity between firms is the industry concerned, so that analysis of the relationship with size is best done industry by industry.

In the early debate in the 1960s some economists claimed to have discovered inverse correlations at least for some industries in several European countries and for Canada (Hamberg, 1966; Morand, 1970; de Melto et al., 1980). However, these results have been disputed as more complete evidence became available. Reviewing all the most recent data, both Cohen (1995) and Symeondis (1996) conclude that R&D spending seems to rise more or less proportionately with firm size above a threshold level:

The most robust finding from the empirical research relating R&D to firm size and market structure is that there is a close positive monotonic relationship between size and R&D which appears to be roughly proportional among R&D performers in the majority of industries or when controlling for industry effects.

(Cohen, 1995, p. 196)

However, there are exceptions to this generalization, both by industry and by country.

In addition to the points made at the end of the previous chapter, it could be postulated that the small firms who do perform R&D would tend to fall into three categories:

1. Firms which have just begun to develop or exploit a new invention. In this case sales could be relatively low in relation to R&D and a very high research intensity could be expected. For example, Genentech in 1994 had an R&D/sales ratio of over 40 per cent. This might tend to fall in the event of successful commercial exploitation of the innovation and growth of the firm and its sales.
2. Highly specialized firms which have a particular expertise, sustained by an intensive research programme in a very narrow field. Here too, research intensity might often be high. For example, some spin-off firms in science parks have R&D/sales ratios between 10 and 20 per cent over fairly long periods.
3. Firms struggling to survive in industries in which new product competition makes R&D increasingly necessary. A very varied management response might be expected in these circumstances, with some firms trying to scrape by with a subthreshold R&D effort, others relying mainly on co-operative research, and still others taking high risks with an ambitious programme.

From much indirect evidence, such as the growth of science parks and the number of spin-off firms around universities it is reasonable to suppose that the number of firms in the first two categories has been increasing fairly rapidly from the 1970s to the 1990s.

If these suppositions are correct, they would account both for the relatively weak correlation between research intensity and size of firm, found in some studies and for the empirical observations of wide inter-industry variations in the strength of this correlation. In the UK, France, Germany and the USA it has been found that in some industries, small or medium-sized firms had higher research intensity than large firms. Even for all industries taken together, although the US figures showed a consistently higher research intensity for firms employing more than 25,000 it was the federal contracts placed with firms that accounted for the greater part of the difference (Table 9.3). Taking company financed R&D only, the difference by size of firm was not great (remembering that we are dealing here only with those firms that do perform R&D).

Turning to variations in research intensity among large firms, Hamberg (1964) and Scherer (1965a, b) found only a weak correlation with size measured in terms of employment or sales, and still less with size measured in terms of assets. Hamberg's sample consisted of 340 large firms

Table 9.3 Funds for R&D as percentage of net sales in R&D performing companies by size of company, USA

Firm size	Total R&D (including federal contracts) as % of net sales			Company funds for R&D (excluding federal contracts) as % of net sales		
	1957	1967	1977	1957	1967	1977
Less than 1,000	1.8	1.7	1.7	1.4	1.6	1.6
1,000 to 4,999	1.8	1.7	1.5	1.2	1.4	1.3
5,000 to 9,999	} 3.9	2.1	1.9	} 1.6	1.6	1.5
10,000 to 24,999		} 5.2	1.8		} 2.3	1.5
25,000 or more			4.2			2.4
All firms	3.4	4.2	3.1	1.5	2.1	2.0

Source: National Science Foundation (1979).

Table 9.4 Concentration of patents, R&D expenditure, and employment and various inventive activity intensity measures for firms with more than 25,000 employees, ranked by employment

Number of firms included	Percentage of all 130 firms			Number of patents per $ bill. sales	R&D as % of sales	Number of patents per $ mill. R&D
	Patents	Employment	R&D			
First 4	9.04	23.98	24.13	11.86	2.69	0.441
First 8	19.89	34.62	38.39	17.98	2.94	0.609
First 12	25.91	40.84	43.87	20.17	2.90	0.695
First 16	35.21	45.98	51.61	20.06	2.50	0.803
First 20	40.71	50.39	54.50	21.41	2.44	0.879
First 30	53.13	59.28	63.88	24.47	2.50	0.978
First 40	58.31	66.25	69.69	23.03	2.34	0.984
First 50	64.81	71.93	75.11	23.55	2.32	1.015
First 75	78.99	83.87	78.75	23.17	2.14	1.085
First 100	91.08	92.77	94.11	22.99	2.02	1.138
All 130	100.00	100.00	100.00	23.03	1.96	1.176

Source: Soete (1979).

from the *Fortune 500* list, while Scherer's sample was 448 firms from the same list. Scherer made the interesting observation that in several industries, research intensity generally rose with size up to sales of $250 million, but began to fall somewhere between $200 million and $600 million. (See also the literature review by Kamien and Schwarz, 1975.) Soete (1979) examined more recent evidence which became available during the 1970s and concluded that the US R&D data did not on the whole support the views of Hamberg and Scherer, although the patent data did still provide some support. Soete maintained that the evidence for the United States showed some tendency for R&D intensity to increase with size of firm, at least in some industries. However, it must be remembered that all the data in Table 9.4 refer to firms with more than 25,000 employees (i.e. very large firms). Evidence in Part One also suggests that the largest firms were sometimes the most research intensive (IG Farben, Standard Oil and Bell).

Thus, summing up the evidence on size of firm and R&D expenditures, which was available in the 1970s:

1. R&D programmes were highly concentrated in all countries for which statistics were available.
2. These programmes were mainly performed in large firms with more than 5,000 employees, but the degree of concentration was significantly less by size of firm than by size of programme.
3. The vast majority of small firms (probably well over 95 per cent) did not perform any specialized R&D programmes.
4. Among those firms which did perform R&D, there was a significant correlation between size of total employment and size of R&D programme in most industries.
5. There was a generally weaker correlation between the relative measure of research activity (research intensity) and size of firm and it was not significant in some industries.
6. In several countries those small firms that did perform R&D had above average R&D intensities.

The most recent evidence for the 1980s and 1990s indicates a reduction in concentration both in terms of size of programme and size of firm; but R&D nevertheless remains more concentrated than employment or sales. Whereas the vast majority of small firms still performs no specialized R&D, the number of R&D performing small firms has increased.

Before attempting to interpret these results it is necessary to consider a little further the relationships between R&D expenditures (inputs into R&D) and R&D output.

9.2   SIZE OF FIRM AND INVENTION

A number of economists has maintained that despite the heavy concentration of R&D expenditure in large firms, it is the small firms that account for most of the important inventions and innovations. As has already been indicated, whereas the measurement of R&D inputs has made significant progress in the past twenty-five years, less progress has been made with measurement of R&D outputs.[2] It was only in 1980 that the OECD devoted a full conference to output (OECD, 1980b). It is generally accepted that the direct output of industrial R&D is a flow of new knowledge and information relating to new and improved products and processes. This may take the form of research reports, technical specifications, operational data and instruction manuals based on experience with pilot plants or prototypes, scientific papers, formulae, oral communications, blueprints, or patents (Table 1.1). No one has found a way to reduce this flow to a common denominator which could be used for inter-firm or inter-industry comparison. The most obvious method would be numbers of inventions and innovations, either unweighted or weighted by some kind of qualitative assessment.

The only statistics of numbers of inventions which are generally available are patent statistics, and ingenious attempts have been made to use these for various forms of comparison, including relative output by size of firm.

However, as already noted in Chapter 5 they are unsatisfactory for a variety of reasons, of which the main one is that firms and industries vary considerably in their propensity to patent. Some firms attach great import- ance to patents and have large departments with a strong interest in patenting activity, which will tend to inflate their inventive output, when measured in this way. Other firms either do not want to bother with patents or prefer to rely on secrecy. There has been a tendency to assume that large firms would have a higher propensity to patent than small firms and that consequently a measure of output of R&D based on patent statistics would understate the contribution of small firms. Since, in the United States, small firms show a much higher number of patents per dollar of R&D expenditure than large firms, this has been claimed as evidence of superior productivity of small firm R&D (see Rothwell and Zegveld, 1982).

However, Schmookler (1966, p. 33), the leading expert on United States patent statistics, presented convincing evidence for the view that, contrary to general belief, large firms in the United States have a lower propensity to patent than small ones. He based this on the empirically demonstrable effects of anti-trust actions on the patent policies of large firms, on the far greater possibilities of pretesting before filing of applications of large firms, and on the greater security of large firms in relation to patent sharing and know-how exchange arrangements. Small firms usually cannot afford not to patent and cannot afford to wait, so that patent statistics tend to exag- gerate the contribution of the smaller firms to inventive output, and that of private individuals. This view was supported in Britain by the work of Pavitt (1982) and the analysis of the workings of the British patent system by Taylor and Silberston (1973).

The other major problem associated with patent statistics is the vari- ability in importance of patents. One way of trying to get round this difficulty is by weighting patents, or by listing major inventions. The difficulty of these methods is that they are very time-consuming, unless they are confined to a small number of really outstanding inventions. In this case the difficulties which arise are those of subjective judgement in selecting the most important inventions, and of rating the relative import- ance of radical primary inventions, compared with the vast multitude of secondary improvement inventions. By far the best known example of this technique is Jewkes' study, which has already been discussed (Jewkes et al., 1958) and which attempted to show that a majority of 70 major twentieth- century inventions were made outside the R&D departments of large firms. The US Department of Commerce study (1967) adopted an essentially similar view of the importance of private inventors and small firms, but with less empirical supporting evidence, and a tendency to confuse invention with innovation. Similar ideas were propounded earlier by Grosvenor (1929) and Hamberg (1966).

Jewkes' analysis may be criticized on the grounds that some important corporate inventions were omitted or, perhaps more justifiably, on the grounds that the contribution of large firms has become much more important since the 1920s. If his list of inventions is broken down, the share of corporate R&D is weak before 1930, but dominant since (Freeman, 1967, 1992). However, after making allowance for these criticisms, it must be conceded that Jewkes and his colleagues have made a strong case for the view that universities, private inventors and smaller firms have made a disproportionately large contribution to the more radical type of twentieth-century inventions. This was also confirmed by our own historical account.

9.3 SIZE OF FIRM AND INNOVATION

However, it does not necessarily follow that, because smaller firms may score better on numbers of patents or numbers of major inventions in relation to their R&D inputs, they are consistently more efficient in R&D performance than large firms. First, it has already been noted that a number of Jewkes' private inventions were in fact developed and brought to market by large corporations. Of the inventions made outside large firm R&D, perhaps about half were innovated in this way. The final aim of industrial R&D is a flow of innovations, so that efficiency in development is just as important as the earlier stages of inventive work. Indeed, it is often very difficult to say who made an invention because of the tangled chain of claim and counter-claim, but it is usually possible to say more precisely which firms made an innovation, in the sense of first launching a new product or process commercially. The relative performance of large firms is apparently better with respect to innovations than with respect to inventions, and Jewkes accepts that their role in development work (which is usually far more expensive) is much more important.

Thus, it may be reasonable to postulate that small firms may have some comparative advantage in the earlier stages of inventive work and the less expensive, but more radical innovations, whereas large firms have an advantage in the later stages and in improvement and scaling up of early breakthroughs. Moreover, there are significant differences between industries in the relative performance of small and large firms. In the chemical industry, where both research and development work are often very expensive, large firms predominate in both invention and innovation. In the mechanical engineering industry, inexpensive ingenuity can play a greater part and small firms or private inventors make a larger contribution. Patent statistics reflect these differences very clearly, and the point is fully confirmed by the results of the project described at the end of this section. However, it must be noted that in the case of computer software, on which patents could until recently not usually be taken out, the major contribution of small new firms may not be reflected in patent statistics.

As we have seen in Part One there are some types of innovation which are beyond the resources of the small firm. The absolute number of components is one factor which will affect this. The extreme case is Apollo XI, for which more than two million components were required, but there are other more mundane complex engineering products for which more than 10,000 components may be needed, such as advanced jet aero-engines, electronic telephone exchanges, large computer systems, nuclear reactors, or some process plant. Large firms also have a comparative advantage where there are several possible alternative routes to success, with uncertainty attached to all of them, but benefits from the simultaneous pursuit of several. Similarly, they enjoy an advantage where large numbers of different specialists are needed to solve a problem or expensive instrumentation is essential.

Table 9.5 Comparative advantage of types of firms in instrument innovation

Innovation process	Established large firm	Recent small firm on second or subsequent products	Entrepreneur, first product
Motivation to innovate	3	1–	1
Ability to have or develop own knowledge, technology	1	3	1
Cost advantages, using outside knowledge	2	3	1
Resources available to penetrate market	1	2–	3
Resources for new product development	1	3	1 or 2
Advantage in costs and speed of prototype and early model manufacture	3	1–	1
Flexibility to adopt new product or technology	3	2	1+
Cost advantage, large series production and marketing	1	2–	3

1 = highest comparative advantages, 3 = lowest comparative advantages

Source: Shimshoni (1970, p. 61).

Probably the greatest advantage of the small firm lies in flexibility, concentration and internal communications. SAPPHO suggested that greater concentration of management effort is important. Efficient coupling of marketing–production and R&D decision-making may be much more easily achieved in the small firm environment. In the discussion of the electronic scientific instrument industry reference has already been made to Shimshoni's work (1966, 1970). He found that new small firms had played a critical part in innovating several key instruments and postulated that their main advantages lay in motivation, low costs, lead time in development work (from speed in decision), and flexibility (Table 9.5). He also concluded that new firms had a major advantage in external economies in the form of technological expertise brought from elsewhere in the R&D system. In his studies of spin-off instrument firms, Roberts also pointed to the critical importance of technological entrepreneurs bringing with them ideas and half-developed new products from a scientific environment in university and government laboratories. Golding demonstrated this mechanism operating within the American semiconductor industry. The exceptionally important role played by new small firms and spin-off firms in the American semiconductor industry has led some observers to conclude that Schumpeter's 'Mark 1' is a more realistic picture of contemporary reality than his 'Mark 2' (i.e. that large firms do not predominate in the process of innovation – see Chapter 1). However, before jumping to this conclusion, it is important to keep in mind the following points: first that the larger corporations (Bell, GE, RCA, AT&T, etc.) did continue to contribute a large share of the key innovations – perhaps as much as half; second, that they accounted for more than half the key process innovations; third, that in

Europe and Japan, both the imitation process and the innovation process were dominated to a much greater extent by the large corporations. Rothwell and Zegveld (1982), while accepting that small firms enjoy some advantages in the innovation process, have also pointed out some of the disadvantages, such as access to finance, ability to cope with government regulations and lack of specialist management expertise.

How far is it possible to test systematically the relative contribution of small and large firms to innovation in various industries and the economy generally? While the evidence is still incomplete and the measurement problems remain formidable, there were several major advances in the 1970s and 1980s and projects carried out in both Britain and the United States enable us to give a fairly definite answer, even if it is not so detailed and precise as we might wish.

A project carried out at the Science Policy Research Unit in 1971 attempted to measure directly the number of innovations made by each of three size categories of firms in many branches of British industry (Freeman, 1971; Townsend et al., 1981). The inquiry was carried out for the Bolton Committee of Inquiry on Small Firms, which defined small firms as those with fewer than 200 employees. The survey covered the period 1945–70, but later surveys, supported by the Research Councils (SERC and SSRC) added new information for the period 1971–83 (Pavitt et al., 1987). Lists of important innovations were obtained from independent sources for each of a large number of different branches of industry. The innovations were then traced to the innovating firms, 90 per cent of whom were able to supply information on their size in terms of total employment at the time of the innovation.

On the reasonable assumption that the branches of industry included in the survey were representative of British industry as a whole, the most important conclusions were as follows:

1. Small firms accounted for about 17 per cent of all industrial innovations made between 1945 and 1983. This may be compared with their share of production and employment, which in 1963 amounted to about 19 per cent of net output and 22 per cent of employment.
2. The share of small firms in innovation was apparently fairly steady (Table 9.6) from 1950 to 1970 but rose quite steeply after this.
3. The share of the largest firms (10,000 employees and over) in the total number of innovations increased in the 1960s and 1970s but declined in the 1980s.
4. Over the entire period firms with a total employment of more than 1,000 accounted for two-thirds of all innovations.

9.4   INNOVATION BY SIZE OF FIRM AND BRANCH OF INDUSTRY

For the period from 1945 to 1970 the analysis by branch of industry showed big variations in the contribution of small firms to innovation. Industries may be classified into two fairly clear-cut groups:

1. Those in which small enterprises made little or no discernible contribution to innovation, either absolutely or relatively. These included

Table 9.6 Innovation share by size of firm in UK Industry 1945–83

	Size of firm					
Time period	*1–199*	*200–499*	*500–999*	*1,000–9,999*	*>10,000*	*No. of innovations*
1945–49	16.8	7.5	5.3	28.3	42.0	226
1950–54	14.2	9.5	4.5	32.2	39.6	359
1955–59	14.4	10.1	9.1	24.9	41.4	514
1960–64	13.6	9.2	6.0	27.8	43.4	684
1965–69	15.4	8.2	8.5	24.7	43.7	720
1970–74	17.5	9.0	6.3	20.7	46.5	656
1975–79	19.6	9.6	7.5	16.2	46.2	823
1980–83	26.3	12.1	4.3	14.9	41.9	396
Number of innovations	774	411	299	1,004	2,020	4,378
Average percentage	17.0	9.4	6.8	22.9	43.1	100

Source: Rothwell and Dodgson (1994).

aerospace, motor vehicles, dyes, pharmaceuticals, cement, glass, steel, aluminium, synthetic resins and shipbuilding (Table 9.7), and (in a special category) coal and gas. In this group small firms accounted for only just over 1 per cent of innovations (6 out of a total of 479), but about 8 per cent of net output in 1963.

2. Those in which small enterprises made a fairly significant contribution to innovation in the industry concerned. These included scientific instruments, electronics, carpets, textiles, textile machinery, paper and board, leather and footwear, timber and furniture, and construction. In this group small enterprises accounted for 103 out of 623 innovations, or about 17 per cent, compared with about 20 per cent of net output in 1963.

If industries are ranked according to the share of small enterprises in the number of innovations for each industry, then this order corresponds fairly well with a measure of concentration based on share of small enterprises in net output (Table 9.7), but the contribution of innovations relative to net output share rises steeply.

In scientific instruments, some types of machinery and paper and board, small enterprises contributed proportionately more than their share of output to innovations. Medium-sized firms (employing 200–999) also contributed substantially to innovation in these industries.

Those industries in which small firms contributed much less than their share of output or nothing at all, correspond broadly to industries of high capital intensity. The major exceptions were aerospace, shipbuilding and pharmaceuticals. In these industries development and innovation costs for most new products are very heavy, although capital intensity is low. Paper and board was again an exception. Although this industry is generally one of relatively high capital intensity, small and medium firms have made important innovations, mainly in speciality products, and in board rather than paper. In these sectors capital intensity is lower.

Table 9.7   Share of small firms in innovations and new output of industries surveyed in UK

1958 SIC MLH number	1958 SIC title of industry	Per cent share of innovation by small firms 1945–70	Number of innovations by small firms 1945–70	Number of innovations by all firms 1945–70	Per cent share of net output by small firms 1963	Value of net output by all firms 1963 (£m)
471–3	Timber and furniture	39	7	18	49	220
351	Scientific instruments	28	23	84	23	154
431–3	Leather and					
450	footwear	26	5	19	32	157
335	Textile machinery	23	15	65	21	65
481–3	Paper and board	20	6	30	15	317
339	General machinery	17	18	108	14	409
332	Machine tools	11	4	38	18	100
411–15 417, 419 492	Textiles, carpets	10	6	63	18	670
364	Electronics	8	13	160	8	320
211–29	Food	8	3	38	16	814
381	Vehicles, tractors	4	3	64	5	733
276	Synthetic resins and plastics	4	2	52	12	77
370	Shipbuilding	2	1	59	10	215
271(1)	Dyes	0	0	22	7	35
272(1)	Pharmaceuticals	0	0	44	12	124
463	Glass	0	0	13	14	96
464	Cement	0	0	18	0[a]	41
383	Aircraft	0	0	52	2	185
321	Aluminium	0	0	16	10[a]	100[a]
311–13	Iron and steel	0	0	68	9	630
101	Coal	0	0	23	0	655
601	Gas	0	0	15	0	216
500, 336 337	Construction, earth-moving equipment and contractor's plant	12	4	33	53	1,931

[a] Estimated.

Source: Freeman (1971).

With this exception it seems to be true that in the capital intensive industries both process and product innovations have been mainly monopolized by large firms. This finding corresponds closely with the conclusions which emerge from Part One. The small firms made their contribution mainly in the field of machinery and instrument innovations, where both capital intensity and development costs are low for many products, and entry costs are low for new firms. This again corresponds closely to the historical account given in Part One. Machinery, instruments and electronics accounted for two-thirds of all the small firms' innovations reported. Although small firms made a significant contribution to innovation in such traditional industries as textiles, leather and furniture, the total number of innovations in these industries was relatively small. An important conclusion from this inquiry is that Jewkes was right in believing that the growth of professional R&D in the large corporation has not eliminated the contribution of small firms to industrial invention and innovation. But Galbraith was right, too, in believing that the larger corporation predominated in contemporary industrial innovation.

Since 1970 the increased share of small firms in total innovation in the UK was primarily in those industries where their contribution was already strong, especially electronics, instruments and computers. Research in the United States showed broadly similar results.

Whereas recent empirical research has demonstrated an increased share of total innovations coming from small firms in the 1970s and 1980s, history matters here too. In the past in the early stages of major new technologies, small firms make a disproportionately large contribution, but as a technology matures a process of concentration tends to take place, dominant designs emerge and lock-in may prevail. (See Figure 6.1 and other work of Utterback, 1993). Acquisitions and mergers are often motivated by the desire to spread the growing R&D costs and to gain control of R&D in competitive firms. There is already in the 1990s evidence of a renewed process of concentration both in ICT industries and in biotechnology, where the acquisition of NTBFs by large chemical and drug firms has been especially notable. Kaplinsky (1983) gave a good example of this process of 're-concentration' in the case of computer-aided design (CAD) firms. One other reservation should be made: the contribution of small firms may be exaggerated in some reports by a failure to distinguish clearly enough between small subsidiaries of large multinational firms or establishments and subsidiaries of domestic firms and the truly independent small firm.

But when all is said and done, there is still impressive support for the view that small firm innovations have genuinely increased their share of the total in the final quarter of the twentieth century. Whereas in previous waves of technical change, such an increase has not been sustained and the long-term trend towards concentration has set in again, the situation is somewhat different now, because of the trend towards networking identified in Chapter 8 and the growth in complexity of technology. These trends may be reinforced by the preference of many engineers and scientists to work in smaller and more intimate organizations. An important finding of Autio (1994) in his research on NTBFs in the Cambridge (Mass.) area, Cambridge (England) and Helsinki area was that the majority did not want to become larger or less specialized. The twenty-first century therefore may see a new form of symbiosis between large and small firms, rather than the concentration trend which we have documented in Part One for the nineteenth and twentieth centuries. This will be an important research topic in the twenty-first century.

9.5 CONCLUSIONS

This chapter has shown that statistical generalizations about size of firm, scale of R&D, inventive output and innovation need to be heavily qualified. Industry matters, technology matters and history matters. The role of small firms in the early stages of the evolution of an industry or a technology can be very different from the later stages. Simplistic generalizations about lower or higher concentration leading to better innovative performance cannot be sustained. True enough that competitive pressures may induce more innovation but these pressures can be as strong in a highly concentrated international oligopoly as in a predominantly small firm industry.

This means that competition policy is not so simple as it at one time appeared. An OECD survey by Symeonidis (1996) gives an admirably succinct summary of the conclusions which emerge from his study of 'Schumpeterian hypotheses':

> The present literature survey suggests that there seems to be little empirical support for the view that large firm size or high concentration are factors generally conductive to a higher level of innovative activity. Of course, once it is recognised that all these variables are endogenously determined, the emphasis shifts from causality to mere correlation. Again there is no evidence of a general positive association between innovation and market structure or firm size, although there are circumstances where a positive association exists. This implies that there is no general trade-off between competition policy and technical progress, although in some R&D-intensive industries a high level of concentration may be inevitable. . . . The range of sustainable levels of concentration in any given industry will depend on a number of industry-specific factors. These include technological characteristics, such as technological opportunity, the average cost of an R&D project, the degree of continuity and predictability of technology and the extent of learning economies; demand characteristics, such as the degree of horizontal product differentiation; and aspects of strategic interaction such as the intensity of price competition.
>
> (Symeonidis, 1966, p. 33)

Finally, it is worth remarking that although much of the literature on this topic refers to the 'Schumpeterian' hypothesis on the supposed advantages of large size for innovation, Schumpeter himself did not formulate such a hypothesis in a clear or unambiguous way. It is true that he did refer (1942) in rather a provocative way to large firm advantages but these can be taken to refer mainly to the fact that only large firms could undertake some very complex types of product and process development. In his earlier work (1912) he spoke mainly of the advantages of inventor–entrepreneurs and small firms and we have ourselves (Chapter 1) indicated the differences between his earlier ('Mark 1') model and his later ('Mark 2') model. Although he certainly pointed to the concentration of R&D in large firms, he also indicated the dangers of bureaucracy in large organizations. Moreover, he did not have the advantages of being able to refer to the statistical evidence which only became available after his death. Consequently, the judgement which is often made that the Schumpeterian hypothesis has been refuted does not do justice either to the historical dimension of his work or to the complexity of the inter-industry and inter-technology differences of which he was aware.

NOTES

1. For definitions of R&D see OECD (1963a).
2. The whole problem of output measurement in R&D is dealt with more fully in UNESCO (1970), Irvine and Martin (1980, 1981), Martin and Irvine (1983), and Proceedings of OECD Conferences on Output Measurement (1980b, 1982); see also the special issue of *Research Policy* on output measurement, vol. 16, nos. 2–4 (1987).

CHAPTER 10

UNCERTAINTY, PROJECT EVALUATION AND INNOVATION

10.1 RISK AND UNCERTAINTY

The power of the giant corporation should not be exaggerated. From everyday observation as well as from Part One, and from the results of SAPPHO and other empirical studies, it is clear that many attempted innovations fail in large firms as well as in small firms. The assertions which are often made about the proportion which fail are rather unreliable for several reasons. Such generalizations are usually based on the experience of one firm or a few firms over a particular period. Moreover, they are usually vague about the criterion of failure. Thus the conventional wisdom of R&D management often refers to a success rate of one project in ten, or even one in a hundred. But everything here depends upon the stage at which such measurements are made. The higher figures often refer to the preliminary selection or screening process by which the less attractive R&D projects or proposals are weeded out before much money has been spent on them, and long before they reach the stage of commercial launch. Shelved research projects or development projects may be regarded as failed innovations in the early stages (Centre for the Study of Industrial Innovation, 1971) but the attrition rate is much higher in the R&D stage than after commercial launch. The SAPPHO project was concerned with attempts which reached this last stage.

Nevertheless, the failure rate is still high when it comes to this stage. This chapter discusses briefly the reasons for the high failure rate and some of the main difficulties confronting the firm in its project selection procedures. Finally, it concludes that it will be difficult to reduce this failure rate by better management of innovation or project selection and control techniques, except for the adaptive and imitative type of project.

This conclusion might appear to be at variance with the findings of SAPPHO and other projects designed to increase our understanding of innovation management. But it is important to recognize fully the limitations of such findings. Even if they are broadly correct in their interpretation of the characteristic pattern of success and of failure this is very far from providing a recipe or formula which will ensure success.

Insofar as the technical and commercial success of other innovators may affect the outcome of each attempt, some failure rate is almost inevitable when there are parallel or competitive attempts. Fuller knowledge about the conditions of success may raise the general standard of management in all attempts but it will not eliminate the possibility of failure where winners and losers are part of the game.

An analogy may be made with the management of football teams. Managers of teams are generally aware of what is needed to win a match. They have a fairly good idea of the pattern of success. So usually have their opponents. But it is by no means so easy to translate the ideal into reality on the field of play for many reasons, including the behaviour of the competition. What can be recognized ex post cannot always be controlled or initiated ex ante. Many of the variables involved are in any case not easy to manipulate.

It is true, of course, that there are some market situations where it is quite possible for several innovators to be successful simultaneously or nearly so. The success of one player does not necessarily mean the failure of another; there are some races where all can have prizes and others which are one-horse races. But even in the case of a monopolist or a socialist system of innovation, failures would persist for three reasons: technical uncertainty, market uncertainty and general political and economic uncertainty, sometimes described as business uncertainty.

The last category applies to all decisions about the future and it is generally assumed that a suitable discount rate applied to estimated future income and expenditure is the appropriate way to handle it in project evaluation. However, it will have bigger implications for innovations than for other types of investment to the extent that innovation projects have a longer time scale before the potential benefits can be realized. These implications are discussed at the end of this chapter.

The other types of uncertainty are specific to the particular innovation project and cannot be discounted, eliminated or assessed as an insurable type of risk. It is true that technical uncertainty can be very much reduced in the experimental development and trial production stages and that is indeed one purpose of these activities. But the outcome of these stages cannot be known before their completion, otherwise the work is not experimental and the activity is not truly innovative. Moreover, even after successful prototype testing, pilot plant work, trial production and test marketing some technical uncertainty still remains in the early stages of the innovation. As we have seen, one of the characteristics of successful innovators is the effort to get rid of bugs in the development stage. But some usually remain even in well-managed innovations, and occasionally they lead to serious setbacks some time after commercial launch. Some very expensive and well-known examples were the Comet jet airliner and Du Pont's Corfam (see Chapter 5).

Technical uncertainty is not merely a matter of 'work' or 'not work', although this is, of course, decisive for success. Indeed, the problem is very rarely reduced to this simple level. Much more usually it is a question of degree, of standards of performance under various operating conditions and at what cost. The uncertainty lies in the extent to which the innovation will satisfy a variety of technical criteria without increased cost of development, production or operation. Several processes for producing indigo dyes synthetically were technically successful without being commercially viable (see Chapter 4). The same was true of many of the hundreds of early cotton-picking machines.

The risk attached to technical innovation differs from normal risks which are insurable. Most economists, following Knight (1965), distinguish

Table 10.1 Degree of uncertainty associated with various types of innovation

1	True uncertainty	Fundamental research
		Fundamental invention
2	Very high degree of uncertainty	Radical product innovations
		Radical process innovations outside firm
3	High degree of uncertainty	Major product innovations
		Radical process innovations in own establishment or system
4	Moderate uncertainty	New 'generations' of established products
5	Little uncertainty	Licensed innovation
		Imitation of product innovations
		Modification of products and processes
		Early adoption of established process
6	Very little uncertainty	New 'model'
		Product differentiation
		Agency for established product innovation
		Late adoption of established process innovation and franchised operations in own establishment
		Minor technical improvements

between measurable uncertainty or risk proper and unmeasurable uncertainty or true uncertainty (see also Shackle, 1955, 1961). Technical innovation is usually classified with the second category. By definition, innovations are not a homogeneous class of events, but some categories of innovation are recognizably less uncertain than others (Table 10.1), and less risky. As Knight recognized, the classification of risk and uncertainty is a matter of degree except in the extremes. Life and fire insurances and other repetitive, calculable risks are usually cited as instances of the first type of risk which can be dealt with in a fairly straightforward manner by the theory of statistical probability, but uncertainty enters in even here. The second type of risk will not normally be assumed by insurance companies or indeed by banks. Special forms of financial institution have, therefore, been developed to handle this kind of uncertainty involving specific judgement in each individual instance.

Even the lower levels of uncertainty illustrated in Table 10.1 are such that only a very small proportion of R&D is financed directly by the capital market. Internally generated cash flow predominates. Where the risk is not borne by the firm or those fairly familiar with the individual project, usually either some type of cost-plus R&D contract is needed, or outright government ownership and finance of the R&D facility.

Numerous attempts have been made to deal with the uncertainty inherent in innovation by substituting subjective probability or credibility estimates for the relatively objective data used in estimating life insurance tables and other insurable risks (for example, Allen, 1968, 1972; Beattie and Reader, 1971). These attempts raise complex philosophical issues which will not be discussed here, but the empirical evidence will be examined to see how firms actually approach innovation decision-making and how far they are capable of statistically based estimation procedures.

It will be argued that the nature of the uncertainty associated with innovation is such that most firms have a powerful incentive most of the time not to undertake the more radical type of product innovation and to concentrate their industrial R&D on defensive, imitative innovations, product differentiation and process innovation. This proposition will be argued on theoretical grounds and the supporting empirical evidence will then be discussed. The distinction between inhouse process innovation and open market product innovation is very important here. Product innovation involves both technical and market uncertainty. Process innovation may involve only technical uncertainty if it is for inhouse application, and, as Hollander (1965) has pointed out, this can be minimal for minor technical improvements.

The general prevalence of uncertainty in R&D decision-making and the development of innovation strategies is clearly evident from the purely theoretical arguments of Keynes and Knight. However, it is also abundantly confirmed from the empirical evidence on firm behaviour. A particularly interesting example of this evidence is the study by Augsdorfer (1996) on *Forbidden Fruit: An Analysis of Bootlegging, Uncertainty and Learning in Corporate R&D*. He investigated the corporate R&D of more than fifty firms in France, Germany and Britain and found the widespread existence of 'bootlegging', i.e. 'under-the-table' covert research projects conducted by R&D personnel. He refers to many authors in the 1980s and 1990s, such as Roberts (1991), who referred to the existence of 'bootlegging' but Augsdorfer was the first to demonstrate not only its very widespread occurrence but also its significance in terms of the prevalence of uncertainty and the consequent extreme difficulty of 'rational' central planning of R&D by management. The tacit acceptance and even encouragement of bootlegging by some managers is another very interesting phenomenon. The existence of bootlegging can sometimes lead to more radical projects being initiated by adventurous individual researchers without the knowledge of management (Augsdorfer, 1994). Pearson (1990) explicitly relates bootlegging to high uncertainty in both means and ends. Another brilliant confirmation of the prevalence of uncertainty in the inventive and innovative process was the empirical research of Scherer (1997) on High Technology start-up companies.

10.2 PROJECT ESTIMATION TECHNIQUES AND THEIR RELIABILITY

Let us now consider the problems confronting the decision-makers in deciding whether to embark on an innovation project in the firm. Basically, they will be concerned, whatever particular selection technique they may favour or even if they operate purely on hunch, to make some estimate of three parameters:

1. The probable costs of development, production, launch and use of marketing of the innovation and the approximate timing of these expenditures.
2. The probable future income stream arising from the sale or use or the innovation and its timing.
3. The probability of success, technically and commercially.

Ideally, the decision-maker would like a complete cash-flow diagram of the future expenditures and income associated with the innovation

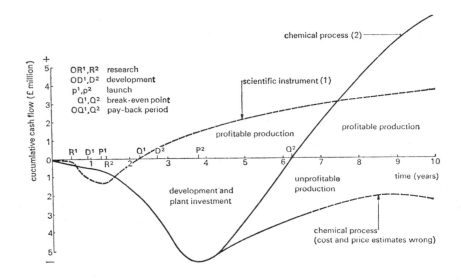

Fig. 10.1 Cumulative cash flow diagram

(Fig. 10.1). Some estimate of the development costs and subsequent launch costs is clearly essential to any kind of assessment of likely profitability, but as we know all too well from the publicized experience of aircraft development, these estimates can often be wildly wrong. Some improvements in estimating techniques can be made and a great deal of effort has gone into this, both in military and civil projects. But it must never be forgotten that estimates can only be really accurate if uncertainty is reduced, and uncertainty can only be significantly reduced either by further research or by making a project less innovative. Those firms who speak of keeping development cost estimating errors within a band of plus or minus 20 per cent are usually referring to a type of project in which technical uncertainty is minimal, for example, adapting electronic circuit designs to novel applications, but well within the boundaries of established technology, or minor modifications of existing designs (categories 5 and 6 in Table 10.1).

This conclusion was strongly confirmed by the empirical work which has been done on project estimating errors at the Rand Corporation (Marschak et al., 1967; Marshall and Meckling, 1962; Mansfield et al., 1971, 1977; Allen and Norris, 1970; Norris, 1971; Keck, 1977, 1970, 1982; Kay, 1979). Mansfield's work is particularly valuable because it permits comparison of different types of innovation. Although it must be conceded that it is very hard to measure just how radical any innovation is, it is nevertheless certain that new chemical entities constitute a more radical class of innovations than alternative dosage forms, and likewise that new products are normally a more radical departure than improved products. The differences in average cost and time overruns, as well as in the variance of estimates, are very striking (Table 10.2).

Table 10.2 Average and standard deviation of ratio of actual to estimated cost by project type and relative size of technical advance, 69 technically completed projects, US Proprietary Drug Laboratory

Project type	Size of technical advance		
	Small	Large and medium	Total
Product improvement			
Average	1.39	1.49	1.41
Standard deviation	1.39	1.64	1.41
Number of projects	28.00	5.00	33.00
New products			
Average	2.21	5.46	2.75
Standard deviation	3.56	5.86	4.11
Number of projects	30.00	6.00	36.00
Actual to estimated time for above			
Product improvement			
Average	2.80	1.74	2.64
Standard deviation	1.28	0.84	1.27
Number of projects	28.00	5.00	33.00
New products			
Average	3.14	3.70	3.24
Standard deviation	1.53	2.19	1.80
Number of projects	30.00	6.00	36.00

Source: Mansfield *et al.* (1971, pp. 102 and 104).

Mansfield's work is also extremely important because it confirms that large errors are not confined to the military sector or the aircraft industry. Moreover, his work on the chemical industry shows that estimating errors cannot be attributed to inexperience, as the firms which he investigated had long experience of project estimation and innovation, and were among the leading R&D performers in the US industry. The results do, however, suggest that there is some trade-off between cost and time, as the average overrun in military projects was much greater with respect to cost than time, whereas the opposite was true of civil projects, both in the USA and the UK. The work of Allen and Norris also suggests that time overruns were greater in research than in development. This trade-off was explored in some depth by Peck and Scherer (1962) in their work on the weapons acquisition process.

In addition to the very large errors involved, the tendency to optimistic bias is notable. This bias is present in other types of investment forecast, but not in such an extreme form. It suggests strongly that the social context of project estimation is a process of political advocacy and clash of interest groups rather than sober assessment of measurable probabilities. This view is confirmed both by historical accounts of individual innovation decision processes and by what little academic research has been done on the subject. Particularly important here was the work of Howard Thomas (1970) on project estimation in two scientific instrument firms. Not only did he find that engineers deliberately made very conservative estimates of development costs, but he also found that they did this in spite of strong financial incentives (including profit-sharing arrangements) to make 'honest' estimates:

> Many engineers in the firm admit quite freely that their estimates of cost and sales volume for projects are often biased in such a way that the resulting return factor estimates appear favourable to the firm. They point out that the procedures themselves are very inaccurate and do not incorporate the technical feeling about a project that an engineer often has, but is not necessarily understood by a finance or marketing man. So the engineer deliberately amends estimates (the means by which evaluations are made) in order to make the return factors acceptable to the firm. They do not do this to make projects personally and technically attractive to them more acceptable to the firm, because they are aware that the firm's financial interests and theirs are in one-to-one correspondence given the profit-sharing and preferential share-purchase plans offered by the firm as part of the remuneration package. Their sole motivation is to make the firm move towards more flexible numerical criteria for differentiating between projects.
>
> (Thomas, 1970)

While it is true that empirical evidence on project estimation is still not as comprehensive as we might wish, it must be regarded as persuasive support for the hypothesis that wide margins of error (with an optimistic bias) are characteristic of the experimental development process. This in itself must make innovation hazardous at least in relation to that part of the decision-making which precedes prototype test and pilot plant work. The evidence on project estimation in former socialist countries also pointed to the same conclusions.

Errors in development cost estimation alone may, of course, be sufficient to bankrupt a firm, if these costs account for a large proportion of its available resources. This occurred in spectacular fashion with Rolls-Royce in the 1970s and several smaller scale examples are cited in Part One. But as we shall see, the market uncertainty is frequently far greater than the technical uncertainty.

Seiler (1965) found that research managements of 100 large US firms rated 'probability of technical success' and 'development costs' as easier to estimate accurately than either 'probability of market success' or 'revenue from sales of product' (Table 10.3).[1] It is easy to think of several reasons why this should be so:

1. The market launch and growth of sales is more distant in time and may be spread over twenty years. A great many things can change during this time. This is partly a question of general business uncertainty relating to the future, but it is also specific to the project so far as it affects forecasts of consumer behaviour.

2. Whereas the development work is largely or entirely under the firm's own control, this is hardly ever true of the market, particularly in a capitalist economy. Economic theory is not capable of predicting the reactions of oligopolistic competitors in the face of innovation by one member of an oligopoly. Nor can the reactions of future customers or the trends of future legislation in relation to new products be safely predicted.

Table 10.3 Percentage of research managements of US firms rating accuracy with which factors affecting R&D can be estimated

	1	2	3	4	5	4+5
Factor	Excellent	Good	Fair	Poor	Totally unreliable	
Cost of research	3.5	27.8	52.2	14.8	1.7	16.5
Cost of development	2.6	38.8	46.6	9.5	2.5	12.0
Probability of technical success	3.5	51.3	39.9	6.3	0.0	6.3
Time to complete research	0.9	18.6	50.4	24.8	5.3	30.1
Manpower to complete research	2.6	34.2	53.5	7.0	2.7	9.7
Probability of market success[a]	3.6	33.6	38.2	14.5	10.1	24.6
Time to complete development	1.8	34.5	41.8	17.3	4.6	21.9
Market life of product[a]	4.6	28.0	29.0	23.4	15.0	38.4
Revenue from sales of product[a]	5.3	36.0	28.9	27.2	2.6	29.8
Cost reduction if R&D succeeds	10.7	57.1	14.3	14.3	3.6	17.9

[a] Assuming success of R&D

Source: Seiler (1965, pp. 177–8).

3. The prediction of future sales revenue and possible profit depends not only on forecasting total quantity which can be sold, but also on forecasting future costs of production, price and price elasticity. This is a formidable undertaking for a product not previously used by consumers.
4. Technological obsolescence may kill a new product or process almost as soon as it has been launched.

The empirical evidence, although unsystematic, confirms what theory suggests. Early estimates of future markets have been wildly inaccurate. As suggested in Part One, the major civil innovations in the past 80 years have been in the electronic industry and in synthetic materials. Almost every major innovation in these two industries was hopelessly underestimated in its early stages, including polyethylene, PVC and synthetic rubber in the materials field, and the computer, the transistor, the robot and numerical control in electronics.

As has already been shown in detail in Chapter 7, one of the most interesting cases is the computer. The early estimates almost all assumed that the market would be confined to a few large-scale scientific and government users. Even firms like IBM, as Watson Junior has confirmed, had no inkling of the potential until several years after electronic computers were in use. Optimistic estimates made in 1955 put the total US computer stock at 4,000 in 1965. The actual figure turned out to be over 20,000. Similarly large errors were made in underestimating the future potential applications of numerically controlled machine tools and robots.

Both of these, together with the examples of polyethylene and PVC, are cases of gross underestimation of future market potential for radical innovations. But there are also examples of gross over-optimism, for example in relation to the fuel cell, the airship, Ardil (the synthetic fibre), various nuclear reactors and the IBM STRETCH computer.

It may be said that the forecasting techniques in use before 1980 when many of these estimates were made, were still very primitive and that there are now much more sophisticated techniques which will reduce these errors. This remains to be seen, but the portents are not encouraging. There can have been few cases where more effort and expertise were devoted to market and cost estimation than in the case of Corfam. A computer model of the world market for hides, leather and shoes was developed, and a prolonged programme of manufacturer and customer trials with Corfam uppers. There are few firms with such an impressive record of product innovation as Du Pont, and they probably knew more about the shoe market than any firm in the industry by the time they launched Corfam. Yet they apparently lost about $100 million on this venture before they withdrew the product from the market.

If we consider the various innovations discussed in Chapters 2 to 7, it is difficult to think of any which worked out as originally expected. The gestation period was often far longer than the pioneers had anticipated (PVC, ammonia, TV, synthetic rubber, catalytic cracking, optical character recognition, indigo) and the development costs were frequently very much higher. The Concorde was probably the most spectacular example of gross underestimation of R&D costs and overestimation of the market in the 1960s and 1970s, with the result that, despite intense efforts by the British and French governments, they lost over £1,000 million and had to subsidize production, sales and airline operation as well as R&D. Even so, only a very small number of aircraft entered service.

10.3 ANIMAL SPIRITS AND PROJECT ESTIMATION

All of this is surprising only to those who believe that some new project evaluation technique or simulation technique would resolve the difficulties which are inherent in the very nature of innovation. Keynes was a great deal wiser. Although he was able to make a fortune on the stock exchange, and to write a treatise on the theory of probability as well as to revolutionize economic theory, he had no illusions about risky investments, whether speculative or innovative:

> Most, probably, of our decisions to do something positive, the full consequences of which will be drawn out over many days to come, can only be taken as a result of animal spirits – of a spontaneous urge to action rather than inaction, and not as the outcome of a weighted average of quantitative benefits multiplied by quantitative probabilities. Enterprise only pretends to itself to be mainly actuated by the statements in its own prospectus, however candid and sincere. Only a little more than an expedition to the South Pole, is it based on an exact calculation of benefits to come. Thus if the animal spirits are dimmed and the spontaneous optimism falters, leaving us to depend on nothing but a mathematical expectation, enterprise will fade and die – though fears of loss may have a basis no more reasonable than hopes of profit had before.
>
> It is safe to say that enterprise which depends on hopes stretching into the future benefits the community as a whole. But individual initiative will only be adequate when reasonable calculation is supplemented and supported by animal

spirits, so that the thought of ultimate loss which often overtakes pioneers, as experience undoubtedly tells us and them, is put aside as a healthy man puts aside the expectation of death.

(Keynes, 1936, pp. 161–2)

The uncertainty surrounding innovation means that among alternative investment possibilities innovation projects are unusually dependent on 'animal spirits'. But animal spirits are feared and distrusted by cautious decision-makers. As a standard textbook on investment decisions put it: 'Management will show a preference for projects with known outcomes over those whose outcomes are uncertain and the value of an investment proposal will be reduced according to its degree of uncertainty? (Townsend, 1969).

Often the substitution of subjectively estimated expected value for an objectively derived probability estimate rests on a false assumption which Keynes exposed, that 'an even chance of heaven or hell is precisely as much to be desired as the certain attainment of a state of mediocrity'. The risk (or uncertainty) aversion of entrepreneurs varies enormously of course, but we are on fairly safe ground in assuming that not many are prepared to gamble their survival on a fifty-fifty chance. Use of subjective probability estimates is usually acceptable only where possible outcomes are not extreme, and there is some repetition of previous experience.

This means that the acceptance of a high degree of uncertainty in innovation is likely to be confined to the following categories:

1. A few small-firm innovators who are ready to make a big gamble, or who are impelled to do so by some threat to their existence.
2. Large-firm innovators who use careful project selection methods but who can afford to adopt a portfolio approach to their R&D, offsetting a few very uncertain investments against a large number of mediocre projects. The size of the very uncertain investments will not usually be such that failure would threaten the continued existence of the firm.
3. Large-firm innovators who are not closely controlled by any formal project selection system and who are able to use corporate resources with a good deal of freedom, and hence impose their subjective estimates or preferences upon the organization.
4. Large- and small-firm innovators who unwittingly accept a very high degree of uncertainty, through 'animal spirits', because the enthusiasm of inventors, entrepreneurs, or product champions leads them on. In some cases (probably the majority) they may not bother to make any sophisticated calculations of the probable return on the investment. In others they may accept grossly over-optimistic subjective estimates of the probable outcome.
5. Government-sponsored innovators who accept very high risks because of urgent national needs (usually war, or threat of war) or a deliberate national science policy strategy, which creates an assured and profitable market in the event of success.
6. Government-sponsored innovators who accept grossly over-optimistic estimates of future returns for other reasons, where failure does not pose a serious threat to the decision-makers, as in the case of Concorde,

where diplomatic and prestige considerations were allowed for nearly twenty years to overrule commercial judgement and commonsense.
7. 'Bootlegging' individual researchers who initiate unofficial projects within a corporate R&D environment. Some of these may later become official projects.

The empirical evidence relating to the use of project selection techniques confirms that the more advanced portfolio methods which have been developed by statisticians and management consultants are seldom used. In the USA, Baker and Pound (1964) found that a few of the techniques had been used occasionally and then discarded in favour of simpler rule-of-thumb methods or discounted cash flow (DCF) calculations (see also Rubenstein, 1966). These methods are strongly biased towards short-term payback and the system in which they are used is frequently project based rather than portfolio based. Their widespread use probably discourages the more radical type of innovation, which would find more favour either in a fairly sophisticated selection system or without any very formal system. A survey in Sweden in 1971 confirmed that only simple quantitative methods were then used in Swedish industry and indicated some reasons for resistance to sophisticated techniques (Naslund and Sellstedt, 1972). Similarly, Olin's survey (1972) concluded that in the European chemical industry project selection remained a pragmatic and intuitive art.

A partial alternative to a quantitative cost benefit or DCF approach is to use a qualitative checklist method of evaluation. A checklist approach has the advantage of being able to take into account many factors which may be difficult to incorporate in a mathematical formula. For example, a critical factor in the success of any R&D project is the enthusiasm and capacity of the project leader and his or her other commitments. Another is the firm's resources of skilled people and accumulated know-how in the field and the possible spin-off from other R&D projects. A third may be the firm's relationship with potential customers and so forth. While all these factors may be taken into account by a research manager or an entrepreneur in calculating probability factors for technical or commercial success, a checklist procedure has the merit of compelling fairly systematic attention to be paid to each point. The actual checklist may be varied in accordance with the peculiar circumstances and characteristics of the firm (or other innovating organization), but the kind of questions that would tend to appear in most checklists would include the following (Dean, 1968; Seiler, 1965):

1. Compatibility with company objectives.
2. Compatibility with other long-term plans.
3. Availability of scientific skills in R&D.
4. Critical technical problems likely to arise.
5. Balance of R&D programme.
6. Interaction with other R&D projects.
7. Competitors' R&D programmes.
8. Size of potential market.
9. Factors affecting expansion of the market.
10. Influence of government regulations and control.

11. Export potential.
12. Probable reaction of competitors.
13. Possibility of licensing and know-how agreements.
14. Possibility of R&D co-operation with consultants or other organizations.
15. Effect on sales of other products.
16. Availability and price of materials needed.
17. Possibilities of spin-off exploitation of innovation.
18. Availability of production skills and equipment.
19. Availability of marketing skills and experience.
20. Advertising requirements.
21. Technical sales and service provision.
22. Effects on company image.
23. Risks to health or life.
24. Probable development, production and marketing costs.
25. Possibility of patent protection.
26. Scale and timing of necessary investment.
27. Location of new or extended plant(s).
28. Attitude of key R&D personnel.
29. Attitude of principal executives.
30. Attitude of production and marketing departments.
31. Attitude of trade unions.
32. Overall effect on company growth.

Some firms may also take into account additional external costs and benefits, such as problems of waste disposal or employment effects, retraining requirements and contributions to research outside the company. The consideration of some environmental effects has become far more widespread in the 1980s and 1990s and indeed has become a legal obligation in some countries (see Chapter 18). Firms may also find even more valuable those types of technique which attempt to foresee and avoid all the conceivable bugs and blockages which could frustrate the future progress of the project (Davies, 1972). Finally, still other types of checklist and portfolio approaches go even wider and relate the solution of the project to the overall strategy of the firm, both with respect to technology and other objectives (see, for example, Patterson, 1996; Kanz and Lam, 1996).

Although the checklist approach permits consideration of many factors which may be disregarded or overlooked in a quantitative analysis, it too has serious limitations. It does not permit easy comparison between alternative projects or ranking of a list of projects, nor does it provide any indication of the likely absolute size of the pay-off. Since most firms have a backlog of projects from which to choose, these are serious defects. Consequently, the ideal method of project selection is probably a combination of a quantitative cost–benefit approach with a qualitative checklist approach. Several such scoring systems have been developed. Figure 10.2 illustrates that originally worked out for project evaluation at Morganite by Hart (1966). Hart's system was based on calculating a project index value, which takes account of estimated peak sales value, net profit on sales, probability of R&D technical success and a time discount factor in relation to future R&D costs according to the formula:

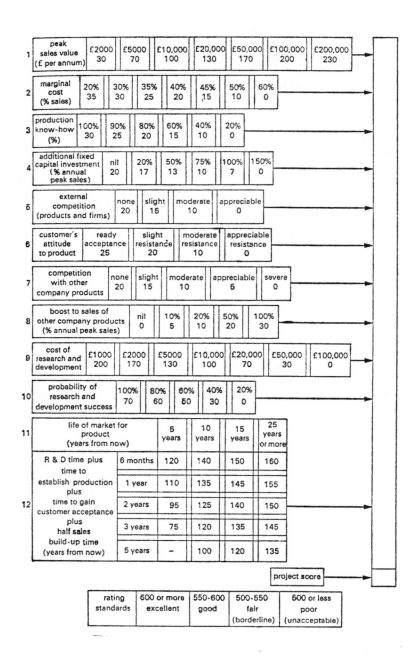

Fig. 10.2 Evaluation chart for product research and development projects
Source: Hart (1966).

$$I \text{ (index value)} = \frac{S \times P \times p \times t}{100\ C}$$

where S = peak sales value £ per annum,
P = net profit on sales (per cent),
p = probability of R&D success on a scale 0 to 1,
t = a time discount factor,
C = future cost of R&D (£).

Estimates for the variables are obtained by answering a checklist of questions. Figure 10.2 shows a score for each possible answer to the question and permits scoring by addition rather than multiplication by using logarithmic functions of the answers. The method can be varied by using different questions and different scores to suit the circumstances of a particular firm.

The advantage of this method is that it permits the firm to take into account such factors as external competition and customer attitudes, yet at the same time ranks projects on some systematic basis, on the assumption of sales growth and high profitability as major company objectives. Another advantage of the technique is that it can be used to involve all departments of the firm in discussion and evaluation, thereby contributing to mobilization and integration of the firm's resources. This is probably the main benefit of any formal evaluation technique. Indeed, some type of formal technique is usually necessary simply to monitor and control the progress of a project. The periodic revision of estimates and reconsideration of projects is essential for effective management of technical innovations.

However, it is apparent that one of the major factors affecting the selection of a project is the balance of work in the R&D department and in the firm as a whole. Consequently, project selection must be related to programming. What management is looking for is a portfolio of projects rather than a series of separate projects. By thinking in terms of a portfolio rather than a project it is possible to select a blend of safe and high risk projects, so that the more long-term and radical advances are not ignored as they would tend to be if selection were based entirely on a scoring system or rate of return system (Kay, 1979, 1982, 1984).

The empirical evidence confirms that industrial R&D is heavily concentrated on the less uncertain types of project. Only a few firms perform any basic research and this accounts for less than 5 per cent of all industrial R&D expenditures in most OECD countries. Several surveys have confirmed that the bulk of R&D expenditures are devoted to minor improvements and quick payback projects, rather than the more long-term radical innovations (FBI, 1961; Schott, 1975; Kay, 1979, 1984; Nelson et al., 1967). Table 10.4 shows the results of an Italian industrial R&D survey based on data supplied by individual researchers. This confirms the involvement in technical service activity and the relatively low proportion of time spent on fundamental research. It must be remembered that in terms of expenditure (as opposed to time of researchers) the share of experimental development would be higher and that of research lower.

Although not directly related to the distribution of R&D expenditures, the study of Gibbons and Johnston (1972) provided additional interesting

Table 10.4 Distribution of the working time devoted by researchers to various activities, broken down by industry, type of researcher and level of responsibility

	Researchers no.	% of time devoted to							
		Individual R&D	Group R&D	Studying and monitoring	Teaching	Organization, administration	Technical services	Advertising, other functions	Other
Food and textiles	187	18	21	11	2	25	16	6	1
Metallurgy	264	11	29	15	2	11	26	7	0
Mechanical eng.	1,669	23	20	10	2	17	20	6	2
Electrical eng.	644	17	24	12	2	22	15	6	1
Electronics telecom.	3,056	20	27	12	4	15	16	6	1
Transport vehicles	1,209	22	23	11	3	11	19	8	2
Pharmaceuticals	2,139	22	28	19	3	13	10	3	0
Chemicals	2,328	17	29	14	2	15	14	7	1
Rubber and fibres	868	18	32	12	1	16	15	6	0
Research companies	1,083	22	33	15	3	10	10	5	0
Other manuf. sectors	1,564	16	22	14	2	8	25	9	4
Total	15,010	20	26	13	3	14	16	6	1
% researchers who perform the activity	—	72	83	81	18	60	55	42	6
Scientists	3,248	33	39	14	1	6	4	3	0
Engineers	11,762	16	23	13	3	16	20	7	2
Junior researchers	5,959	27	26	14	2	8	15	5	3
Project leaders	7,154	16	29	13	3	15	17	7	0
Directors	1,808	10	18	13	4	27	18	9	1

	Researchers no.	Distribution of R&D time (%)						Type of researcher		
		Funda-mental research	% of researchers who per-form the activity	Applied research	% of researchers who per-form the activity	Experi-mental development	% of researchers who per-form the activity	Scientists (%)	Engineers (%)	Average age
Food and textiles	187	9	53	34	94	57	94	0	2.4	40.4
Metallurgy	264	2	18	49	94	49	94	1.4	2.0	42.2
Mechanical eng.	1,669	10	41	56	100	35	83	6.9	6.0	37.0
Electrical eng.	644	8	28	55	88	37	67	4.1	6.7	31.8
Electronics telecom.	3,056	8	30	41	86	51	85	10.1	20.8	36.2
Transport vehicles	1,209	8	44	51	91	41	82	12.4	11.3	35.9
Pharmaceuticals	2,139	17	62	55	94	28	68	31.2	15.9	36.2
Chemicals	2,328	11	43	43	86	45	83	11.9	14.8	40.5
Rubber and fibres	868	10	39	51	93	39	80	6.0	3.9	38.3
Research companies	1,083	10	47	53	93	38	77	13.8	10.1	35.1
Other manuf. sectors	1,564	8	35	43	98	49	96	2.3	6.0	38.4
Total	15,010	11	43	49	91	41	79	100.0	100.0	37.4
% researchers who perform the activity	—							—	—	—
Scientists	3,248	16	57	69	98	15	58	100.0	—	35.4
Engineers	11,762	9	39	43	89	48	86	—	100.0	37.8
Junior researchers	5,959	12	46	49	88	38	76	61.5	33.7	34.1
Project leaders	7,154	9	40	48	93	43	82	34.2	51.4	38.1
Directors	1,808	11	45	50	93	39	83	4.3	14.9	43.4

Source: Sirilli (1982).

Table 10.5 Distribution of R&D activities in Italy

	Percentage
Expansion of range without technical modification	17
New application of existing product	2
Standard product with new specification appropriate to particular application (e.g. portable)	29
Conforming to new standard (e.g. metric)	2
Standard product made easier to use	6
Standard product, new marketing	2
Standard product, new design	2
Products developed outside UK	23
New products, involving technical change and developed by UK firms	18

evidence. They attempted to derive a random sample of innovations by listing all new product announcements which appeared in UK technical journals on a selected date in 1971. They found that when they examined the list of 1,317 products, after eliminating 258 duplications, 32 process or service innovations, and 16 of non-industrial origin, the remaining 1,000 new products broke down as shown in Table 10.5.

When they came to examine in greater detail a sample of the last 18 per cent, which were those relevant for their purpose, they found that half of this restricted sample could be described as modifications of existing products of the company.

There are grounds for believing that firms are more ready to attempt radical innovations in relation to their own processes than in relation to their products. This includes, of course, the adoption and modification of the product innovations of the capital goods industries. The market uncertainty is very much reduced with inhouse process innovations as the firm controls the application. For similar reasons, much more radical product innovations may be expected in response to an assured market (whether government or otherwise) than on a competitive market. This was clearly evident in the development of radar and synthetic materials as well as in military aircraft.

If the process which is developed cannot be used by the firm itself, then the uncertainty is, of course, greater. This difference accounts largely for the respective approach to process innovation of the chemical plant contractor, as compared with the chemical firms themselves. As we have seen in Chapter 4, the contractors, who cannot use the process themselves but have to face an extremely uncertain market, tend to concentrate on improvements in design and scaling up. They usually attempt completely new processes only in association with a chemical or oil firm. The chemical and oil firms, on the other hand, have a strong incentive not only to make improvements in their inhouse processes, but to explore radically new processes for use in their own establishments, hoping that they can keep the innovation sequence largely under their own control. The introduction of such a new process may be used to lower production costs and increase profitability by comparison with competitors, or to lower product prices and expand markets. In some cases introduction may be retarded to preserve the profitability of existing investment. If the process can ultimately be licensed to other firms, this may be regarded as an additional bonus, but the successful marketing (licensing) of the process is not usually essential, as it is in the case of the contractor-originated process. Indeed, there may be a deliberate preference for secrecy and not licensing.

Research in the 1980s and 1990s has demonstrated that potential appropriability of the revenues from a new product or process plays an important role in project evaluation but that this varies very considerably between industrial sectors. This emerged both from the systematic Yale survey of innovations (Levin, 1988; Levin *et al.*, 1987) and from the later even more comprehensive PACE Survey (Arundel, 1995) in several European countries.

10.4 R&D BUDGETING AND THE STRATEGY OF THE FIRM

By undertaking R&D work mainly with a relatively low degree of uncertainty (Table 10.1), the firm is in effect using its R&D budget as a form of insurance against the risks of technical change. Or, as Arrow (1962) puts it, 'the Corporation acts as its own insurance company'. Management often actually bases its R&D budget on a percentage of sales calculation. This insurance premium varies in different branches of industry depending upon the intensity of technological competition, but the level of expenditure is often fairly uniform among many firms in each branch. Although management cannot calculate accurately the return on any individual project or piece of R&D, it has learnt from experience and from observation of competitors that this 'normal' level of R&D spending will probably help it to survive and grow. However, there is room for a variety of alternative strategies and these are discussed more fully in the next chapter. Some firms may spend much more heavily on R&D than is usual for their industry branch and follow a high-risk offensive strategy. Others may try to get by with very little R&D, or none at all, relying on other sources of competitive advantage. Threshold factors complicate the budgeting problem still further, and the factors discussed in Chapter 7 illustrate some of the dangers of R&D budgeting by an industry average. Naslund and Sellstedt (1972) produced evidence to show that in Swedish industry many firms allocated funds to R&D on an ad hoc project basis, rather than as a stable regular budget. But Kay (1979) argues persuasively that these differences in behaviour may be largely a function of size of firm. The Swedish firms were mainly small ones but the larger European and American firms typically follow a more long-term budget strategy.

Between industries wide variations in research intensity continue to exist, and it may be postulated that they are attributable on the one hand to historical circumstances (new technological opportunities), and on the other to the varying pressures of competition. In an industry in which new processes or new generations of products emerge every ten years or so, a moderately high level of research intensity would be necessary to avoid obsolescence of the product range or excessive costs (drugs, instruments, machinery, vehicles). Although the individual firm in such an industry might increase profitability for a few years by cutting back on R&D, this would be at the expense of long-term profitability and survival. A very low level of R&D activity, or none at all, would be a viable strategy in those branches of industry where technological obsolescence is not a problem, or where changes in product range are mainly fashion based. An extra-ordinarily high level of R&D activity might be necessary for survival in industries such as aircraft and electronics where an artificial stimulus to obsolescence derives from military R&D and procurement as well as rapid changes in world science and technology. In the extreme case, if it works, it's obsolete. The fact that one-third of the net output of the aircraft industry in the United States and the United Kingdom was actually R&D for long periods can only be explained in these 'Alice in Wonderland' terms. Thus the main determinant of research intensity is the branch of industry, and the ranking of industries by intensity is similar in all industrialized countries. Nelson (1991) and Pavitt (1984) are among the authors who have most strongly emphasized the importance of this variation by industry.

The outcome of the individual projects with a high degree of technical and market uncertainty cannot be precisely foreseen, either by the firms or by anyone else. Otherwise they would make fortunes more easily and enjoy a high and relatively stable rate of growth. But within an industry branch it is much more likely that someone will succeed in making the big advances, even though we cannot predict exactly which firm. Thus the uncertainty attached to R&D would lead one to expect a stronger statistical association between R&D spending and the growth of an *industry* than between R&D spending and the growth of the individual *firm*. This is in fact what the empirical evidence does show.

The most research intensive industries are, by and large, those with the highest growth rate (Fig. 7.1), whereas industries with little R&D are on the whole relatively slow-growing or stagnant. But within a fast-growing, research intensive industry, such as electronics or pharmaceuticals, there is not such a strong association between high growth and research intensity by firm. Some of the empirical data suggest a weak correlation and some suggest none at all. This result could be expected not only on grounds of the high degree of uncertainty surrounding the outcome of expensive offensive projects in any individual firm, but also on grounds of externalities which continually arise in R&D. Many firms in an industry may benefit from the technical progress made in only one or two or in a different industry altogether. The whole electronics industry benefited from Bell's work on semiconductors, but only a small part of this benefit was recovered by Bell in the form of licence and know-how payments or indeed in sales. Whereas the patent system strengthens the possibility of appropriating the benefits of knowledge gained through inhouse R&D, it cannot and does not prevent the diffusion of this information through a variety of channels, particularly the movement of people as in the American semiconductor industry.

A much stronger association might be expected between firm growth and some combined measure of R&D with Scientific and Technological Services (STS). This would capture the important productivity growth attributable to minor productivity improvements of the type described by Hollander. The empirical results of Katz confirm this hypothesis and so too does some work on the distribution of qualified personnel and growth of firms.

Firms recognize the need to perform R&D in order to stay in business or retain their independence, but there is no recipe for successful innovation.

This is one of the main factors contributing to 'higgledy-piggledy' or 'Tolstoyan' patterns of growth. Among the characteristics of successful innovation are the capacity of the innovator to couple efficient R&D with knowledge of the market requirements. But this is more obvious ex post than ex ante. Burns and Stalker in a classic study (1961) showed the internal difficulties within the firm in achieving the necessary degree of integration of these functions. Since innovation is often a complex of events extending over several years, the coupling process is a continuous one and is liable to be severely strained by internal problems within the firm as well as by extraneous events. The process is one of groping, searching and experimenting and even the best laid plans may come to grief. A firm with an efficient R&D set-up is more likely to survive, but it is by no means sure to do so. Even its own innovations may increase the general instability and uncertainty, so that they will often be unwelcome within the firm itself. Project SAPPHO showed that typically there was opposition to an innovation within the firm both on commercial and technical grounds. A greater degree of opposition on commercial grounds was quite strongly associated with failure of the innovation.

The existence of conflict in relation to R&D decision-making and the uncertainty inherent in the process mean that selection and forecasting procedures are not always what they appear from formal descriptions of the methods. A great deal has been written about various modes of technological forecasting and their application in American industry (Jantsch, 1967). Such TF techniques can undoubtedly be very useful and Bright (1968) in particular demonstrated their value in company strategy in identifying new technological opportunities and threats. However, as with other management techniques, the reality differs from the impression given by enthusiastic advocates. In view of the importance of recognizing what really happens in industry, and distinguishing this from idealized abstract concepts, the conclusions of one survey are quoted at some length:

Since some companies could probably benefit from formal technological forecasting but do not practice it, we searched within the organization for factors that inhibit its use. We found these common management-oriented obstacles to the use of this technique:

1. *Failure to integrate technological forecasting into the organization's regular plans.* Whereas most managers support the viewpoint that the most critical factor in implementing any forecasting technique is its integration into a long-range planning program, including the selection of research projects and the allocation of resources consistent with overall corporate objective, this is most often not the case in practice.

More typical is the experience of the executive who was transferred into the advanced planning group of his company with the task of instituting a formal forecasting program to simplify the planning process. . . . There had been no attempt to apply forecasting to the technological future of the company's major product line, and hence his efforts had had no impact on planning.

In another company, one individual with a technical background developed an interest in sophisticated technological forecasting and, with the support of the corporate vice-president for research, had been developing descriptive reports for more than a year. In addition to preparing reports on techniques, he also addressed the R and D planning process, with special attention to the problems of integrating technological forecasting with planning. Yet we found no evidence that anyone was using these techniques for decision making contrary to reports by Jantsch on the same organization. The company's efforts represent the work of one man who had hopes for the future, but has met with little success to date in selling his ideas.

2. *Failure to objectively select research and development projects.* In most of the companies we studied, the planning and control of R and D expenditures appears haphazard at best. An objective, factual assessment of the economic benefits, direct and indirect, of R and D investments seems to be very much the exception. While part of this is a natural outgrowth of the inevitable uncertainty of the task and the necessary flexibility and informality that characterize most research activities, it also represents management's failure to deal adequately with the planning and control process. The R and D project-selection process observed was primarily one of 'advocacy', based on the personal interests of researchers, the pet projects of key administrators, and a variety of other criteria which could be at odds with the strategic interests of the company.

In one company major R and D decisions are determined by internal power dynamics, which has led to a considerable amount of 'hobby work', or unauthorized research on pet projects. In another, funds are allocated by function or discipline on the basis of advocacy and power, even though it is recognized that individual product allocations might provide a more solid basis for planning. Despite the need for an objective cost-justification of research projects, we found little incentive among R and D decision makers for either planning or forecasting of technology.

3. *Failure to understand the role of sophisticated management techniques.* A further aspect of managerial resistance to technological forecasting (and to other management techniques) results from a fear of the unknown, a concern that decision-making prerogatives are being pre-empted, and/or the fear that systematic decision-making techniques may uncover incorrect decisions made in the past. In addition, the adoption of sophisticated forecasting is likely to further complicate the planning task rather than simplify it.

4. *Failure of top management to support forecasting efforts.* The support of top management is a requisite for many major changes, but we found few top managers supporting technological forecasting, and none initiating it. Initiation generally came from one man with the right background, interest and motivation, but without the influence necessary to establish his technological forecasting ideas.

The staff-line barrier is another aspect of this problem, manifested by a corporate staff trying to sell the technique to a divisional planning group, which, in turn, is asked to sell it to the divisional management.

5. *Failure of divisional management to look far enough ahead.* A final management impediment to technological forecasting is the short time perspective of line decision makers in profit-controlled divisions. The pay-off from technological forecasting is often in the long run, and, as one director of technology planning noted, 'The big corporation has no memory for the long-term investment'.

(Dory and Lord, 1970; Roberts, 1968)

Many other empirical surveys could be cited to confirm these conclusions, notably Olin (1972). They support the general arguments of Nelson and Winter (1977), Dosi (1984, 1988), Downie (1958), Gold (1971, 1979) and

Marris (1964) on the theory of the firm. Downie was concerned to explain why it was that the process of concentration in more efficient firms did not proceed more rapidly and more 'rationally' since big inter-firm differences in efficiency were clearly apparent. His explanation was in terms of 'unexpected' success of innovations in firms which had fallen behind. Thus the 'innovation mechanism' offset the efficiency 'transfer mechanism', constantly changing the relative position of competitive firms. Marris was arguing that growth maximization was a more realistic explanation of firm behaviour than profit maximization. He postulated, however, that such growth policies were subject to a profits constraint. Insofar as R&D is regarded mainly as a force contributing to growth and survival, its spread may be associated with the type of professional management attitudes which he identified as characteristic of the modern corporation, but in so far as they are unable to estimate likely profitability, it will remain higgledy-piggledy, with the profit/survival constraint sometimes weeding out the unlucky as well as the inefficient.

Moreover, the empirical evidence confirms that decision-making in relation to R&D projects or general strategy is usually a matter of controversy within the firm. The general uncertainty means that many different views may be held and the situation is typically one of advocacy and political debate in which project estimates are used by interest groups to buttress a particular point of view. Evaluation techniques and technological forecasting, like tribal war dances, play a very important part in mobilizing, energizing and organizing.

Although a defensive R&D strategy may be regarded as the typical response of the firm in research intensive branches of industry, it is by no means the only possible response. Particularly in countries where science is in any case underdeveloped, such a strategy may not be a realistic possibility. Even in the case of firms in advanced economies, managements of some firms may prefer a strategy of imitation, dependence or even suicide. These may be the only realistic alternatives where strong military or civil demands from government provide a powerful artificial stimulus in particular sectors of the world market, or where multinational corporations enjoy overwhelming advantages in scale of R&D, dynamic economies and market power.

The possibilities open to the firm will be considerably affected by national innovation policies. From the historical account in Part One it was evident that the growth of synthetic materials in Germany and of electronics in the USA were intimately related to government policies. By greatly reducing both the technical and the market uncertainty, governments provided a very powerful stimulus to industrial innovation. The profitability constraint (or profit maximizing behaviour for those who prefer this assumption) means that the time horizon of most firms in their decision-making is relatively short. This inevitably militates against long-term strategies, so that the advocates of long-term R&D policies in the firm will usually be at a disadvantage, unless they have external support of this kind. This is of great importance in considering such problems as pollution, energy conservation and resource depletion (see Part Four).

For this reason (the uncertainty and long-term nature of radical R&D), it must be expected that there will be a tendency in a private capitalist economy to underinvestment in long-term research and innovation, in spite of the potential advantages which the individual firm may gain. This underinvestment will be greatest in fundamental research and the more radical types of innovation (Nelson, 1959; Arrow, 1962). It is largely for this reason that in capitalist and the former socialist economies alike, governments finance most fundamental research and a certain amount of radical innovation (see Chapter 16). Conversely, there may well be overinvestment in short-term R&D associated with product differentiation and brand image. In conditions of oligopoly the firm strives to reduce market uncertainty by differentiating the market for its own products through a combination of advertising and minor technical changes. It does not necessarily follow, however, as Arrow and other economists have suggested, that a more perfect innovation system would separate R&D from the firm. The evidence from SAPPHO as well as most other innovation studies confirms the view that the coupling entrepreneurial function of the firm can be most efficiently performed if it is active in R&D itself. Successful innovation depends on combining technical with market knowledge. There is also the negative evidence that those socialist economies which initially separated industrial R&D from the enterprise were generally revising these policies in the 1970s in the direction of greater emphasis on enterprise-level R&D. The debates on the 'core competences' of firms (Teece, 1986; Prahalad and Hamil, 1990) generally led to the conclusion that there were major advantages for those firms which were able to integrate their manufacturing know-how with their R&D. Some economists explained this in terms of transaction cost theory (Williamson, 1975, 1985) but sociological explanations also carry conviction (Mowery 1983).

The overall picture which emerges from this survey of uncertainty and project selection in relation to innovation is rather more Tolstoyan than the neoclassical theory of the firm tends to assume. Most firms are unable to make very rational calculations about any one project, because of the uncertainty which is inherent in the process, because they lack the information necessary for rational behaviour and because they lack the time and the inclination to get it or to use very complex methods of assessment. This means that growth is higgledy-piggledy and that no one foresees very clearly the outcome of their own or their competitors' behaviour. If anyone doubts this let them consider the behaviour of the firms involved in the United States and European computer industries between 1950 and 1990, or in the radio industry between 1900 and 1930. Nevertheless, the social benefits and costs arising from this untidy innovative process can be very great. We now turn to a consideration of the various strategies which are open to the firm in the face of this degree of technical and market uncertainty, and in Part Four we consider the problem from the standpoint of national policy.

NOTE

1. One can only agree with Mansfield's ironic comments on the apparent optimistic self-deception of US research managements in believing that 'good' estimates could be made of many of these factors. The point here, however, is the relative accuracy.

CHAPTER 11

INNOVATION AND THE STRATEGY OF THE FIRM

11.1 THE RANGE OF INNOVATION STRATEGIES

Even though the survival and profitability constraints are obviously of the greatest importance in explaining firm behaviour, we conclude from Chapter 10 that rational profit maximizing behaviour (or growth maximizing) is seldom possible in the face of the uncertainties associated with individual innovation projects. This is not to deny that neoclassical short-run theory is a valuable, precise, abstract model of firm behaviour, but it means that this model has limited relevance, and that other ways of interpreting and understanding innovative behaviour are needed (Nelson and Winter, 1977, 1982; Dosi *et al.*, 1988). One possible approach to such a theory (and it is no more than a first approach) is to look at the various strategies open to a firm when confronted with technical change. Such an approach does not look to an equilibrium which is never attained, but does take into account the historical context of any industry in a particular country. This chapter classifies some possible strategies, and discusses them in relation to R&D, and other innovative activities of the firm.

Any classification of strategies by types is necessarily somewhat arbitrary and does violence to the infinite variety of circumstances in the real world. The use of such ideal types may nevertheless be useful for purposes of conceptualization, just as the use of the concepts of extrovert and introvert is useful in psychology. In practice there is an infinite gradation between types, and many individuals possess characteristics of both types. More-over, individuals (and firms) do not always behave true to type. Finally, people and firm strategies are always changing, so that generalizations which were true of a previous decade will not necessarily be true of the next. For example, information and communication technology is leading to many changes in the behaviour of firms, especially with respect to their external networks of relationships and their collaboration with other firms (Coombs *et al.*, 1996; Hagedoorn and Schakenraad, 1992).

Any firm operates within a spectrum of technological and market possibilities arising from the growth of world science and technology and the world market. These developments are largely independent of the individual firm and would mostly continue even if it ceased to exist. To survive and develop it must take into account these limitations and historical circumstances. To this extent its innovative activity is not free or arbitrary, but historically circumscribed. Its survival and growth depend upon its capacity to adapt to this rapidly changing external environment and to change it. Whereas traditional economic theory largely ignores the complication of world science and technology and looks to the market as the environment, changing technology is a critically important aspect of the environment for firms in most industries in most countries.

Within these limits, the firm has a range of options and alternative strategies. It can use its resources and scientific and technical skills in a variety of different combinations. It can give greater or lesser weight to short-term or long-term considerations. It can form alliances of various kinds. It can license innovations made elsewhere. It can attempt market and technological forecasting. It can attempt to develop a variety of new products and processes on its own. It can modify world science and technology to a small extent, but it cannot predict accurately the outcome of its own innovative efforts or those of its competitors, so that the hazards and risks which it faces if it attempts any major change in world technology are always present.

Yet not to innovate is to die. Some firms actually do elect to die.[1] Firms which fail to introduce new products or processes in the chemical, instruments or electronics industries cannot usually survive, because their competitors will pre-empt the market with product innovations, or manufacture standard products more cheaply with new processes. Consequently, if they wish to survive despite all their uncertainties about innovation, most firms are on an innovative treadmill. They may not wish to be offensive innovators, but they can often scarcely avoid being defensive or imitative innovators. Changes in technology and in the market and the advances of their competitors compel them to try and keep pace in one way or another. There are various alternative strategies which they may follow, depending upon their resources, their history, their management attitudes, and their luck (Table 11.1).

They differ from those which are normally considered in relation to the economist's model of perfect competition, since two of the assumptions of this model are perfect information and equal technology. Both of these assumptions are completely unrealistic in relation to most of the strategies we are considering, but they are perhaps relevant for the traditional strategy which may be followed by firms producing a standard homogeneous commodity under competitive conditions. Such firms can concentrate all their ingenuity on low-cost efficient production and can ignore other scientific and technical activities or treat them as exogenous to the firm. Some products are still produced under conditions which may sometimes approximate to traditional competitive assumptions but they are only at one end of the spectrum. The traditional strategy is essentially noninnovative, or insofar as it is innovative it is restricted to the adoption of process innovations, generated elsewhere but available equally to all firms in the industry. Agriculture, building and catering are examples of industries which in some respects approximate to these assumptions, although all three are now quite strongly affected by information technology and by organizational innovations.

We consider six alternative strategies, but they should be considered as a spectrum of possibilities, not as clearly definable pure forms. Although some firms recognizably follow one or other of these strategies, they may change from one strategy to another, and they may follow different strategies in different sectors of their business.

Table 11.1 Strategies of the firm

Strategy	Inhouse scientific and technical functions within the firm									
	Fundamental research	Applied research	Experimental development	Design engineering	Production engineering quality control	Technical services	Patents	Scientific and technical information	Education and training	Long-range forecasting and product planning
Offensive	4	5	5	5	4	5	5	4	5	5
Defensive	2	3	5	5	4	4	4	5	4	4
Imitative	1	2	3	4	5	3	2	5	3	3
Dependent	1	1	2	3	5	2	1	3	3	2
Traditional	1	1	1	1	5	1	1	1	1	1
Opportunist	1	1	1	1	1	2	1	5	1	5

Range 1–5 indicates weak (or non-existent) to very strong.

11.2   OFFENSIVE STRATEGY

An 'offensive' innovation strategy is one designed to achieve technical and market leadership by being ahead of competitors in the introduction of new products.[2] Since a great deal of world science and technology is accessible to other firms, such a strategy must either be based on a special relationship with part of the world science–technology system, or on strong independent R&D, or on very much quicker exploitation of new possibilities, or on some combination of these advantages. The special relationship may involve recruitment of key individuals, consultancy arrangements, contract research, good information systems, personal links, or a mixture of these. But in any case the technical and scientific information and knowledge for an innovation will rarely come from a single source or be available in a finished form. Consequently the firm's R&D department has a key role in an offensive strategy. It must itself generate that scientific and technical information and knowledge which is not available from outside and it must take the proposed innovation to the point at which normal production can be launched. A partial exception to this generalization is the new firm which is formed to exploit an innovation already wholly or largely developed elsewhere, as was the case with many scientific instrument innovations. The new small firm is a special category of offensive innovator. The remarks here apply primarily to already established firms, but we may recall the conclusion of Chapters 8 and 9 that the importance of the new small innovating firm is related to the reluctance and inability of many established firms to adopt an offensive strategy.

The firm pursuing an offensive strategy will normally be highly research intensive, since it will usually depend to a considerable extent on inhouse R&D. In the extreme case it may do nothing but R&D for some years. It will usually attach considerable importance to patent protection since it is aiming to be first or nearly first in the world, and hoping for substantial monopoly profits to cover the heavy R&D costs which it incurs and the failures which are inevitable. It must be prepared to take a very long-term view and high risks. Examples of such an offensive strategy which have been considered in Part One are RCA's development of television and colour television, Du Pont's development of nylon and Corfam, IG Farben's development of PVC, ICI's development of Terylene, Bell's development of semiconductors, Houdry's development of catalytic cracking, and the UK Atomic Energy Authority's development of various nuclear reactors. It took more than ten years from the commencement of research before most of these innovations showed any profit, and some never did so.

The extent to which an offensive strategy requires the pursuit of inhouse fundamental research is a matter partly of debate and partly of definition. From a narrow economic point of view it is fashionable to deride inhouse fundamental research, and to regard it as an expensive toy or a white elephant. Certainly it can be this, and the advice of many economists and management consultants to leave fundamental research to universities has a kernel of good sense, but it is too narrow. Certainly some of the most successful offensive innovations were partly based on inhouse fundamental research. Or at least the firms who were doing it described it as such, and it could legitimately be defined as research without a specific practical end in view (the definition of applied research). However, it was certainly not completely pure research in the academic sense of knowledge pursued without any regard to the possible applications. Perhaps the best description of it is oriented fundamental research or background fundamental research. A strong case can be made for doing this type of research as part of an offensive strategy (or even in some cases as part of a defensive strategy).

The straightforward economic argument against inhouse fundamental research holds that no firm can possibly do more than a small fraction of the fundamental research which is relevant, and that in any case the firm can get access to the results of fundamental research performed elsewhere. This oversimplified 'economy' argument breaks down because of its failure to understand the nature of information processing in research, and the peculiar nature of the interface between science and technology. There is no direct correspondence between changes in science and changes in technology. Their interaction is extremely complex and resembles more a process of mutual scanning of old and new knowledge. The argument that 'anyone can read the published results of fundamental scientific research' is only a half-truth. A number of empirical studies which have been made in the Unites States indicate that access to the results of fundamental research is partly related to the degree of participation (Price and Bass, 1969; Steinmueller, 1994). In trying to answer the question: 'Why do firms do basic research with their own money?', Rosenberg (1990) described basic research investment as a 'ticket of admission' to scientific and knowledge-building networks. Many case studies of innovation show that direct access to original research results was extremely important, although the mode of access varied considerably (Illinois Institute of Technology Research Institute, 1969; Langrish et al., 1972; Wilkins, 1967; Gibbons and Johnston, 1974; Industrial Research Institute Research Corporation, 1979; Hounshell and Smith, 1988). More systematic studies by Mansfield (1991) of 76 major American firms showed that most of them believed that a significant proportion of their new products and processes introduced between 1975 and 1985 could not have been developed without the results of fundamental university research in the fifteen years prior to the innovations.

Inhouse fundamental research was obviously important in some of the cases considered in Part One (e.g. nylon and polyethylene), and its role in relation to Bell's discovery and development of the transistor is discussed in a classic paper by Nelson (1962). It was also important in a significant proportion of the American case studies, for example in GE and Dow. The results of SAPPHO, although not strongly differentiating between success and failure on the basis of fundamental research performance, did suggest a marginal advantage to fundamental research performers (Science Policy Research Unit, 1972). It may sometimes be a matter of hair-splitting as to whether research is defined as background, oriented basic or applied research. It must always be remembered that all schemes of classification are to some extent arbitrary and artificial.

Price and Bass (1969) attempted to measure the relative importance of direct participation as one of the modes of access to original research. They classified 244 coupling events in 27 innovation case studies. A coupling event is one which links developments in basic science with technological

Table 11.2 Frequency of use of coupling method

Category of coupling	Suits and Bueche	Frey and Goldman	Tannenbaum (MAB)
Indirect[a]	8	5	25
Passive availability[b]	28	17	43
Direct participation[c]	38	18	40
'Gatekeeper'[d]	14	2	6
All 'coupling events'	88	42	114

[a] No direct dialogue between originators and users of new scientific knowledge.
[b] Scientists are open to approach but do not initiate a dialogue. Technologists request assistance.
[c] Includes inter-disciplinary teams, exchanges and consultants.
[d] Gifted individuals assigned the specific function of promoting communication between scientists and engineers.

Source: Price and Bass (1969).

advances. The results shown in Table 11.2 indicate that direct participation was involved in 40 per cent of the events, and passive availability of scientists outside firms was also very important. It is not unreasonable to postulate that here too the effectiveness of communication is to some extent a function of the degree of involvement in basic research.

Most of these studies relate to innovations made by firms which would probably be classified as offensive, and tend to confirm the view that inhouse oriented fundamental research combined with monitoring activities and consultancy are important modes of access to new knowledge for firms pursuing such a strategy. Price and Bass conclude that:

1. Although the discovery of new knowledge is not the typical *starting point* [our italics] for the innovative process, very frequently interaction with new knowledge or with persons actively engaged in scientific research is essential.
2. Innovation typically depends on information for which the requirement cannot be anticipated in definitive terms and therefore cannot be programmed in advance; instead key information is often provided through unrelated research. The process is facilitated by a great deal of freedom and flexibility in communication across organizational, geographical and disciplinary lines.
3. The function of basic research in the innovative process can often be described as meaningful dialogue between the scientific and technological communities. The entrepreneurs for the innovative process usually belong to the latter sector, while the persons intimately familiar with the necessary scientific understanding are often part of the former.

(Price and Bass, 1969, p. 804)

These findings are extremely important, because it has often been concluded from individual case studies that technical innovations bear no relation to basic research or the advance of scientific knowledge. The results of the US Department of Defense Project Hindsight (Sherwin and Isenson, 1966) and of the Manchester Queen's Award study (Langrish *et al.*, 1972) were often wrongly construed in this way, because they suggested that most of the new products were based on an 'old' science. Any major innovation will draw on a stock of knowledge, much of which is old in this sense. But the capacity to innovate successfully depends increasingly on the ability to draw upon this whole corpus of structured knowledge, old and new (Steinmueller, 1994).

The availability of external economies in the form of a highly developed scientific and technological infrastructure is consequently a critical element in innovative efficiency. Although these external economies are to some extent worldwide, and to this extent it makes sense to talk of a world stock or pool of knowledge, access to many parts of it is limited. Cultural, educational, political, national and proprietary commercial barriers prevent everyone from drawing freely on this stock as well as purely geographical factors. The ability to gain access to it is an important aspect of R&D management and bears a definite relationship to research performance and reputation. Pavitt's inter-country comparisons of innovative performance (1971, 1980) also bear out this conclusion and so too does the study by Gibbons and Johnston on the interaction of science and technology (1974).

We may conclude, therefore, both from the results of Price and Bass and other empirical studies and from our own survey, that the performance of fundamental research, while not essential to an offensive innovation strategy, is often a valuable means of access to new and old knowledge generated outside the firm, as well as a source of new ideas within the firm. While ultimately all firms may be able to use new scientific knowledge, the firm with an offensive strategy aims to get there many years sooner. Even if it does not conduct oriented fundamental research itself it will need to be able to communicate with those who do, whether by the performance of applied research, through consultants or through recruitment of young postgraduates or by other means. This has very important implications for training policy as well as for communications with the outside scientific and technological community.

However, although access to basic scientific knowledge may often be important, the most critical technological functions for the firm pursuing an offensive innovation strategy will be those centred on experimental development work. These will include design engineering on the one hand, and applied research on the other. A firm wishing to be ahead of the world in the introduction of a new product or process must have a very strong problem-solving capacity in designing, building and testing prototypes and pilot plants. Its heaviest expenditures are likely to be in these areas, and it will probably seek patent protection not only for its original breakthrough inventions but also for a variety of secondary and follow-up inventions. Since many new products are essentially engineering systems, a wide range of skills may be needed. Pilkington's were successful with the float glass process and IG Farben with PVC, largely because they had the scientific capacity to resolve the problems which cropped up in pilot plant work and could not be resolved by rule of thumb. The same is even more true of nuclear reactor development work.

There has been a great deal of confusion and misunderstanding over expenditure on R&D in relation to the total costs of innovation. It became fashionable to talk of R&D costs as a relatively insignificant part of the total costs of innovation – at most 10 per cent. This view is not supported by any empirical research and is based on a misreading of a US Department of

Commerce report frequently quoted and requoted. The small amount of empirical research which has been done on this question indicates that R&D costs typically account for about 50 per cent of the total costs of launching a new product in the electronics and chemical industries. As in so many aspects of industrial innovation it was Mansfield and his colleagues (1971, 1977) who got down to the hard task of systematic empirical observation and measurement, rather than plucking generalizations from the air. Their results were confirmed on a larger scale by the Canadian surveys of industrial R&D and by German work (OECD, 1982).

This is not to minimize the importance of production planning, tooling, market research, advertising and marketing. All of these functions must be efficiently performed by the innovating firm, but its most important distinguishing feature is likely to be its heavy commitment to applied research and experimental development. As we have seen, this was characteristic of IG Farben, Du Pont, GE, RCA, Bell and other offensive innovators. In the case of the new firm established to launch a new product, the inventor–entrepreneur is the living embodiment of this characteristic.

However, in order to succeed in its offensive strategy the firm will not only need to be good at R&D, it will also need to be able to educate both its customers and its own personnel. At a later stage these functions may be socialized as the new technology becomes generally established, but in the early stages (which may last for some decades) the innovating firm may have to bear the brunt of this educational and training effort. This may involve running courses, writing manuals and textbooks, producing films, providing technical assistance and advisory services and developing new instruments. Typical examples of this aspect of innovation are the Marconi school for wireless operators, the BASF agricultural advisory stations, the ICI technical services for polyethylene and other plastics, the IBM and ICL computer training and advisory services, UKAEA's work on isotopes, and technical education of the consortia and the CEGB. As we have seen, many observers (e.g. Brock, 1975) believed that the efficient provision of these services was the decisive advantage of IBM in the world computer market at the time of its greatest dominance.

The offensive innovator will need good scientists, technologists and technicians for all these functions as well as for production and marketing of the new product. This means that such firms are likely to be highly education intensive in the sense of having an above average ratio of scientifically trained people in relation to their total employment. The generation and processing of information occupy a high proportion of the labour force, but whereas for the traditional firm this would represent a top heavy and wasteful deployment of resources, these activities are the life-blood of the offensive innovating firm. It is the conversion of information into new knowledge of products and processes which is its most important feature.

11.3   DEFENSIVE INNOVATION STRATEGY

Only a small minority of firms in any country are willing to follow an offensive innovation strategy, and even these are seldom able to do so consistently over a long period. Their very success with original innovations may lead them into a position where they are essentially resting on their laurels and consolidating an established position. They will in any case often have products at various stages of the product cycle – some completely new, others just established and still others nearing obsolescence. The vast majority of firms, including some of those who have once been offensive innovators, will follow a different strategy: defensive, imitative, dependent, traditional, or opportunist. It must be emphasized again that these categories are not pure forms but shade into one another. The differences assume particular importance in relation to industry in the developing countries, but they are also important in Europe and the United States.

A defensive strategy does not imply absence of R&D. On the contrary a defensive policy may be just as research intensive as an offensive policy. (The example of Sloan's approach to R&D at GM described in Chapter 6 is a good illustration.) The difference lies in the nature and timing of innovations. The defensive innovators do not wish to be the first in the world, but neither do they wish to be left behind by the tide of technical change. They may not wish to incur the heavy risks of being the first to innovate and may imagine that they can profit from the mistakes of early innovators and from their opening up of the market. Alternatively, the defensive innovator may lack the capacity for the more original types of innovation, and in particular the links with fundamental research. Or they may have particular strength and skills in production engineering and in marketing. Most probably the reasons for a defensive strategy will be a mixture of these and similar factors. A defensive strategy may sometimes be involuntary in the sense that a would-be offensive innovator may be outpaced by a more successful offensive competitor.

Several surveys (Nelson et al., 1967; Schott, 1975, 1976; Sirilli, 1982) have shown that in all the leading countries, most industrial R&D is defensive or imitative in character and concerned mainly with minor improvements, modifications of existing products and processes, technical services and other work with short time horizons. Defensive R&D is probably typical of most oligopolistic markets and is closely linked to product differentiation. For the oligopolist, defensive R&D is a form of insurance enabling the firm to react and adapt to the technical changes introduced by competitors. Since defensive innovators do not wish to be left too far behind, they must be capable of moving rapidly once they decide that the time is ripe. If they wish to obtain or retain a significant share of the market they must design models at least as good as the early innovators and preferably incorporating some technical advances which differentiate their products, but at a lower cost. Consequently, experimental development and design are just as important for the defensive innovator as for the offensive innovator. Computer firms which continued to market valve designs long after the introduction of semiconductors could not survive. Chemical contractors which attempted to market a process which was technically obsolescent could not survive either. The defensive innovator must be capable at least of catching up with the game, if not of leap-frogging. This means that in industries such as semiconductors and software, all innovators must be extremely agile since the life of each new generation of components and products is so short (Hobday, 1994).

In an interesting study of the computer market, Hoffmann (1976) maintains that IBM has mainly followed a defensive innovation strategy, although with some offensive elements, while Sperry Rand (Univac) pursued a more consistently offensive strategy and Honeywell an imitative strategy. Since IBM spent far more on R&D than Sperry Rand in absolute terms, this illustrates the point that the defensive innovator may well commit greater scientific and technical resources than the offensive innovator. A certain amount of slack may be necessary in order to cover many new possibilities and to retain the flexibility needed to move very fast in catching up with the technical advances first introduced by competitors. However, even with heavy R&D spending a defensive strategy may lead to a firm being outflanked by more agile competitors. The classic case of an R&D-intensive firm being outmanoeuvred in this way was IBM's late development of the personal computer. In the end, they had to do this using external sources and an R&D team which stood outside their main R&D facilities (Chapter 7).

Patents may be extremely important for the defensive innovator but they assume a slightly different role. Whereas for the pioneer patents are often a critical method of protecting a technical lead and retaining a monopolistic position, for the defensive innovator they are a bargaining counter to weaken this monopoly. The defensive innovators will typically regard patents as a nuisance, but will claim that they have to get them to avoid being excluded from a new branch of technology. The offensive innovators will often regard them as a major source of licensing revenue, as well as protection for the price level needed to recoup R&D costs. They may fight major legal battles to establish and protect their patent position (RCA with television, ICI with polyethylene, La Roche with tranquillizers, EMI with their scanner, Telefunken with PAL), and typically their receipts from licensing and know-how deals will far exceed expenditure. (In 1971, ICI had receipts of £13 million and expenditure of £3 million.)

The defensive innovators will probably find it necessary to devote resources to the education and training of their customers as well as their own staff. They will also usually have to provide technical assistance and advice and these functions may be just as important for the defensive as for the offensive innovators. On the other hand, advertising and selling organizations, the traditional weapons of the oligopolist, will probably be more important, and to some extent technical services to customers will be bound up with this. The oligopolist may well attempt to use a combination of product differentiation and technical services to secure a market share not attainable by sheer originality (Brock, 1975; Hoffmann, 1976).

Both the offensive and the defensive innovator will be deeply concerned with long-range planning, whether or not they formalize this function within the firm. In many cases this may still often be the vision of the entrepreneur and his immediate associates, but increasingly this function, too, is becoming professionalized and specialized, so that product planning is a typical department for both offensive and defensive innovators. However, the more speculative type of 'technological forecasting' is more characteristic of the offensive innovator, and as we have seen in Chapter 10, still has considerable affinities to astrology or fortune-telling. It should probably still be regarded as a kind of sophisticated war dance to mobilize a faction in support of a particular project or strategy, but increasingly serious techniques have been developed (Bright, 1968; Beattie and Reader, 1971; Encel *et al.*, 1975; and Jones, 1981).

The defensive innovator, then, like the offensive innovator, will be a knowledge intensive firm, employing a high proportion of scientific and technical personnel. Scientific and technical information services will be particularly important, and so will speed in decision-making, since survival and growth will depend to a considerable extent on timing. The defensive innovators can wait until they see how the market is going to develop and what mistakes the pioneers make, but they dare not wait too long or they may miss the boat altogether, or slip into a position of complete dependence in which they have lost their freedom of manoeuvre. R&D will be geared to speed and efficiency in development and design work, once management decides to take the plunge. Such firms will sometimes describe their R&D as advanced development rather than research.

Most commonly, the large multi-product chemical or electrical firm will contain elements of both offensive and defensive strategies in its various product lines, but a defensive strategy is more characteristic of firms in the smaller industrialized countries, which cannot risk an offensive strategy or lack the scientific environment and the market.

The strategy which a firm is able or willing to pursue is strongly influenced by its national environment and government policy. Thus, for example, European firms since the war have generally been unable or unwilling to attempt offensive innovations in the semiconductor industry and their role has been almost entirely defensive (Chapter 7 and Hobday, 1991). French chemical firms have often followed a defensive strategy while German chemical firms have often been offensive. The complex interplay of national environment and firm strategy cannot be dealt with in detail here. But it is important to make the simple but fundamental point that many firms in the offensive group are United States firms, while most firms in the developing countries are imitative, dependent or traditional, with Europe in an intermediate position. In the eighteenth and early nineteenth centuries many British firms were following offensive innovation strategies, although without formal R&D departments. An over-simplified interpretation of Japanese experience since 1990 would be in terms of the movement of an increasing proportion of firms from traditional to imitative strategies, and then to defensive and offensive innovations. Japanese national policy has been designed to facilitate this progression. The extent of this shift is clearly observable in the statistics of the Japanese technological balance of payments since the war. In the early post-war period Japanese firms were spending far more on buying foreign licences and know-how than they were receiving from the sale of their own technology. At this time it was customary to regard the Japanese as 'superb imitators' and the long-term elements in their strategy were often overlooked. During the 1970s and 1980s on the new contracts which they signed, Japanese firms received more from the sale of their own technology than they paid out.

Hobday (1995) has shown in his research on the innovative strategies of East Asian firms in the 'Tigers' that there has been similar progression in their strategies from dependence, imitation and joint projects with foreign firms to increasingly independent innovations. National policies have tried to facilitate this (see Chapter 12). A technology policy of this sort involves a gradual change in the mix of STS in the direction of a more R&D-intensive mix. The type of R&D also changes from adaptive to increased originality, but it may require a long period in which most enterprises follow a dependent or imitative strategy, while slowly strengthening their technical resources, on the basis of a carefully conceived long-term national policy, involving protection of 'infant technology' as well as the build-up of a wide range of government-supported STS. The main elements of this long-term national strategy have been well described by Allen (1981) and are analysed further in Chapter 12. The precise balance of STS must vary with the size, resource endowment and historical background of each country. But in many developing countries STINFO (Science and Technical Information Services), survey organizations, standard institutes, technical assistance organizations and design-engineering consultancy organizations capable of impartial scrutiny and feasibility studies for projects involving imported technology are all of critical importance. They can provide the essential science and technology infrastructure which enables the STS at enterprise level to function effectively, despite the inevitable limitations in trained scientific and technical personnel. Only a few enterprises will gradually be able to develop first an adaptive and later an original innovative capacity. However, even in the United States the vast majority of firms are traditional, dependent or imitative in their strategies. We now turn to a consideration of these alternatives.

11.4 IMITATIVE AND DEPENDENT STRATEGIES

The defensive innovators do not normally aim to produce a carbon-copy imitation of the products introduced by early innovators. On the contrary, they hope to take advantage of early mistakes to improve upon the design, and they must have the technical strength to do so. At least they would like to differentiate their products by minor technical improvements. They will try to compete by establishing an independent patent position rather than simply by taking a licence, but if they do take a licence it will usually be with the aim of using it as a springboard to do better. However, their expenditure on acquisition of know-how and licences from other (offensive and defensive) firms may often exceed their income from licensing. For the imitative firm it will always do so.

The imitative firm does not aspire to 'leap-frogging' or even to keeping up with the game. It is content to follow some way behind the leaders in established technologies, often a long way behind. The extent of the lag will vary, depending upon the particular circumstances of the industry, the country and the firm. If the lag is long then it may be unnecessary to take a licence, but it still may be useful to buy know-how. If the lag is short, formal and deliberate licensing and know-how acquisition will often be necessary. The imitative firm may take out a few secondary patents but these will be a byproduct of its activity rather than a central part of its strategy. Similarly, the imitative firm may devote some resources to technical services and training but these will be far less important than for the innovating firms, as the imitators will rely on the pioneering work of others or on the socialization of these activities, through the national education system. An exception to this generalization might be in a completely new area (e.g. in a developing country) when neither imports nor the subsidiary of an innovating firm have opened up the market. The enterprising imitator may aspire to become a defensive innovator, especially in rapidly growing economies. This will mean an upgrading of STS and the strengthening or commencement of R&D activities, leading often to joint ventures or collaborative agreements with foreign or domestic firms. Examples of this are discussed in Part Three.

The imitator must enjoy certain advantages to enter the market in competition with the established innovating firms. These may vary from a captive market to decisive cost advantages. The captive market may be within the firm itself or its satellites. For example, a large user of synthetic rubber, such as a tyre company, may decide to go into production on its own account. Or it may be in a geographical area where the firm enjoys special advantages, varying from a politically privileged position to tariff protection. (This was the typical situation in many developing countries in the period of import substitution and still today in many cases). Alternatively or additionally, the imitator may enjoy advantages in lower labour costs, plant investment costs, energy supplies or material costs. The former are more important in electrical equipment, the latter in the chemical industry. Lower material costs may be the result of a natural advantage or of other activities (e.g. oil refineries in the plastics industry). Finally, imitators may enjoy advantages in managerial efficiency and in much lower overhead costs, arising from the fact that they do not need to spend heavily on R&D, patents, training, and technical services, which loom so large for the innovating firm. The extent to which imitators are able to erode the position of the early innovators through these advantages will depend upon the continuing pace of technological change. The early innovators will try to maintain a sufficient flow of improvements and new generations of equipment, so as to lose the imitators. But if the technology settles down, and the industry becomes mature, they are vulnerable and may have to innovate elsewhere. Du Pont's decision to move right out of the rayon industry despite their technical strength is a good example of strategic planning of this kind. Several other more recent cases in the chemical industry are discussed in Quintella (1993) in his book on strategic management. In some industries and technologies (but by no means all) the growth of an industry may be represented as following a cyclical pattern – a product cycle from 'birth' to 'maturity'. Sometimes the pace of technical change and the rapid succession of new generations of products may delay 'maturity', while in others apparently 'mature' industries may be rejuvenated. However, in those cases where it is a useful approximation to the growth pattern Hirsch (1965) has summarized the characteristics of the product cycle which may permit imitators to compete (Table 11.3 and Fig. 11.1). The extent to which they are actually able to do so, particularly in developing countries, is strongly influenced by institutional factors and government policies (see Chapters 12 and 15).

Unless the imitators enjoy significant market protection or privilege they must rely on lower unit costs of production to make headway. This will usually mean that in addition to lower overheads, they will also strive to be more efficient in the basic production process. They may attempt this by

Table 11.3 Characteristics of the product cycle

Characteristics		*Cycle phase*	
	Early	*Growth*	*Mature*
Technology	Short runs Rapidly changing techniques Dependence on external economies	Mass production methods gradually introduced Variations in techniques still frequent	Long runs and stable technology Few innovations of importance
Capital intensity	Low	High, due to high obsolescence rate	High, due to large quantity of specialized equipment
Industry structure	Entry is know-how determined Numerous firms providing specialized services	Growing number of firms Many casualties and mergers Growing vertical integration	Financial resources critical for entry Number of firms declining
Critical human inputs	Scientific and engineering	Management	Unskilled and semi-skilled labour
Demand structure	Sellers' market Performance and price of substitutes determine buyers' expectations	Individual producers face growing price elasticity Intra-industry competition reduces prices Product information spreading	Buyers' market Information easily available

Source: Hirsch (1965).

production factors	product cycle phase		
	new	growth	mature
management	2	3	1
scientific and engineering know-how	3	2	1
unskilled labour	1	2	3
external economies	3	2	1
capital	1	3[a]	3[a]

Fig. 11.1 The relative importance of various factors in different phases of the product cycle.

Note: The purpose of the blocks is simply to rank the importance of the different factors, at different stages of the product cycle. The relative areas of the rectangles are not intended to imply anything more precise than this.
[a] Considered to be of equal importance.

Source: Hirsch (1965).

process improvements, but both static and dynamic economies of scale will usually be operating to their competitive disadvantage, so that good adaptive R&D must be closely linked to manufacturing. Consequently, production engineering and design are two technical functions in which the imitators must be strong. Even if they are making carbon copies under licence, the imitators cannot afford to have high production costs unless they have high tariff protection. They will also wish to be well-informed about changes in production techniques and in the market, so that scientific and technical information services are another function which is essential for the imitator firm. The information function is also important for the selection of products to imitate and of firms from which to acquire know-how. It is clear that in all of this the would-be imitator in the typical developing country may be severely handicapped by local circumstances, unless national policies are carefully designed to facilitate technical progress.

If national policies and firm strategies are strongly oriented towards catch-up, for example, with respect to training, education, finance for investment and the import of technology, then it may be possible for latecomers to turn their lateness into a competitive advantage. Gerschenkron (1962) developed this theory of latecomer advantages mainly with respect to scale of plant in the steel industry in the nineteenth century. He pointed to the importance of financial institutions which could bear the costs of heavy investment and argued that given this 'social capability' latecomer firms could enjoy the advantages of the existence of an established world market and the availability of skills and technologies which could be imported quickly and more cheaply than those which were available to the early innovators. Jang-Sup Shin (1996) has extended Gerschenkron's analysis with the example of the South Korean steel industry and the South Korean semiconductor industry.

The examples cited in Chapter 2 on the British cotton industry show, however, that the British offensive innovators in the late eighteenth and early nineteenth century were able to reinforce their competitive advantages by economies of agglomeration and networking advantages, as well as by organizational and marketing innovations. In the end, they were overtaken by catch-up firms in latecomer countries, but it was a very extended process. The balance is a delicate and complicated one and the extent to which latecomer firms and countries can overcome their disadvantages by determined and intelligent catch-up strategies is one of the key questions discussed in Part Three, especially in the chapters on national systems of innovation (Chapter 12) and development (Chapter 15).

A dependent strategy involves the acceptance of an essentially satellite or subordinate role in relation to other stronger firms. The dependent firm does not attempt to initiate or even imitate technical changes in its product, except as a result of specific requests from its customers or its parent. It will usually rely on its customers to supply the technical specification for the new product, and technical advice in introducing it. Most large firms in industrialized countries have a number of such satellite firms around them supplying components, or doing contract fabrication and machining, or supplying a variety of services. The dependent firm is often a subcontractor or even a sub-subcontractor. Typically, it has lost all initiative in product design and has no R&D facilities. The small firms in capital intensive industries are often in this category and hence account for rather few innovations (see Chapter 9). For the special role of such firms in the Japanese economy, see Clark (1979); Sako, (1992); Womack et al., (1990).

The dependent small subcontract firm may, however, also seek to upgrade its technology and in some instances its major customers may help it to do so. In Chapter 9, we have seen that the most dynamic small firms are the so-called NTBFs (new technology based firms). These will often be offensive innovators in very specialized niche markets. However, small subcontract firms may also move from a dependent status to the category of innovative firms by the upgrading of their specialized knowledge in a narrow field. They may also lessen their dependence by enlarging their customer network once they strengthen their own innovative competence.

As we saw in Chapters 6 and 7, the Japanese automobile and electronic industries are often cited as examples of changing subcontractor relationships. The large assembly firms may extend technical assistance to their 'first-tier' suppliers, lend them engineers, and collaborate with them in upgrading the specifications of the components and materials which they supply. The same strategy is being increasingly imitated in Europe and North America and indeed some large firms in both retailing and manufacture already pursued it decades ago. Marks and Spencer is a good example of this. The case of Chrysler is another interesting example. For a long time this firm was distinctly the weakest of the three top US automobile firms and was actually saved from bankruptcy by the federal government. It deliberately attempted to emulate the Japanese strategy of working closely with suppliers in order to emerge from its difficulties. From 1989 to 1996 it reduced the number of suppliers from 2,500 to 1,140 but established a new relationship of collaboration in design and manufacturing with the remaining companies. Dyer (1996), who studied the supplier companies, described this transformation as the creation of an 'American *Keiretsu*' and showed that Chrysler's profit per vehicle leapt from an average of $250 in the 1980s to a record for all US automobile firms of $2,110 per vehicle in 1994. From being the least profitable, Chrysler became the most profitable of the US auto firms in the 1990s and the supplier firms were able to make numerous suggestions for improvements, many of which were implemented. Although the main benefits in this case appear to have gone to Chrysler, there were clearly also gains for suppliers and a changed form of dependence based more on mutual trust and co-operation.

The pure dependent firm is in effect a department or shop of a larger firm, and very often such firms are actually taken over. But it may suit the larger firm to maintain the client relationship, as subcontractors are a useful cushion to mitigate fluctuations in the work load of the main firm. In the 1980s and 1990s, there has been a fairly strong worldwide trend towards outsourcing by large firms of activities which were once performed inhouse. The dependent firm may also wish to retain its formal independence as the owners may hope they will ultimately be able to change their status by diversification or by enlarging their market. They may in any case prize even that limited degree of autonomy which they still enjoy as a satellite firm. In spite of their apparently weak bargaining position, they may enjoy good profits for considerable periods, because of low overheads, entrepreneurial skill, specialized craft knowledge or other peculiar local advantages. Even if they are 'squeezed' pretty hard by their customers, they may prefer to endure long periods of low profitability rather than be taken over completely. Although bankruptcies and take-overs may be common, there is also a stream of new entries.

11.5 TRADITIONAL AND OPPORTUNIST STRATEGIES

The 'dependent' firm differs from the 'traditional' in the nature of its product. The product supplied by the 'traditional' firm changes little, if at all. The product supplied by the 'dependent' firm may change quite a lot, but in response to an initiative and a specification from outside. The traditional firms sees no reason to change its product because the market does not demand a change, and the competition does not compel it to do so. Both lack the scientific and technical capacity to initiate product changes of a far-reaching character, but the traditional firm may be able to cope with design changes which are essentially fashion rather than technique. Sometimes indeed, this is its greatest strength.

Traditional firms may operate under severely competitive conditions approximating to the perfect competition model of economists, or they may operate under conditions of fragmented local monopoly based on poor communications, lack of a developed market economy, and pre-capitalist social systems. Their technology is often based on craft skills and their scientific inputs are minimal or non-existent. Demand for the products of such firms may often be very strong, to some extent just because of their traditional craft skills (handicrafts, restaurants and decorators). Such firms may have good survival power even in highly industrialized capitalist economies. But in many branches of industry they have proved vulnerable to exogenous technical change. Incapable of initiating technical innovation in their product line, or of defensive response to the technical changes introduced by others, they have been gradually driven out. These are the 'peasants' of industry.

An industrialized capitalist society includes some industries which are predominantly traditional, and others characterized by rapid technical innovation. It has been argued that an important feature of the twentieth century has been the growth of the research intensive sector. But it is a matter of conjecture and of policy as to how far this change may continue. It is a complex process, since sometimes the very success of a technical innovation may lead to standardized mass production of a new commodity with little further technical change or research for a long time. Usually, however, the industries generated by R&D have continued to perform it, so that the balance has gradually shifted towards a more research intensive economy, and a higher rate of technical change. It is the contention of this book that this has been one of the most important changes in twentieth-century industry, but it must be seen over a long time perspective.

This change is now extending to service industries and may prove to be an even more important structural change in the economies of the twenty-first century, both in the presently industrialized countries and in the developing countries. Mainly as a result of the pervasiveness of ICT, some service industries, such as the financial services, entertainment and infor-mation services are becoming both more capital intensive (through their heavy investment in computers and communication equipment) and more research intensive (through their employment of software and electronic engineers). The telecommunication industry which was already fairly R&D-intensive and capital intensive is now even more so and is rapidly extend-ing its linkages to the world of entertainment and multimedia services. Moreover, the old boundaries between manufacturing and services, which were already being bridged in the 1950s and 1960s are now often com-pletely eroded.

It was always hard to classify a firm like IBM, either to services or to manufacturing since it was extremely strong in both. As the share of software and consultancy in the total output of the computer industry has been rapidly increasing, IBM (and similar computer firms) has become more and more like a service firm, even though manufacturing remains a vitally important activity. The same tendency can be seen in the case of firms manufacturing telecommunication equipment: Ericsson, one of the most successful European firms in this industry, now employs fewer than 10 per cent of its employees in manufacturing. Most of its personnel are engaged in software, design of systems, R&D, worldwide consultancy and marketing, technical services, management and networking. It subcontracts a great deal of manufacturing. Smart (1996) has described how even a classical manufacturing firm like General Electric in the United States is moving into service areas as a matter of deliberate strategy. Manufacturing accounted for 56 per cent of total revenue in 1990 but this had declined to 44 per cent by 1995 and was projected to decline still further to 33 per cent by the year 2000. Financial services, medical services, consultancy and 'after-market services' were all expanding rapidly.

An even more extreme example is that of Benetton, which most people would think of as a firm belonging to the clothing industry. But, as Belussi (1993) and others have shown, Benetton actually has hardly any employees in manufacturing, except in experimental operations designed to keep abreast of new technological developments. Almost all manufacturing is subcontracted to a network of small firms in North-Eastern Italy, whereas Benetton itself concentrates on design and worldwide marketing through hundreds of franchised retail outlets all over the world. ICT is extremely important for Benetton since the daily sales data from these retail outlets are processed and co-ordinated by a computerized warehouse near Venice and form the basis for the manufacturing orders to the subcontractors.

All these examples are illustrative of the fundamental changes affecting the economy through the worldwide diffusion of ICT. Firms everywhere are being obliged to rethink their strategies as a result. Of course, this rethinking may not take the form of deliberate sophisticated new management strategies. Only in a minority of rather large firms will sophisticated management tools such as technological forecasting be deployed. The latest management fashions, such as re-engineering or technological audits, lean production or bench-marking will be more common. But in the vast majority, the rethink will take the form of chief executives, entrepreneurs and other managers responding to ideas or pressures from competitors, from the media, from suppliers and other external sources of information and knowledge, and adapting their own hunches or visions of the future accordingly. The shift in thinking related to the diffusion of the ICT techno-economic paradigm is everywhere apparent (see Chapter 17).

This shift has been less the result of any conscious central government strategy (although government policies have increasingly tended to favour this change) than the outcome of a long series of adaptive responses by firms to external pressures at home and abroad, and of attempts to realize the dreams of inventors. The efforts of firms to survive, to make profits and to grow have led them to adopt one or more of the strategies which have been discussed. But the variety of possible responses to changing circumstances is very great, and to allow for this element of variety one other category should be included, which may be described as an opportunist or niche strategy. There is always the possibility that entrepreneurs will identify some new opportunity in the rapidly changing market, which may not require any inhouse R&D, or complex design, but will enable them to prosper by finding an important niche, and providing a product or service which consumers need, but nobody else has thought to provide. Imaginative entrepreneurship is still such a scarce resource that it will constantly find new opportunities, which may bear little relation to R&D, even in research intensive industries.

Those firms which adopt a strategy of offensive or defensive innovation have gradually learned how to innovate. But there is no recipe which can ensure success and intense controversy still surrounds the important ingredients. The fact that they are often innovating on a world market increases the uncertainty which they confront, and has often led to the involvement of government to subsidize R&D, to create appropriate infrastructures and to diminish market uncertainty. Economic policy inevitably becomes enmeshed with policy for science and technology. These problems are particularly acute for the developing countries and are discussed further in Part Three.

11.6 CONCLUSIONS

In Part One it was argued from historical evidence that the professionalization of the R&D process was one of the most important social changes in twentieth-century industry. In Part Two it has been argued that the requirements of successful innovation and the emergence of an R&D establishment within industry have profoundly modified patterns of firm behaviour. This means that it is no longer satisfactory (if it ever was) to explain behaviour exclusively in terms of response to price signals in an external environment, and adjustment towards an equilibrium situation. World technology is just as much a part of the firm's environment as the world market, and the firm's adaptive responses to changes in technology cannot be reduced to predictable reactions to price changes. This makes things difficult for economists. It means that they must pay much more attention to engineers and to sociology, psychology and political science. Economists have an elegant theory which is confronted with a very untidy and messy reality. Their theory was and is an important contribution to the explanation and prediction of many aspects of firm behaviour, but it is not self-sufficient and attempts to make it so can only lead to sterility.

The chapters in Part Two offer little support to those theories of the firm which have postulated either perfect knowledge or optimizing behaviour with respect to the future. A more sophisticated modern defence of these theories is the 'as if' version. It is conceded that firms are incapable of making the type of calculations about the future, which are assumed in the neoclassical story, but it is argued that the outcome is nevertheless the same because competition ensures the survival and growth of those firms who have behaved 'as if' they could. However, as Hodgson (1992) and Winter (1986) have convincingly argued, this story is barely more credible than the original version. Neither biological evolution, nor the evolution of firms and industries, leads to optimality.

Far more plausible, in the light of the evidence in Parts One and Two, are those theories of the firm, such as those of Nelson and Winter (1974, 1982) and of Dosi *et al.* (1988) which take full account of bounded rationality, imperfect information, market and technical uncertainty. Moreover, any satisfactory theory of the firm must also take account of the variety of behaviour in different industrial sectors and over different historical periods.

The late Edith Penrose (1959) pointed economics in this new direction with her 'resource-based' theory of the firm, with various combinations of 'competences' and accumulated skills and knowledge. Most recent theorizing has followed Teece (1986) in his development of these concepts in relation to various functions within the firm, i.e. competence in R&D, manufacturing and marketing. At one extreme this can lead to the notion of the 'hollow corporation' which subcontracts all manufacturing. However, the examples which have been discussed above do not point unequivocally in this direction. Even Benetton recognized the importance of retaining some minimal competence in manufacturing, if only to check the work of subcontractors and to avoid being overtaken by sudden new developments in technology.

An interesting and original alternative to this type of thinking is that advanced by Christensen (1995). He proposes a conceptual distinction between four generic categories of assets for innovation: (1) scientific research assets; (2) process innovative assets; (3) product innovative application assets; (4) aesthetic design assets. The last of these is too often forgotten in theorizing about competence but is of the greatest importance in many industries and services. Whereas innovation may sometimes depend on only one or two of these four assets, more commonly a constellation of assets has to be mobilized. However, these assets may be located in a variety of organizational settings, which are moreover often regrouping. Christensen's theory indicates many exciting avenues of research both for economists and for sociologists and organization theorists.

The discussion in this chapter is not intended as an alternative theory of firm behaviour. Such a theory requires a greater integrative effort in the social sciences. But it is intended to indicate the kind of issues which must be embraced by any theory which seeks to explain the firm's innovative and adaptive responses to technological change, as well as to price changes in its factor inputs and the market for its products. There are encouraging indications that social scientists from several disciplines, including economists, are beginning to tackle the development of a more comprehensive and satisfactory theory of the firm (MacKenzie, 1990; Stirling, 1994).

NOTES

1. Metcalfe's study (1970) on Lancashire cotton firms showed that a large number were not willing to purchase a simple new piece of equipment (a size box), even though it cost less than £100, and the payback period was clearly demonstrated by the Research Association and the manufacturers to be less than one year. Mansfield et al.'s study (1972) of the adoption process of numerically controlled machine tools in the American tool-and-die industry similarly showed that many firms did not intend to adopt, 'even when firm owners granted that the lack of numerical control would soon be a major competitive disadvantage'. Mansfield estimated the median payback period in this case as five years and suggests that in many firms in this category the owners were close to retirement.
2. The new product may, of course, be a process for other firms.

PART TWO REVIEW ARTICLES, LITERATURE SURVEYS AND KEY REFERENCES

Afuah, A. (1977) *Innovation Management: Strategies, Implementation and Profits*, Ann Arbor, University of Michigan Press.

Archibugi, D., Cesaretto, S. and Sirilli, G. (1987), 'Innovative activity, R&D and patenting: the evidence of the survey on innovation diffusion in Italy', *Science Technology and Industry Review*, no. 2, pp. 135–50.

Barnett, C.K. (1997) 'Organisational learning theories: a review and synthesis of the literature', *Academy of Management Review*.

Carroll, G.R. and Teece, D.J. (eds) (1996) *Industrial and Corporate Change*, special issue on Firms, Markets and Organisations, vol. 5, no. 2, pp. 203–645.

Cohen, W. (1995) 'Empirical studies of innovative activity', in Stoneman, P. (ed.) *Handbook of the Economics of Innovation and Technological Change*, Oxford, Blackwell.

Cohen, W.M. and Levin, R.C. (1987) 'Empirical studies of innovation and market structure', in Schmalensee, R. and Willig, R. (eds) *Handbook of Industrial Organisation*, Amsterdam, North Holland.

Coombs, R., Saviotti, P. and Walsh, V. (eds) (1992) *Technological Change and Company Strategies*, London, Academic Press.

Coombs, R., Richards, A., Saviotti, P. and Walsh, V. (eds) (1996) *Technological Collaboration: The Dynamics of Industrial Innovation*, Cheltenham, Elgar.

Cyert, R.M. and March, J.G. (1963) *A Behavioural Theory of the Firm*, London, Prentice-Hall.

DeBresson, C. (1996) *Economic Interdependence and Innovative Activity*, Cheltenham, Elgar.

Dodgson, M. (1993) 'Organisational learning: a review of some literatures', *Organisation Studies*, vol. 14(3), pp. 375–94.

Dosi, G. (1988) 'Sources, procedures and micro-economic effects of innovation', *Journal of Economic Literature*, vol. 36, pp. 1126–71.

Duysters, G. (1995) *The Evolution of Complex Industrial Systems: The Dynamics of Major IT Sectors*, Maastricht, UPM.

Freeman, C. (1994) 'The economics of technical change: a critical survey', *Cambridge Journal of Economics*, vol. 18, pp. 463–514.

Gaynor, G.H. (ed.) (1996) *Handbook of Technology Management*, New York, McGraw Hill.

Gjerding, A.N. (1996) *Technical Innovation and Organisational Change*, Aalborg, Aalborg University Press.

Gold, B. (1979) *Productivity, Technology and Capital: Economic Analysis, Managerial Strategies and Government Policies*, Lexington, Lexington Books.

Granstrand, O. and Sjølander, S. (1992) 'Managing innovation in multi-technology corporations', *Research Policy*, vol. 19, no. 1, pp. 35–61.

Håkansson, H. and Snehota, I (eds) (1995) *Developing Relationships in Business Networks*, London, Routledge.

Jewkes, J., Sawers, D. and Stillerman, R. (1969) (2nd edn.) *The Sources of Invention*, London, Macmillan.

Kodama, F. (1996) *Emerging Patterns of Innovation: Sources of Japan's Technological Edge*, Cambridge, MA, Harvard Business School Press.

Leonard-Barton, D. (1995) *Wellsprings of Knowledge: Building and Sustaining the Sources of Innovation*, Cambridge, MA, Harvard Business School Press.

Loveridge, R. and Pitt, M. (eds) (1990) *The Strategic Management of Technological Innovation*, New York, Wiley.

Miyazaki, K. (1995) *Building Competencies in the Firm: Lessons from Japanese and European Opto-electronics*, Basingstoke, Macmillan.

Pavitt, K.L.R. (1990) 'What we know about the strategic management of technology', *California Management Review*, Spring, pp. 17–26.

Pearson, A.W., Stratford, M.J.W., Wadee, A. and Wilkinson, A. (1996) 'Decision support systems in R&D Project Management' in Gaynor, G.H. (ed.) *Handbook of Technology Management*, New York, McGraw Hill.

Penrose, E. (1959) *The Theory of the Growth of the Firm*, Oxford, Blackwell.

Quintella, R.H. (1993) *The Strategic Management of Technology*, London, Pinter.

Rothwell, R and Gardiner, P. (1988) 'Re-innovation and robust design: producer and user benefits', *Journal of Marketing Management*, vol. 3, no. 3, pp. 372–87.

Sakakura, S. and Kobayashi, M. (1991) 'R&D Project Management in Japanese Research Institutes', *Research Policy*, vol. 20, no. 6, pp. 531–59.

Scherer, F.M. (1992) *International High Technology Competition*, Cambridge, MA, Harvard University Press.

Scherer, F.M. (ed.) (1994) *Monopoly and Competition Policy*, Library of Critical Writings in Economics, Cheltenham, Elgar.

Steinmueller, E. (1994) 'Basic research and industrial innovation', in Dodgson, M. and Rothwell, R. (eds) *Handbook of Industrial Innovation*, Cheltenham, Elgar.

Swann, P. (ed.) (1993) *New Technologies and the Firm*, London, Routledge.

Symeonidis, G. (1996) *Innovation, Firm Size and Market Structure: Schumpeterian Hypotheses and Some New Themes*, Economics Department Working Paper no. 161, Paris, OECD.

Teece, D. (ed.) (1987) *The Competitive Challenge: Strategies for Industrial Innovation and Renewal*, Cambridge, MA, Ballinger.

Utterback, J.M. (1993) *Mastering the Dynamics of Innovation*, Boston, Harvard Business School Press.

von Hippel, E. (1988) *The Sources of Innovation*, Oxford, Oxford University Press.

Whiston, T.G. (1992) *Managerial and Organisational Integration: The Integrative Enterprise*, London, Springer.

Whittington, T. (1993) *What is Strategy and Why Does it Matter?*, London, Routledge.

Williamson, O.E. and Winter, S.G. (eds) (1993) *The Nature of the Firm: Origins, Evolution and Development*, New York, Oxford University Press.

Winter, S.G. (1988) 'On Coase, competence and the corporation', *Journal of Law, Economics and Organisation*, vol. 4, pp. 163–80.

PART THREE

MACRO-ECONOMICS OF INNOVATION: SCIENCE, TECHNOLOGY AND GROWTH AND GLOBALIZATION

INTRODUCTORY NOTE

The growth and development of nations has always been closely linked with access to and the effective exploitation of science and technology. As was illustrated in Part One, recent history has been dominated by the rise of science-related industry and the diffusion of many new products, typified by the growth of the chemical industry (Chapters 4 and 5) and development of automobiles (Chapter 6) and of electronics (Chapter 7). The broad consequence of such developments has been to underscore the dependence of advanced economies on the successful use of new technologies, and the extent to which this is partly dependent on indigenous scientific and technological capabilities.

In Part Two, using this historical evidence, we attempted to generalize about the conditions for success or failure in the innovative efforts of firms. We followed this with an analysis of the relationship between size of firm, R&D and innovation and then with a discussion of the ways in which firms attempt to evaluate their innovative projects. Finally, in Chapter 11 we discussed the strategies of firms confronting an uncertain market and changing technology. This leads in naturally to the discussion of the national and international environment in which they innovate.

We start this third part with a chapter setting out the many interactions between various institutions dealing with science and technology as well as with higher education, innovation and technology diffusion in the much broader sense. These national interactions, whether public or private, have become known in the literature as 'national systems of innovation'. As Chapter 12 illustrates, a clear understanding of such national systemic interactions provides an essential bridge when moving from the micro- to the macro-economics of innovation. It is also essential for comprehending fully the growth dynamics of science and technology and the particularly striking way in which such growth dynamics appear to differ across countries.

Understanding these growth dynamics was undoubtedly one of the main driving forces behind the political debate about 'technological gaps' between the United States and Europe in the early 1960s. The rapid growth of international trade, international investment and technological transfers between national economies showed how quickly industrial leadership could shift from one country or region to another, within the framework of an increasingly interdependent world economy. Japan and the Soviet Union, one inside and the other outside the mainstream of world economic competition, illustrated quite opposite sides of the technology/innovation/economy coin. The Japanese economy appeared particularly successful in bringing technology to the marketplace and in building up the scientific and the technological capabilities needed to sustain the process. In contrast, the largest parts of the scientific and technological efforts which the ex-Soviet Union spent over the same period – substantially more than Japan – were directed to space and defence, whereby the centrally planned, defence-oriented closed economic system provided little incentive to enterprises to diffuse new technologies out of these sectors or to innovate outside these sectors.

Already in the 1960s, the thesis that technological disparities and 'gaps' between countries were at the core of the different growth performance had become popular among economists and many European policy-makers. The recognition of the importance of the contribution of science and technology to economic growth was exemplified in the frantic search for explanations of the 'residual', that part of growth which could not be explained by labour and capital accumulation. As Chapter 13 illustrates growth theory dominated much of the economic literature in the 1960s. With the economic crises of the 1970s and the 'disappearance' of the 'residual', the subject slowly disappeared from the academic agenda of the economics profession. Alternative explanations more in the Schumpeterian tradition and later on in the 'new' growth theory tradition were formulated and growth theory became again in the late 1980s at the centre of economic theory and policy debate. Today, one is even witnessing an empirical revival of neoclassical growth theory; the residual having more or less 'disappeared' and growth becoming once again something which can supposedly be explained by and large in terms of labour and capital accumulation.

But while the search for explanations of growth seems to have come full circle over the last 40 years, the world has witnessed some fundamental changes with respect to the extent to which countries have relied upon foreign sources to realize their growth performance. The interaction between science and technology and international trade and investment has become increasingly recognized. One of the most crucial variables of direct influence on both technological and growth performance is of course international trade. The old parable with which we start Chapter 14 shows the need to break out of the traditional neoclassical, 'factor-endowment' explanations of trade flows and to bring in both older, more classical explanations as well as the 'new' trade explanations which have so much influenced the thinking of economists over the last two decades. But the international opening up of our economic systems does not of course end with exports and imports of goods and commodities.

Already before the First World War, the foreign investment of what later became known as multinational corporations, led to a considerable international diffusion of technological knowledge (see, for example, the case of electrification in Chapter 3). In the 1960s, multinational corporations primarily located in the USA and often competing in the most technologically advanced sectors of industry, developed new strategies for the international location of R&D capabilities, production capacities and market penetration. These strategies led to a more substantial international migration of production from the USA to Europe as the new technologies moved through the R&D, innovation and marketing phases of the product cycle. Such processes of internationalization have accelerated in the 1980s and 1990s and become much more widespread and diversified with the appearance of the so-called network corporation, a phenomenon more complex than the multinational with overseas subsidiaries, based as it is on inter-company alliances of technical, production, financial and marketing competences across national boundaries. More recently the term 'techno-globalism' has been launched to express the further trend towards the internationalization of technology based on 'global' or 'multi-domestic' firms' competitiveness rather than just on a nation's international competitiveness.

But as Chapter 15 illustrates, such processes have, at least up to now, not led to world convergence in growth and development performance at the world level. Underdevelopment, while less general and widespread than thirty years ago, is still one of the main challenges of our times. The promises of technology seem still not to have reached millions of people in Africa, Latin America and Asia. This is why we conclude Part Three with a return to the many issues of development and the potentiality of technology diffusion and indigenous technological development.

CHAPTER 12

NATIONAL SYSTEMS
OF INNOVATION

12.1 INTRODUCTION: FRIEDRICH LIST'S 'NATIONAL SYSTEM'

In Part One, our historical account demonstrated the special characteristics of the British economy, technology, culture and political system, which enabled that country to forge ahead of the rest of the world during the Industrial Revolution (Table 12.1). In Chapter 3 we pointed in a similar way to some special features of the United States national system which enabled that country to overtake Britain and in its turn to forge ahead of other countries (Table 12.2). In later chapters, we pointed to some special features of the German and Japanese economies which enabled them to achieve very high growth rates in the twentieth century. All of these illustrate the point that the national environment can have a considerable influence in stimulating, facilitating, hindering or preventing the innovative activities of firms. In this chapter we examine the concept of the 'national system of innovation' and explore in particular the contrasting systems respectively in the former Soviet Union and in Japan. We also analyse the contrast between countries in Latin America and East Asia in their efforts to catch up with the leading industrial countries. This account leads on to the more formal and generalized discussion of technology and economic growth in the following chapter.

As we have already seen in Part One, Friedrich List criticized the classical economists for giving insufficient attention to science, technology and skills in the growth of nations. His book on *The National System of Political Economy* (1841), might just as well have been called 'The National System of Innovation'. The main concern of List was with the problem of Germany overtaking England and for underdeveloped countries (as Germany then was in relation to England), he advocated not only protection of infant industries but a broad range of policies designed to accelerate or to make possible, industrialization and economic growth. Most of these policies were concerned with learning about new technology and applying it. He clearly anticipated many contemporary theories on 'national systems of innovation' (Lundvall, 1992; Nelson, 1993; Mjøset. 1992).

After reviewing the changing ideas of economists about development in the years since the Second World War, the World Bank (1991, pp. 33–5) concluded that it is intangible investment in knowledge accumulation, which is decisive rather than physical capital investment, as was at one time believed. The Report cited the 'new growth theory' (Romer, 1986; Grossman and Helpman, 1991) in support of this view but the so-called 'new' growth theory has in fact only belatedly incorporated into neo-classical models the realistic assumptions which had become commonplace

Table 12.1 Characteristics of British national system of innovation, eighteenth and nineteenth centuries

---

- Strong links between scientists and entrepreneurs.
- Science has become a national institution, encouraged by the state and popularized by local clubs.
- Strong local investment by landlords in transport infrastructure (canals and roads, later railways).
- Partnership form of organization enables inventors to raise capital and collaborate with entrepreneurs (e.g. Arkwright/Strutt).
- Profits from trade and services available through national and local capital markets to invest in factory production especially in textiles.
- Economic policy strongly influenced by classical economics and in the interests of industrialization.
- Strong efforts to protect national technology delay catching up by competitors.
- British productivity per person about twice as high as European average by 1850.
- Reduction or elimination of internal and external barriers to trade.
- Dissenters' academies and some universities provide science education. Mechanics trained in new industrial towns on part-time basis.

Table 12.2 Characteristics of US national system of innovation, late nineteenth and twentieth centuries

---

- No feudal barriers to trade and investment; slavery abolished 1865; capitalist ideology dominant.
- Railway infrastructure permits rapid growth of very large national market from 1860s onwards.
- Shortage of skilled labour induces development of machine intensive and capital intensive techniques (McCormick, Singer, Ford).
- Abundant national resources exploited with heavy investment and big scale economies (steel, copper, oil).
- Mass production and flow production as typical US techniques.
- Strong encouragement of technical education and science at federal and state level from 1776 onwards.
- US firms in capital intensive industries grow very large (GM, GE, SO, etc.) and start inhouse R&D.
- US productivity twice as high as Europe by 1914.
- Major import of technology and science through immigration from Europe.

---

among economic historians and neo-Schumpeterian economists (see Chapter 13). Indeed, it could just as well have cited Friedrich List (1841), who in criticizing a passage from Adam Smith said:

> In opposition to this reasoning, Adam Smith has merely taken the word *capital* in that sense in which it is necessarily taken by rentiers or merchants in their book-keeping and their balance sheets. . . . He has forgotten that he himself includes (in his definition of capital) the intellectual and bodily abilities of the producers under this term. He wrongly maintains that the revenues of the nation are dependent only on the sum of its material capital.
>
> (List, 1841, p. 183)

and further:

> The present state of the nations is the result of the accumulation of all dis-coveries, inventions, improvements, perfections and exertions of all generations

which have lived before us: they form the intellectual capital of the present
human race, and every separate nation is productive only in the proportion in
which it has known how to appropriate those attainments of former generations
and to increase them by its own acquirements.

(Ibid., p. 113)

List's clear recognition of the interdependence of tangible and intangible
investment has a decidedly modern ring. He argued too that industry
should be linked to the formal institutions of science and of education:

There scarcely exists a manufacturing business which has no relation to physics,
mechanics, chemistry, mathematics or to the art of design, etc. No progress, no
new discoveries and inventions can be made in these sciences by which a hundred
industries and processes could not be improved or altered. In the manufacturing
State, therefore, sciences and arts must necessarily become popular.

(Ibid., p. 162)

It was thanks to the advocacy of List and like-minded economists as well
as the long-established Prussian system, that Germany developed one of
the best technical education and training systems in the world. This system
not only was, according to many historians, (e.g. Landes, 1969; Barnett,
1988; Hobsbawm, 1968) one of the main factors in Germany overtaking
Britain in the latter half of the nineteenth century, but still a century later
was the foundation for the superior skills and higher productivity of the
German labour force (Prais, 1981) in many industries. Many British policies
for education and training for over a century can be realistically viewed as
spasmodic, belated and never wholly successful attempts to catch up with
German technological education and training systems.

Not only did List anticipate these essential features of current work on
national systems of innovation, he also recognized the interdependence of
the import of foreign technology and domestic technical development.
Nations should not only acquire the achievements of other more advanced
nations, they should increase them by their own efforts. Again, there was
already a good model for this approach to technological learning in
Prussia: the acquisition of machine tool technology. It was British engineers
(especially Mawdslay) and mechanics who were responsible for the key
innovations in machine tool technology in the first quarter of the nineteenth century. This technology was described by Paulinyi (1982) as the
'Alpha and Omega of modern machine-building' because it enabled the
design and construction of metal-working precision machinery for all other
industries. Those involved attempted to maintain a considerable degree of
secrecy, but its importance was recognized by the Prussian government,
which took decisive steps to acquire the technology, despite the fact that
the British government was attempting to ban the export of machine tools
(with the imposition of heavy fines for contravention).

The Prussian government, which had set up technical training institutes,
(Gewerbe-Institut) made sure that they received imported British machine
tools for reverse engineering and for training German craftsmen, who then
disseminated the technology in Germany industry (Paulinyi, 1982). British
craftsmen were also attracted to Prussia as much of the technology
depended on tacit knowledge. (Three out of four of the leading machine

tool entrepreneurs in Britain at that time had themselves spent years with Mawdslay in his workshop.) The transfer of technology promoted and co-ordinated by the Prussian state was highly successful: the German machine tool industry and machine-building proved capable of designing and manufacturing the machinery necessary to make steam locomotives in the 1840s and 1850s. This set Prussia (later Imperial Germany) well on the road to overtaking Britain. Although he did not cite this particular example, List therefore was not talking in a purely abstract way about industrialization and technology transfer but about a process which was unfolding before his eyes.

Not only did List analyse many features of the national system of innovation which are at the heart of contemporary studies (education and training institutions, science, technical institutes, user–producer interactive learning, knowledge accumulation, adapting imported technology, promotion of strategic industries, etc.), he also put great emphasis on the role of the state in co-ordinating and carrying through long-term policies for industry and the economy. Here, as often, he took issue with Jean-Baptiste Say, his favourite target in his polemics with the classical school, who had argued that governments did not make much difference, except in a negative way. The role of the Prussian state in technology catch-up was summed up by Landes (1969) as follows:

> Only the government could afford to send officials on costly tours of inspection as far away as the United States; provide the necessary building and equipment; feed, clothe, house, and in come cases pay students for a period of years. Moreover, these pedagogical institutions were only part – though the most important part – of a larger educational system designed to introduce the new techniques and diffuse them through the economy; there were also non-teaching academies, museums, and, most important perhaps, expositions.
>
> Finally, the government provided technical advice and assistance, awarded subventions to inventors and immigrant entrepreneurs, bestowed gifts of machinery, allowed rebates and exemptions of duties on imports of industrial equipment. Some of this was simply a continuation of the past – a heritage of the strong tradition of direct state interest in economic development. Much of it, in Germany particularly, was symptomatic of a passionate desire to organize and hasten the process of catching up.
>
> In so far as this promotional effort stressed the establishment of rational standards of research and industrial performance, it was of the greatest significance for the future.
>
> (Landes, 1969, p. 151)

The United States was of course even more successful than Germany in overtaking Britain in the second half of the nineteenth century and List had learnt a great deal from his residence there and especially from Hamilton's (1791) *Report on Manufactures*. The widespread promotion of education (though not of industrial training) was even more remarkable in the United States than in Germany. However, the abundance of cheap, accessible materials, energy and land together with successive waves of immigration imparted to the United States national system some specific characteristics without parallel in Europe. The proactive role of the state was greater in Germany while foreign investment played a greater role in the United States.

Although List anticipated many features of the contemporary debate about national systems of innovation (even though his terminology was different), it would of course be absurd to imagine that he could have foreseen all the changes in the world economy and national economies over the next century and a half. In particular, he did not foresee the rise of inhouse professionalized R&D in industry, still less the rise of multi-national (or transnational) corporations (TNCs), operating production establishments in many different countries and increasingly also setting up R&D outside their original base. These are major new developments which deeply affect the whole concept of national systems. In the next section, this chapter will first discuss the rise of R&D and the types of comparison of national systems to which this has led. The third section will discuss some contrasting national systems and a fourth section will discuss the role of TNCs and the ways in which they may affect the performance of national economies in different continents.

12.2 THE RISE OF PROFESSIONAL R&D

In the twentieth century it became tempting to analyse a 'national system of innovation' in terms simply of its R&D and education system. In this section, we shall try to show how and why this rather narrow quantitative approach was abandoned in favour of a much wider qualitative analysis embracing a wide range of social institutions.

As we have seen in Chapters 4 and 5, it was in Germany that the major institutional innovation of the inhouse industrial R&D department was introduced in 1870. Product and process innovation by firms took place of course for more than a century before that but it was the German dyestuffs industry (Beer, 1959) which first realized that it could be profitable to put the business of research for new products and development of new chemical processes on a more regular, systematic and professional basis. Hoechst, Bayer and BASF have continued and strengthened this tradition down to the present day when their R&D laboratories now employ many thousands of scientists and engineers. Undoubtedly such discoveries and innovations as synthetic indigo, many other synthetic dyestuffs and pharmaceuticals and the Haber–Bosch process for fertilizers were the main factors in estab-lishing the German chemical industry's leading position in the world before and after the First World War. When the three companies merged in 1926 to form the giant IG Farben Trust they further reinforced their R&D, as shown in Chapter 5 and made many of the key innovations in synthetic materials, fibres and rubbers (PVC, polystyrene, urea-formaldehyde, Buna, etc.).

From their origins in the chemical and electrical industries gradually during the latter part of the nineteenth century and the first half of the twentieth century, specialized R&D laboratories became characteristic features of most large firms in manufacturing industry (although not of the vast majority of small firms or of service industries) (Mowery, 1980, 1983; Hounshell, 1982; Hughes, 1989). This change in industrial behaviour and the growth of government laboratories, of independent contract research institutes and university research impressed many observers and led to the comment by a leading physicist that the greatest invention of the nine-teenth century was the method of invention itself. A great many inventions

Table 12.3 Estimated gross expenditure on research and development as a fraction of GNP (GERD/GNP ratio), 1934–83

	1934	1967	1983	1983 civil R&D only
USA	0.6	3.1	2.7	2.0
EC*	0.2	1.2	2.1	1.8
Japan	0.1	1.0	2.7	2.7
USSR	0.3	3.2	3.6	1.0

* Estimated weighted average of 12 EC countries.

Sources: Estimates based on Bernal (1939) adapted to 'Frascati' definitions (1963), OECD statistics, and adjustments to Soviet statistics based on Freeman and Young (1965).

had of course been made for centuries or indeed for millennia before 1870 but the new professional R&D labs seemed like a giant step forward. This perception was powerfully reinforced in the Second World War. Science was already important in the First World War – more important than most people realized at the time – but it was the Manhattan Project and its outcome at Hiroshima which impressed on people throughout the world the power of science and especially, as it seemed, Big Science. Many other developments on both sides, such as radar, computers, rockets and explosives resulted from large R&D projects, mobilizing both government, industrial and academic engineers and scientists.

It was therefore hardly surprising that in the climate which existed after the Second World War, the prestige of organized, professional R&D was very high. The proposals made by a visionary physicist (Bernal, 1939) to increase R&D in Britain and other European countries by an order of magnitude seemed absurdly utopian at the time but this was in fact achieved in the new political climate after the Second World War. A similar rapid expansion occurred in all industrial countries (Chapter 1 and Table 12.3) and even in Third World countries there was a trend to establish research councils, national R&D labs and other scientific institutions, to do nuclear physics and in some cases to try and make nuclear weapons (e.g. Argentina, India, Brazil, Israel, Yugoslavia). It was hardly surprising either that a simplistic linear model of science and technology 'push' was often dominant in the new science councils that advised governments. It seemed so obvious that the atom bomb (and it was hoped nuclear power for electricity) was the outcome of a chain reaction: basic physics → large-scale development in big labs → applications and innovations (whether military or civil). The 'linear model' was specifically endorsed in the influential Report of Vannevar Bush (NSF, 1945), *Science, the Endless Frontier* (see Stokes, 1993).

This meant that the R&D system was often seen as almost the only source of innovations – an impression that was reinforced by the system of measurement which was adopted, first by the National Science Foundation in the United States and later during the 1950s and 1960s by all the other OECD countries. This was standardized by the so-called Frascati Manual (OECD, 1963a) and despite the fact that the authors pointed out that technical change did not depend just on R&D but on many other related activities, such as education, training, production engineering, design and quality control, nevertheless R&D measures were very frequently used as a surrogate for all these activities which helped to promote new and improved products and processes. Furthermore, the importance of all the feedback loops from the market and from production into the R&D system was often overlooked or forgotten. The simple fact that the R&D measures were the only ones that were available reinforced these tendencies.

Their effect could be seen in many national reports as well as in the 'Science Policy Reviews' conducted by the OECD in its member countries in the 1960s and 1970s. The admirable aim of these reviews, like those of member countries' economic policies which still continue and on which they were modelled, was to produce a friendly but independent and critical assessment of each country's performance by an international comparative yardstick. In practice they concentrated mainly on the formal R&D system and technical education. This was of course still quite useful but it meant that the 'national system' was usually defined in rather narrow terms. Much research on invention and innovation had amply demonstrated that many factors were important for innovative success other than R&D. However, the practical difficulties of incorporating these factors in international comparisons were very great. 'League table' comparisons of R&D were much easier and more influential. For similar reasons purely quantitative measures of R&D and education were often used in models later developed for the 'new growth theory' (Chapter 13).

Gradually, during the 1950s and 1960s the evidence accumulated that the rate of technical change and of economic growth depended more on efficient diffusion than on being first in the world with radical innovations and as much on social innovations as on technical innovations. This was reflected in the change of emphasis in various OECD reports (OECD, 1963b, 1971b, 1980a, 1988, 1991, 1992) and in the introduction of country reports on 'innovation'. Basic science was of course still recognized as being very important but much more was said about technology and diffusion than hitherto.

Although various OECD reports are a convenient record of changing ideas and policies for science and technology, they rarely originated these changes. The OECD documents summed up and reflected recent experience and changes in the member countries and disseminated what were thought to be the lessons of this experience. The OECD was also, however, more ready than most international organizations to involve independent researchers so that their reports also embody some input from academic research on technical change as well as from industrial R&D management sources. Comparisons with Japan were especially influential after it joined the OECD in the 1970s.

12.3   SOME CONTRASTING FEATURES OF NATIONAL SYSTEMS OF INNOVATION IN THE 1970S AND 1980S

As empirical evidence and analysis began to accumulate about industrial R&D and innovation in Japan, the United States and Europe, it became increasingly evident that the success of innovations, their rate of diffusion and the associated productivity gains depended on a wide variety of other influences as well as formal R&D. In particular, incremental innovations came from production engineers, from technicians and from the shop floor. They were strongly related to different forms of work organization (see especially Hollander, 1965).

Furthermore, many improvements to products and to services came from interaction with the market and with related firms, such as subcontractors, suppliers of materials and services (see especially von Hippel, 1976, 1988; Lundvall, 1985, 1988b, 1992; Sako, 1992). Formal R&D was usually decisive in its contribution to radical innovations but it was no longer possible to ignore the many other contributions to and influences upon the process of technical change at the level of firms and industries.

Not only were inter-firm relationships shown to be of critical importance but, as shown in Part Two, the external linkages within the narrower professional science–technology system were also shown to be decisive for innovative success with radical innovations (NSF, 1973; Gibbons and Johnston, 1974). Finally, research on diffusion revealed more and more that the systemic aspects of innovation were increasingly influential in determining both the rate of diffusion and the productivity gains associated with any particular diffusion process (see especially Carlsson and Jacobsson, 1993). The success of any specific technical innovation, such as robots or CNC depended on other related changes in systems of production (Fleck, 1993). As three major new 'generic' technologies (information technology, biotechnology and new materials technology) diffused through the world economy in the 1970s, 1980s and 1990s, systemic aspects of innovation assumed greater and greater importance.

At the international level two contrasting experiences made a very powerful impression in the 1980s both on policy-makers and on researchers: on the one hand the extraordinary success of first Japan and then South Korea in technological and economic catch-up; and on the other the collapse of the socialist economies of Eastern Europe.

At first in the 1950s and 1960s the Japanese success was often simply attributed to copying, imitating and importing foreign technology and the statistics of the so-called technological balance of payments were often cited to support this view. They showed a huge deficit in Japanese transactions for licensing and know-how imports and exports and a correspondingly large surplus for the United States. It soon became evident, however, as Japanese products and processes began to outperform American and European products and processes in more and more industries (Chapters 6 and 7), that this explanation was not adequate even though the import of technology continued to be important. Japanese industrial R&D expenditures as a proportion of civil industrial net output surpassed those of the United States in the 1970s and total civil R&D as a fraction of GNP surpassed the USA in the 1980s (Table 12.3). The Japanese performance could now be explained more in terms of R&D-intensity, especially as Japanese R&D was highly concentrated in the fastest growing civil industries, such as electronics. Patent statistics showed that the leading Japanese electronic firms outstripped American and European firms in these industries, not just in domestic patenting but in patents taken out in the United States (Patel and Pavitt, 1991, 1992; Freeman, 1987).

Table 12.4 Contrasting national systems of innovation, 1970s

Japan	USSR
High GERD/GNP ratio (2.5%). Very low proportion of military/space R&D (<2% of R&D).	Very high GERD/GNP ratio (c.4%). Extremely high proportion of military/space R&D (>70% of R&D).
High proportion of total R&D at enterprise level and company financed (approx. two thirds).	Low proportion of total R&D at enterprise level and company financed (<10%).
Strong integration of R&D, production and import of technology at enterprise level.	Separation of R&D, production and import of technology and weak institutional linkages.
Strong user–producer and subcontractor network linkages.	Weak or non-existent linkages between marketing, production and procurement.
Strong incentives to innovate at enterprise level involving both management and workforce.	Some incentives to innovate made increasingly strong in 1960s and 1970s but offset by other negative disincentives affecting both management and workforce.
Intensive experience of competition in international markets.	Relatively weak exposure to international competition except in arms race.

However, although these rough measures of research and inventive activity certainly did indicate the huge increase in Japanese scientific and technical activities, they did not explain how these activities led to higher quality of new products and processes (Grupp and Hofmeyer, 1986; Womack et al. 1990); to shorter lead times (Graves, 1991; Mansfield, 1988) and to more rapid diffusion of such technologies as robotics (Fleck and White, 1987; Mansfield, 1989). Moreover, the contrasting example of the (then) Soviet Union and other East European countries showed that simply to commit greater resources to R&D did not guarantee successful innovation, diffusion and productivity gains. It was obvious that qualitative factors affecting the national systems had to be taken into account as well as the purely quantitative indicators.

Some major differences between the two national systems of Japan and the Soviet Union as they were functioning in the 1970s are summarized in Table 12.4. The most striking contrast of course was the huge commitment of Soviet R&D to military and space applications with little direct or indirect spin-off to the civil economy. It has now been shown that the desire to keep pace with the USA in the superpower arms race led to about three-quarters of the massive Soviet R&D resources going into defence and space research. This amounted to nearly 3 per cent of GNP, so that only about 1 per cent remained for civil R&D. This civil R&D/GNP ratio was less than half of most West European countries and much smaller than the Japanese ratio (Table 12.3).

Nevertheless, it could have been far more productive if the social, technical and economic linkages in the system and the incentives to efficient performance had been stronger. The Soviet system grew up on the basis of separate research institutes within the academy system (for fundamental research), for each industry sector (for applied research and development) and for the design of plant and import of technology (the project design organizations) (Barker and Davies, 1965; Amann *et al.*, 1979). The links between all these different institutions and enterprise-level R&D remained rather weak despite successive attempts to reform and improve the system in the 1960s and 1970s. R&D at the enterprise level remained very weak in civil industry.

Moreover, there were quite strong negative incentives in the Soviet system retarding innovation at enterprise level (Gomulka, 1990), such as the need to meet quantitative planned product targets. Thus, whereas the integration of R&D, production, and technology imports at firm level was the strongest feature of the Japanese system (Baba, 1985; Takeuchi and Nonaka, 1986; Freeman, 1987), it was very weak in the Soviet Union except in the aircraft industry and other defence sectors. Finally, the user–producer linkages which were so important in most other industrial countries were very weak or almost non-existent in some areas in the Soviet Union.

There were some features of their national systems in which both countries resembled each other, and both did of course enjoy high economic growth rates in the 1950s and 1960s. They had (and still have) good education systems with a high proportion of young people participating in tertiary education and strong emphasis on science and technology. They also had methods of generating long-term goals and perspectives for the science–technology system, but whereas in the Japanese case the long-term 'visions' were generated by an interactive process involving not only MITI and other government organizations but also industry and universities (Irvine and Martin, 1984), in the USSR the process was more restricted and dominated to a greater extent by military/space requirements.

A similar sharp contrast can be made between the national systems of innovation typically present in Latin American countries in the 1980s and those in the 'Four Tigers' of East Asia (Table 12.5) and especially between two 'newly industrializing countries' (NICs) in the 1980s: Brazil and South Korea (Table 12.6). The Asian countries started from a lower level of industrialization in the 1950s and lower GDP *per capita* (Table 12.7), but whereas in the 1960s and 1970s the Latin American and East Asian countries were often grouped together as very fast growing NICs, in the 1980s a sharp contrast began to emerge (Table 12.8): the East Asian countries' GNP grew at an average annual rate of about 8 per cent, but in most Latin American countries, including Brazil, this fell to less than 2 per cent, which meant in many cases a falling *per capita* income.

There are of course many explanations for this stark contrast. Some of the Asian countries introduced more radical social changes, such as land reform and universal education than most Latin American countries and clearly a structural and technical transformation of this magnitude in this time was facilitated by these social changes. In the case of Brazil and South Korea it is possible to give some more detailed quantitative indicators of some of these contrasting features. As Table 12.6 shows the contrast in educational systems in the 1980s was very marked as well as enterprise-level R&D, telecommunication infrastructure and the diffusion of new technologies (see Nelson, 1993, for more detailed comparisons and Villaschi, 1993, for a detailed study of the Brazilian NS). More recently, the performance of the Brazilian economy improved after the 'lost decade' of the 1980s (Ferraz *et al.*, 1992) but still remained in strong contrast to the case of Korea.

Table 12.5 Divergence in national systems of innovation in the 1980s

East Asia	Latin America
Expanding universal education system with high participation in tertiary education and with high proportion of engineering graduates.	Deteriorating education system with proportionately lower output of engineers.
Import of technology typically combined with local initiatives in technical change and at later stages rapidly rising levels of R&D.	Much transfer of technology, especially from the USA, but weak enterprise level R&D and little integration with technology transfer.
Industrial R&D rises typically to >50 per cent of all R&D.	Industrial R&D typically remains at <25 per cent of total.
Development of strong science–technology infrastructure and at later stages good linkages with industrial R&D.	Weakening of science–technology infrastructure and poor linkages with industry.
High levels of investment and major inflow of Japanese investment and technology with strong yen in 1980s and 1990s. Strong influence of Japanese models of management and networking organization.	Decline in (mainly US) foreign investment and generally lower levels of investment. Low level of international networking in technology. Recovery of volatile portfolio investment in 1990s but less recovery of long-term direct investment.
Heavy investment in advanced telecommunications infrastructure.	Slow development of modern telecommunications.
Strong and fast growing electronic industries with high exports and extensive user feedback from international markets.	Weak electronic industries with low exports and little learning by international marketing.

Table 12.6 National systems of innovation, 1980s: some quantitative indicators for Brazil and South Korea

Various indicators of technical capability and national institutions	Brazil	South Korea
% age group in third level (higher) education	11 (1985)	32 (1985)
Engineering students as per cent of population	0.13 (1985)	0.54 (1985)
R&D as per cent GNP	0.7 (1987)	2.1 (1989)
Industry R&D as % total	30 (1988)	65 (1987)
Robots per million empl[t]	52 (1987)	1,060 (1987)
CAD per million empl[t]	422 (1986)	1,437 (1986)
NCMT per million empl[t]	2,298 (1987)	5,176 (1985)
Growth rate electronics	8% (1983–87)	21% (1985–90)
Telephone lines per 100 (1989)	6	25
Per capita sales of telecommunication equipment (1989)	$10	$77
Patents (US) (1989)	36	159

Table 12.7 Starting levels for industry, Latin America and Asia, 1955

	Ratio of manufacturing to agricultural net product	$ net value of manufacturing per capita
Argentina	1.32	145
Brazil	0.72	50
Mexico	1.00	60
Venezuela	1.43	95
Colombia	0.42	45
South Korea	0.20	8
Thailand	0.28	10
India	0.30	7
Indonesia	0.20	10

Source: Maizels (1963).

Table 12.8 Comparative growth rates, 1965–89

GDP % pa	1965–80	1980–89
East Asia	7.5	7.9
South Asia	3.9	5.1
Africa (sub-Sahara)	4.0	2.1
Latin America	5.8	1.6

GDP per capita % pa	1965–80	1980–89
East Asia	5.0	6.3
South Asia	1.5	2.9
Africa (sub-Sahara)	1.1	−1.2
Latin America	3.5	−0.5

Source: World Bank (1991).

12.4   GLOBALIZATION AND NATIONAL SYSTEMS

It has been argued that a variety of national institutions has powerfully affected the relative rates of technical change and hence of economic growth in various countries. The variations in national systems which have been described are of course extreme contrasting cases. Nevertheless, they have certainly been important features of world development in the second half of the twentieth century and they point to uneven development of the world economy and divergence in growth rates. Moreover, differences in national systems were also important between Japan, the Unites States and the EC and between European countries themselves, as the major comparative study between more than a dozen national systems of innovation illustrated (Nelson, 1993). The comparative study of Ireland with other small countries by Mjøset (1992) also demonstrated this point and the comparison of Denmark and Sweden by Edqvist and Lundvall (1993) showed that big differences exist between neighbouring countries which superficially appear very similar in many ways. Moreover, Archibugi and

Pinta (1992) demonstrated the growing pattern of specialization in technology and trade and Fagerberg (1992) showed the continuing importance of the home market for comparative technological advantage.

However, the whole concept of national differences in innovative capabilities determining national performance has been recently challenged on the grounds that transnational corporations (TNCs) are changing the face of the world economy in the direction of globalization. For example, Ohmae (1990) in his book *The Borderless World* argues that national frontiers are 'melting away' in what he calls the ILE (interlinked economy) – the triad of USA, EC and Japan, now being joined by NICs. He accepts that before the 1980s national policies and national differences were important, especially in Japan, but today this ILE is becoming 'so powerful that it has swallowed most consumers and corporations, made traditional national borders almost disappear, and pushed bureaucrats, politicians and the military towards the status of declining industries' (p. xii).

Against this, Michael Porter has argued that:

> Competitive advantage is created and sustained through a highly localised process. Differences in national economic structures, values, cultures, institutions and histories contribute profoundly to competitive success. The role of the home nation seems to be as strong or stronger than ever. While globalisation of competition might appear to make the national less important, instead it seems to make it more so. With fewer impediments to trade to shelter uncompetitive domestic firms and industries, the home nation takes on growing significance because it is the source of the skills and technology that underpin competitive advantage.
>
> (Porter, 1990, p. 19)

Even in the case of global network industries, such as telecommunications, Davies (1996) and Hulsink (1996) have argued that the nation base has become even more important in global competition. In addition to Porter's argument, Lundvall (1993) points out that if uncertainty, localized learning and bounded rationality are introduced as basic and more realistic assumptions about micro-economic behaviour, rather than the traditional assumptions of perfect information and hyperrationality, then it must follow that local and national variations in circumstances may often lead to different paths of development and to diversity rather than to standardization and convergence.

At first sight, the activities of multinational corporations might appear to offer a powerful countervailing force to this local variety and diversity. The largest corporations in the world, whether their original domestic base was in Europe, the United States, Japan or elsewhere, have often been investing in many different new locations. This investment, even though initially it may have been in distribution and service networks, or in production facilities has more recently also included R&D. While the greater part of the 1980s' investment has been within the OECD area and in oil-producing countries and could be more accurately described therefore as 'triadization' rather than 'globalization', it has also flowed, even though very unevenly, to other countries of the Third World and there is now a trickle to the former socialist group of countries.

As Harry Johnson (1975) long ago pointed out, in this sense the multi-nationals do indeed unite the human race. Since the basic laws of physics, chemistry, biology and other sciences apply everywhere there is an under-lying unified technology which can in principle be applied anywhere with identical or very similar results. Insofar as large 'global' TNCs are able to sell their products and services worldwide and to produce them in many different locations, they can and do act as a very powerful agency tending towards the worldwide standardization of technology and output. As the model developed by Callon (1993) indicates, the diffusion process can tend to enhance similarities between adopters.

Even in the case of consumer goods where it might be reasonable to suppose that there would continue to be wide variations in consumer tastes, we are all sufficiently familiar with such products as Coca-Cola and such services as those provided by McDonald's to recognize the reality of such global production and distribution networks, offering standardized products and services worldwide. Is it not realistic to suppose that an ever-larger proportion of world production and trade will take this form? Supporting such a view are not only the obvious examples of hotel chains, soft drinks, canned beer, tourist agencies and credit cards but theoretical economic arguments based on static and dynamic economies of scale in production, advertising, marketing, design and finance, as well as the ability of large multinationals to take advantage of surviving differences between nations in costs of capital, labour, energy and other inputs.

However, it would be unwise to assume that these tendencies are the only or even necessarily the strongest tendencies within the world economy. Nor are they so unequivocally desirable that they should be promoted by both national and international economic policies. In fact the arguments for preserving and even encouraging diversity may sometimes outweigh the shorter term advantages of the scale economies derived from flows of investment. In fact both processes (global standardization in some areas but increasing diversity in others) co-exist.

While there are certainly some products and services, such as those already mentioned, where there is indeed a demand which is global in nature and where local variations in taste, regulation, climate and other circumstances can be largely or wholly ignored, there are far more products and services where such variations certainly cannot be overlooked without dire consequences. Innumerable cases leap to mind where climate con-ditions affect the performance of machines, instruments, vehicles and materials and even more examples are obvious in relation to variations in national standards, specifications and regulations. While it is true that international standardization is a countervailing force though the activities of the International Standards Organization (ISO) and many other bodies attempting to achieve harmonization of technical standards, it is also true that the experience of the European Community over the past 20 years demonstrates the extreme difficulties attending this process in many areas (as well as the feasibility in others). And all this still does not take into account the cultural aspects of the problem which deeply affect such areas as food, clothing and personal services.

So far we have been discussing mainly the case of established products and pointing to some factors which limit global standardization even in the simplest cases. Advocates of a strong globalization hypothesis would of course accept most of these points, although they might argue that some of them will constantly tend to diminish as the media, travel, education and international organizations all exert their long-term influence. Rothwell (1992) has pointed to the 'electronification' of design as an important factor facilitating the internationalization of design and R&D. It can be argued further that local variations can easily be dealt with inside the framework of the global strategies of the multinational corporations. Indeed, globalization of R&D has already led to local adaptation and modification of products to meet national variations, as a normal and almost routine activity of TNCs. Some companies such as Honda go one step further and claim to have a strategy of diversity in worldwide design which goes beyond the simple modification of a standard product to the idea of local variation at the design stage in several different parts of the world. However, the vast majority of Japanese-based TNCs remain essentially Japanese companies with international operations rather than truly international companies and the same is true of US and most other TNCs in relation to their home environment (Hu, 1992). Most R&D activities of TNCs are still overwhelmingly conducted in the domestic base of the company and are heavily influenced by the local national system of innovation. Moreover, ownership and control still remain overwhelmingly based on the domestic platform.

The statistics are rather poor but analysis of all the available data and cross-checking with the patent statistics (Patel and Pavitt, 1991; Patel, 1995) suggests that the R&D activities of US companies outside the USA amount to less than 10 per cent of the total, while those of Japanese companies are much lower – less than 2 per cent – though rising rather rapidly (Table 12.9). The picture in Europe is more complex both because of the development of the European Community and the Single European Market, and because of the existence of several technically advanced small countries where the domestic base is too small for the strong MNCs which are based there (Netherlands, Sweden, Belgium, Switzerland). A larger part of national R&D activities in these countries and most other parts of Europe is undertaken by foreign multinationals and their 'own' TNCs perform much more R&D abroad than is the case with the USA or Japan. Only a small part of total world R&D is conducted outside the leading industrial countries and only a very small part of this is financed by TNCs.

Qualitative analysis of the transnational activities of corporations shows that most of it is either local design modification to meet national specifications and regulations or research to facilitate monitoring of local science and technology. The more original research, development and design work is still overwhelmingly concentrated in the domestic base, although there are important exceptions in the drug industry and electronics industry where specialized pools of scientific ability play an important role.

These examples of local growth in innovative capability and of R&D performance in the subsidiaries of TNC parents in Third World countries should on no account be ignored, as they may foreshadow major long-term changes in the next century (Reddy and Sigurdson, 1994). The fact that they have occurred in the drug and electronics industries, i.e. two of the fastest growing industries in the world economy, is particularly important.

Table 12.9 Japanese R&D facilities in Europe

Country	1990	1991	1992 (start of year)	1993	1994
UK	24	47	72	79	83
France	10	17	29	33	34
Germany	14	28	38	46	53
Netherlands	3	4	6	12	16
Belgium and Luxemburg	4	8	10	17	16
Ireland	1	2	2	5	7
Spain	11	15	23	20	26
Italy	3	8	8	12	14
Denmark	0	1	1	1	1
Portugal	0	0	0	0	0
Greece	0	0	0	0	0
Total EU	70	130	189	225	250
Finland	0	0	1	1	0
Norway	0	0	1	1	1
Sweden	1	1	3	5	5
Austria	0	0	2	5	3
Switzerland	1	3	3	3	4
Iceland	0	1	1	0	1
Total Europe	72	135	200	240	264
(inc. over previous year)		88%	48%	20%	10%

An interesting example is that of the Malaysian electronic industry analysed by Ariffin and Bell (1997). Based on in-depth interviews with 25 subsidiaries of foreign TNCs they show that although there is great diversity in the nature of parent–subsidiary relationships and linkages, two-thirds of the sample had moved into innovative activities drawing heavily but not exclusively on learning links with their parent companies. One-third of the sample firms had entered into collaborative innovation projects with their parents or sister firms. United States and European TNCs were more likely than Japanese TNCs to deepen their technology investment and shift into higher technology products. Siemens Penang was selected in 1991 as their worldwide optosemiconductor centre, while Motorola Penang increased the number of R&D engineers from 5 in 1976 to 130 in 1995.

This example in Malaysian electronics indicates that the local national environment and policies can both facilitate and induce changes in TNC behaviour, as well as changes in the behaviour of nationally owned firms, contrary to some previous theory. Other examples could be cited from Singapore, Taiwan and Korea. In general, the Asian electronics industry, with its high degree of interdependence in components, materials, capital goods and software and its rapid change of product and process generations, indicates the possibility that the type of shift envisaged by some prophets of globalization may indeed occur on a more substantial scale than has hitherto been the case in the worldwide distribution of R&D and other innovative activities. The ideas of Perez and Soete (1988) on 'windows of opportunity' for catching up countries are highly relevant here. Their suggestion that information and communication technology may be more easily transferable and offer entirely new opportunities for catch-up are further developed in Chapter 15.

As long as we are dealing with a static array of products and discussing only minor variations to adjust to local consumer tastes and environments, then the standardization arguments, the globalization arguments and even some of the simplifying neoclassical assumptions about perfect information are at the borderlines of credibility and usefulness. But once we leave this world and enter the dynamic world of radical innovations, both technical and organizational, and of extremely uneven and unequal access to new developments in science and technology, then the whole picture is transformed. More realistic assumptions and a more realistic vision are essential if economic theory is to be of any help in policy-making. The variations in national systems are an essential feature of this more complete vision, illustrated by the example of East Asia and now increasingly South Asia.

Lundvall (1993) points out that, even in the case of continuous incremental innovation in open economies, the drive towards standardization is limited. Geographical and cultural proximity to advanced users and a network of institutionalized (even if often informal) user–producer relationships are an important source of catch-up capability as well as diversity and comparative advantage, as is the local supply of managerial and technical skills and accumulated tacit knowledge. While Lundvall accepts that TNCs might locate in 'national strongholds' in order to gain access to the fruits of this interactive learning process, he points out that it is not always simple to enter such markets because of the strength of the non-economic relationships involved. Competing standards for the global market may be important weapons in such situations as well as other forms of product differentiation and quality improvement.

When it comes to radical innovations the importance of institutional variety and localized learning is even greater. Posner's (1961) theory of technology gaps and imitation lags is of fundamental importance here (see Chapter 14). It may be many years before imitators are capable of assembling the mix of skills, the work organization and other institutional changes necessary to launch into the production and marketing of entirely new products. The study by Ariffin and Bell (1997) which has been cited makes throughout a valuable distinction between learning to produce and learning to innovate, while Hobday (1995) emphasizes the role of learning to export.

It is of course true that in the global diffusion of radical innovations, TNCs may have an extremely important role. They are in a position to transfer specialized equipment and skills to new locations if they so wish and to stimulate and organize the necessary learning processes. They are also in a position to make technology exchange agreements with rivals and to organize joint ventures in any part of the world. It is for this reason that many governments in Europe as well as in the Third World and the ex-socialist countries have been anxious to offer incentives to attract a flow of inward investment and associated technology transfer from firms based in Japan and the USA.

However, such efforts will meet with only limited success unless accompanied by a variety of institutional changes, as in some Asian countries, designed to strengthen autonomous technology capability within the importing countries. This is especially true of those generic technologies which have been at the centre of the worldwide diffusion process over the past two decades. Here it is essential to emphasize the interdependencies between innovations and between technical and organizational innovations. A theory of technical change which ignores these interdependencies is no more helpful than a theory of economics which ignores the interdependencies of prices and quantities in the world economy. Hobday's (1995) study of the various ways in which Asian firms have organized and reorganized is particularly valuable in this connection.

Perez (1983) has pointed out that the social and institutional framework which is hospitable to one set of technologies will not be so suitable for a radically new technology. Whereas incremental innovations can be easily accommodated, this may not be the case with radical innovations which by definition involve an element of creative destruction. When we are talking about large clusters of radical innovations combined with rapid processes of incremental innovation, then the problems of structural and social adjustment can be very great. This is quite obvious when we consider such aspects as the change in management techniques and skill mix which are called for, but it also applies to many other types of institutional change in standards, patents, new services, new infrastructure, government policies and public organizations.

It is in this context that the concept of national systems of innovation assumes such great importance and in the light of this approach it is not surprising that the recognition of the scope and depth of the computer revolution, which was accelerated by the microprocessor in the 1970s, has been followed by a growing recognition of the importance of organizational and managerial change ('multi-skilling', 'lean production systems', 'downsizing', 'just-in-time' stock control, worker participation in technical change, quality circles, continuous learning).

The diffusion of a new techno-economic paradigm is a trial and error process involving great institutional variety. There are evolutionary advantages in this variety and considerable dangers in being locked in too early to a standardized technology or organization. A technological monoculture may be more dangerous than an ecological monoculture. Even when a technology matures and shows clear-cut advantages and scale economies it is important to retain flexibility and to nourish alternative sources of radically new technology and work organization.

National and international policies thus confront the need for a sophisticated dual approach to a complex set of problems. Policies for diffusion of standard generic technologies are certainly important and these may sometimes entail the encouragement of inward investment and technology transfer by MNCs. But also important are policies to encourage local originality and diversity.

12.5 CONCLUSIONS

This chapter has attempted to show that historically there have been major differences between countries in the ways in which they have organized and sustained the development, introduction, improvement and diffusion of new products and processes within their national economies. These differences can perhaps be most easily demonstrated in the case of Britain in the late eighteenth and early nineteenth centuries when it achieved leadership in world technology and world trade and could temporarily claim to be 'the workshop of the world' (Chapter 2).

Historians (von Tunzelmann, 1993; Mathias, 1969) are generally agreed that no single factor can explain this British success. It can be rather attributed to a unique combination of interacting social, economic and technical changes within the national economic space. It was certainly not just a succession of remarkable inventions in the textile and iron industries, important though these were. Among the more important changes were the transition from a domestic 'putting out' system to a factory system of production, new ways of managing and financing companies (partnership and later joint stock companies), interactive learning between new companies and industries using new materials and other inputs as well as new machinery; the removal of many older restrictions on trade and industry and the growth of new markets and systems of retail and wholesale trade; a new transport infrastructure, a hospitable cultural environment for new scientific theories and inventions and certainly not least important – the dissemination and widespread acceptance of economic theories and policies which facilitated all these changes. It was the British Prime Minister who said to Adam Smith: 'We are all your pupils now.' The benefits from foreign trade and in some cases piracy and plunder also played their part, especially in the process of capital accumulation, but it was the mode of investment of this capital within the national economy, rather than the simple acquisition of treasure or luxury expenditure which was a decisive impulse to economic growth.

Of course, many of these changes also took place within other European countries but there were distinct and measurable differences in the rate and direction of institutional and technical change, the rate of economic growth, of foreign trade and of standards of living, which were fairly obvious at the time as well as to historians since. This can be seen not only from the writings of economists but from novelists and travellers. It was therefore hardly surprising that Friedrich List and other economists on the continent of Europe were concerned to develop theories and policies which would help them to understand the reasons for British commercial supremacy and enable Germany (and other countries) to catch up. This chapter has attempted to show that in this endeavour List anticipated many contemporary ideas about 'national systems of innovation', including the crucial importance of technological accumulation through a combination of technology imports with local activities and proactive interventionist policies to foster strategic 'infant' industries.

In the second half of the nineteenth century, new developments in the natural sciences and in electrical engineering led to progressive entrepreneurs and reformers realizing that in the new and fastest growing industries, learning by doing, using and interacting in the old British ways had to be accompanied or replaced by more professional and systematic processes of innovation and learning. The organizational innovation of the inhouse R&D department put the introduction of new products and processes on a firmer foundation while new institutions and departments of higher education and secondary education provided the new more highly qualified scientists, engineers and technicians. This chapter has argued that it was in Germany and the United States that these institutional innovations began and made the greatest impact on national systems in the second half of the nineteenth century and early twentieth century.

The rapidly widening gap between a small group of industrialized countries and the rest of the 'underdeveloped' world (Durlauf and Johnson, 1992; Dosi et al., 1992) as well as the 'forging ahead', 'catching up' and 'falling behind' (Abramovitz, 1986) among the leaders clearly called for some explanation of why growth rates differed so much. It is analysed in the next chapter on technology and growth. The brave simplifying assumptions of neoclassical economics might lead people to expect convergence rather than divergence in growth rates (perfect information and costless, instant transfer of technology). Nor did formal growth theory and models provide much help since most of the interesting phenomena were confined to a 'residual' which could not be satisfactorily disaggregated and because of the interdependencies involved (Nelson, 1973).

Many economic historians and proponents of what have now become known as 'national systems of innovation' would claim that the differences were due to varying types of institutional and technical change which may be the subject of qualitative description, even though difficult to quantify. This chapter has argued that the oversimplification of quantitative R&D comparisons was an inadequate method in itself and has attempted to show by the examples of Japan and the former Soviet Union, and of the East Asian and Latin American NICs that institutional differences in the mode of importing, improving, developing and diffusing new technologies, products and processes played a major role in their sharply contrasting growth rates.

Finally, the chapter discussed the controversial issue of globalization and its bearing on national systems of innovation. It is ironical that just as the importance of technology policies and industrial policies has been increasingly recognized alike in OECD and in developing countries, the limitations of national policies are increasingly emphasized and the relevance of national systems increasingly questioned (see, for example, Humbert, 1993). The global reach of transnational corporations, the drastic cost reductions and quality improvements in global telecommunications networking and other rapid and related changes in the world economy must certainly be taken into account in any satisfactory analysis of national systems (Chesnais, 1992; Freeman and Soete, 1994; Edqvist, 1997).

It is tempting at first sight to follow Ohmae (1990) and to discard national economies and nation-states as rapidly obsolescent categories. The speed of change and the difficulties of focusing analysis may be illustrated by the confusion in terminology. Ohmae maintains that nation-states are losing their power and their influence both 'upwards' and 'downwards', on the one hand to supranational institutions (the EC, NAFTA, UN organizations etc., as well as transnational companies) and, on the other hand to subnational (or 'infranational') provincial, urban or local authorities and organizations (the disintegration of federal and centralized states, the growing importance in some areas of local government agencies and even of various forms of tariff-free zones, and of 'Silicon Valleys'). Unhappily, at least in the English language, the same word 'regions' has to be used to describe both processes – the very large 'regional' trading blocs such as NAFTA, or the emerging East Asian 'region' and the much smaller subnational 'regions'. Some terminological innovation is needed here. It is undoubtedly important to keep track of both and of their interaction with national systems.

The work of geographers as well as economists (e.g. Storper and Harrison, 1991; Saxenian, 1991; Scott, 1991; Lundvall, 1992; Antonelli, 1994) has convincingly demonstrated the importance of subnational regions for network developments and new technology systems. They have argued that local infrastructure, externalities, especially in skills and local labour markets, specialized services and, not least, mutual trust and personal relationships have contributed greatly to flourishing regions. It should not be forgotten, however, that 'regional systems of innovation' and 'economies of agglomeration' have always underpinned national systems from the beginnings of the Industrial Revolution (Arcangeli, 1993). Marshall (1890) already stressed the importance of what were then known as 'industrial districts' where 'the secrets of industry were in the air' (Foray, 1991). Piore and Sabel (1984) have especially underlined the importance of these regions in many parts of Europe both in the nineteenth century and again today. A very interesting question for research is how far information and communication technology will underpin or undermine these regional economies (Chapter 17).

The vulnerability of national economies to external shocks is also by no means a new phenomenon of the last decade or two, even though the liberalization of capital markets and international flows of trade and investment combined with computerization and new telecommunications networks may have increased this vulnerability. Small and distant nations were already affected by shocks from the City of London under the gold standard and the Popular Front government in France suffered just as severely from the 'flight of capital' in the 1930s as the socialist government of France in the 1980s.

This chapter has argued that nation-states, national economies and national systems of innovation are still essential domains of economic and political analysis, despite some shifts to upper and nether regions. Michael Porter (1990) could even be right in his contention that the intensification of global competition has made the role of the home nation more important, not less. From the standpoint of developing countries, national policies for catching up in technology remain of fundamental importance. Nevertheless, the interaction of national systems both with 'subnational systems of innovation' and with transnational corporations will be increasingly important, as will be the role of international co-operation in sustaining a global regime favourable to catching up and development. These sometimes conflicting and sometimes converging trends will surely be one of the most exciting areas of research in the next century. Just as exciting will be the study of 'sectoral systems of innovation' affecting specific industries or clusters of industries and sometimes performing better than the national systems (see for example Carlsson and Jacobsson, 1991).

CHAPTER 13

TECHNOLOGY AND ECONOMIC GROWTH

Economists have long recognized the importance of science and technology for long-term economic growth and productivity. As we saw in Chapter 1, both Smith and Marx viewed inventions and innovations as the most dynamic element in the growth of capitalist economies, interacting with capital accumulation, scale economies and expanding markets. Growth theory has traditionally recognized the crucial role of knowledge accumulation in the growth process. Without technological change, capital accumulation will not be sustained – its marginal productivity declining – and the *per capita* growth rate of the economy will inexorably tend towards zero. The inventions of new machines and intermediate goods provide the opportunities for new investment. Thus, as has been highlighted in the various cases described in Part One, The Rise of Science-related Technology, the introduction, diffusion and continuous improvements of new products and processes has been one of the major factors behind the efficiency gains in those sectors over the post-war period.

In this chapter, we turn to the most salient macro-economic implications of science and technology. The literature on this subject is huge by any standard and has witnessed a dramatic revival in the 1980s and 1990s. We refer the reader to the numerous overview articles which have been published recently on this subject and are referred to at the end of this Part (see among others Amable, 1994; Nelson, 1994; Verspagen, 1993; Schneider and Ziesemer, 1995).

We start this chapter with a brief empirical overview of the long-term pattern of growth. As we already highlighted in the previous chapter, long-term growth performance has been strikingly uneven over time and between countries, raising questions as to the simple convergence hypothesis popular in many of the growth models developed in the 1950s and 1960s. In the second section, we turn to some of the underlying economic explanations for growth starting from the old, neoclassical elegant equilibrium growth models and moving to the more recent 'new' growth models which have attempted to encapsulate some of the more 'endogenous' features of technological change. In the third section, we then turn to some of the structural Schumpeterian explanations for growth based on the emergence of major new technologies with pervasive macro-economic impacts. Despite the work of many economists over the last fifty years, the actual long-run growth performance across countries in the world remains, in the words of the *Economist* (1996), to some extent 'a mystery'.

Table 13.1 Estimates of pre-industrial per capita GNP (at 1960 US dollars and prices)

	Period	GNP per capita
Countries now developed		
Great Britain	1700	160–200
USA	1710	200–260
France	1781–90	170–200
Russia	1860	160–200
Sweden	1860	190–230
Japan	1885	160–200
Countries now less developed		
Egypt	1887	170–210
Ghana	1891	90–150
India	1900	130–160
Iran	1900	140–220
Jamaica	1832	240–280
Mexico	1900	150–190
Philippines	1902	170–210

Note: It is probably safe to assume nearly stationary per capita incomes in the years which precede the estimates, so that, for example, the figure of Ghana 1700 should not have been significantly different from that of Ghana 1891 (if anything it could have been higher before slave trade). Clearly the figures should be taken as rough estimates and it must also be remembered that many historians and sociologists would argue that per capita incomes are not comparable over very long periods because of structural changes in the economy, qualitative change in the composition of output and many institutional changes. According to some economists too, comparisons can be useful between countries only during certain phases of growth and not over very extended periods. See also the discussion on measurement problems in the transition to the information society in Chapter 17.

Source: Bairoch (1981).

13.1 ON FACTS AND FIGURES: LONG-RUN PHASES OF GROWTH

As many economic historians have argued and contrary to popular belief, in the long run the pattern of growth at the world level seems to have been characterized by an 'explosion' of diverging growth trends. It can be reasonably argued that most countries in the world had much more similar per capita income levels before the Industrial Revolution than today. As Tables 13.1 and 13.2 show, before the Industrial Revolution the income gap between the poorest and richest country was relatively small, between 1.5 and 2. Over the current industrialization century that ratio increased dramatically. If anything, the dominant pattern of economic growth over the last two centuries has been one of a fast-increasing differentiation among countries and growth divergence (see Box 13.1).

Long-term growth and technological convergence
of the now-industrialized countries

Not surprisingly, a large number of contributions from economic historians focused on these diverging long-term growth paths. In line with the

Table 13.2 Estimates of trends in per capita GNP (1960 US dollars and prices, 1750–1977)

Year	Developed countries		Third World		Gaps	
	(1)	(2)	(3)	(4)	(5)	(6) Ratio of most developed to least developed
	Total ($bn)	Per capita	Total ($bn)	Per capita	= (2)/(4)	
1750	35	182	112	188	1.0	1.8
1800	47	198	137	188	1.1	1.8
1830	67	237	150	183	1.3	2.8
1860	118	324	159	174	1.9	4.5
1913	430	662	217	192	3.4	10.4
1950	889	1054	335	203	5.2	17.9
1960	1394	1453	514	250	5.8	20.0
1970	2386	2229	800	380	7.2	25.7
1977	2108	2737	1082	355	7.7	29.1

Source: Bairoch (1981, pp. 7–8), as quoted in Dosi *et al.* (1992).

Box 13.1   Some 'stylized facts' (SF) on growth patterns

SF1   Economies have grown over the past two centuries probably faster than during any previous period in recorded history.

SF2   But they have grown at different and variable rates (sometimes negative for particular periods and particular countries – Argentina, other less developed countries, USSR, etc. – or for many countries for particular periods, deep recessions).

SF3   The long-term patterns for the whole set of countries show an increasing differentiation, highlighted by a secular increase in the variance in per capita income.

SF4   Catching up with forging ahead has been relatively rare (Britain overtaking Holland in the eighteenth century; USA, Germany and others overtaking Britain in the late nineteenth and twentieth centuries; Japan overtaking almost everyone in the late twentieth century).

SF5   progress in catching up has been more widespread (Western and Central Europe in the nineteenth century; Scandinavia and Italy in the twentieth century; East Asian countries in the late twentieth century; the EEC catching up with the USA during the 1960s and 1970s).

SF6   Falling behind has also been a rather frequent phenomenon (many less developed countries in the 1970s and 1980s; a few countries falling behind after a considerable spurt of catching up – compare the 1950s with the 1980s in Latin America and Eastern Europe; Britain experiencing a long-term relative decline).

SF7   One can hardly identify, in general, persistent features of national growth patterns just conditional on initial performances (e.g. 'all laggard countries will tend to grow faster' or 'those that have grown faster will grow faster also in the future'). Closer inspection of particular economies or groups of them does appear to show long-term persistence (e.g. Japan or, conversely, Britain), but the causes of the phenomenon are plausibly country specific rather than a common feature of the world economy.

Source: Dosi *et al.* (1992)

arguments advanced in Chapter 12, many authors have attempted to explain these in terms of differences in countries' success in using science and technological advances more efficiently or better than other countries. It was Abramovitz (1986) who introduced first the distinction between 'forging ahead', 'catching up' and 'falling behind' in economic growth. As we discussed in more detail in Part One, there are only two major examples of 'forging ahead' (see Chapters 2 and 3 for the UK in the nineteenth century and USA in the twentieth century), while there are often examples of forging ahead in particular industries (see, for example, Chapters 4 and 5 for the case of the chemical industry in Germany, or Chapter 7 for Japanese consumer electronics).

When it comes to the phenomenon of 'catching up' in aggregate growth rates, there are many more examples of European countries in the twentieth century and an increasing number from Asia. Both in forging ahead and in catching up, economic historians and economists have recognized the crucial role of technical change but few have made any intensive study of those institutions which promote this change. Abramovitz was one of the few who did and he introduced the notion of 'social capability' to describe those institutions which could achieve the requisite combination of institutional and technical change.

As we have seen in Chapter 12, this approach is characteristic of recent work on 'national systems of innovation'. Circumstances are obviously different in a small island economy, such as Mauritius, or in a service and entrepôt trading centre, such as Hong Kong, from a continental giant economy such as China or India. Such differences have led some economists and historians to eschew generalizations and aggregate growth models and insist on the variety of national circumstances. However, whereas the national system approach tries to explain the performance of each country in terms of its special features, growth theory attempts to generalize about all countries or at least large groups of countries. The sheer variety of culture, of political institutions, of economic and social structures, of scientific and technical institutions, of policies and of their interdependencies makes this an inherently difficult project. Wars, fluctuations in mobility of labour and capital, ease of technology transfer and operations of multinational corporations add to the difficulties. Nevertheless, it is obviously desirable to attempt generalizations. In practice, such generalizations may prove to be valid for only one phase of economic growth and for a particular group of countries, for example, West European catch-up with the United states after the Second World War, or Latin American and East European 'falling behind' in the 1980s. Nevertheless, as Hahn (1987) has argued, growth modelling has a heuristic value of this kind, even if it yields somewhat restricted results, demonstrating periods of convergence and divergence in particular areas, or even negative results.

As one of the few authors who took time and effort to collect historical data for a sample of now developed countries Maddison (1982, 1987) identified a number of quite distinct 'phases' of economic growth (see also Landes, 1965; Cornwall, 1977; Abramovitz and David, 1994; von Tunzelmann, 1995). Enlarging the economic data collected by Maddison with technological performance data, Pavitt and Soete (1981) found strong evidence of a relationship between those phases of economic growth and different patterns of innovative performance. Thus, and following the different Maddison phases,[1] they concluded as follows. The period 1890 to 1913 appears to have been characterized by four features:

1. What could be called 'entrepreneur-driven' output growth: the countries that grew more were those with a higher innovative strength which was ready to be exploited at the beginning of the period.
2. International technological convergence: late-coming countries tended to catch up in terms of innovativeness, but this technological upgrading did not imply directly higher growth.
3. International convergence in terms of GDP per capita via other processes, not linearly linked with technological innovativeness (described by Abramovitz as 'social capability' and by Freeman (1995) as 'congruence' of technology, culture, politics and economy).
4. In general, relatively loose links between technological variables and growth rates: that is, each nation's pattern of economic growth was structurally rather different with respect to the role of technological accumulation.

The period 1913 to 1929 on the other hand brings to the forefront:

1. A direct link between rates of change in innovativeness and income growth: innovative activities start being directly incorporated in the process of accumulation.
2. The tendency toward catching up and technological convergence continues, but in terms of per capita income growth one observes a divergence process.
3. The international uniformity of the links between technology and growth on the other hand increases.

In the period of the Great Depression and the Second World War:

1. The relationship between rates of change, innovativeness and growth remains strong: a sign of a transition to a phase of development where innovation, investment, productivity growth, and competitiveness become more directly linked.
2. Divergence in both technological capabilities and per capita income occurs.
3. The international structural uniformity – as defined above – in terms of the 'fine tuning' between technological regimes and patterns of growth becomes very low.

The period of high growth after the Second World War (1945–75) appears to be the only period where one can detect a high degree of structural uniformity, that is a relatively unified regime of growth in which technological activities play an important role:

1. Technological and income catching-up are correlated in a 'virtuous circle' which consists of imitation, learning, growing innovativeness, growing labour productivity, and high rates of investment.

2. Foreign competitiveness is now clearly fundamental to growth since it expands growth via the so-called foreign trade 'multiplier' and constrains it via the foreign trade balance.
3. Manufacturing output growth is the main engine of aggregate income growth.

The period after 1975 is generally acknowledged to correspond to a new phase: one in which the regime of international growth of the 'golden age' period – 'les trentes glorieuses' in Fourastier's terms – has come to an end, and with it the stability in the 'fine tuning' between technology and economic growth (Boyer, 1988). For an increasing number of the previously 'lagging behind' developed countries, the possibilities of large-scale imitation of US levels of technology became exhausted. As the long-standing association between rates of increase in innovative activities and output weakened, it was not apparently replaced by any new 'entrepreneur-driven' mechanism whereby high national levels of innovative activity were associated with subsequently favourable trends in economic performance.

Post-war growth at world level: how growth rates differed

As highlighted above the development path of the by now industrialized countries appears to be dominated by convergence. Most empirical econometric contributions have shown how over the post-war period in the OECD area the least developed OECD economies have increased per capita GDP at the highest rate. However, broadening the analysis to country samples larger than just the OECD raises many questions about any simple (linear) relationship between per capita income and subsequent growth rates. Again there exists by now a voluminous literature attempting to explain 'why growth rates differ' to put it in Denison's original terms.[2]

As an illustrative way of getting some feeling for the relation between growth rates and technology in the world, Figure 13.1, taken from Verspagen (1993) summarizes some of the available evidence on growth rates by trying to detect regularities in growth performance across countries, in relation to the initial country's development gap with the leading country in the post-war period: the USA.

Verspagen calculated, as illustrated in Figure 13.1, the initial (1960) values of 114 countries' per capita income gap (G, defined as the logarithm of per capita income in the USA over the per capita income in the country) against its motion over the 1960–85 period.[3] Each point corresponds to a particular country, classified into 'oil exporters', 'newly industrializing country', 'developed market economy' and 'other' following the traditional World Bank or UN definitions. The lines drawn in the figure are simple (linear) regression lines for the different subsamples of countries. A positive change of the income gap implies convergence or catching up; a negative change divergence or falling behind. Hence, positively sloped regression lines are consistent with convergence.

As noted by Verspagen, if there is any systematic pattern in the total cloud of points in Figure 13.1 it is the variance, which grows bigger the larger the per capita income gap becomes. Thus, countries close to the

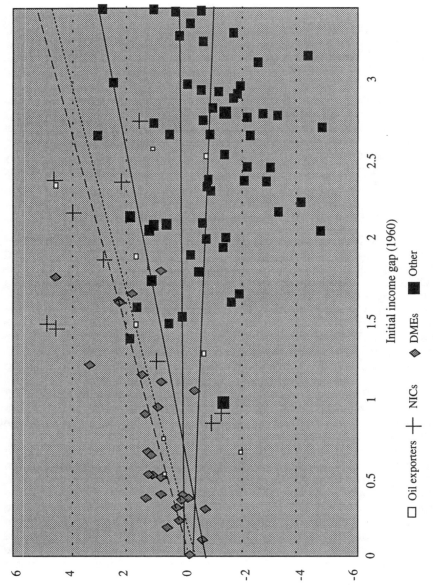

Fig. 13.1 Annual percentage decrease of the income gap, 1960–86

world economic and technological frontier (as measured by the perform-ance of the USA) show smaller (absolute) growth rate differentials relative to this frontier than those further away from it.

However, the results in Figure 13.1 indicate that there is a clear dicho-tomy between convergence and divergence at the world level. Countries facing the largest gaps in the 1960s (the least developed countries) have also experienced increases of the gap (a negatively sloped regression line). They have seen their income *per capita* lag even further behind, which is exactly opposite to what the convergence or catching up hypothesis predicts.

13.2 ON MODELLING TECHNICAL CHANGE AND GROWTH

It is only fair to say that no significant body of economic thought, whether classical, neoclassical, Keynesian or structuralist, has ever suggested that the influence of technical change was unimportant. Economists of all persuasions have always accepted that technical innovation was one of the most important, if not the most important source of dynamism in capitalist economies. However, this superficial appearance of agreement should not obscure the differences in relation to the analytical framework used by various economists in their approaches to the problems of technology and growth. These differences were clearly apparent once attempts were made to model the growth process.

From old to new growth theory

Recent years have seen an explosion of renewed interest in growth modelling. This renewed interest is to some extent a reflection of the importance of the change of perspective which has been adopted con-cerning the sources of growth with the advent of a class of 'new' growth models. The new approaches stress the particular endogenous role played by technical progress highlighting the determinants of such progress. First, we turn to a brief review of the older, more traditional growth models. Whereas 'technical progress' could be regarded as a 'residual' in many older growth models, it now enters centre stage.

The 'old' neoclassical growth model (Solow, 1956, 1957) was charac-terized by a set of largely traditional neoclassical assumptions: a constant returns production function – a hypothesis compatible with perfect com-petition; capital and labour as the two production factors, whereby capital could be accumulated, labour not and 'optimizing' behaviour on the part of individual agents in the economy. Probably in response to some of the fundamental questions raised by Harrod (1948) with respect to the inherent instability of growth, failing to achieve 'full employment', Solow laid the foundation of neoclassical growth theory and illustrated how a stable full employment growth path could be obtained in a relatively simple neo-classical growth setting. Both output and the capital stock would grow along a balanced growth path at a rate equal to the sum of population growth and the rate of (labour-augmenting) technological progress. Without population growth and without such a labour-augmenting or constant rate of technological progress, growth would vanish due to the decreasing marginal returns on the only accumulated factor, capital. Accumulating capital would bring about decreasing returns which would act as disincentive for investment in the long run. Only the two exogenous influences (population growth and technology) could save growth, hence the long-term equilibrium growth rate became exogenous and by and large independent of economic influences.

In more formal terms, output (Y) can be assumed to be produced at constant returns in a production function with two factors, capital (K) and labour (L):

$$Y = f(K,L)$$

An exogenous trend of technical progress as well as a constant growth rate of the population can be added to this specification. An equation representing the accumulation of capital, whereby the latter is homogenous, the same output may be used for investment or consumption purposes, can also be added to the model. Agents are confronted with allocation decisions concerning investment or consumption.

The growth process can then be represented as follows: the variation in the level of capital (K) depends on the level of output (Y), since Y and K are linked by a functional relationship: the savings ratio (s). Since the marginal productivity of K in the production of Y decreases with the level of K, the higher this level is, the less capital contributes to increasing production. As a result, capital accumulation becomes more and more difficult, eventually leading to zero growth in the long run. This result stems from the specification of the savings behaviour and is a typical neoclassical result. While elegant and illuminating, the neoclassical growth model does not explain long-term growth. Growth of per capita income in the long term can only exist if there is an exogenous trend due to technical progress. Otherwise, growth in Solow's basic model is limited to what has been called 'transitory' dynamics.

There are other aspects of Solow's model that are also worth noting. The first concerns the influence of savings on the capital/worker ratio. The level of the savings ratio influences the capital/worker ratio, but not the equilibrium growth rate, exogenous by definition. At the equilibrium point, in fact, all savings are used to build up capital for new incoming workers, not for existing workers. As a result, policies affecting the level of savings will only have a level effect. A transitory effect will occur when the economy converges towards the equilibrium level of capital per worker, but this will not affect the growth rate.

Second, in an international framework, all countries must converge towards the same level of capital and income per capita under the hypothesis that all agents have the same tastes, that is, they have the same savings rate. As such, poor countries will 'catch-up' to rich countries. The observed facts of the last decades (Figure 13.1) do not seem to lend much support to this finding.

To obtain 'unceasing' growth, one must consider an external factor that increases the productivity of inputs over time. Exogenous technical progress T can then be integrated into the production function:

$$Y = f(K,L,T)$$

but it must be assumed that T is not a production factor like K or L. The assumption of constant returns in the production function would mean that the marginal returns of capital to technology would have to be decreasing, preventing them from contributing to growth in the long run unless additional specific assumptions are introduced.

The effect of technical progress could be twofold: (i) there is a direct increase in productivity; (ii) an increase in the return to capital, leading to additional investment and thus extra income. Accumulation of capital is a direct consequence of technical change, which is the only source of growth. As technical progress is not explained, growth is still exogenous in this model.

This weakness can be overcome by attempting to endogenize technical change as Kaldor (1957) and Arrow (1962) did by focusing on learning effects as a source of technology improvement. Other 'older' endogenous growth models adopt similar approaches, where technical progress is a byproduct of production, investment or human capital.[4] The breakthrough in the formal modelling of such 'endogenous' growth features came with the advent of 'new' growth models some ten years ago.

New growth models

As compared with the traditional neoclassical growth model (Solow, 1956, 1957) where technical progress is a simple time trend, new growth models consider determinants of technical change, by which in essence they come up with an endogenous determination of the sources of growth. Perpetual growth is now made possible by the presence of increasing returns to scale or externalities, which guarantee that marginal productivity in the accumulation of factors does not go to zero when these factors are accumulated. As Amable (1994) notes, the 'new' growth models 'adopt a particular specification of accumulation: a certain quantity of resource produces a given percentage increase of a factor, not a given quantity'.

Many authors have attempted to classify such endogenous growth models. For present purposes, the classification proposed by Amable (1993) based on the sources of growth is probably the most interesting one. A number of 'increasing' sources of growth might indeed exist.

1. A first source of endogenous growth lies in investment in a certain factor. Romer (1986) for instance, considers a relatively simple and traditional growth model not restrained, however, by constant returns, but with economies of scale which are external to the firm.
2. A more directly obvious source of growth is technological innovation, itself dependent on the amount of resources devoted to R&D and other knowledge generating activities. In a model again originally put forward by Romer (1990), capital is now not a homogeneous good, but consists of a set of different intermediate goods. New, intermediate inputs are discovered when R&D resources are devoted to a search process. A contrasting framework is presented by Aghion and Howitt (1992) where such innovation can also consist of a number of 'creative destructions'

rather than just new additions to the range of available inputs to production. Each new input now replaces the preceding one, thus ending also the monopoly power attached to it.

3. The accumulation of human capital is another source of endogenous growth. In the model developed by Lucas (1988), individuals as with investment in Romer (1986), accumulate human capital in a context of increasing returns. The higher the average level of human capital, the higher the productivity of each worker in the production of the final good.

4. Finally, growth may also be realized through public goods and infrastructure; one may think today of communication networks, information services, etc. Such goods increase the productivity of private factors. The possibility simultaneously to use these goods by a large number of agents makes them public goods in the traditional sense of the word. They are produced by social institutions financed through taxes. Public goods policy will be of crucial importance for the provision of these goods.

The previous typology illustrates the particular role of investment – in physical, R&D, human and public capital – in the relationship between technological change and 'endogenous' economic growth. It is important to see such explanations of growth in their dynamic demand-led context. In terms of an individual firm's decision to invest and in more qualitative terms, it might be described as follows. Before developing new growth opportunities, e.g. through the introduction of a new product, the firm will generally look for opportunities in line with its know-how. After evaluating demand trends for new and improved products, and in order to adapt its production programme accordingly, it will undertake when necessary research and development efforts (intangible investment) which will augment the firm's stock of technological knowledge. The latter, however, can only become effective if followed up with physical investment. Depending on successful marketing, this dual investment decision, initiated itself by the search for new opportunities, will eventually lead to the firm's expansion, taking into account the 'creative destruction' with respect to reducing demand for older products. At the level of the production process, the virtuous cycle can be seen to start with a strategic decision to reorganize whereby the company expects an improvement in its competitivity/price (or its competitivity/flexibility).

The corresponding managerial innovation is dependent on information on the potential of new types of equipment. Again, this will generally but not always lead to physical investment which, in turn, will be translated into an increase in productivity, and subsequently of the firm's market share.

At the aggregate macro-economic level this virtuous cycle of intangible investment, learning, physical investment and market pressures (both in costs and demand), described above at the level of the individual firm, highlights the particular importance of intangible investment for 'endogenous' growth, to embrace the positive externalities of technological change. Growth in investment is led by market growth and market size as in the classical contributions of Smith (1776), or Young (1928), or the more Keynesian contributions of Kaldor (1957). The 'post-war' liberalization of trade has certainly further facilitated market expansion in terms of geographical size, purchasing power and diversity. This continuous market widening and quality improvement trend has offset any inherent tendency for individual product market saturation (cf. also Pasinetti, 1991). The extent and nature of this market growth has led to vigorous competition, both in terms of product differentiation and cost reduction, causing a speeding up of technological and organizational innovation in terms of potential productivity gains and adaptation to more diversified demand. Furthermore, this speeding up has required an increasing number of researchers and research resources within the science-based sectors and many 'high tech' user sectors. Because of the increased complexity and diversity of science and technological capabilities, this has resulted in the growth of networks, alliances, technological and scientific exchanges, that is to a growth in the 'externality' effects of technology.

As described in the previous chapter, the latter now go well beyond the innovating firm or sector as well as beyond the innovating country. The intensive production of the latest technological novelties tends to accelerate the acquisition of new equipment, thereby increasing physical investment. This increase may be further enhanced by macro-economic liberalization: the liberalization of capital markets and the internationalization of financial markets and investment, which are, as we discuss in more detail in the following chapters, closely connected to the internationalization of technology and the globalization of enterprises. New investment results in productivity gains and a wide range of new or differentiated products. Following the classical 'increasing returns' arguments (cf. Young, 1928; Verdoorn, 1949, 1980; Kaldor, 1980; Buchanan and Yoon, 1994), this increase in production generates additional revenue that creates additional demand for other goods and services and new opportunities, closing the virtuous growth circle to the first trend described above.

Despite the insights such a view of the interaction between technology, investment and growth contains, particularly with respect to some of the increasing return features of technological change, it will be obvious from the brief description above that new growth models remain practically by definition rather schematic presentations of the complexity of the interaction between technology and growth (Nelson, 1994). Other more qualitative analyses focus on some of the structural features associated with growth and have blossomed over the post-war period. It is to these contributions that we turn next.

How old is new growth theory?

First of all, it should be emphasized that in these new growth models the conventional production factors, capital and labour, do not play the role we are used to from conventional (neoclassical) models. This holds both for capital and labour and is, of course, due to the fact that many of these models are meant for analytical rather than empirical purposes. With respect to labour, for example, the 'new growth theory' makes a distinction between labour and human capital. In most models human capital is the major input in the research (R&D) sector. It is then (for simplicity) assumed that the same factor (human capital) is also employed in the consumption goods sector. In practice, however, there is a big difference between (the quality of) labour employed in the R&D sector and labour employed elsewhere. Thus, the 'increasing returns to labour/human capital' effect is limited to the extent that it is indeed a larger employment of researchers that causes the increase in the growth rate. If, for (institutional) reasons, such as a lack of schooling or rigidities in the labour market, a larger resource base would not lead to a larger employment in the research sector, the scale effect would not be present in the model. This also points to a possible explanation for the often undesired (by the authors of the models themselves) 'prediction' of most of the models that countries with a larger resource base (in terms of population growth) would grow faster than countries with a constant or declining population.

Capital too in the usual sense is absent in most of the 'new' growth theories.[5] It is either assumed that physical capital is absent, and only investment in knowledge matters (Romer, 1986) or, in more sophisticated models, that there is only an intermediate good which, contrary to physical capital in the usual sense, does not accumulate. In the latter case, it is assumed that improvements in the quality of these intermediate goods are generated in the research sector and improve productivity in the consumption goods sector. Moreover, again for simplicity, most models assume either that there is a homogeneous intermediate good, or that all intermediate good classes are affected by technological change to the same extent.

This leads to the conclusion that more investment in R&D leads to a higher increase in the quality of the full range of intermediate goods and thus, via an increase in investment, to a higher growth rate. However, once some Schumpetarian features of 'creative destruction' are allowed for, as in Aghion and Howitt (1990), the simple fact that the addition to intermediate goods will also lead through a so-called 'business stealing effect' to some intermediate goods becoming obsolescent. Interestingly in this case, the previous conclusions no longer hold. Increases in R&D could lead to lower growth.[6]

With respect to technology, one of the main arguments in the 'new' growth models is that technological change is a good which has to be produced in the same sense as other goods, but also contains some typical features of a public good (Romer's concept of 'shareability'). One characteristic of such goods is that in a free market economy, there will be an inherent tendency for underinvestment in the production of it. Another characteristic is that the level of output of this good has an influence on the growth rate of the economy (increasing returns to scale). At the empirical level, this latter conclusion has been the basis for estimating correlations between investment and the growth rate of output, or marginal productivities of capital in so-called 'unrestricted' (i.e. leaving open the possibilities for increasing returns to scale ex ante) production functions. These methods, which are of course partly motivated by the availability of data, implicitly adopt the assumption found in most 'new' growth models that capital (or intermediate goods) are homogeneous (De Long and Summers, 1991).

If one does not aim for formal econometric modelling estimations, the empirical conclusions from the new growth theories lose much of their power. Most of the conclusions that one would draw from such analyses can also be drawn from analyses based on other models. Take, for example, the well-known correlations between so-called 'total factor productivity' and R&D outlays, as estimated by, for example, Griliches (1984) on a number of occasions. These correlations would fit the picture drawn from 'new' growth theories rather well. The same argument holds for the relation between the number of patents, which could be interpreted as an indicator for the number of varieties of an intermediate good and economic growth. Such empirical experiments introducing directly a technology variable in the analysis might improve to a greater extent one's insight into the nature of the relation between technological change and economic growth, than some of the present empirical exercises.

The scope for further integration of some of the ideas of new growth theory with a more detailed analysis of the differentiated role of science and technology in specific sectors is in our view a promising avenue for further research in this area. If, for example, we recognize that in the production of the good, we call GDP both inputs which are not so much subject to (quality) changes due to technological change and inputs which are subject to such changes do play a role, a logical extension of 'new' growth theory might consist in separating out those inputs in the production function in which technological change plays a small role (call these inputs low tech) from those in which technological change plays a large role (call these high tech). The simple aggregate dynamic scale effects put forward by the 'new growth theories' would not necessarily hold in the case of investment being directed towards low tech investment goods. By contrast, for high tech inputs, the scale effects are more likely to hold an improvement in quality of these goods (via resources applied in the R&D sector) resulting in more investment in it, and thus having a positive influence on the growth of productivity in the aggregate economy. The soundest basis for interpreting the stylized framework of new growth theories with reference to the real world seems therefore to allow for a distinction between high tech and low tech inputs.[7]

It is not difficult to see how the integration of technological investment, in the form of intangible investment, particularly within the more disaggregated framework set out above, leads one also to reassess in a number of ways some of the traditional macro-economic policy questions. Thus, once the more realistic picture of increasing returns with which many features of technological change can be associated is introduced in macro-economic growth analysis, it becomes very clear that policies with respect to technology (e.g. R&D subsidies) cannot be equated with other traditional static efficiency improving micro policies, such as competition policy, but do have a significant dynamic growth impact. In the more recent open economy growth models (Grossman and Helpman, 1990), this policy conclusion becomes, not surprisingly subject to the relative efficiency with which new technologies are 'produced' in one country compared to another. We come back to this 'internationalization' aspect in the next chapter.

13.3 A COMPLEMENTARY NEO-SCHUMPETERIAN VIEW OF GROWTH

Not necessarily opposed, but rather complementary to the 'new' growth tradition, the more descriptive contributions in the Schumpeterian tradition, often based on detailed case studies of the emergence and diffusion of particular new technologies, bring to the forefront the complexity of the mechanisms governing the development and diffusion of technological change. The hypothesis discussed in Chapter 1 that technological development is characterized by the existence of 'technological paradigms' which embody powerful heuristics and determine a relatively ordered pattern of technological change – so-called technological 'trajectories' – allows one also to introduce some of the disruptive, creative capital destruction features of technological change. Within such a Schumpeterian framework, one may consider changing paradigms and 'normal' technological progress within existing paradigms as describing trends in discontinuous versus continuous technological development.

The latter distinction, crucial to Schumpeter, is of course central to many of the recent concerns with respect to the apparent lack of macro-economic evidence of productivity gains resulting from the application of new technologies and the more general concern about the slow diffusion of so-called 'generic technologies' (ICT and biotechnology in particular).

We discuss below some of the possible explanations for this so-called Solow productivity paradox. Solow pointed out that although computers could be seen everywhere, the actual measured productivity gains from their use appeared to be very small or even non-existent. The explanation most emphasized brings to the forefront the importance of various diffusion barriers including human resources, organizational and management practices and more broadly the surrounding social, economic and institutional environment.

The analysis starts here from the recognition that clusters of radical technical innovations do also lead to major disruptions not just in the production sphere but also in the broad social, institutional and organizational sphere. As a consequence it is far from paradoxical that the economic and social potential of such technologies will only be realized over fairly long historical periods and that during this 'learning' process changes in management strategies as well as in the institutional environment will influence the success of enterprises and national economies. From such a perspective, such new generic technologies require also fundamental changes in societal attitudes and institutions, without which their diffusion and potential productivity gains will not be realized.

As a case in point, it has been argued by many authors that information technology is so pervasive and has so many new features that it is the latest hurricane in Schumpeter's successive 'creative gales of destruction'. Whereas incremental innovations do not give rise to major problems of structural adjustment, the introduction of such a radically new technology system gives rise to many such problems. It does so because – and the list is not exhaustive – it requires a redesign and new configuration of the capital stock, a new skill profile in the labour force, new management structures and work organization, a new pattern of industrial relations and a new pattern of institutional regulation at national and international level, for example, in relation to the global telecommunications network or traded information services.

The profound nature of this transformation is now generally recognized. As the *Economist* (1986) put it: 'The factory is being re-invented from scratch. Traditional production lines are being ripped apart to make room for flexible "make-anything" machinery . . . long narrow production lines with men crawling all over them – a feature of manufacturing everywhere since the early days of the car-making dynasties – are being . . . replaced with clusters of all-purpose machines huddled in cells run by computers and served by multi-fingered robots. The whole shape of the industrial landscape is changing.'

Similar points could and have been made about the 'office of the future'. The personal computer, the word processor, data banks, electronic mail and many other new information services are bringing about an equally dramatic transformation in the pattern of service activities, many of which are becoming capital intensive for the first time and in some cases R&D-intensive (see Chapter 17). Such pervasiveness of a cluster of new technologies brings, of course, to the forefront the question of the diffusion process across sectors of the economy. With a pervasive technology such as IT, this is one of the main problems of structural adjustment. It is hardly surprising that in contrast to the computer industry, which has been able to make the most effective use of its own technology, most manufacturing and service industries which have little previous experience with this radically new technology should experience innumerable difficulties and 'snarl-ups' as they attempt to use it. They frequently lack the necessary skills as well as the management competence. It is no longer a question of an incremental improvement in an established trajectory with which they were familiar but of a break with the past.

In John Diebold's book *The Advent of the Automatic Factory*, written in 1952, he showed astonishing foresight about these problems. Diebold pointed to the new skills that would be required, especially in design and maintenance, on an enormous scale. He also pointed out that computerization involved the 'design of the whole range of products as well as processes'. This would not be possible without changes in 'structures of most firms to facilitate the flow of information between R&D, design, production and marketing'. This in turn would require changes in 'the structure of management to facilitate horizontal movement of people and information'. The simple availability of computer hardware was only the most elementary first step and the whole process would be a matter of decades not years.

Events have proved Diebold completely right. It was only in the 1980s that computerized automation in the form of FMS, CIM, CAD, etc. started to take off on a significant scale and undoubtedly it has a very long way to go. This applies *a fortiori* to computerization in the service sector. In this complex diffusion process it is necessary to adapt computer technology (especially software) and telecommunication technology to the specific needs of each sector and each enterprise. All managers of information systems and of software consultancy firms would confirm that this is so. For example, Lockett (1987) in his study of information technology innovations in a large multinational company pointed out that an intense dialogue between the provider and the user was an essential condition of the successful implementation of IT innovations. An experimental R&D approach was necessary.

These findings confirm earlier innovation research in the 1960s and 1970s (Chapter 8) that the understanding of user needs was the most important condition for successful innovations. However, in the case of information technology this requirement of 'user friendliness' is exceptionally important and attended by special problems. In the first place, the user often finds great difficulty in conceptualizing and specifying the precise nature of the requirement. Second, the provider rarely has the possibility of understanding this requirement immediately from the outside. Third, the technology itself is still extremely fluid. Finally, the introduction of IT frequently brings with it the need for reorganization of management itself and/or production systems.

The historical analogy with the diffusion of electric power drawn by Freeman (1987) and David (1991) is particularly revealing here. The key technical innovations were made between the 1860s and 1880s. With the establishment of effective generating and transmission systems in the 1880s and 1890s (which also involved a wave of social, legal and organizational innovations) the emphasis shifted to diffusion of applications of electric power throughout the economy both in industry and households. In this phase, although technical innovations continued at a high rate and big improvements were made in the available hardware, the key problem was the organizational change in firms, change in skills, factory lay-out and in attitude of engineers, managers and workers. Warren Devine (1983) described this change in words which might also be applied to the diffusion of computerized automation (see Chapter 3).

The analogy between electrification and computerization is a useful one because it illustrates the long time scales involved in any type of technical change which affects a very broad range of goods and services, where many products and processes have to be redesigned to realize the full potential of the new technology. Perez (1983, 1989) has pointed out that the dominant technologies of a particular period actually become an integrated web since they are developed together to take advantage of a specific range of new technical and economic opportunities, whether related to electricity and cheap abundant steel, or to electronic computers, cheap integrated circuits and universal cheap telecommunications. As the techno-economic paradigm crystallizes and develops, the interrelated advantages become more and more apparent and the logic of the new system comes to seem self-evident. A flow of new technical and organizational innovations helps to resolve the main problems and bottlenecks. But in the early stages there are great difficulties and risks in adopting the new equipment because it is not yet an integrated system, and adoption involves a piecemeal trial and error process of learning by doing and using.

These considerations mean that it would be quite unreasonable to expect sudden increases in productivity in sectors far removed from the new technology. Such increases could be expected to emerge only after a fairly long process of familiarization, of developing customized equipment, new skills and new management attitudes and structures. One may stress here the cross-sectoral aspect of the diffusion process, the learning by doing and by interacting (Lundvall, 1988a, b) and the time lags involved in moving from the potential productivity gains of a radical new technology system to their actual realization.

The point will be clear. In the process of the translation of potential productivity gains to actual productivity gains, the focus on just physical investment will not be enough. There will be crucial 'learning' and other adaptation lags, which might well explain why the diffusion process has been slow and uneven between various countries as we saw in Chapter 12.

In his review of old and new growth theory, Nelson (1996) pointed out that most of the ideas embraced in the new growth theory were actually advanced by Abramovitz in 1952 or by other economic historians even earlier. He poses the question: 'Why such a lag in incorporation of under-standings long held by scholars of economic growth into formal growth theory? . . . What if anything is gained when understandings which are widely held are incorporated in formal theories?'

In answering these questions he argues that what he terms 'appreciative theorising' proceeded for the most part quite independently of formal growth modelling. Formal theorizing failed to keep up with appreciative theorizing and lent it little help. Our own survey leads us to endorse his conclusions: that economic research yields the best results when appreciative and formal theorizing interact effectively and that the new growth theory offers better possibilities for such a positive interaction.

NOTES

1. Many authors have referred in this context to the notion of 'conditional convergence'. See e.g. Barro (1991) and Mankiw *et al.* (1992). Maddison (1995) has further enlarged his analysis to other developing countries, focussing more explicitly on the conditions for convergence to emerge.
2. Among which are Abramovitz (1979, 1986), Baumol (1986), Dowrick and Nguyen (1989).
3. Data are taken from Verspagen (1994); see this paper for more details on the construction of the growth rates.
4. See, among others, Uzawa (1965); Phelps (1966).
5. An exception being Scott (1989).
6. Since then, starting with Grossman and Helpmann (1990), many authors have focussed on such growth models called 'Schumpeterian' only because they include such features as 'creative destruction'. For an overview see Aghion and Howitt (1994).
7. See in particular some of the econometric work on R and D spillovers e.g. Griliches (1980).

CHAPTER 14

INNOVATION AND INTERNATIONAL TRADE PERFORMANCE

14.1 INTRODUCTION: ON STATICS AND DYNAMICS IN INTERNATIONAL COMPETITIVENESS

In Part One we have mentioned many examples of strong export performance associated with success in technical and organizational innovation (the British cotton and pottery industries, the German synthetic dyestuffs and plastics industries, the Japanese automobile and consumer electronics industries). Such examples, although they certainly accord with popular impressions of the time, lack any systematic underpinning in terms of trade theory. Such a theory should of course take into account imports as well as exports, the advantages and disadvantages of trade specialization, the role of natural resources and so forth. In this chapter, therefore, we attempt to relate the wealth of empirical findings on the connections between innovation and trade performance to the general economic theory of international trade.

The importance of technological innovation in determining long-term international competitiveness and trade performance has received increasing attention over the last two decades from both international trade economists and policy-makers. As in the case of the new growth theory discussed in the previous chapter, the main reason for this increased policy recognition appears at first sight to be the new recognition that knowledge creation is characterized by significant growth externalities and that at the international level strategic technology policy intervention might be justified, even required in particular cases, thus breaking down nearly twenty-five years of what some have called an economists' cartel agreement in support of all trade liberalization.

The chapter starts with a brief account and assessment of old and not so new trade theory based, among other things, on the analysis presented in Dosi *et al.* (1990). The somewhat pervasive notion of strategic trade policy and its various forms of implementation is introduced. In the next section, the basis of the national focus of such policies is questioned. Strategic national policy discussions appear increasingly out of tune with the internationalization level at which most of the supposed beneficiaries of such policy intervention operate – large often multinational firms. As many authors have emphasized, most explicitly Ostry (1995), it appears somewhat paradoxical that now that the scope for national strategic trade and technology policy is being recognized by both international trade theorists and national policy-makers, the effectiveness and possibility for implementation of such national policies appears to have been reduced.

Once upon a time, international trade theorists like to tell each other,[1] there was a paradise where everybody lived efficiently, producing and trading whatever was demanded in the most efficient combination. Then an angel came and stamped on each person's forehead a different colour, you could say a national flag, allowing him or her to produce and trade only with capital and land with the same colour. The diaspora which followed led to large differences in efficiency across the world, with a huge world welfare loss. Since that unhappy moment, international trade theorists – by definition economists with a world rather than national welfare vision – have been trying to show how to get back to this paradisiacal situation.

The first main direction of analysis, going back to the classical economists, tried to show how, despite a country's poor efficiency there could nevertheless be gains in welfare by specializing in those products/industries in which it was relatively most efficient. Such gains were by and large based on the principles of division of labour applied to an international world. The neoclassical extension of this line of analysis introduced more formally 'factor endowments' in explaining a country's comparative advantage and established a number of crucial links with factor price equalization, income distribution and growth. In terms of our parable, it could be said that trade theory established how, through God's invisible hand, free trade would undo the angel's ill-doings and re-establish paradise all over the world, despite the national differences in 'factor endowments'.

There is no doubt that international trade has been one of the main engines of growth in the post-war period. With the continuous liberalization of trade, world trade flows increased over the period 1950 to 1975 by more than 500 per cent, compared with an increase in world output of only 200 per cent. However, despite the success of institutions such as GATT (now WTO) in developing a stable, liberal and non-discriminatory trade system which came to dominate larger and larger areas of manufacturing, international trade theorists started to question the theoretical basis underlying such trade flows.

First, empirical trade analysts found it increasingly difficult not to be surprised by the number of trade flows which did not fit 'pure' trade theory explanations. These findings conveniently described as a 'paradox' (the so-called Leontieff paradox) seemed evidence of the limits and limited value of pure trade theory in a world dominated by imperfect competition phenomena. The unease with the existing theoretical trade framework became a standard opener of many trade analyses. For example, Bhagwati, writing some thirty years ago, put it as follows:

> The realistic phenomena . . . such as the development of new technologies in consumption and production involve essentially phenomena of imperfect competition for which, despite Chamberlain and Joan Robinson, we still do not have today any serious theories of general equilibrium. . . . Unless therefore we have a new powerful theoretical system . . . we cannot really hope to make a dent in the traditional frame of analysis.
>
> (Bhagwati, 1970, p. 23)

Hufbauer (1970, p. 192) in a more ironical tone reviewed his empirical 'paradoxical' results by observing that they could 'as yet offer little to compare with Samuelson's magnificent (if misleading) factor-price equalisation theorem'.

The second set of queries related more directly to the success (or failure) of 'traditional trade theory' in quantifying the gains from trade. Much to the surprise of many policy-makers, trade analysts came up with rather low estimates of gains from trade following trade liberalization and, for example, the creation of free trade markets.[2] The estimated gains actually excluded some of the most important and obvious gains from opening up to trade related to imperfect competition; for instance, the efficiency gains associated with reaping scale economies or the consumer gains associated with the availability of greater product variety, which the pure trade model did not and could not include. As the quote from Bhagwati suggested, it became only a matter of time before a 'new' line of analysis was developed.

This second main 'new' direction of analysis, developed over the last ten years and associated with the names of Grossman, Helpman and Krugman, started from a fundamentally different assumption: that many if not most economic activities are characterized by increasing rather than decreasing returns. From this perspective gains from trade are in the first instance the result of the scale economies each national economy can achieve through free trade, whether the size of Luxemburg or the USA. These gains are actually far more significant than traditional trade theory would lead one to believe. Many empirical studies within the 'new' trade theory tradition have thus pointed towards the significance of such gains. This was done most clearly by Smith and Venables (1988) with respect to the further harmonization of the European Community's internal market and by Harris (1984) with respect to the Canada–USA free trade agreement.

In terms of our parable, it could be said that just as in large nations where particular activities have been concentrated in particular locations – Krugman's favourite example being mushroom production in Pennsylvania – 'paradise' for the world as a whole will be achieved by bringing resources together, whether motor car manufacturing in Japan or ceramic tiles in Italy. Again 'free trade' will be essential for achieving this, this time though primarily because it allows consumers to reap the 'imperfect competition' advantages (scale economies and product differentiation) of such trade. Under this approach the advantages accruing to the region or country from the agglomeration of a particular set of activities are not so much of importance compared to the advantages to every world consumer of the efficient exploitation of world economies of scale.

However, setting out from such a radically different starting assumption, 'new' trade theory has also led to a plethora of 'new', sometimes perverse, sometimes similar, theoretical results with respect to some of the basic trade theorems. The most controversial normative result from a traditional trade perspective has undoubtedly been the illustration by Brander and Spencer (1983, 1985) that free trade might in some cases of market structure no longer yield world maximum welfare gain, but that a 'strategic' trade policy might be justified and actually needed. As Dixit pointed out in his contribution to Krugman's book on strategic trade policy:

Recent research contains support for almost all the vocal and popular views of trade policy that only a few years ago struggled against the economists' conventional wisdom of free trade. Now the mercantilist arguments for restricting imports and promoting exports are being justified on grounds of 'profit sharing'. The fears that other governments could capture permanent advantage in industry after industry by giving each a small initial impetus down the learning curve now emerges as results of impeccable formal models. The claim that one's own government should be aggressive in the pursuit of such policies because other governments do the same is no longer dismissed as a *non sequitur*.

(Dixit, 1986).

The discussion surrounding strategic trade policy has brought to the forefront many features which appear at first sight to reflect better the industrial reality with which both policy-makers and businessmen are confronted in many sectors. Particularly with respect to analyses of technical change and international trade, this discussion seems to offer a better theoretical framework to discuss appropriate policies covering the whole spectrum of trade, industrial and technology policies. The importance of monopoly rents, profit sharing and strategic trade manipulation seem of particular relevance to many high technology industries. Furthermore, the actual emergence of these new theories on the US academic scene occurred at a time of increasing fear in the USA of the Japanese challenge in trade and technology (Mowery and Rosenberg, 1989).

However, the emergence of strategic trade concepts also brings to the forefront some of the dynamic features associated with technological change, in particular its cumulativeness. Compared to the previous set of 'new' trade theories, the alternative Schumpeterian trade theories, as reviewed in Dosi *et al.* (1990), put a greater emphasis on the dynamics of increasing returns, particularly those associated with production technology and innovation. In terms of the opening parable, to the extent that technological development and growth are irreversible processes, there is no possible return to 'paradise'. Like virginity, once lost, it is lost for ever. There is rather a historically driven development process with no particular connotation of optimality. Dynamic features closely associated with technical change such as learning and the accompanying feedback effects (both positive and negative) are, from this perspective, the main explanatory factor behind the very different development and specialization patterns (some virtuous, some vicious) which countries appear to get 'locked into'.

As emphasized in many locational theories, the reasons for such differentiated 'path-dependent' development processes have to do with the way industrialization locations get selected early on and, by appropriating the available agglomeration economies, exercise some 'competitive exclusion', to use Arthur's (1988a, b) term, on other locations. The example of the British cotton industry cited in Chapter 2 is an excellent one. In other words, from a dynamic technology perspective, it does matter whether a region or country is specialized in mushroom production or silicon chips. Few authors in the 'new' trade theory have yet fully integrated these dynamic features in their trade models[3] and policy conclusions.

Is there no 'normative' world paradise to be attained in this last vision of the world? In terms of unifying overall principle, the answer to that question can indeed only be no. The normative counterpart of any dynamic and more evolutionary analysis brings to the forefront the crucial role of history, of manmade interventions, of institutions, of particular international investment decisions of multinational corporations; of the whole spectrum of individual and collective decisions made in a complex system such as the international economic environment.

More than in the second direction of analysis, this third approach brings to the forefront the possible trade-off between policies directed towards static allocative efficiency and dynamic growth efficiency. Once concepts such as increasing returns are introduced, there appears to be nothing in the mechanism leading to static allocative efficiency that would also necessarily guarantee the fulfilment of the criteria of dynamic efficiency. As highlighted in Dosi *et al.* (1990):

> In the static neo-classical 'pure' trade world the theorem of comparative advantage will operate in its purest form: each trading partner gains from trade since it gets more commodities from abroad than it would otherwise be able to manufacture domestically without forgoing any production and consumption of the commodities in which it specialises. The same could be said with respect to the static interpretation of the early 'new' trade contributions: as in the traditional case trade gains in the true meaning of static allocative efficiency are typically of a 'once-and-for-all' nature. By contrast, once some of the dynamic economies of scale associated with 'strategic' products and industries are introduced one is confronted with the *possibility of* significant trade-offs between statics and dynamics.
>
> (Dosi *et al.*, 1990, p. 202)

This point has been highlighted by many authors (see e.g. Chenery, 1959) long before contributions to trade theory brought it to the attention of policy-makers in a coherent and formalized way. Indeed, if different commodities or sectors present significant differences in their 'dynamic strategic potential', for example, in terms of economies of scale, technical progress, learning-by-doing, etc., international specializations which are efficient in terms of static comparative advantage criteria may either generate in the long run virtuous or vicious circles of technological advance/backwardness.[4]

14.2   INNOVATION AND INTERNATIONAL COMPETITIVENESS

Interest in differences in technology as a basis for trade grew partly in response to the Leontieff paradox. Whereas traditional classical factor proportions theory would have led one to expect that the United States would perform best in capital intensive goods, Leontieff (1953) showed empirically that in fact it did rather better in labour intensive commodities. Superior technology was presented as a potential explanation for this pattern of US trade. The first theoretical contribution was the technology gap model of Posner (1961), who observed that the production of new products and processes conferred a temporary monopoly advantage on the producing country and could provide a basis for trade not founded on differences in natural endowments. However, the diffusion of the innovation over national boundaries would undermine the monopoly advantage. Only by continually producing new innovations and thus sustaining a 'technology gap' could a country maintain a comparative advantage in new products over time. This theory emphasized the ability to appropriate the benefits from innovations and the monopoly advantages they confer on the innovator. The product cycle approach (Chapter 11) considers a similar dynamic framework for the international location of production based on the characteristics of goods as they mature. Maturity brings standardization over time and leads to the diffusion of production from the advanced innovating country, which has an advantage in the production of non-standardized products, to less technologically advanced countries.

These theoretical perspectives on trade and technology are distinct from attempts to incorporate differences in technology into the mainstream neoclassical Heckscher–Ohlin framework, by the way they treat technology as a concept. They have expanded the two-factor endowment model (labour and capital) to include an additional endowment of 'knowledge' or intellectual capital. This expansion of the Heckscher–Ohlin framework, the so-called neo-factor-endowment approach, also frequently involves the subdivision of labour into endowments of skilled and unskilled labour, with the former clearly related to technological superiority. The outcome is that a country with a relatively large endowment of knowledge will have a comparative advantage in producing knowledge intensive goods, a prediction consistent with that from the technology gap approach. However, in the neo-factor-endowment approach knowledge is perceived as a static endowment to the economy, not as a dynamic process involving innovation followed by diffusion as in the technology gap or product cycle models. The neo-factor-endowment approach is static, and thus neglects the dynamic implications of differences in 'knowledge'.

While the technology gap and product cycle models include some relevant features of innovation, such as monopoly advantage, they also lack a more complete treatment of the innovation process. In the technology gap model the benefits of innovation are appropriable in the short run and lead to monopoly power, but the dynamic implications of such benefits are not addressed. In what they term a 'neo-Schumpeterian' approach, Dosi et al. (1990) observed that the cumulative benefits of innovation, due to its firm-specific nature and the ability partially to exclude others from its benefits, can lead to the firm-, sector- and country-specific technological advantage accumulating over time, causing 'virtuous' and 'vicious' cycles of development. Again, the examples of the British cotton industry and the German plastics industry cited in Part One provide clear-cut historical illustrations. Technology is characterized as embodying specific, local, often tacit, and only partly appropriable knowledge which can explain the geographical concentration of innovation over time. This perpetuation of technology gaps over time is in clear contradiction to the process of convergence between countries which relies on the diffusion of innovation from the advanced to the less developed countries and the subsequent erosion of technological advantage, as for instance in the traditional Solow growth model (see Fagerberg, 1994). Dosi et al. (1990) consider the outcome, in terms of divergence or convergence, as depending on the balance between factors favouring the accumulation of technological advantage over time and opportunities for countries to 'catch-up' to the technology frontier. The latter include imitation and technology transfer through the diffusion of innovation.

The 'new' trade theory, by incorporating endogenous technical change into neoclassical models of trade, predicts many of the same results. For instance, in Krugman (1986, 1989, 1990), learning by doing leads to the accumulation of knowledge over time and the lock-in of countries to certain specialization and growth paths. In this approach, technology is frequently conceptualized as consisting of 'blueprints' or inventions, the benefits of which can be partially appropriated by the innovating firm, but which also contribute to a pool of collective knowledge. The role of spillovers of knowledge between countries is emphasized along with the public good aspects of innovation. This conceptualization of innovation separates it from the neo-Schumpeterian approach, which places more emphasis on the cumulative nature of technological change.

While these different theoretical approaches to trade and technology all stress different aspects of the innovation process, the actual empirical differences between them are not so clear. Partly due to a lack of precise technology indicators, many of the empirical papers have focused on other approaches to the technology–trade relation, such as the importance of innovation in influencing non-price competitiveness. One empirical framework is to include innovation in an export demand model (for instance, Greenhalgh, 1990, for the UK and Magnier and Toujas-Bernate, 1994, for a number of OECD countries), as a proxy for the quality of goods, along with price and income factors. Growth in market shares relies not just on the ability to compete in terms of price (or unit labour costs) but also in terms of technology, or product quality and in the creation of new products and markets. The results from these models of competitiveness indicate an important role for non-price factors in positively influencing trade performance. For the UK, Greenhalgh (1990) found that this relationship varied considerably according to the sector. While the trade performance of over half the sectors benefited from innovation, either their own or from other sectors, some important innovating sectors did not. Likewise, Magnier and Toujas-Bernate (1994), for five OECD countries, found innovation (proxied by R&D expenditure) to be an important factor in affecting market shares in the long run, again with 'significant national and sectoral disparities' emphasizing the importance of considering the relationship at the sector level. Amable and Verspagen (1995) confirmed these results using patents as the proxy for innovation and allowing country and sector specific effects. Amendola et al. (1993), in a country level model of the dynamic determinants of international competitiveness, confirmed that differences in innovation have important long-run consequences for trade performance. In conclusion, it can be argued that a substantial literature has emerged stressing the importance of differences across countries in the rate of accumulation of technological advantage over time for international competitiveness (see Wakelin, 1997).

In the next sections of this chapter we discuss, albeit briefly, some of the international implications of 'strategic' trade and technology policies. There is, as we will try to argue below, a significant role for policy-making, both in terms of the need for a more harmonized and more coherent set of national industrial, technology, competition and trade policies and in terms of the need for international rule-based systems going beyond trade and including industrial and technology policy.

From this perspective there is a significant paradox in the emergence of such 'domestic' strategic industrial policies and their theoretical foundations at a time when the 'domestic' firms at which such policies are aimed are becoming increasingly global and multinational, involved in so-called strategic alliances, and increasingly sourcing on an international scale 'strategic' science and technology inputs. It is as if such firms have become cause, victim and beneficiary of some of the increased trade friction following the widespread implementation of strategic trade policy.

As the repetitive yet rather differentiated use of the word strategic in the above sentences illustrates, it might be useful to discuss some of the various conceptual definitions applied with respect to strategic industries.

14.3 STRATEGIC INDUSTRIES AND POLICIES: AN ATTEMPT AT CLARIFICATION

From an analytical point of view, it seems useful to consider within this context three rather different definitions of 'strategic': a technological one; a trade one; and an industrial one.

The first probably minimalistic definition of 'strategic' can actually be found in its analogy with the military interpretation of the term, whereby long-term access is the main reason for justifying the strategic interest and readiness for extra support costs. Access to some products or technologies contains from this perspective a 'strategic' advantage. Such a military notion of strategic is probably most clearly reflected in the attempt over a long period to prevent the export of 'strategic' high tech products to Eastern European countries. The purpose here was clearly twofold, a military one – which need not be discussed here – and an economic one, closely related to the essential role of certain high tech products as inputs in both capital and final consumer goods.

However, it is not immediately obvious why high tech products would fall under the category of strategic products, certainly not when compared to some scarce natural resource (e.g. oil), where world stocks are concentrated in one or a few countries. To the extent that high tech products are continuously subject to 'creative destruction' through the emergence of new inventions and innovations, and that knowledge is difficult to contain within firms let alone countries, new scientific and technological breakthroughs and the international diffusion of technology are likely to be major factors in quickly rendering obsolete such strategic high tech products. If one thinks of the costs of developing 'strategic' capabilities in, for example, microelectronics technology, it will be obvious that the continuous improvements in performance by the leading firms might render the costs of strategic support policies in this area often prohibitive and certainly highly risky. The apparently successful case of technological 'leap-frogging', South Korea, which succeeded in developing a technological capability in the production of VLSI chips over a short period, probably illustrates more the particularly 'fitting' institutional surroundings, as well as the haphazard luck of good fortune, than any careful 'strategic' policy consideration of costs and benefits. Thus, the entry of Samsung and Gold Star in wafer-based IC production a couple of years before Japanese firms were forced to raise chip prices following American anti-dumping suits and import tariff measures, allowed those firms quickly to reap high returns on their investments. The 'financial patience' of the *chaebols* in pouring in resources with little financial shareholder accountability and the close US–Korean ties both in terms of trade flows, as well as in terms of the training of Korean scientists and engineers in the USA, were all crucial exceptional factors in the Korean 'strategic' leap-frogging story (see also Chapter 7).

However, as this case illustrates well, the 'strategic' argument involved in the case of high tech products is one which is in the first instance based on the cumulative, learning and dynamic increasing returns features of technological advance in this area. For many technologies, so the strategic policy argument runs, access to an interaction with a wider national technological capability might be essential for future technological success and the successful transfer and effective use of technology in other sectors of the economy. The high tech products which fall under this first heading are 'strategic' in that they have an out-of-proportion importance in terms of their 'pervasiveness'; for example, they are essential 'raw material' or intermediate technological input in many capital and final consumer products, and there are strong cumulative and increasing return features involved in the development of such technologies. National and supranational technology policies have very much focused on such products. One may think of the VLSI, Sematech and Jessi support programmes in Japan, the USA or Europe. At the same time the term 'strategic' has often been used to justify support policies in particular high tech areas which did not really satisfy the 'pervasiveness' criteria: one may think of nuclear energy, HDTV or the European aerospace programmes.

The second notion of 'strategic' increasingly used in the policy arena is the one most closely related to new trade theory, and very much identified with the paper by Brander and Spencer (1983). The argument here is a straightforward economic one based on the notion of increasing returns. These are, however, more directly associated with the actual production of many products which are being traded internationally. The resulting international concentration of production of particular goods in some regions/ countries and not in others raises the possibility for 'strategic' intervention, that is the initial stimulus to get the static production increasing returns underway within the region/country before any other region/country did the same. The problem here is that if everybody were to develop such 'strategic' policies, no one would reap the benefits of the scale and agglomeration economies which in theory justified such policies. From a dynamic point of view, however, the picture becomes more complicated. The regional or national externalities linked to the strategic product or sector could have a significant impact on growth, apparently justifying in a more systematic way policy support for such strategic sectors.

Trade and industrial support policies for some particular sectors which differ very much from country to country could be said to fall into this category. One may think of the European support policies for the aerospace industry, the French TGV initiative, etc. The product or sectoral focus of industrial policy is here clearly dictated by notions of the region or country's comparative or potential comparative advantage. The main practical implementation problem relates to the delineation of such sectors. No-one would probably include any longer the iron and steel sector under the heading of 'strategic', although many did so at one time. It is clear that both in theory and practice the static and dynamic economies of scale have been and still are significant in this sector.[5] Gerschenkron (1962) identified it as the main illustration for his theory of latecomer advantages in scale of plant.

The third and probably broadest notion of 'strategic' underpins directly the *raison d'être* of some industrial policies. It can be best described with reference to the French notion of *'filières'*. From a national perspective some sectors have such essential forward and backward linkages both in terms of material and knowledge inputs and outputs that they have become 'strategic' to the country. The French automobile industry is probably the best illustration of such a sector. One in ten Frenchmen has been estimated to be linked to the production of French motor cars. In this very broad interpretation a sector can be said to have become strategic because of its widespread infiltration of the whole economy through the large amount of vertical linkages. It is obvious that this last 'broadest' interpretation of the term 'strategic' can easily become a very defensive definition. Here too analogy with some military use of the term 'strategic' seems appropriate (e.g. a 'strategic' withdrawal behind lines which can be better defended or from where a new attack can be launched). In case of substantial import penetration, for example, the domestic sector might need to be protected temporarily for national strategic reasons. The additional costs of doing so are again justified in dynamic terms: if lost, the costs in developing such a widespread new *filière* or re-entering the sector could well be substantially higher.

A variety of arguments can thus be put forward in favour of 'strategic' policy: primarily technology inspired (our first case); trade inspired (our second case); or industrially inspired (our third case). Let us now turn to some of the limits of such policies, less from a government's implementation perspective than from the actual international effectiveness of such national policies.

14.4   DOMESTIC STRATEGIC POLICIES FOR MULTINATIONAL FIRMS

Opposed to the new theoretical insights which point to the possible justification for strategic 'domestic' industrial and technology government support action, there is the rather paradoxical reality that the domestic firms which would display some of the main industrial or technological characteristics justifying such government support action have become increasingly rare.

The growth and emergence of the transnational corporation is of course not a new or recent feature. There exists now a vast literature providing detailed insights into the reasons underlying the growth of transnational firms.[6] The actual internationalization of production has been just one feature of the more general 'internationalization' pattern of growth in trade and flows of capital and technology which has been characteristic of the post-war stable, liberal trading system. While many international investments might have been initially inspired by protectionist fears and a profound desire to secure access to large markets, the post-war growth in the internationalization of production has been generally of a complementary nature rather than a substitute for international trade flows. One could even go a step further: it is primarily the internationalization of production over the last three decades which has led to catching up and rapid technological diffusion of 'best practice' production techniques and products from the USA to a large number of OECD countries, and thus to the convergence of income levels, rather than just the actual international trade flows as traditional trade theory would lead one to believe.[7] Perhaps even more remarkable has been the relocation of Japanese MNCs in the 1990s, partly as a result of the high exchange rate for the yen. The ratio of VCRs produced by Japanese firms outside Japan in 1994 was 53 per cent and that of colour TVs 78 per cent.

One of the main reasons why this has been the case, is the fact that foreign investment was never limited just to production, but often included maintenance, engineering and development activities. This is of course not surprising. Important differences might exist between domestic and foreign user requirements. As we saw in Chapter 12, foreign regulations, standards and other procurement specifications will in all likelihood be rather different from the home-base country of the firm; not to speak of other tastes and economically induced differences between the foreign and home country. These factors have undoubtedly led many multinational firms to set up or take over research and development laboratories in foreign countries, sometimes directly linked to their production subsidiaries, sometimes not.[8]

Since the Second World War this trend has accelerated.[9] This process has been most evident in the case of multinational firms with their 'home base' in small OECD countries (one can think of the Dutch, Swiss and Swedish multinationals). In these countries the need for more international sourcing of relevant scientific and technological inputs is of course felt much more rapidly. There is indeed no reason why only the 'home' basis should be relied upon in terms of the provision of well-qualified scientists and engineers. However, evidence on the international location of R&D activities remains rather limited. The available US Department of Commerce data point to the fact that over the 1970s and 1980s there has been a gradual increase (from about 6 to 10 per cent) in the foreign share of total R&D expenditure of US MNCs (see Freeman and Hagedoorn, 1992).

Table 14.1, based on available 'official' OECD data, indicates for the EC as a whole the trends in the domestic and foreign company financed R&D for the period 1981 to 1989. The figure suggests that the importance of foreign R&D remains small and has only been gradually increasing to some 8.5 per cent of total company financed R&D. There are reasons to believe, however, that such official data may underestimate the amount, particularly in some of the small European countries such as Belgium, the Netherlands, Sweden or Switzerland. In Table 14.1 the official R&D data are compared with evidence based on US patent data gathered by Patel and Pavitt (1991). The table illustrates not only the likely underestimation of the foreign owned/controlled share of total company financed R&D, particularly in small countries, but also illustrates the extreme diversification in international R&D location pattern between OECD countries, and between Japan, the USA and the large European countries in particular.

Table 14.1   Foreign controlled domestic technology compared to nationally controlled foreign technology (based on R&D 1988 and US patenting 1981–6)

Home country	Foreign company financed R&D (as % of total R&D expenditure, 1989)	US patenting from foreign firms (as % of country's total US patenting)	US patenting by national firms (as % of country's total US patenting)
Belgium	n.a.	45.7	16.5
France	13.5	11.8	3.8
FR Germany	3.0	11.5	8.5
Italy	7.8	11.2	3.0
Netherlands	4.4	9.5	73.4
Sweden	2.3	5.4	16.7
Switzerland	2.0	12.5	27.8
UK	16.2	22.3	24.5
EC/Western Europe	8.4	7.4	9.3
Canada	18.8	28.1	12.5
Japan	0.1	1.2	0.5
USA	13.7	4.2	4.4

Sources: US, Science and Engineering Indicators (1991); OECD, Science and Technology Indicators; Patel and Pavitt (1991).

With respect to the small(er) countries, the patent data reported in Table 14.1 illustrate, how such countries can sometimes be characterized by either a strong presence of foreign MNCs' R&D activities (Belgium, Canada, Switzerland and, to broaden the size a little, the UK), or by large domestic MNCs with a strong R&D presence abroad (the Netherlands, Sweden, Switzerland and the UK). National technology support policies in those countries, whether strategic or not, have long been faced with the growing discrepancy between the effectiveness of such national support compared to its international foreign impact. The difference between technology support policies in various small countries is rather revealing in this context. In countries such as Belgium or Canada, the policy focus is very much on improving the physical and human infrastructure and attracting foreign MNCs' investment in production and R&D. In countries such as the Netherlands, the policy is aimed at small and medium-sized firms in an effort to 'free' themselves from the dominance of the large domestic MNCs. The latter are pushed in their demands for domestic R&D support to the higher EC policy level. These differences in national policy reflect responses to the basic question: Given the international sourcing and leakage effects, why should nations have a (strategic) technology support policy, and how can policy-makers assess the 'good citizenship' of foreign firms? Singapore is a good example of strong national policies attracting MNC R&D and other technical activities which harmonize with local objectives.

In more recent times this debate can no longer be said to be confined to such smaller countries, or to the foreign R&D labs set up or acquired by US and European MNCs over the post-war period.[10] The examples given in Chapter 12 illustrate this point. A more fundamental 'globalization' trend, involving a much wider set of international exchanges including strategic alliances, networks of scientific and technological information and

R&D subcontracting, has emerged and grown rapidly, particularly between the so-called triad firms.[11] Such a globalization trend is not contradicted by the evidence of Porter (1990) or Patel and Pavitt (1991) concerning the strongly 'national' home basis of the competitive advantage of such emerging global firms. In line with the arguments set out in section 14.1, it is indeed the national virtues which first create the opportunity to cross borders. However, this emphasis, exactly as in the case of strategic trade theory, while bringing to the forefront some of the essential 'verités' of the nature of international competitiveness and post-war trade flows, does underscore the new emerging trend of globalization and networking between such firms, as they become increasingly international in their marketing, distribution, R&D and technology sourcing, as reflected, for example, in the number of strategic alliances.

The growth in strategic alliances and networks – reviewed in depth by Chesnais (1992), Hagedoorn (1991) and Mowery (1992) – between these 'global' firms raises three clear issues with respect to the strategic policy discussion.

First, one may raise some questions as to the nature of such alliances. Are they a new, more or less permanent feature of the new global network economy closely related to the complexity of science and technology and the need for international sourcing of and access to science and technology, or are they a temporary feature – the first step in the emergence at world level of oligopolistic cartels in sectors dominated by static and dynamic economies of scale?

Obviously, the answer to this question is not just one or the other. The literature on strategic alliances points to a wide variety of more or less technology inspired motives for strategic inter-firm co-operation. These range from research risk-sharing to seeking access and entry to foreign markets (Hagedoorn and Schakenraad, 1992). The present-day trend towards globalization, alliances and networking involves a far greater share of world production, including production of component suppliers; investment, including intangible investment; and access to markets and other firms' tacit knowledge, including mergers. It cannot be viewed independently from the increase in a trend towards world oligopoly in many sectors dominated by economies of scale.

Consequently, a first major policy issue is whether a global competition policy is needed and how it should be implemented. Without entering this debate,[12] it will be clear that the existence of a supranational form of international competition policy aimed at counteracting the emergence of worldwide cartels between global firms would directly undermine one of the reasons for strategic government support, namely that based on arguments of dependence on 'foreign' monopoly pricing.

The second major policy issue, which the discussion on networking and globalization brings to the forefront within the context of strategic technology policies, is the possible impact and inducement effects on international localization and networking. Indeed, are strategic alliances and technology networks truly based on the needs of large global firms for international exchanges to improve resource allocation and obtain more dynamic innovation and faster spread of best practice techniques, or are they motivated by the desire to take advantage of various domestic 'strategic' support policies? From this perspective, the attempt of global firms to become 'multinational' and present themselves as 'good domestic citizens' in as many countries as possible is also the result of the growing importance of national strategic support schemes, which could provide major competitive advantages to domestic 'national' competitors (Mowery and Rosenberg, 1989).

Again an adequate response to these questions can only consist of emphasizing both features. The aim of many of the largest firms is increasingly directed towards global strategies which find a balance between reaping some of the industry-specific scale advantages of global markets, yet exploiting the often geographically determined diversity of consumers and production factors. The large multinational firm's organization and its production technology often provide the necessary flexibility to confront this diversity. The decentralization of its production units, marketing and even research, together with a diversification of subcontractors, will enable it to take full advantage of global access, including access to government-sponsored scientific and technological knowledge. At the same time, the precise location of such a firm's plant will depend heavily on the local surrounding environment. Whereas the locational choice will often depend on the availability of local skills, infrastructure access to knowledge and local government support, the firm itself will of course contribute not only to local output and employment but also to the long-term development and growth of the region regarding skills, training, access to knowledge, local suppliers' know-how and networks. These often scarce factors constitute precisely the 'externalities', increasing return growth features of long-term development, and explain why regional/local authorities have always been keen on providing incentives for foreign firms to invest and locate in their particular region.

The contribution of multinational firms to such locally created advantages raises a number of interesting policy paradoxes. At the national, or supranational (e.g. the EU) level, there will be major concern, particularly at the technological end, about national 'strategic' support policies flowing to such 'foreign' firms. Attempts will be made to exclude such firms from national (including here also the EU) sponsored strategic policies. However, at the local site level there will be increasing rivalry concerning the services offered to firms, with little interest in the domestic or foreign origin of such firms – to the region most of the firms will be foreign. Such rivalry will often result in a multiplicity of 'new' growth sites, 'science parks' or 'technopoli' being set up with few developing the necessary size to reach some of the essential scale externalities and dynamic growth features and all increasing the cost of communicating and interacting.[13]

The third policy issue with respect to the increasing trend to networking and strategic alliances, and of particular relevance to the discussion surrounding strategic policy, relates to access to technology for those firms/countries not belonging to the networks. As indicated above, one has witnessed particularly over the 1980s a rapid growth in 'strategic' technology alliances by companies on an international scale. This has occurred primarily between firms from the so-called triad: the USA, Japan and the EU. It has been estimated that over 90 per cent of all technology agreements and alliances are made between companies from the USA, Japan and the EU (Hagedoorn and Schakenraad, 1991; Freeman and Hagedoorn, 1992).

This geographically concentrated network of strategic alliances raises major issues about access for those countries/companies not belonging to the existing networks. In the absence of an international regulatory framework, it is likely that such technology networking will increase inequality of access to technology and investment. Such possibility of 'exclusion' is characteristic of processes of increasing returns and learning. There is here an emerging need for broad international principles such as reciprocity concerning access to technology networks, including the particular preoccupations of the NICs and developing countries.[14]

14.5 CONCLUSIONS

'New' trade theory and, in its footsteps, 'new' growth theory[15] have brought a great deal of economic realism back into economic analysis: realism with respect to the way firms operate in industries dominated by economies of scale and imperfect competition; consumers consume differentiated commodities and look continuously for variety; and countries grow and trade not on the basis of decreasing returns or 'given' factor endowments, but on the basis of often historically grown 'externalities', based on absolute cost and technology 'created' advantages. In doing so, trade theory has returned to the forefront the importance of the gains from free trade to world consumers, yet at the same time has opened up a Pandora's box about the possibility of governments intervening to set in motion the virtuous circle of growth, international competitiveness and technology accumulation.

To some extent, these 'new' insights highlight the point made by many businessmen and national policy-makers long before: that international specializations which are clearly efficient in terms of static comparative advantage criteria may well not be so in the long run, because of sectoral differences in dynamic growth potential. However, what this rather old debate, dressed up in its new 'strategic' policy clothes, has brought to the forefront is that arguments about the existence of possible national policy trade-offs do also have an international price. That price is probably least in terms of the actual subsidies spent on the 'strategic' sector; it is possibly highest when it leads to retaliatory trade and pricing action as, for example, typified in the 1986 Semiconductor Agreement between the USA and Japan, with its high cost to world consumers.

In this chapter, the emphasis has been on another international feature of strategic policy, namely the fact that the domestic firms (for whom such 'strategic' domestic policies could be developed on the basis of the 'new' found theoretical wisdom) have grown increasingly global and are involved in 'strategic' alliances with foreign firms, possibly as a result of strategic foreign policies. This increasing globalization trend raises a number of important policy issues, not least those with respect to the level at which policy should be implemented. It is obvious that global, network or multinational firms increasingly call into question the meaning of many national policies. In many cases such firms might be just as good 'citizens' as nationally owned firms; in other cases, they will not be. It is difficult, if not impossible, for governments to draw lines here: the result will generally be total inclusion or total exclusion of 'foreign'-owned firms in national policies.

Now that technology policy is again high on the policy agenda of most developed countries, it might sound frustrating to policy-makers that its effectiveness seems to become undermined by the very actors on whom it has been traditionally focused: the large, nationally owned, R&D performing firms. But perhaps this is exactly where technology policy has gone wrong and needs to be reoriented away from national prestige and the technological nurturing of its leading firms, and more towards those less footloose features of the process of technical change: the local network of small and medium-sized firms; the higher education, training and basic research infrastructure; the local institutional set-up. The importance of this broader concept of technology policy which is increasingly recognized in the notion of 'national systems of innovation' is not being questioned by our analysis, rather to the contrary.

NOTES

1. The parable is from Paul Krugman, who quotes Paul Samuelson.
2. See in particular the estimates with respect to the enlargement of the EC in the 1970s and later on with respect to the formation of the Single European market.
3. For more detailed analysis see Dosi et al. (1990).
4. See also Cimoli and Soete (1992) and Fagerberg et al. (1994).
5. This also explains why in the 1950s and 1960s both in the EC (ESCC) and Japan these sectors were at the core of much 'strategic' policy intervention even though the word strategic was not yet used.
6. For recent overviews see in particular Dunning (1988, 1989) and Narule (1996).
7. This emerges, for example, from the crucial importance in EU imports and intra-EU trade of high tech sectors such as electronics, of foreign multinationals intra-firm trade. The latter not only accounts for more than 50 per cent of total EU imports in sectors such as computers, electronic consumer goods and electronic components, it also accounts for more than 50 per cent of total intra-EU trade in those sectors. In other words, it is in the first instance (foreign) MNCs which appear to exploit most of the trade advantages of the Single European Market. See Grupp and Soete (1993).
8. Just after or even before the Second World War, one can think of ITT, GM and Ford in Europe, or North American Philips in the USA, IBM in Europe including Switzerland, the takeovers of Pathé in France by Kodak and of the Belgian Gevaert by Agfa.
9. There now exists a voluminous literature on this subject. See among others Cantwell (1989); Casson (1991); Pearce and Singh (1991); Granstrand et al. (1992). See also Chapter 12.
10. This view is broadly in line with the arguments put forward by Patel and Pavitt (1992) and Duysters and Hagedoorn (1993).
11. For an overview of the evidence, see Freeman and Hagedoorn (1992).
12. For a clear outspoken view on this issue, see Ostry (1990).

13. For early discussions of some of those locational features of agglomeration effects see David (1984), Arthur (1989) and Murphy *et al.* (1989).
14. This issue has been emphasized in particular in the OECD so-called TEP report (OECD, 1991).
15. Just as 'new' growth theory incorporated many ideas which were already commonplace among economic historians and Schumpeterian economists (see Chapter 13), so 'new' trade theory returned in some respects to an 'older' tradition in trade theory, which has been ably analyzed by Reinert (1994, 1997). He points out that already in 1613, Antonio Serra clearly expounded a theory of increasing returns and 'Schumpeterian mercantilism.' (Reinert, 1997, p. 29).

CHAPTER 15

DEVELOPMENT AND THE DIFFUSION OF TECHNOLOGY[1]

15.1 INTRODUCTION

As illustrated in previous chapters, the importance of 'foreign' technology and its international diffusion has been historically a well-recognized factor in the industrialization of both Europe and the United States in the nineteenth century, and even more strikingly of Japan in the twentieth century. That importance emerges stronger daily from the evidence with regard to the rapid industrialization of the so-called 'newly industrializing countries', sometimes also called 'dynamic Asian economies', such as South Korea, Taiwan, Singapore, and more recently China, Malaysia, Thailand and Indonesia. Developing countries as a group have reached growth rates during the 1980s and 1990s which have been substantially higher than those of the developed countries. As noted by the IMF, the growth rates of the developed (OECD) countries no longer represent the 'engine' of world growth, in contrast with the 1950s and 1960s. At the same time many developing countries, particularly in Africa, have continued to lag behind and see their growth in GDP systematically outperformed by population growth, hence reducing the per capita wealth available.

The enormous difficulties experienced by most developing countries in their effort to industrialize do not lend great support to a set of arguments based at first sight on historical analogy and a relatively simple mechanistic view of the growth process. Following the arguments set out in Chapter 12, this chapter will focus directly on the technological issues which underlie the process of development. It aims to provide insights into how technologies evolve and diffuse and under what conditions a process of 'effective' technological catching up can take place.

There is a voluminous literature on the subject of international technology diffusion and technological 'catching up' which has been a focal point of research for economic historians (e.g. Gerschenkron, 1962; Landes, 1969; Hobsbawm, 1968; Rosenberg, 1976). We do not intend to review this literature here. Suffice to say that there appears to be some 'convergence' between contributions based on in-depth historical studies of countries in the production and use of particular technologies (see, in particular, Ames and Rosenberg, 1963; Habakkuk, 1962; von Tunzelmann, 1978) and the more recent international trade and growth models based on imitation and 'catching up' discussed in the previous chapters. That convergence clearly puts the emphasis back on the historical institutional framework within which the process of imitation/technological catching up takes place, including the role of historical accidents, the importance of 'developmental' constraints, whether primarily economic (such as lack of natural resources) or more political in nature, the role of immigration (see Scoville, 1951) and other 'germ carriers', and the crucial role of governments (see Yakushiji, 1986; Wade, 1990; Lall, 1995).

From this perspective, the international diversity in growth performance of countries demonstrated the importance of path-dependent development, with many bifurcations and possibilities of 'locked-in' development, whereby some industrialized locations got 'selected' early on. Such locations may then exert some competitive 'exclusion' (Arthur, 1988) of other places by appropriating the available agglomeration economies. Indeed, it is the increasing returns associated with industrialization and development which make the conditions of development so paradoxical. Previous capital is needed to produce new capital, previous knowledge is needed to absorb new knowledge, skills must be available to acquire the new skills and a certain level of development is required to create the agglomeration economies that make development possible. In summary, it is to some extent within the logic of the dynamics of the system that the rich get richer and the gap remains and widens for those left behind.

All development policies have in one way or another been geared to breaking out of this vicious circle. Most have concentrated on tackling the investment and infrastructure locational questions, with some but relatively less direct attention to the knowledge and skills constraint. The question which will be tackled in this chapter is whether these constraints are always equally formidable or whether their intensity varies in time with some increasing and some decreasing, thereby opening windows of opportunity to escape the vicious circle. It is clear that the international diffusion of technology provides a crucial ingredient in the debate on technological 'locking out' of underdevelopment. This was also a central argument in some of the neo-technology accounts of international trade reviewed above, where comparative advantage would shift with the further international diffusion of the technology to 'less developed' countries. Through the 'use' of imported technologies these countries would acquire some comparative industrialization advantage in low tech mature products and industries. At first sight, the choice of such products as a point of entry might appear to be the only one to initiate a development process and establish a process of economic growth. However (and leaving aside for the moment all aspects of technological 'blending' and other user-initiated technological change) insofar as mature products are precisely those that have exhausted their technological dynamism, there will also be risks of getting 'trapped' in a low wage, low skill, low growth, development pattern. Technological catching up will only be achieved through acquiring the capacity for creating and improving as opposed to the simple 'use' of technology. This means being able at some stage to enter either as imitators or as innovators of new products or processes. Under what conditions would this be possible?

To answer this question, the particular role of technological change as a disruptive process with changes in direction and deep structural transformations needs to be far better understood, as argued in Part One. The notion of technological change as a global more or less continuous process underlies in many ways the traditional way development is viewed: a cumulative unidirectional process, a race along a fixed track, where catching up is merely a question of relative speed. As Perez (1988) in particular has emphasized, speed no doubt is a relevant aspect, but history is full of examples of how successful overtaking has been primarily based on running in a new direction.

In this chapter the focus will be on the specific conditions under which technological catching up and imitation can take place using as an example the most pervasive set of new technologies presently diffusing: microelectronics (Chapter 7). This will allow us to highlight the importance of some of the specific disruptive characteristics of these new technologies with respect to international economic growth and catching up.

The chapter starts with a brief exploration of some of the methodological insights gained through diffusion models. We then discuss in general terms some of the conditions for imitators to enter and effectively catch up. In a third section we discuss some of the implications of the analysis for international trade and growth theory. In the last part we discuss some of the policy implications. Throughout the analysis we do not refer specifically to any particular group of countries, although the case of the East Asian economies, discussed in Chapter 12, is clearly of some relevance here.

15.2 TECHNOLOGY DIFFUSION AND INDUSTRIAL GROWTH

There is a striking level of methodological similarity between the traditional epidemic innovation diffusion model and some of the models of industrial growth and economic development introduced in the 1930s by Kuznets (1953) and Schumpeter (1934) among others. This is in many ways not surprising. The concepts of 'imitation' and 'bandwagons', so crucial to the diffusion literature, have been and are still central in many of the more structural accounts of economic growth,[2] where the S-shaped diffusion pattern is similar to the emergence and long-term rise and fall of industries. An attempt at linking the two theories was made in Freeman *et al.* (1982). Here it was precisely the notion of 'clusters' of innovations including the follow-up innovations made during the diffusion period which were linked to the rapid growth of new industries, and in the extreme case could even provide the ingredients of an upswing in overall economic growth.

In the more restrictive diffusion terminology, this could be viewed as an 'envelope' of diffusion curves of a set of closely interrelated clusters of innovations, which occurring within a limited time span could tilt the economy in the early diffusion phase to a higher rate of economic performance. Another similarity with diffusion models can be found in Rostow's theory of the stages of economic growth (1960), again with a distinct S-shaped pattern of take-off, rapid growth with the 'drive to maturity', and slower growth with the 'age of high mass-consumption' and standardization. Rostow phases contain many of the S-shaped development patterns assumed to exist for new products, as typified in the marketing and subsequent international trade literature on the 'product life cycle'. Such an argument was also put forward in the mid-1960s by Hirsch (1965) who showed how the relative importance of certain production factors would change over the different phases of the product cycle (see Chapter 11, Figure 11.4). Hirsch, and after him Vernon (1966) and many other proponents of the product life cycle trade theory, illustrated how such changes would shift comparative advantage in favour of less developed countries as products reached the maturity phase.

Within the development literature of the 1960s, particularly the 'dependencia school', such views and particularly Rostow's theory were heavily criticized. The mechanistic, quasi-autonomous nature of the process of economic growth assumed by Rostow was even branded as 'ahistorical'. Interestingly though, the critique on the mechanistic nature of Rostow's growth model finds its reflection in the most recent diffusion literature, criticizing the 'mechanistic, atheoretical' nature of the S-shaped 'epidemic' technology diffusion models.

The more recent diffusion contributions also provide a number of interesting insights into some of the broader industrial growth theories mentioned earlier. The first area of critique on the 'standard' diffusion model has led to the application of 'probit analysis' to develop alternative models of inter-firm diffusion. Probit analysis was already a well-established technique in the study of the diffusion of new products among individuals.[3] The central assumption underlying the probit model is that an individual consumer (or firm) will be found to own the new product (or adopt the new innovation) at a time when their income (size) exceeds some critical level. This critical or tolerance income (or size) level represents the actual tastes of the consumer (the receptiveness of the firm) which can be related to any number of personal or economic characteristics. Over time, though, with the increase in income and assuming an unchanged income distribution, the critical income will fall with an across-the-board change in taste in favour of the new product, due both to imitation, more and better information, bandwagon effects, etc.

The relevance of the probit model for industrial growth theory is self-evident. A 'critical' income per capita level is a concept which can be introduced in a straightforward manner in Rostow's theory of the stages of economic growth or in Abramovitz's concept of 'absorptive capacity'. Replacing the concept of individuals by 'countries', different behaviour between countries in their growth performance can be explained and expected. Considering both the extreme variation in a country's ability to take risks and 'assess new innovations' (the variation in consumer tastes in the probit model), and the extreme levels of income inequality at the world level, it should come as no surprise that industrialization at the worldwide level (diffusion) has been slow, and that many poor countries, even with the fall over time in the 'critical income' industrialization level, have not reached the stage of take-off.

A second criticism against the standard diffusion model relates to its static nature and the way the diffusion process is reduced to a pure demand-induced phenomenon. Metcalfe (1981, 1982) in particular has emphasized the limits of the standard model in this area. There are, as many detailed studies of the 'innovation process' have indicated, plenty of reasons for expecting both innovation and its surrounding economic environment to change as diffusion proceeds.

At the technological end, one may expect significant improvements to the innovation to occur, as diffusion proceeds further. These incremental developments can be either more or less autonomous or induced by the diffusion process: for example, through user feedback information as well as through the wider application of the innovation to new users, requiring better performance and/or more precisely defined quality characteristics. As diffusion proceeds and the specific user's demands become more stringent, it can be expected that the effective use of 'scientific knowledge' in improving the performance, quality and reliability of the innovation will increase substantially.

From an economic perspective too, the static, demand-focused nature of the standard diffusion model is questionable. In the model developed by Metcalfe (1981, 1988), the price of the new innovation is no longer a constant or evolving along a particular time path, but is determined by the process of diffusion. In addition, supply of the innovation is limited by productive capacity, whose rate of increase depends on the profitability of producing the innovation. The pattern also depends on the capacity of the innovators to educate the users (see, for example, the illustrations in Chapters 7 and 8). A typical Schumpeterian 'scenario' of an entrepreneur–innovator emerges, his temporary reward consisting of the initial monopoly profits which are gradually competed away as imitation takes place and the 'innovative potential' is exhausted. At the same time though, as the rate of return to the innovation supplier(s) falls, 'the associated reductions in price increase the profitability of adopting the new inno- vation', which is further diffused. Finally, the diffusion path depends also on systemic factors, such as availability of materials, new capital goods, components and so forth.

Once the importance of these systemic supply factors is fully recognized, it also becomes more apparent how past investment in the 'old', established technology can slow down the diffusion of the new innovation; past investment not just in physical capital but also in human capital, even 'intellectual' capital (to use the expression of Friedrich List). The importance of past investment in, and existing commitment to, the technology which is being displaced, in slowing down the diffusion of the new technology, points also towards the phenomenon of inter-technology competition and to questions of 'locked-in' technological development (Arthur, 1988a, b). New technology will initially compete against existing technology on disadvan- tageous terms. As Rosenberg (1976) observed, the diffusion of steam power in the last century was significantly retarded by a series of improvements to the existing water power technology which further prolonged the economic life of the old technology. The process of a 'dying' technology is indeed slow, with the old technology firms often living on past fully recovered investment and sometimes being able to underprice the innovation- adopting firms. This process has been described as the 'sailing ship effect' since this was an obvious example in relation to the competition between sailing ships and steamships in the nineteenth century.

15.3   INTERNATIONAL TECHNOLOGY DIFFUSION AND CATCHING UP

The implications for the international diffusion of technology and the potential for technological catching up are significant, at least from this methodological perspective. There is of course no reason not to expect that the vast majority of new technologies will originate primarily from within the technologically most advanced countries. However, there are good reasons to expect that the internal national diffusion of such major new technology might well be hampered by the various factors mentioned earlier, the new technology competing (in its diffusion) on disadvantageous terms. Thus the possible previous investment outlays in the existing technology, combined with commitment from the management, the skilled labour force and even the 'development' research geared towards improving it, might all hamper the diffusion of the new technology to such an extent that it will diffuse more quickly elsewhere, in a country uncommitted to the old technology, both in terms of actual production and investment. At the same time, as diffusion proceeds, some of the resultant crucial incremental innovations, for example, from user-feedback information, will further shift the technological advantage to the country in which the new technology is diffusing more rapidly.

The industrialization in the nineteenth century of Germany, France, the USA and a number of smaller European countries gives some illustrations of this view. The dramatic change in fortune in the UK's position from an absolute technological leadership, producing in the mid-nineteenth century more steam engines than the whole of the rest of the world put together and accounting for more than half of all world exports of manufacture, is a powerful illustration of this phenomenon. It pinpoints the significant advantages of 'late industrializers', in terms of catching up with present technological leaders, as well as in terms of acquiring foreign technology at a more 'competitive' price. As we saw in Chapter 12, in recent times this has been most obvious in the case of Japan in the 1960s and 1970s, and South Korea in the 1980s, in their rapid industrialization where world 'best-practice' productivity levels were achieved over a very short time in steel, cars and electronic consumer goods, and in the most recent years computers, largely on the basis of initially imported technology.

Jang-Sup Shin (1996) showed the similarity between the South Korean latecomer advantages in steel production and those described by Gerschenkron (1962) for European latecomers a century earlier. Nevertheless, the scarcity of such successful examples illustrates how 'non-automatic' and exceptional such processes of effective technological catching up and leap-frogging have been. The use of foreign, imported technology might appear as a straightforward 'industrialization' short cut, but the effective assimilation of foreign technology is actually difficult and complex. The crucial question will be the country's, and the domestic firm's 'absorptive capacity'. (See Radosevic, 1996, for a more complete discussion of technology transfer at the firm level.)

As illustrated in Perez and Soete (1988) a central problem in the previous discussion on the international diffusion of technology is the almost automatic interchange of the words 'use' in the narrow diffusion sense and 'entry' in the broader technology sense. Technological catching up implies in the end the effective 'use' of foreign technology with the aim of mastering and improving it. The emphasis is on the acquisition and the assimilation of technology. From an industrializing country or from any other lagging position this can in the first instance only be done through the 'use' of foreign technology.

Diffusion theory gives us hints as to why such international diffusion is not automatic and what the reasons are for delays in 'adopting'. The probit model emphasizes the threshold level below which no 'adoption' is likely to take place. Below this level, practically by definition, there will be no acquisition of technology, learning or catching up, let alone technological leap-frogging. When the probit model is applied to the diffusion of a new consumer product, the notion of a threshold determined by income level and taste is fairly straightforward. However, when referring to the diffusion of production technology, the concept of a threshold for entry as opposed to use is considerably more complex.

15.4   TECHNOLOGICAL EVOLUTION, AND COST OF ENTRY

The most unrealistic of the assumptions in the early diffusion model relates to the once and for all static nature of the technologies. When a product or process is first introduced it is almost inevitably in a relatively primitive form and is submitted to successive incremental improvements which either reduce its cost of production and/or increase its quality, performance, reliability or whatever other aspect is important to the users or can contribute to enlarge the market. Such improvements could well follow what Nelson and Winter (1977) have termed a 'natural' trajectory and Dosi (1982) a 'technological' trajectory. As in the product life cycle model, the path of such successive incremental innovations from introduction to maturity of any particular technology, could be represented in the familiar S-shape fashion. Improvements are achieved slowly at first, then accelerate and finally slow down again, according to Wolff's law of diminishing returns to investment in incremental innovations.

This means that the imitator does not always enter the 'same' technology as the innovator. Nor do later imitators enter at the same point in the technology's evolution or trajectory as earlier ones. All these improvements have a cost and they all imply the generation of additional innovation-bound knowledge and experience. This implies that cost of entry curves vary in time. A reasonable assumption would be that they constantly shift upwards as they now cover the cost of the original investment plus all subsequent investment in incremental improvements. However, this is not necessarily so. As Hirsch (1965) observed, requirements for entry vary in importance through the various phases of the product cycle. In our terminology, this would imply that both the various components of the cost of entry vary in relative importance, as well as the minimum threshold levels as technologies evolve over the various phases of the product life cycle. Figure 15.1 illustrates the four phases of the product cycle.

Phase I is the period of first introduction where the focus is on the product itself. It has to perform its function adequately and break successfully into the market. It is a learning process for designers, plant engineers, management, workers, distributors and consumers. It is the world of the Schumpeterian entrepreneur. Since original design and engineering are involved, the minimum scientific and technological threshold (S) is likely to be high, whereas the relevant skills and experience (E) threshold might be

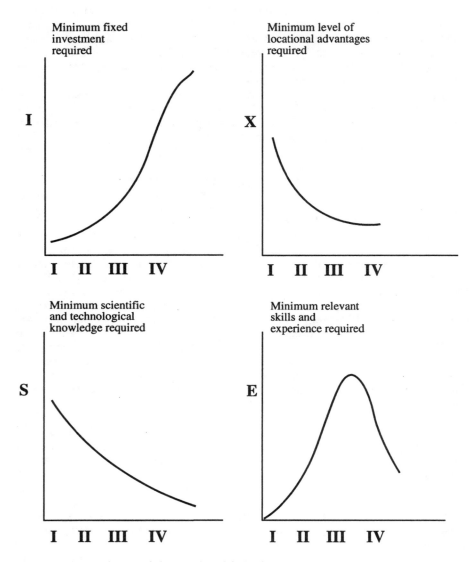

Fig. 15.1 Four phases of the product life cycle

Source: Perez and Soete (1988).

low. The level of locational advantages required (X) can be crucial and relatively high for successful introduction. Finally, the initial investment costs (I) are likely to be low if not always in absolute terms at least relative to what they will become as the technology evolves.

Phase II is the period of market growth. Once the product is basically defined and its market has been tested as clearly capable of growing, the focus shifts to the process of production. Plant design becomes important and successive improvements are made to both the product and the process of production to achieve the optimal match between the two, in order to increase output and productivity. Materials and shape might be changed to lower costs and increase efficiency or respond to market demand. Plant organization is gradually optimized and the most appropriate equipment chosen or specified. It is the world of the production engineer and the marketing manager. As the scientific and technical problems are gradually solved and their solution is embodied in both product and production equipment, the threshold for imitators decreases. But the threshold in terms of required skills increases rapidly as experience accumulates within the producing firm in relation to the product, the process of production and successful marketing. Locational and infrastructural economies of the sort generated by the innovation itself grow at the expense of the producers, so later entrants could find the relevant infrastructure more available than earlier ones. The cost of Investment (I) is now higher than before as optimal plant size has grown and more sophisticated and better adapted equipment has been incorporated to handle the larger volumes.

By **Phase III** all the main conditions have been clearly established. Market size and rate of growth are well known, the relationship between product and process has been 'optimized' in an engineering sense and the direction of further incremental innovations to increase productivity is clearly seen. The focus is now on managing firm growth and capturing market share. Scale-up of both plant and firm are the characteristics of this phase. As it proceeds, many firms that were successful in the previous two phases might be eliminated. The actual capital costs and the management skills required to stay in the race in Phase III can be formidable. This is therefore no time for new entrants. The Knowledge (S) component of entry costs is by now relatively low but the Experience (E) and Investment (I) components are at their highest and growing. Locational advantages become less important by comparison with the internalized economies that successful firms have accumulated in market and financial power by this time. Furthermore, going back to the price a firm would charge for selling the technology, one could say that in Phase I the price can tend to infinity due to an interest in monopolizing the technical information of the Knowledge (S) sort, but in Phase III it can again become relatively high in order to monopolize markets, through keeping the now much greater Experience (E) within the firm.

Finally, in the maturity stage **Phase IV** both the product and its process of production are standardized. Further investment in technological improvements results in diminishing returns. Since factor inputs are established and fixed, the advantage in costs of production goes to the firm or locality that can make the greatest comparative saving in any of them. This might lead the established firms to relocate some of their own plants even from the end of Phase III. But it can also lead them to concentrate on other innovations and to turn the technology acquired in the previous phases into a commodity, i.e. being willing to sell it at a discretionary price in the form of licences and 'know-how' contracts. This practice could eventually result in a buyer's market if there are competing suppliers. Thus, in the final or maturity phase of a technology the threshold of entry comes further down even though the actual costs of entry may still be high. The previous knowledge requirements are now very low because they are almost totally embodied in the product and the equipment. The required skills are well codified and at a price can be purchased, though their real acquisition for efficient production may not be guaranteed without enormous efforts on the part of the buyer. The relevant locational advantages continue to be important; those relating to the education of input suppliers and consumers are at their highest. Finally, fixed investment costs are much higher than in Phase I but suppliers are available who have the experience and know the specifications for all the necessary equipment.

Given the appropriate circumstances, Phases I and IV provide the 'easiest basis to attain' threshold conditions for new entrants, but under radically different costs and requirements. With little capital and experience, but with scientific and technical knowledge plus an adequate provision of locational advantage or compensatory 'help' an innovator or imitator can enter the market at the early stages. Entry at Phase IV depends on traditional comparative and locational advantages. But it requires considerable amounts of investment and technology purchase funds. An important difference between the two entry points, is that entry at Phase I does not guarantee survival in the race. Much further investment and technology generation efforts are required as competitors advance along the improvement path. A maturity entrance appears relatively safer as long as the product in question is not substituted by a newer one in the market. Profits will depend on how many other new producers struggle for a share at this stage.

This then appears to support both the view put forward by product life cycle trade theory, illustrated in the success of export-led industrialization strategies achieved on the basis of manufacturing mature traditional products, and the apparently contradictory early entry 'events' of a number of industrializing countries in such technologies as digital telecommunications, electronic memory chips or biotechnology products. The latter 'early entry' phenomenon is again, and as already discussed in the previous section, further supported by much historical evidence with regard to the industrialization of many presently industrialized countries.

15.5 TECHNOLOGICAL CATCHING UP: FROM PRODUCT LIFE CYCLES TO TECHNO-ECONOMIC PARADIGMS

One of the main shortcomings of the product cycle theory framework set out and used above is that it assumes that all products are independent of one another. Every new product is seen as a radical innovation, the successive improvements to it and to its production process are the incremental changes which bring it to maturity, after which the next product is seen as a radical departure destined to follow a similar evolution. In fact, and as discussed at greater length in the chapters in Part One, products build upon one another and are interconnected in technology systems. Each product cycle develops within a broader family which in turn evolves within an even broader system.

There are two reasons why the notion of technology systems is more relevant for development strategies than that of single product cycles. One is that the knowledge, the skills and the experience required for the various products within a system are interrelated and support each other. The other is that the analysis of technologies in terms of systems allows the identification of those families of products and processes which will shorten the time of learning and catching up as well as provide a wide scope for development and growth.

Going back to the various entry phases of the previous section, one finds that the requirements for entry in new products (Phase I) are relatively low in managerial ability and capital which would make them ideal for industrializing countries if it were not for the other two factors: the need for high levels of locational advantages and for scientific and technical knowledge. Assuming that some overall government action could eventually compensate for the lack of locational and infrastructural advantages, let us concentrate on the type of barrier created by scientific and technical knowledge.

In the industrialized countries, truly new technology systems do not necessarily originate in the most powerful, large, experienced firms, as shown in Part One. They often involve small firms started up by entrepreneurs with advanced university training in specialized areas, such as has been seen in microelectronics and biotechnology, or revolutionary new ideas as in the case of Henry Ford (Chapter 6). Much of the knowledge and skills which will later be required for the growth phase of the system and for subsequent products is developed within these firms as they evolve and either grow or are absorbed by large firms or simply disappear. Much of the knowledge required to enter a technology system in its early phase is in fact public knowledge available at universities. Many of the skills required must be invented in practice. As the system evolves it generates new knowledge and skills which are increasingly of a private nature and are generally not willingly sold to competitors anywhere. Gradually, as the system approaches maturity, both the knowledge and the skills tend to become public again or are willingly sold at a price.

This implies that given the availability of well-qualified university personnel, a window of opportunity could open for relatively autonomous entry into new products in a new technology system in its early phases. This partly explains the cases of innovation outside the main technologically leading nations or firms mentioned earlier on. The problem now becomes whether the endogenous generation of knowledge and skills will be sufficient to remain in business as the system evolves. And this implies not only constant technological effort but also a growing flow of investment. Development is not about individual product successes but about the capacity to establish interrelated technology systems in evolution, which generate synergies for self-sustained growth processes.

If we follow the taxonomy put forward by Freeman and Perez (1988), it will be clear that the technology systems discussed here are in turn the elements of a larger whole which also evolves in time from an early phase through growth to maturity. The 'life cycle' of such a techno-economic paradigm is composed of a series of interrelated technology systems. It is clear that it is the interconnection between technological systems which generates and diffuses knowledge, skills and experience which are of widespread and general use. In this view, the present transition period is identified with a change in techno-economic paradigm and will affect the whole range of technology systems which evolved and matured under the previous paradigm. Most of them will be profoundly transformed as the new information intensive, flexible, systemic, micro-electronics-based paradigm propagates across the productive system. Mature industries reconvert, mature products are redesigned, new products and industries appear and grow, giving rise to new technology systems based on other sorts of relevant knowledge and requiring and generating new skills and new locational and infrastructural advantages.

What this means for developing countries is that during periods of paradigm transition a temporary window of opportunity is open for entry into new industries. An illustrative example is provided by the entry of the East Asian countries into micro-electronics despite the intense US/Japanese competition (Hobday, 1995). The speed with which, for example, US technology was transferred abroad mainly to a number of South East Asian industrializing countries, sometimes within less than a year of the original innovation, is a reflection of that pressure.

The competing away of the monopoly profits in these new industries appears at least from an international point of view faster and even less subject to national control than in previous historical cases. Multinational productive capital is of course more mobile and much of the observed technology transfer abroad is purely a reflection of the need for multi-national firms to 'internalize' internationally their scientific and techno-logical knowledge. One reason for this relates of course to the specific difficulties in legally 'appropriating' electronic inventions and innovations (see e.g. the debate about the protection of 'software' knowledge). The failure of the (international) patent system to provide effective protection for inventions and innovations in electronics has meant, however, that imitation, the competing away of the innovation/monopoly profits, and the 'swarming' effect have all proceeded much more rapidly, both nationally and internationally.

Second, microelectronics insofar as it is characterized by a radically different resource saving potential might well be more 'appropriate' in its application as well as production to the situation of a large number of semi- or newly industrializing countries. In the terminology introduced in the previous section, the locational advantages of these countries could well be higher than with regard to many previous technologies. In particular, the specific capital-saving potential of information technology seems to be of relevance to countries where growth has been hampered by general capital shortage.

Third, it is generally accepted that the so-called 'deskilling' effects of microelectronics might be particularly severe in relation to a wide variety of highly specialized technical (mechanical and electrical) skills, which form at this moment probably the major specific human capital bottleneck in most semi- or newly industrializing countries. Consequently, the 'wrong' skills and 'irrelevant' experience could well amount to a more significant bottleneck and diffusion retardation factor in the advanced countries than in many industrializing countries.

The potential for technological catching up in microelectronics remains, however, subject to many of the various threshold levels and the entry cost components mentioned earlier on. Locational and infrastructural advantages do not fall from heaven, nor does a particular country's endowment in scientific and technical personnel and skills. They result from the previous history of development, plus natural resources, plus social, cultural and political factors. And, depending on the nature of the new paradigm, these can be excellent, very good, bad or hopelessly inadequate in any particular country. Furthermore, taking advantage of the new opportunities and of favourable conditions requires the capacity to recognize them, the competence and imagination to design an adequate strategy and the social conditions and political will to carry it through.

The real chances of advance for any particular country will be very large or very small depending on all the factors mentioned, but they will also be affected by the ultimate shape taken by the socio-institutional framework at the international level. Our main point is that the present period has been and continues to be particularly favourable for attempting a leap in development of whatever size is possible. And this demands a complete reassessment of each country's conditions in the light of the new opportunities.

15.6 CONCLUSIONS: ON THE NEED FOR COMPLEMENTARY INDIGENOUS CAPACITIES

The discussion in this chapter has shown the crucial role of knowledge diffusion (Figure 15.1) in the catch-up process. We conclude the chapter by pointing to the implications for scientific research, for R&D and for STS.

The import of foreign technology is often discussed in terms of two equally impracticable extremes. On the one hand, a position of complete autarchy in science and technology and of striving to be completely independent in every single branch of research and development would be ruinously expensive and almost impossible for all but the largest superpowers. The mechanisms for the international transfer of technology are of the greatest importance for policy-makers in the developing countries. Every country stands to gain enormously from international interchange and division of labour in world science and technology.

On the other hand, an international division of labour in science and technology which is so one-sided that it leaves large areas virtually denuded of independent scientific capacity is equally unacceptable. Even on the narrowest economic grounds it is highly inefficient, and is only recommended by those who have had no practical contact with the problems of technology transfer. Simply to assimilate any sophisticated technology today and operate it efficiently requires some independent capacity for R&D, even if this is mainly adaptive R&D.

Not just in agriculture but also in manufacturing, the variety of local conditions is so great that simple 'copying' is often ruled out. Thus in many countries the capacity to receive technology from outside imperatively requires some independent indigenous science base. To solve the innumerable local problems of soil, materials, environment, skills and climate requires that the indigenous base should grow and flourish. Of course, a variety of other scientific and technological services (STS) are needed as well as R&D, such as information services, consultancy services, geological surveys, project surveys, testing facilities and training organizations. For a long time, these may employ more engineers and scientists than R&D itself.

What is desirable on economic and technological grounds is even more so on cultural and political grounds. While some scientific and technical capacity is necessary for assimilation of the results of foreign research and technical progress, it is undoubtedly possible to get by with a smaller commitment than that made in the superpowers or in several West European countries. Obviously the size of a country has a very great bearing on this question and will affect the degree of specialization which is necessary. Heavy reliance on imported technology is an inescapable necessity for most countries in the world. The economic consequences of this situation are perhaps not too serious, but the political and cultural consequences are very great. One must therefore expect that the smaller countries, as well as the developing countries, will lay increasing stress on equitable international arrangements for access to world science and technology. The attempt to establish more expensive 'autarchic' R&D is to some extent a defensive reaction against the political dangers of potential lack of access. Only in proportion to the growth of mutual trust, and a genuinely international policy, will the achievement of a more equitable and mutually beneficial international division of labour in science and technology be possible. Such a division must in any case be based in principle on all countries contributing to as well as drawing from the world stock of knowledge (see Chapter 16).

The implications of this are complex for technology and multinational corporations but relatively clear for fundamental science. The greatest significance of fundamental research is that it provides a multipurpose general knowledge base on which to build a wide range of scientific and technical services. Every country without exception requires such a base, even if only on a very small scale. Without it there cannot be any independent long-term cultural, economic or political development. One of the main objectives of world policy for science and technology should be to build and sustain an indigenous scientific capacity throughout the developing world. The Canadian International Development Research Centre was an important step towards the reorientation of world science and aid policies in this direction (IDRC, 1972) and has since been imitated by other agencies and foundations. Regrettably, the amount of 'aid' for investment projects and for the strengthening of autonomous science and technology, as advocated by IDRC, has been cut back in the 1990s. In Part Four, we follow through on this discussion of policy, which has been initiated here, starting with policies for basic scientific research at a national and global level.

NOTES

1. The greater part of this section is based on Perez and Soete (1988). We are grateful to Carlota Perez for helpful comments and suggestions.

2. Many more formal structural growth models have since then been developed, which are not discussed here. See amongst others: Dosi *et al.* (1992); Landesmann and Goodwin (1994); Silverberg *et al.* (1988); Silverberg and Lehnert (1994); Silverberg and Verspagen (1995) and Soete and Turner (1984).
3. The application of probit models in economics began with Farrell (1954) and Tobin (1955). Although Tobin's example of durable goods did not at first find favour with editors of journals, it is now the most cited reference.

PART THREE   REVIEW ARTICLES, LITERATURE SURVEYS AND KEY REFERENCES

Antonelli, C. (1995) *The Economics of Localised Technological Change and Industrial Dynamics*, Dordrecht, Kluwer.

Archibugi, D. and Michie, J. (eds) (1997) *Technology, Globalisation and Economic Performance*, Cambridge, Cambridge University Press.

Cantwell, J. (1995) 'The globalisation of technology: what remains of the product cycle model?', *Cambridge Journal of Economics*, vol. 19, pp. 155–74.

Cappeleu, A. and Fagerberg, J. (1996) *East Asian Growth: A Critical Assessment*, Oslo, NUPI.

de la Mothe, J. and Paquet, G. (eds) (1996) *Evolutionary Economics and the New International Political Economy*, London, Cassell.

Dore, R. (1987) *Taking Japan Seriously: A Confucian Perspective on Leading Economic Issues*, London, Athlone Press.

Dosi, G., Pavitt, K. and Soete, L. (1990) *The Economics of Technical Change and International Trade*, Brighton, Wheatsheaf.

Fagerberg, J. (1987) 'A technology gap approach to why growth rates differ', *Research Policy*, vol. 16, no. 2, pp. 87–99.

Fagerberg, J. (1995) *Convergence or Divergence: The Impact of Technology*, Working Paper no. 524, Oslo, NUPI.

Fagerberg, J., Verspagen, B. and von Tunzelmann, N. (eds) (1994) *The Dynamics of Technology, Trade and Growth*, Aldershot, Elgar.

Freeman, C. (1991) 'Networks of innovators, a synthesis of research issues', *Research Policy*, vol. 20, no. 5, pp. 499–514.

Hagedoorn, J. and Schakenraad, J. (1990) 'Strategic partnering and technological cooperation', in Freeman, C. and Soete, L. (eds) *New Explorations in the Economics of Technical Change*, London, Pinter.

Hobday, M. (1995) *Innovation in East Asia: The Challenge to Japan*, Aldershot, Elgar.

Jang-Sup Shin (1996) *The Economics of Late-comers, Catching-up, Technology Transfer and Institutions in Germany, Japan and S. Korea*, London, Routledge.

Kaplinsky, R. (1990) *The Economies of Small: Appropriate Technology in a Changing World*, London, IT Publications.

Krugman, P. (1995) 'Technological change in international trade', in Stoneman, P. (ed.) *Handbook of the Economics of Innovation and Technological Change*, Oxford, Blackwell.

Lazonick, W. (1991) *Business Organisation and the Myth of the Market Economy*, Cambridge, Cambridge University Press.

Lundvall, B.-Å. (ed.) (1992) *The National System of Innovation*, London, Pinter.

Maddison, A. (1982) *Phases of Capitalist Growth*, Oxford, Oxford University Press.

Mjøset, L. (1992) *The Irish Economy in a Comparative International Perspective*, Dublin, National Economic and Social Council.

Narula, R. (1996) *Multinational Investment and Economic Structure*, London, Routledge.

Nelson, R. (1996) *The Sources of Economic Growth*, Cambridge, MA, Harvard University Press.

Nelson, R. (ed.) (1993) *National Innovation Systems: A Comparative Analysis*, New York, Oxford University Press.

Niosi, J. (ed.) (1991) *Technology and National Competitiveness: Oligopoly, Technological Innovation and International Competition*, Montreal, McGill University Press.

Perez, C. (1988) 'New technologies and development', in Freeman C. and Lundvall, B-Å. (eds.), *Small Countries Facing the Technological Revolution*, London, Pinter, pp. 85–97.

Reinert, E. (1994) 'Symptoms and causes of poverty: underdevelopment in a Schumpeterian system', *Forum for Development Studies*, nos. 1–2, pp. 71–109.

Reinert, E. (1997) 'The role of the State in economic growth', *Working Paper 1997–5*, Centre for Development and Environment, University of Oslo.

Ruigrok, W. and van Tulder, R. (1995) *The Logic of International Re-structuring*, London, Routledge.

Scott, M. F. (1989) *A New View of Economic Growth*, Oxford, Clarendon Press.

Soete, L. (1989) 'The impact of technological innovation on international trade patterns: the evidence reconsidered', *Research Policy*, vol. 16, nos. 2–4, pp. 101–30.

Strange, S. (1988) *States and Markets*, New York, Blackwell.

Vernon, R. (1966) 'International investment and international trade in the product cycle', *Quarterly Journal of Economics*, vol. 80, pp. 190–207.

Verspagen, B. (1993) *Uneven Growth between Interdependent Economies: An Evolutionary View on Technology Gaps, Trade and Growth*, Aldershot, Avebury.

Verspagen, B. (1992) 'Endogenous innovations in neo-classical growth models: a survey', *Journal of Macro-Economics*, vol. 14, no. 4, pp. 631–62.

Wade, R. (1990) *Governing the Market: Economic Theory and the Role of Government in East Asian Industrialisation*, Princeton, Princeton University Press.

PART FOUR
INNOVATION AND PUBLIC POLICY

INTRODUCTORY NOTE

Part Three of this book has illustrated the variety of ways in which the national environment, or 'national system of innovation' may influence the innovative activities of firms. Differences in national systems can go a long way towards explaining differences in growth rates over the past two centuries. However, Chapters 14 and 15 in particular showed that the international global environment exerts an increasing influence on the behaviour of firms. Even in the largest countries the world economy, world science and technology increasingly affect the innovative activities of all firms, large and small. Public policies too are heavily affected by the international environment. This is obvious in the case of war or the Cold War, but it is also true in the case of policies which are motivated primarily by economic goals or social objectives.

In this final part, which deals with public policies for science, technology and innovation, we outline historically in Chapter 16 the main trends in public expenditures for R&D and related activities since the Second World War. It is quite impossible to deal with all aspects of this complex topic and our account is necessarily restricted to a few key themes. We do not go into any depth in the discussion of military R&D and military innovations and their management, although our account certainly shows that they were extraordinarily important over the whole half-century, above all in the early post-war period. We endeavour to show that the military R&D system had a powerful influence on civil R&D, most obviously in the larger countries but also in many smaller countries.

Neither was it possible to discuss the immense variety of national systems of procurement, fiscal incentives, subsidies and so forth. We have concentrated rather on some of the major common trends in the industrialized countries. Part Three demonstrated the importance of variety; in this part, we illustrate some of the strong tendencies towards policy convergence in the last fifty years. This convergence has arisen partly because science and technology have an intrinsically universal quality, but also partly because competition in the global economy compels countries to heed the behaviour of competitors and to adopt what appear to be the more successful techniques. Fashion also plays its part.

Whereas in the 1940s and 1950s military goals dominated in the larger countries, during the 1960s and 1970s economic policy objectives such as the growth of productivity began to take precedence, even in countries with strong military commitments.

International organizations, especially the OECD, provided a forum in which some of the main policy problems were analysed and debated. A brief historical survey shows that during the 1970s a general consensus developed that a change was necessary in priorities for innovation policy.

Whereas in the 1950s and 1960s public subsidies for both military and civil R&D projects had been widely employed, the value of these subsidies was now increasingly questioned and economists generally argued that public expenditures in support of innovation should be mainly oriented into four areas:

1. Fundamental research, mainly in universities.
2. Generic technologies and their diffusion, especially ICT.
3. Industries whose structure prevented the effective performance of research and development at firm level. Agriculture was the typical example but the case for technical consultancy and research services to support small and medium-sized firms (SME) in many industries was widely advocated and implemented.
4. Infrastructural investment in STS as, for example, in bibliometric services, data banks and other information services.

Chapter 16 presents the main economic arguments for public support of basic research, which have been almost universally accepted in the second half of the twentieth century and introduces the similar arguments which have been used to justify public expenditures in support of generic technologies. By far the major part of such expenditures has been to promote the development and diffusion of information and communication technology (ICT) in the 1980s and 1990s. This technology has had such pervasive effects throughout the world economy that it merits some special consideration, especially in relation to its employment effects, which have been at the centre of the public debate. We have therefore devoted Chapter 17 entirely to the consideration of ICT, employment and related issues of the 'information society'.

In Chapter 18 we return to a consideration of wider public policy issues and we look further forward into the twenty-first century and the next millennium. No one can accurately forecast the future but we take the view that environmental policy issues are likely to become of overwhelming importance during the next fifty years. We suggest that this calls for a radical new look at policies for science and technology, as proposed in the so-called 'Maastricht Memorandum' (Soete and Arundel, 1993) involving a systems approach.

Finally, in the concluding chapter we discuss the political and social dimensions of technology policy. 'Technology assessment' has had a chequered history; the original Office of Technology Assessment (OTA) in the United States has closed but in several European countries TA is developing in new directions. Especially in the Netherlands 'constructive technology assessment' offers a new and original approach. We argue that in one form or another technology policy is nothing but continuous and constructive technology assessment and that this necessarily involves political, social and ethical considerations as well as economics.

CHAPTER 16

ASPECTS OF PUBLIC POLICY FOR SCIENCE, TECHNOLOGY AND INNOVATION

16.1 INTRODUCTION

As we have seen in Part Three, some public support for basic research activities has actually proved to be essential even in less developed countries whose main concern is with the import, imitation, assimilation and improvement of technologies already available from outside. At least some minimal research activity in universities and public laboratories, together with the education and training of some postgraduate students abroad is essential simply to gain points of entry and depth of understanding. At later stages, with the gradual improvement of facilities and the upgrading of standards, the science system in such countries may make an increasing contribution to the advance of world science, but world scientific publication is still dominated by a few countries (Table 16.1).

The acceptance of a utilitarian argument for the public support of basic scientific research predates the Industrial Revolution itself. Jacob Schmookler (1965) in his paper on 'Catastrophe and Utilitarianism in the Development of Basic Science' pointed out that 'cultural' arguments for the support of science have been less influential than arguments of competitive military and economic advantage. Although private individuals and societies were often motivated by curiosity and the desire to advance knowledge, utilitarian motives have tended to predominate when it comes to public funds. This is not to belittle or to denigrate the motives of those wealthy individuals or patrons who have supported 'science' rather in the same way as they have supported 'art' or 'music'; nor is it to deny the argument of many educationists that science should be taught in the curriculum of every school and university as an essential part of the heritage of human civilization. It is simply to recognize that those motives identified by Schmookler have often proved more effective in generating funds for science than the more general appeals for the advancement of knowledge.

As we shall attempt to show in this chapter, it was the clear demonstration just before, during and after the Second World War that basic science could confer enormous military and economic advantages, which led to an order of magnitude increase in its endowment in many countries and to the arrival of what became known as 'Big Science' (Price, 1963).

In this chapter we first of all discuss the public funding of basic research and then turn to the more controversial question of public funding of other types of R&D. This leads on to a more general discussion of priorities in

Table 16.1   Shares in world scientific publications

	1983	1993
European Union	26.8	29.6
EFTA	4.5	4.6
CEE	10.8	4.8
NAFTA	40.9	40.6
Japan	6.9	8.2
DAE	0.3	1.6
Other Asia	3.0	3.7
Australia/NZ	2.8	2.8
Other America	0.9	1.1
Other	3.1	3.0

Source: OST from ISI data.

science and technology policy as they are reflected in expenditures on R&D and other scientific and technical activities. The changing priorities of the past half century are briefly reviewed and some of the likely priorities for the next half century are introduced. In this book we do not go into any detail on methods of selection between alternative projects in basic science – that is beyond the scope of a book on innovation – but see Irvine and Martin (1984), Martin and Irvine (1989).

16.2   PUBLIC FUNDING OF RESEARCH

The first clear and forceful advocacy of a national science and technology policy based on public support for research is usually attributed to Francis Bacon (1627). In *The New Atlantis* he advocated the establishment of a major research institute ('Salomon's House') which would use the results of scientific expeditions and explorations all over the world to establish the 'knowledge of causes, and secret motions of things'. It seems unlikely that in addition to his other publications, Bacon also wrote Shakespeare's plays, but it is certain that he was an eminent statesman in the early seventeenth century and that his publications were influential. Probably it was Bacon and Newton who, more than any other two individuals, helped to develop a cultural and political climate in Britain highly favourable to science in the seventeenth and eighteenth centuries. The actual endowment of science by grants for the Royal Society from 1662, for expeditions, prizes and other ventures, was still on a minute scale compared with the subventions to which we have grown accustomed in the twentieth century. It was nevertheless significant in terms of public recognition at the highest level. Some of the Italian city states provided more generous support and at an earlier stage but the cultural, religious and political problems in Italy prevented the development of a scientific culture so uniquely favourable to the Industrial Revolution as was the case in Britain in the eighteenth century (Jacob, 1988). It should be noted that in those days it was not customary to make a clear distinction between basic and applied research.

Before the closing decades of the nineteenth century the sums spent on supporting academies of science, scientific societies or university research were very small indeed. Governments also spent small amounts of money on armaments research, often in publicly owned ordnance factories, and on a few regulatory functions. Although the sums were small, their strategic importance was considerable in the context of the time. For example, ordnance factories and arsenals often led the way in the advancement of metal-working technology and naval shipyards were important for the shipbuilding industry as a whole (Reinert, 1993; Modelski and Thompson, 1993). As we have seen in Chapters 3 and 6, the Springfield Armoury pioneered the technology of interchangeable parts while Frederick Taylor's early work on steel alloys was sponsored by the US Navy. The Italian arms factory owned by Beretta, which has been described by historians as the 'oldest industrial dynasty in the world' (Jaikumar, 1988), led the introduction of many new technologies in Italy. Some of Maudslay's early work on machine tools between 1800 and 1810 was undertaken for the Royal Navy and Woolwich Arsenal, notably the 45 machines of 22 types for the Portsmouth blockmaking plant, which could make 130,000 ships' pulley blocks a year – more than enough for the entire requirements of the Navy. This was the first large-scale plant employing machine tools for volume production, even earlier than the 'American system of production' described in Chapter 6 (Corry, 1990).

Towards the end of the nineteenth century, with the rise of new chemical and electrical technologies, governments extended their range of scientific and technical activities. Laboratories for introducing and regulating industrial standards (metrology), such as the Bureau of Standards in the United States, the Physikalische und Technische Reichsanstalt in Germany and the National Physical Laboratory in Britain, assumed important responsibilities for research as well as and in support of their regulatory functions and other government laboratories were established to support civil research.

The First World War stimulated the growth of government support for scientific research and development in many countries as, for example, in the establishment of the Department for Scientific and Industrial Research (DSIR) in Britain in 1915 (McLeod and Andrews, 1970; Andrews and Poole, 1972). This continued as an agency in support of civil science down to the 1960s, controlling a variety of government laboratories and funding university research. Similar developments occurred in many countries but it was the Second World War and the Cold War which led to the greatest surge of government funding of R&D. This took the form of massive support for some huge projects, of which the most famous was the 'Manhattan Project' to design and develop nuclear weapons. It was followed by the establishment in many countries of large institutions for R&D in the military and civil applications of nuclear energy. These were by no means the only Big Science laboratories; there were others, especially for military purposes and of course for space exploration. However, for a long time, some governments had begun to increase their support for 'Little Science', i.e. a large number of research projects carried out by small groups of researchers or by individual researchers or postgraduate students. After the Second World War this type of (mainly university) research received a relatively generous increase in funding in many countries side by side with the growth of Big Science. Big Science itself did not actually begin with the Second World War or the Manhattan Project but had already taken off in the 1930s. Even though the US Department of Agriculture was at that time spending more on research than the Department of Defence (will those days ever return?) and even though 'Small Science' was still by far the predominant mode, yet the first Big Science laboratories had emerged well before the Second World War. The most important was undoubtedly the Lawrence Berkeley Laboratory – the Radiation Laboratory at the University of California in Berkeley (Galison and Hevly, 1992).

Most of the main features of wartime and post-war Big Science had already emerged in this laboratory in the 1930s. Among them were: hierarchical organization and management structures and committees; strong interaction between basic science, technology and engineering (and frequently other disciplines too); complex patterns of extra-mural industrial and government funding and influence and the national and local prestige derived from large instruments and large-scale funding as well as from scientific achievements. In the case of the Radiation Laboratory there were also the problems associated with strong autocratic leadership, which arose in some later Big Science projects, although by no means all. Whether autocratic, bureaucratic or democratic, the working environment and career pattern for young scientists entering Big Science labs undoubtedly differed considerably from traditional university small science. It was nevertheless attractive to many as is evident from the fact that nearly a quarter of the Laboratory's staff in the 1930s were supported by their own resources (both PhD students and post-Docs), compared with 20 per cent by the state, 25 per cent by intra-mural funds and 27 per cent by other external funds.

External sources of funding were of course vital for the expensive capital equipment which was the main feature of Big Science from the standpoint of economics and administration. At Berkeley, the Rockefeller Foundation, the state and the federal government contributed the lion's share, although private individuals and corporations played some part. Both the Second World War and the Korean War played an important role, not in initiating government expenditures on science but in habituating government agencies to the regular disbursement of very large sums for R&D at many centres.

Not surprisingly the huge increase of public funding for research did not go unchallenged. That it was able to survive budget cutting and to continue into the 1980s in many countries was due to a combination of factors. Undoubtedly, one of the main ones was the very strong support of military agencies in many countries, but especially in the superpowers. The examples of nuclear bombs and radar served to convince even the most sceptical that weapons technology now depended very heavily on basic science and especially on physics. In past centuries the armed forces had sometimes been relatively conservative about new weapons but after the Second World War they generally became enthusiastic not only about new weapons projects, but also about basic research in universities and government laboratories which they perceived as underpinning new military technology. Thus the United States Air Force, the Office of Naval Research and other military agencies were among the main sponsors of basic research in the Unites States in the 1950s and 1960s. Their support meant that the claims of the scientific community for more generous endowment were much more favourably received than before the war.

In the Soviet Union and other East European countries the links between the academy institutions (supporting basic research) and the military–industrial complex were even closer and there too public funding of basic research expanded very rapidly, although not concentrated in universities as in Western Europe and the United States.

The economic arguments in favour of public funding of research, although they had been advanced in rather general terms long ago, were not actually elaborated in depth until some time after the Second World War. It was Richard Nelson (1959) and Kenneth Arrow (1962) who wrote the classic papers, which made public expenditure on basic research, as on education, acceptable to most economists, including those most sceptical about public expenditures. The principal point which they made was that private expenditures would tend to be lower than the economically and socially desirable level if left to market funding. Basic research is by definition truly uncertain; the researchers do not know who, if anyone, will benefit from its outcome. Consequently, it is unlikely that firms will finance much if any basic research because they do not know which industries or firms will be able to appropriate a return from the investment (see Chapter 10).

This attitude is heavily reinforced by the time scale of much basic research, which is often very long term. As we have seen in Chapter 10, firms usually have rather a short-term or medium-term perspective for their R&D investments. They hope for a return within a few years or a decade at most and discounted cash flow calculations bias decision-making strongly in this direction. Consequently, very few firms will think it worth while to finance research which may take twenty or thirty years to yield an outcome which is itself uncertain.

The view that firms would be unlikely to finance much basic research was generally borne out by the empirical evidence, as we have seen in Parts One and Two. Certainly, there are examples of industrial support for long-term basic research. Some of them, as in the case of Bell Laboratories (Chapter 7), were facilitated by the peculiar form of regulation of the telecommunications industry. Since the deregulation and reorganization of AT&T, funds for basic research were drastically reduced in the 1990s. In a few other cases, especially in the chemical industry, very large firms such as Bayer, Hoechst, BASF, ICI and Du Pont (Chapter 5) were ready to support 'oriented' basic research, which because of the huge spread of their product portfolios was more likely to yield results of interest to them. Still other firms, as we have also seen in Part Three, were obliged to do a little basic research in order to gain access to and understanding of the results of research conducted elsewhere or to recruit good scientists. Empirical studies in many countries supported the view that firms in a variety of industries needed continuous contact or involvement with university research for successful innovation (see, for example, Senker and Faulkner (1994); Faulkner et al., (1995); Mansfield et al., 1977; Mansfield, 1980).

Further support for the arguments in favour of public funding of basic research came from industry which reported that recruitment of university graduates with an up-to-date knowledge of scientific instruments and of mathematical and computer techniques, derived from recent university research, was often as important or more important than the results of the basic research itself (Pavitt, 1990). The most obvious (and unanticipated) example of the benefits of long-term basic research came from biology, which for decades had been the Cinderella of the natural sciences. The results of biological research from Crick and Watson onwards have led to the extremely rapid growth of biotechnology with myriad potential applications. These unexpected developments meant that firms in the chemical industry had to resort quickly to a variety of techniques to understand and assimilate the results of university research. These included hiring many university biologists as consultants, recruiting them to industrial R&D departments, or financing research in university departments (Faulkner, 1986; Sharp, 1991; Martin and Thomas, 1996).

Finally, the social benefits of publicly funded basic and applied research are much wider than the competitive advantage of firms or the growth of the economy. Obvious examples are research on public health problems and on the environment. As Pavitt (1996) has pointed out, it is highly unlikely that private firms would have financed the early research on BSE or on cancer and smoking. It is even less likely that they would or could finance research on the holes in the ozone layer or life on Mars. We simply do not know what will be the results of research in astronomy or in particle physics and the instrumentation costs are extremely heavy. Yet society may deem it worthwhile to incur these costs even though no private firm would undertake them.

This classical view of the externalities accruing from basic research and of the case for substantial public funding went almost unchallenged in the 1960s and 1970s but has come under attack from Milton Friedman and more recently, from a scientist, Terence Kealey (1996). In his book *The Laws of Scientific Research*, Kealey attempted to show, like many others before him, that the so-called 'linear model' of R&D is fallacious. But whereas most historians, engineers and social scientists have argued that it should be replaced by an 'interactive' model which points to the strong interactions between basic research, applied research, development, production and markets, Kealey argued that basic research comes last of all. Only when society is very rich can it afford basic research. He derived his 'reversed' linear model from the historical evidence that he assembled, which is, however, strongly disputed by historians (e.g. David, 1997).

The policy conclusion drawn by Kealey was that all public funding of basic research should be withdrawn. He believed that it would be substituted by increased funding from private industry, private individuals and private foundations. The kernel of good sense in this argument is that a pluralistic pattern of funding is to be preferred to single source funding.

The uncertainties in basic research, the subjective bias in decision-making and the ideological element mean that it is often desirable to have alternative sources. However, the cessation of all public support would probably have disastrous long-term consequences in most countries and it is highly unlikely that private sources would compensate for the collapse of public investment. This does not mean to say, of course, that all decisions on public expenditures for research have been wise ones. On the contrary, a good argument can be made that quite large sums have been misallocated (as was also the case with some private research). In the next section we turn to other spheres of public expenditure on science and technology and give some examples of such misallocation.

16.3 PUBLIC INVESTMENT IN R&D FOR INDUSTRY

In Part One, we attempted to analyse the main features of technical change since the Industrial Revolution. However, even though some additional empirical data have been cited in Parts Two and Three, it remains true that several important sectors of industry have been largely ignored. These fall into two main groups:

1. Some research intensive industries, such as aircraft and nuclear weapons, where there has been a very heavy government involvement, both in procurement and in R&D itself.
2. Many consumer goods industries and services mostly of low research intensity, such as clothing, furniture, food, consumer durables, construction, distribution, financial and social services.

Much of the argument of the book is relevant to all sectors of industry. For example, the discussion on size of firm in Chapter 9 and methods of project evaluation in Chapter 10 is as relevant to the food industry as it is to electronics and chemicals. But there are some features of these two groups of industries which raise new issues which have as yet hardly been discussed. These are primarily issues of public policy.

In the USA, UK and France the aircraft industry accounted for more than a quarter of total industrial R&D expenditure during most of the post-war period. By far the greater part of this was financed with public money, although the development work was carried out in industry. The recognition in the 1960s that it was no longer necessary or desirable to give such a high priority to the development of new types of military aircraft was often followed by increased programmes of public expenditure for civil aircraft development. In some cases this was no doubt justifiable in terms of civil, economic and social benefits, particularly, for example, in reducing the cost of air travel and some of the disamenity effects such as noise. But there was little evidence that the supersonic transport (SST) could be justified in economic or welfare terms; nor was there any readiness by the manufacturers to finance the main part of the development costs. No private entrepreneurs would accept this risk, but a very high proportion of public expenditure on civil R&D in the UK and France went to the Concorde and other aircraft projects in the 1960s. Only in the 1970s did this tail off with the completion of Concorde, but it continues to this day on a smaller scale.

This is not to say that the principle of public support for civil, industrial R&D including aircraft development is wrong. Indeed it can be justified on a small scale on economic grounds, but a priori there is no more reason why such support should go to the aircraft industry rather than to the railways, telecommunications, the computer industry or machine tools. If a balance is struck between the social and environmental costs and benefits involved, all of these might be stronger candidates, even taking into account technological spin-off to other industries. Indeed, the spin-off effects of advanced energy-saving land transport systems could quite possibly be greater than for air transport, as well as the direct benefits to consumers. The conclusion is difficult to escape that the preferential treatment of R&D in this industry was due less to any considered assessment of transport or communication needs than to habit, the continuing power of a lobby and the associated military and prestige elements. It was certainly not due to any sophisticated project evaluation techniques.

So far as the American experience is concerned, Eads and Nelson (1971) have argued that the attempts to emulate military 'crash' programmes in such civil technologies as SST and nuclear reactors was misconceived and a considerable waste of resources was the result. Government expenditures, although very important, should have concentrated on applied research and early experimental development. It is in the area of fundamental research and enabling technologies that the economic case for public finance and public laboratories is strongest. For those stages of development directly linked to the introduction of commercial products or systems, it is much more likely that waste will be avoided if this is carried out by the enterprise and largely or entirely at the firm's expense. It is clear that there are very grave dangers in major government subsidies to firms to cover their development costs. If government subsidies to enterprises are used at all, then thorough public discussion of priorities is essential.

Whether or not one accepts the basic theme of Galbraith's *New Industrial State*, or the arguments advanced in Parts One and Two of this book, it would be difficult to deny that the military–industrial complex is a reality which very much affects firm behaviour, at least in a few industries. The scale and complexity of modern technology have been carried to extreme limits in research, design and development for military aircraft, missiles and nuclear weapons. The large-scale participation of governments and the peculiar nature of the military market mean that the process of advocacy in project selection, which as we have seen is present in all R&D policy-making, becomes overtly political at the national level. The 'lobby' and the 'corridor padder' are more important in this type of decision-making than elaborate calculations of return on investment. Indeed such calculations may often be used purely as a gimmick to provide a pseudo-rationalistic method of manipulating the political process. Clearly national policies of a non-economic nature have predominated in determining the innovative performance of the aircraft industry, both military and civil, and the same is true for several other industries closely linked to aircraft (Peck and Scherer, 1962; Peck, 1968).

These criticisms of some public subsidies to aircraft and nuclear R&D do not invalidate all arguments for public expenditures in support of the development efforts or the applied research of firms. That would be to throw out the baby with the bath-water à la Terence Kealey. A case where public expenditure on applied research and development has clearly been in the public interest and has led to many beneficial results for consumers as well as producers is that of agricultural research.

In this case, it is the structure of the industry which has led in almost every country to the heavy involvement of the public sector in agricultural R&D. Family farms are too small to finance their own R&D, nor do they usually possess the requisite scientific knowledge. For these reasons, as well as the strategic importance of food production, governments have financed most of the applied agricultural research. Indeed, this was one of the greatest success stories of technical change in the American economy from the 1860s to the 1960s. The situation is changing now with the concentration into large farms (agribusiness), which in some cases now are able to perform their own R&D, and with the advent of genetic engineering and the greatly increased involvement of chemical firms.

These two very different cases of aircraft and agriculture illustrate well the argument of Nelson and Winter (1977) that it is essential to consider the specific features of each industry in the analysis of technical change. Another example of a very different type is that of consumer goods and services.

16.4 INNOVATION IN CONSUMER GOODS AND SERVICES

The evidence of Part One has shown the very great technical achievements of innovators in capital goods, components and materials in the twentieth century. The buyers of capital goods and intermediate products, as of weapon systems, are often scientifically and technically sophisticated customers. They are their own 'customers' for many process innovations, and when they are purchasing outside they often have greater technical sophistication than their suppliers (for example, the chemical firm selecting a pump or a filter). They may often use their purchasing power to commission technical innovations. Buyers in these markets are concerned with genuine technical characteristics and may lay down stringent technical performance specifications. They are less likely to be impressed by product differentiation, and advertising plays a less important part than with consumer goods while technical services to customers play a much more important part.

Most innovation case studies agree that those innovators who take considerable trouble to ascertain the future requirements of their customers are on the whole more successful. The SAPPHO comparisons of success and failure in innovation (Chapter 8) showed that most failures were associated with either neglect of market requirements or relatively poor understanding of the customer's needs. To this extent the argument here is at variance with Galbraith's assumption of 'producer sovereignty' in imposing innovation on the market.

These results are superficially encouraging insofar as they indicate that at least in capital goods the market has been effective in stimulating the types of innovation which match real customer needs and potentially, if not always in practice, social welfare. There are certainly also examples of consumer goods which confirm the SAPPHO conclusions on user needs. Take, for example, the Danish plastic toy 'Lego'. The firm took an enormous amount of trouble to ascertain the needs and preferences of the users (in this case, mainly children). It was rewarded by the most successful sales and export performance of any toy in the world. On the only occasion when the firm introduced a new product without its usual exhaustive prototype tests with users, the innovation was a failure. There are many other examples of great consumer benefits from household products and from drugs. Moreover, most consumers have benefited from the rise in living standards made possible by the productivity advances due to technical innovation in capital goods, materials and communication systems.

Nevertheless, generalizations about the benefits of technical innovation do need to be heavily qualified outside the area of capital goods. Most innovation studies have been concerned with the more spectacular breakthrough innovations and have hardly considered the type of annual model changes which are more characteristic of many consumer products. There are reasons for believing that buyers in these areas are far less capable of making sound technical judgements than in the capital goods area. They may typically have rather poor sources of information and lack the capacity to make any serious technical assessment.

A glance at the distribution of industrial R&D expenditures in any country shows a heavy concentration in capital goods and materials. Consumer industries have very little and much of these limited R&D and other scientific inputs are used for product differentiation in oligopolistic markets, and in the closely related activity of planned obsolescence. The growth and welfare implications of this kind of industrial R&D are dubious. Fisher *et al.* (1962) showed the very high costs of annual model changes in the US car industry and the 'planned obsolescence' in car exhaust pipes was strongly criticized in a UK government report (Report of the Committee on Corrosion and Protection, 1971, p. 128). Obviously it was this aspect of technical innovation which Galbraith had in mind in developing his critique of 'producer sovereignty'. This critique is much more relevant to consumer goods than to capital goods.

This raises the more general problem of the economic theory of the market and the direction of innovation. Theoretically, the ideal consumer market is supposed to provide consumers with the power to choose between an array of alternatives. Possessing 'perfect' information they are free to choose the 'best buy' for price and quality, thus compelling suppliers to adapt their output to consumer needs through the competitive mechanism. This was the idea of 'consumer sovereignty'. Of course, no economist ever imagined that reality would ever quite correspond to the ideal. Consumers would never really be 'perfectly' informed and very frequently there would be some form of collusion between suppliers or other elements of monopoly. But in some primary commodity markets, fruit and vegetable markets, and some markets for capital goods the reality has sometimes approximated quite closely to the ideal abstract model.

Economists have devoted a great deal of attention to the problems of maintaining consumer sovereignty (see, for example, Knox, 1969). Much of the theory of monopoly and the critique of advertising has been concerned with the erosion of consumer choice. In many countries anti-monopoly and consumer protection legislation attempts to restrict or reverse the powerful tendencies towards concentration of producers and greater producer sovereignty (Heath, 1971). We are not concerned here so much with this general problem but with the specific consequences of technical innovation on consumer markets.

There are three main ways in which technical innovation may diminish consumer sovereignty that the normal type of anti-monopoly legislation does little to affect:

1. The theory of consumer choice is essentially static. The consumer supposedly chooses from the existing array of goods and services. But in areas where technical change is important this array has been determined by choices of R&D project or innovation decisions many years before. The critical element lacking in consumer sovereignty is therefore the power to influence the future array of goods and services. Apologists for the present state of affairs maintain that consumers do in fact exert this influence indirectly because producers are concerned to anticipate their wishes in order to make a profit. Up to a point this is true, as we have seen in the case of many products, but the possibility exists that the innovation decision-makers will impose their preferences and choices, rather than those of the consumer. This power is not unlimited, as the example of the Ford Edsel showed, but it is a serious problem.
2. The theory of consumer choice implies perfect information about the available array of products or services. If we are thinking of a shopper looking at the prices and quality of vegetables on a dozen different stalls in a street market this model may not be too far from the truth. But it breaks down where any degree of technical sophistication or product differentiation comes into the picture, as with cars, television and other consumer durables, and increasingly with a great variety of processed food and chemical products. Shopping on the Internet will complicate this problem even further. This market situation is essentially one of completely unequal access to technical information, as all the consumer associations are well aware. The unsatisfactory nature of the repair and maintenance services for consumer durables in all industrial countries is one illustration of this problem. Despite the fact that there are thousands of firms competing for the business, consumer dissatisfaction is chronic and well founded.
3. Consumers cannot possibly be aware of the various long-term side effects of a multitude of individual choices about many products. The individual who buys a car did not and could not work out the long-term consequences for the urban environment of millions of similar decisions. Nor indeed did the suppliers. Yet the social costs may well be so great that they negate the private benefits to most consumers, as Mishan (1969) long ago argued. The problems of waste disposal in relation to plastics and nuclear power are other examples of long-term social costs inadequately considered.

These three defects help to explain why the nominally sovereign consumers may often feel powerless and frustrated. The problem is often one of adequate articulation of a felt need. But it may often also be that producers are almost exclusively oriented to product differentiation and brand image, rather than to imaginative technical innovation or to social needs. For example, efforts to design safer cars and a relatively pollution-free car were very obvious social needs, but the R&D effort which went into improving safety or preventing pollution was negligible for a long time. Interestingly enough, the stimulus came not from the R&D of producers but from outside critics, such as Ralph Nader, and public regulation. The same is even more true of designing a generally more satisfactory land transport system, particularly in congested urban areas.

Several other examples may be cited to illustrate the extent to which innovators and designers may neglect the interests of users and simply pursue their own fashions and enthusiasms. The example of housing is particularly striking in Britain. In the 1960s there was a fashion among architects and town planners for high-rise flats ('high' blocks of flats may be defined for this purpose as blocks with six or more stories). The extraordinary feature of this particular technological fashion was that such flats were much more expensive than conventional housing or low-rise flats throughout this period (McCutcheon, 1972). (Estimates vary between 1.3 and 1.8 times the cost per sq. metre.) Moreover, this was a period of financial stringency when local authorities were under constant pressure from central government (and from electors) to prune their expenditures and cut back on their services. Yet in spite of this almost all major cities put up thousands of high-rise flats throughout the 1960s.

This might have been justifiable if there had been overwhelming evidence of consumer preference for high-rise living, from sociological surveys and/or a clear readiness to pay the much higher rents on an economic basis. No such evidence was ever produced. The sociological surveys were inconclusive but most of them showed, if anything, a dislike of high flats among council tenants, particularly old people and those with children.

These examples demonstrate the extent to which the values and preferences of designers and innovators may be imposed on the consumer, whether through private firms or public authorities. This does not imply malevolence or contempt for the consumer. On the contrary, in every case, the innovators believed that they were acting in their best interests. It is only a special illustration of a problem which has long been familiar to political scientists. The separation of research, development and design into specialized functions with their own ethos, fashions, interests and enthusiasms inevitably carried with it this danger of lack of social accountability. In theory again the competitive market mechanism ought to be able to perform this function automatically. But it has been argued that the marketplace mechanism, which theoretically was supposed to ensure correspondence between consumer wishes and supply, no longer performs this function adequately, if it ever did so in some sectors. This means that increasingly the political mechanism must restore the lost consumer sovereignty which the autonomous market mechanism cannot assure.

It might have been expected that socialist societies would have been able to make social innovations, which would link the public R&D system more closely to consumer needs. But the evidence available does not justify this conclusion, possibly because they were poor countries attempting to industrialize rapidly, and in the case of the Soviet Union and China, to compete militarily with other great powers; probably also because of the absence of civil liberties.

There seems to be no reason in principle why the more enterprising consumer associations should not go into the business of specifying desired technical performance parameters in the same way as capital goods buyers. They might do this either in association with public authorities or on their own, depending upon circumstances. Ultimately this could even lead to the award of R&D contracts and procurement contracts. Some chain stores and department stores already act in this way on behalf of their customers and, in closing this loop, the civil sector would only be learning the social technique which was learnt by the military sector a generation ago, and by capital goods innovators long ago.

The 'customer–contractor' principle of the UK Rothschild Report (1971) can be of great value if it is interpreted in this sense – as an assertion of the need for all R&D organizations to operate with a sense of social responsibility, and a way of improving the coupling mechanism between the scientific community and the public. However, the Report tended to assume a one-to-one correspondence between the interests and preferences of government departments and the interests and preferences of the public on whose behalf they are acting. This is of course very far from being the case. Departments may often share the technological fashions and preferences of designers without checking back to the actual wishes of consumers. (This was in fact the situation both in the case of aircraft development and high-rise flats).

One must certainly hope that the executive branch will be responsive to the known wishes and needs of the population, and will be active in trying to ascertain these preferences where they are not known. But the experience of politics over several thousand years has shown conclusively that this cannot be relied upon (Armytage, 1965; Lenin, 1917). Methods are therefore essential which can ensure that the executive branch is subjected to a continuous process of critical review and control. Local and spontaneous initiatives are important but so too is parliament, and the strengthening of parliamentary process in relation to science and technology is essential. It is this type of problem which has reinforced the moves towards some type of 'technology assessment' under parliamentary control. Even though the original Congressional American Office of Technology Assessment (OTA) was closed down in the 1990s as part of the Gingrich programme of budget cuts, the idea survives in Europe and is likely to resurface in some form or other elsewhere. It is further discussed in the final chapter of this book.

16.5 CHANGING PRIORITIES FOR SCIENCE AND TECHNOLOGY

The political process is involved not only in the public expenditure on basic research and military R&D, or in support of applied research and development in various industries, or in compensating for the deficiencies of this R&D, but also in establishing priorities between all these various objectives. There is considerable unwillingness to think in these terms. The philosophy expressed in the Rothschild Report (1971) is still strong. According to this Report, which was accepted and implemented by British governments, national priorities for R&D cannot ever be established. This can take place only at the departmental level, where each government department must behave as a customer contracting for the R&D which it requires and acting autonomously.

As the example of the aircraft industry has shown, both in Europe and the USA, actual priorities were established even if they were implicit rather than explicit. These real priorities can be recognized by examination of the actual distribution of R&D expenditures. One of the most characteristic features of the pattern of public expenditures in the 1950s and 1960s was the massive scale of nuclear, military and space programmes. This was

Table 16.2 The change of percentage shares of military, space and nuclear R&D expenditure as a proportion of total public R&D expenditure during the 1960s

Country	1960–61				1969–70			
	Defence	Space	Nuclear	Total	Defence	Space	Nuclear	Total
USA	68.7	9.1	10.7	88.5	48.7	23.2	6.5	78.4
Canada	23.2	—	21.2	44.4	11.2	1.4	19.5	32.1
Belgium	6.0	—	24.3	30.3	2.0	6.0	14.8	22.8
UK	64.5	0.5	14.7	79.7	40.4	3.7	11.5	55.6
Norway	8.6	0.4	16.5	25.5	7.1	1.2	8.3	16.6
Japan	5.6	—	7.6	13.2	2.2	0.7	7.4	10.3
Sweden	49.0	0.1	23.9	73.0	28.3	1.5	9.4	39.2
Netherlands	5.0	0.2	11.7	16.9	4.5	2.9	10.5	17.9
France	41.5	—	27.5	69.0	30.7	6.7	17.8	55.2

— = not available.

Source: OECD Statistics (1971).

particularly true of the USA, UK and France, within the OECD, and of the Soviet Union and China outside.

Only in smaller countries and Japan did these outlays fall below 10 per cent of publicly financed R&D (Table 16.2).

Since the USA and the medium-sized European countries accounted for a very high proportion of total R&D in the OECD area, the proportion of all public R&D expenditures going to national security and prestige types of R&D was well over 75 per cent in the early 1960s. This was equivalent to nearly half of R&D expenditure of all kinds (public and private) in the OECD area in the 1960s. During the 1970s the pattern changed considerably (Table 16.3) for a variety of reasons. Already during the 1960s the US government's expenditure on NASA was falling after the successful moon landings, and the policies of international détente also facilitated some reduction in the relative scale of military expenditures. In Britain, however, the very high level of military spending was still further increased in the 1970s.

A classification of R&D expenditures by public goals must always be attended by great difficulties, since the same programme may be supported by different agencies for a variety of motives and, moreover, these motives are changing over time. To attempt such classification is nevertheless highly desirable, since the debate on public priorities and the desirable direction of change must be informed by knowledge of the approximate scale of allocations.

There is an important sense in which the grouping together of military, nuclear and space programmes, at any rate in the 1950s and 1960s, may be justified. All of them depended almost exclusively on public funds during this period; all of them were heavily if not exclusively influenced by considerations of national security and prestige; and all of them involved special institutions to control and operate the programmes, usually of the 'big' variety. Moreover, they became identified in the public mind with a definite kind of science policy. The development of this public image of 'big science and technology' has had very important social consequences. The priority accorded to these projects was so great that not without some

Table 16.3 Total specific government R&D funding by socio-economic objective (percentage distribution)

	USA			Japan[a]		Germany			UK			France		
	1971	1975	1980	1975	1979	1971[b]	1975	1980	1971[b]	1975	1980	1971[b,c]	1975	1980
Defence	52.2	50.8	47.0	3.8	3.6	21.3	17.6	14.2	46.2	52.9	59.4	38.0	32.6	40.9
Space	19.2	14.5	14.4	11.8	9.3	9.4	6.8	6.0	1.9	2.5	2.3	7.0	6.1	5.0
Civil aeronautics	3.1	1.6	1.6		—	3.6	2.6	2.3	14.5	8.2	3.4	7.0	6.7	2.4
Defence and aerospace	74.9	66.9	63.0	15.6	12.9	34.3	27.0	22.5	62.6	63.6	65.1	52.0	45.4	48.3
Agriculture	1.9	2.2	2.2	22.2	18.4	3.1	3.0	2.6	2.9	4.8	4.5	4.0	4.2	4.3
Industrial growth nec	0.6	0.4	0.4	17.7	13.9	8.6	9.1	11.7	4.6	3.1	3.4	7.0	8.9	7.6
Agriculture and industry	2.5	2.6	2.6	39.9	32.3	11.7	12.1	14.3	7.5	7.9	7.9	11.0	13.1	11.9
Production of energy	3.6	7.1	11.8	12.8	17.8	16.4	16.8	20.1	7.5	7.1	7.3	8.0	9.4	8.5
Transport, telecommunications	1.6	1.8	1.1	3.2	2.2	0.9	2.3	2.9	0.9	0.7	0.7	{ 6.0	3.2	3.2
Urban and rural planning	0.4	0.5	0.4	1.0	1.9	0.8	1.8	2.1	1.2	1.7	1.1		1.6	1.5
Earth and atmosphere	1.5	2.0	2.0	1.4	1.9	2.3	2.8	3.9	0.3	0.8	0.9		3.3	3.3
Energy and infrastructure	7.1	11.4	15.3	18.4	23.8	20.4	23.7	29.1	9.9	10.3	10.1	15.0	17.5	16.5
Environment protection	0.9	0.9	1.1	2.6	2.5	0.5	1.6	2.8	0.2	0.6	0.9	{ 3.0	0.9	1.2
Health	8.7	11.9	11.9	5.1	4.3	4.4	5.2	5.7	1.9	2.3	1.8		4.4	5.0
Social development and services	2.6	2.0	2.2	2.0	1.5	6.7	7.7	5.4	0.7	1.2	1.2	1.0	1.2	1.4
Health and welfare	12.2	14.8	15.2	9.7	8.3	11.6	14.5	13.9	2.8	4.1	3.9	4.0	6.5	7.6
Advancement of knowledge nec[d]	3.3	4.3	3.9	13.6[e]	20.2[e]	22.0	22.7	20.2	17.2	14.1	13.0	19.0	17.0	15.2
Total specified R&D funding	100.0	100.0	100.0	100.0	100.0	100.0	100.0	100.0	100.0	100.0	100.0	100.0	100.0	100.0

[a] Government intramural only, except for Advancement of Knowledge and Industrial Development.
[b] Not strictly comparable with following years.
[c] Rough OECD estimate.
[d] Excludes public general university funds throughout and also excludes basic research supported by US mission-oriented agencies. An 'adjusted' US figure might be about 15 per cent in 1980.
[e] Total university receipts from government for specified projects including those for other objectives.

nec = not elsewhere classified.

Source: OECD (1981).

Table 16.4 Three phases of science and technology policy

1940s to 1950s	1960s to 1970s	1980s to 1990s
Manhattan Project	Economic growth	Generic technologies
V1, V2 rockets	Productivity	Materials technology
Military aircraft	Civil aircraft	Biotechnology
	Nuclear power	Market competition
		ICT
Science advisory councils	Science and technology Councils and ministries	Science, technology and industry ministries
Physicists	Physicists, chemists	'Hard sciences'
Chemists	Economists	Biology, Ecology
	Engineers	Social Sciences
		Economics
Weapon systems	Economic growth	Structural change
Basic science	Weapon systems	Environment
Government labs	Industrial R&D	Networks
	University expansion	Weapon systems
Radical innovations	Incremental innovations	Diffusion
Big expansion of science and technology expenditures	Continued but slower expansion of expenditures	Levelling off and sometimes reduction of expenditures
V. Bush (1946) *Science, the Endless Frontier*	OECD (1963a) *Science, Economic Growth and Government Policy* I. Svennilson*	OECD, (1979) *Science and Technology in the New Economic context* R. Nelson* and J.-J. Salomon*
J.D. Bernal (1939) *The Social Function of Science*	OECD (1971b) *The Brooks Report* H. Brooks*	OECD (1991) *Technology, Economy and Productivity Programme* F. Chesnais
		J.Ziman (1994) *Prometheus Bound: Science in a Steady State*
		OECD (1990) *Sundqvist* Report*

* Chairman or main authors of group or committee reports.

justification, sections of public opinion have tended to accept 'big science and technology' as the image of science and technology in general.

Since the Second World War, policies for science and technology in the OECD countries have gone through several different phases (Table 16.4). In the immediate post-war period attitudes were still heavily influenced by the experiences of the war itself and of course by the beginning of the Cold War. During this first period the emphasis was very heavily on the supply side of the science–technology system, and especially on building up strong R&D capability. In the 1960s and 1970s, this gave way increasingly to a more balanced approach which recognized the dangers of a one-sided reliance on the supply side in R&D policy and put greater emphasis on the general economic environment affecting technical change and on the innovation process as a whole.

Finally, in the most recent years, there have been increasing attempts to integrate both these approaches and to link up policies for science and technology with policies for industry and for the economy generally. These differences of emphasis should not be exaggerated; elements of the 'supply' and the 'environment' approaches were of course present in most countries all the way through. Table 16.4 shows in a very over-simplified and schematic way three phases of policy for science and technology in the main OECD countries since the 1940s.

The tremendous success of the Manhattan Project (the development of the atomic bomb), the radar programme and military aircraft during the Second World War had convinced governments that an enormous invest-ment in science and technology could produce an astonishing payoff in purely military terms. They appeared, at least superficially, to justify the view that the assembly of large R&D teams with generous financial support could solve many difficult and complex problems. The thought is often expressed that if human beings can use science and technology to get to the moon, it should not be beyond the wit of human beings to solve some of our more urgent terrestrial problems, such as urban transport. Although naïvely formulated, this thought nevertheless contains an important truth. What science and technology can achieve is partly a question of the social priorities and goals set for research (Nelson, 1977).

One of the consequences of the successful development of the A-bomb was to give immense prestige and weight to the nuclear and aircraft lobbies[1] in national decision-making for R&D. The fashion set in the 1940s dominated world R&D expenditures in the next decade. In the early 1950s at the height of the Cold War, the powers which then led the nuclear weapons race were devoting more than half their national R&D resources and more than 90 per cent of their public expenditure to these objectives. It should be remembered that the R&D programmes were effective in the narrow sense in which they were conceived. Although attended by fantastic cost overruns, they nevertheless produced successive generations of ever more sophisticated weaponry. Moreover, those who queried the extra-ordinarily high priority given to these programmes were often met with the 'spin-off' argument according to which they benefited technology and economic progress in general. The demonstration effect of the Manhattan Project not only served to justify a succession of 'big technology' projects in the countries which originated it, but also led to a wave of competitive and imitative efforts. The primacy of nuclear physics in civil research, and of nuclear energy programmes in national fuel research, cannot be entirely dissociated from the socio-political consequences of the awesome achieve-ments of nuclear weapons. The first de facto science policy institutions in many countries were government nuclear research organizations (Dedijer, 1964).

The launching of Sputnik in 1957 had perhaps even greater reper-cussions. Although it probably owed its successful development to the overall priority given to rocket-type delivery systems for nuclear weapons, its importance obviously transcended these objectives. It led immediately to a further massive competitive increase in US public expenditures on R&D, and to major changes in policy for science and technology in many other countries. It impelled the American president to set as the major national priority for US technology the prestige objective of putting a man on the moon by 1970. This objective was triumphantly achieved, and few would question the magnitude of the technical achievement.

An overemphasis on 'big science and technology' was not the only weakness of the science policies of the 1950s. There was a general over-estimation of the importance of R&D and a relative neglect of other scientific and technical services, and of those other activities which are essential for successful innovation and efficient technical change through-out the economy. With a few exceptions, such as Nelson, Carter and Williams, economists had largely neglected these topics and knew little or nothing about R&D or innovation. Among leading economists only Schumpeter had given invention and innovation pride of place in his models of the behaviour of the economic system and even he took little interest in the policy implications of his analysis, whether for government or industry. This did not mean, however, that economists were unmindful of the claims of science. On the contrary, as we have seen they were generally quite ready to treat science and technology with generosity. Formally speaking, economic theory of almost all kinds recognized that technical progress was the mainspring of economic growth, even though it showed little disposition to delve into its mysteries.

The benign neglect of the economists, the ignorance of the politicians and many managers, and the self-interest of the R&D and military establish-ments combined to create favourable conditions for the extraordinarily rapid growth of expenditures on R&D, both by government and industry. The 1940s and 1950s were the golden age of R&D expansion when very few questions were asked about the efficient use of funds and the general assumption was that increased expenditure could only do good. This was not quite so naïve as it may sound today as the evidence of a very high rate of return on R&D investment was quite strong both in the military and the civil spheres (Schott, 1975; Mansfield et al. 1971, 1977). A fairly rapid expansion of many R&D budgets had a good economic justification.

Our evaluation of the first period of headlong expansion after the war should therefore by no means be purely negative. But it gave way to a period of slower growth or even contraction and to a method of managing R&D and other technical activities which was far more cost conscious and far more concerned with cost-effectiveness. There were many reasons for this transition and generally industry was ahead of government in making it, for obvious reasons. The pressures of the marketplace, of competition and of profitability compelled many firms to review their R&D projects and programmes much more critically. But for governments too there were strong pressures in the same direction. The public embarrassment of huge cost and time overruns on big weapons projects began to be noticed by legislatures and by public opinion already in the 1950s. Moreover, the very fact of the huge expansion carried with it the demand for greater public accountability. As long as R&D budgets were minute they could escape without much scrutiny, but once they were measured in millions and billions, a higher degree of management attention was almost inevitable.

Consequently, during the 1960s and 1970s two new tendencies became apparent in most OECD countries: first a slowing down in the rate of growth of R&D budgets and second far greater interest in the results of R&D. At the same time, academic researchers as well as industrial managers began to take a serious interest in the whole process of industrial innovation and technical change, as well as in the narrower problems of R&D project evaluation and programming.

Paradoxically the first effects of this recognition in the 1960s tended to be a reinforcement of the 'supply-side' policies of the 1950s, a rather indiscriminate build-up of R&D facilities and of the education system. In this early period the technological and productivity gaps between Europe and the United States were the main focus of concern. The gap with Japan in consumer electronics and automobiles only opened up in the 1970s and 1980s, although it was foreshadowed in a few industries, such as shipbuilding, much earlier. The US lead in the 1950s was often wrongly attributed primarily to the scale of military, space and nuclear R&D and to the government policies affecting these technologies (e.g. Servan-Schreiber, 1965).

However, during the 1960s with the increasing sophistication of economic and political analysis, the 'spin-off' arguments of the military and big science lobbies became increasingly discredited. The US–European productivity gap was steadily closing and studies like that of Eads and Nelson (1971) demonstrated, both in terms of economic theory and of practical examples, the wastefulness of much 'big technology', government-financed development. A series of studies, notably by Carter and Williams (1957, 1959a, 1959b, 1964) in the UK and by Hollander (1965) in the USA, pointed to the great importance of other scientific and technical activities, as well as R&D in the industrial innovation process. Thus, although projects such as Concorde and fast breeder reactors still received lavish financial and political support throughout the 1960s and indeed throughout the 1970s, they were subjected to more and more searching examination and criticism and even the expenditure of NASA after the triumph of the moon landings was drastically reduced. Scarcely a single economist could be found in Britain or the United States to lend support to government subsidies for the SST. In part, of course, these more stringent financial approaches simply reflected a general tendency to curtail the growth of government expenditures, which was gathering force in the 1960s and even more in the 1970s and 1980s.

The concentration on the economic growth objective did not go completely unchallenged even though it predominated in most countries in the 1960s (OECD, 1971b). It was questioned on a variety of grounds but especially on two: whether it could be sustained in the long term because of the exhaustion of materials and energy resources; and whether the pollution associated with it might not endanger the very existence of the human race. Neither of these questions was completely new but they attracted great attention in the early 1970s and this had a considerable influence on policies for science and technology throughout the OECD area. They were brought into focus by the debate on *The Limits to Growth* (Meadows *et al.*, 1972) and the counter-arguments advanced in such critiques as *Thinking about the Future* (Cole *et al.*, 1973). One of the outcomes of the debate was an acceptance by most governments and by public opinion that environmental policies should have much greater weight in decision-making, and that considerable scientific and technical efforts should be devoted to the attainment of higher environmental standards and the prevention of some of the more serious pollution hazards (see Chapter 18). Although there has been some backsliding more recently, these remain important influences on the allocation of resources for science and technology and on policy generally, and as we have seen (Table 16.3), did result in a significant reallocation of resources in the 1970s.

Another outcome of the debate was a recognition that if long-term economic growth were to be sustained over a long period, then it could only be through a high rate of technical change in the use of materials, energy and the stock of capital. Critics of the MIT work demonstrated that new discoveries, the use of lower grade ores, economy in use of materials, substitution processes, recycling and social changes could combine to avert the catastrophic scenarios of collapse of the world economy in the next fifty years. However, all of this would be possible only if technology policies were successful and integrated with economic policies, and responded fairly rapidly to new developments (see Chapter 18).

This particular lesson was brought home forcefully by the OPEC crisis of 1973 which confronted the OECD countries with the need to develop urgently alternative sources of energy and at the same time to adopt much more efficient policies for conservation. Both of these requirements also had major implications for priorities in R&D and the introduction of completely new programmes to cope with the situation, such as work on solar energy, geothermal energy, biomass and so forth. Thus the second phase of post-war policies for science and technology in the OECD area ended with a growing recognition that the old objectives of sustained economic growth would be difficult to maintain in the 1980s without some reorientation of science and technology and without a far closer integration of STP with economic, industrial and energy policies. The problem of technical change and employment came increasingly to the forefront and this is dealt with in the next chapter of this book.

The second stage of STP in the OECD countries thus featured a less naïve approach to the problems of R&D and innovation. It was marked by the establishment of separate ministries for science policy or technology policy or both in a number of countries or by considerably strengthened groups at cabinet level in others. At the same time statistics of R&D became generally available and were put on an internationally comparable basis, largely through the efforts of the OECD. This organization also initiated a series of national scientific policy 'reviews' in each of the member countries, which heightened awareness and understanding of the problems.

The achievements of the second stage of STP should not be overrated, any more than those of the first stage. Although policy-makers for science and technology became aware of the high opportunity cost of military R&D and some other types of big science and technology, they were unable or unwilling to achieve very much in the way of reorientation of these expenditures. During the 1960s, military expenditures were falling in most countries as a proportion of total R&D expenditures, but in the 1970s and 1980s they began to creep up again.

Even where resources were transferred or new resources were invested in programmes which were likely to have a more direct and beneficial effect on economic efficiency and economic growth, there were great difficulties in organizing and administering such programmes and considerable doubts about their value. Although new governmental structures were set up in many countries to deal with STP problems, quite often they did not endure for very long and there was a bewildering succession of new committees, councils, agencies, ministries etc.

The reasons for these failings go deep to the heart of the economic, social and cultural structures and traditions of most OECD countries. They concern the relationships between government and industry in a market economy, between the various functions of government in departmental systems of administration, and between the various disciplines in an educational and career system organized on disciplinary lines. They also involve the attitude of the labour force and management to technical change and the responsibility for introducing and managing it.

Finally, what President Eisenhower had described as the 'military–industrial' complex continued to enjoy a special status throughout the second stage of science and technology policy. Indeed this continued even after the end of the Cold War into what we have called the 'third stage'. In the third stage of policy for science and technology some new solutions have been attempted for some of these problems but it could not be said that any of them have been wholly effective.

The theoretical understanding of the problems of government policies and the methods of achieving their objectives has become far more sophisticated during the last quarter century (see, for example, Stoneman, 1987, 1995; Teubal, 1987; Teubal and Steinmueller, 1982; Metcalfe, 1995; Pavitt and Walker, 1976). However, at the same time the problems have become more difficult as the period of rapid growth and full employment gave way to one of deep structural change and high levels of unemployment in the 1980s and 1990s. The collapse of the East European centrally planned economies and the end of the Cold War created the possibility for a drastic reduction in the military expenditures on R&D, which had been such a dominant feature of the international pattern ever since the 1940s. At the same time public concern with a variety of environmental issues, as well as international agreements, obliged governments to devote greater resources to research on some environmental and public health hazards.

These changes have found some reflection in the changing patterns of public R&D expenditures from the second to the third phase. However, they are much less than might have been expected from the dramatic political and economic changes in the 1980s and 1990s. Military and economic objectives continue as the dominant ones in all the major countries (Table 16.5). In fact the share of military R&D actually increased substantially in the United States in the 1980s, from 54 per cent in 1980 to nearly 60 per cent of a much larger budget in 1991. In the UK the share also remained extremely high at around 45 per cent. In both the USA and UK the military share was actually higher in 1991 than in the late 1960s and even by 1994 it was still higher than in that period. The classification of government expenditures is not easy and those in Tables 16.2, 16.3 and

Table 16.5 Public R&D expenditures by principle objectives in various OECD countries

Billions PPPS	USA			Japan			Germany			UK			France		
	1981	1991	1994	1981	1991	1994	1981	1991	1994	1981	1991	1994	1981	1991	1994
Total OB	33.7	65.9	68.3	7.1	10.5	13.1	7.5	14.0	15.3	6.5	7.8	8.5	7.3	14.2	13.5
Defence	18.4	39.3	37.8	0.4	0.6	0.8	0.7	1.5	1.3	3.0	3.4	3.8	2.8	5.1	4.5
Civil	15.3	26.6	30.5	6.7	9.9	12.3	6.8	12.5	14.0	3.5	4.4	4.7	4.5	9.1	9.0
Economic objectives[1]	5.5	5.9	6.9	2.3	3.3	3.8	2.4	3.1	3.2	1.3	1.3	1.4	1.7	3.0	2.0
Health, environment[2]	5.4	11.6	13.5	0.4	0.6	0.8	1.0	1.6	1.9	0.5	1.0	1.2	0.7	0.9	1.0
Non-oriented	1.3	2.6	2.7	0.5	0.9	1.2	0.8	2.1	2.1	0.4	0.4	0.4	1.1	2.2	2.4
Civil space[3]	3.1	6.5	7.4	0.4	0.7	1.0	0.3	0.8	0.9	0.1	0.2	0.3	0.3	1.2	1.4
University funds[4]	0.0	0.0	0.0	3.1	4.4	5.5	2.5	4.7	5.9	1.0	1.5	1.5	0.7	1.8	1.9

[1] Agriculture, forestry, industrial development and energy
[2] Environmental protection, health, social development services, earth and atmosphere
[3] Civil space programmes
[4] Advancement of knowledge, excluding USA

Source: MSTI Database, July 1996.

16.4 are not completely consistent. Nevertheless, the broad trends are relatively clear.

Both in the USA and in the UK, there has been a significant rise in the share of public expenditures devoted to environmental, health and related objectives during the 1980s and early 1990s but this is not so evident elsewhere. The share in German expenditure has been steady at about 13 per cent of the total, whereas the British share only reached this point in 1994.

The rise in the share of military expenditures in the 1980s, and to a lesser extent in environmental expenditures, cannot be explained in terms of economic analysis. Clearly political parties, interest groups and lobbies determined this reallocation. As already argued in Part Two and elsewhere in this book, even within the firm, where economic considerations predominate, interest groups and a process of political advocacy play an important role in decision-making for R&D. This inescapable involvement of political as well as economic features leads to some further fundamental conclusions in the final chapter of this book.

NOTE

1. The expression 'lobby' is used here, and throughout the book, not in a pejorative sense, but in the normal sense of political science, to describe an interest group with a distinctive set of attitudes, which attempts to inform and influence government policy.

CHAPTER 17

THE INFORMATION SOCIETY AND EMPLOYMENT

17.1 NEW TECHNOLOGIES: JOB CREATION AND DESTRUCTION

Ever since Ricardo's famous remarks in 1821 and the ensuing debate, economists have recognized the two-edged nature of technical change: that it both destroys old jobs and creates new ones. In general, economists have argued that the job creation effects have in the long run outstripped the job destruction effects, albeit accompanied by a steady reduction in working hours throughout the nineteenth and twentieth centuries. Nobody has claimed, however, that 'compensation' is automatic, painless or instantaneous. As Ricardo pointed out, the new jobs may not match the old ones either with respect to skill or to location. Where the mismatch is severe and/ or prolonged, economists speak of 'structural unemployment' and the problems of 'structural adjustment' although the precise borderline between 'structural' and the more usual everyday 'frictional' unemployment is not always easy to define precisely. Nevertheless, the existence of some fairly severe problems of structural unemployment from the 1970s to the 1990s is now universally recognized, as has become obvious from the rapid increase and high rates of 'long-term' unemployment or non-employment in most developed countries in the 1990s.[1]

Schumpeter (1939) gave a new twist to the whole debate with his conception of 'successive industrial revolutions' when new technologies were diffusing through the productive system. Whether or not they accept Schumpeter's long wave ideas, few economists or engineers today would deny the enormous worldwide impact of information and communication technology (ICT). In fact, many commentators go even further and suggest that ICT is ushering in an entirely new era or 'post-industrial' society. Everyone would today accept that the extraordinary reduction in costs associated with microelectronics in successive generations of integrated circuits, of telecommunications and of electronic computers (Table 17.1) is having great effects on almost every branch of the economy, whether in primary, secondary or tertiary sectors. Earlier new technology systems such as steam power or electricity had similar pervasive effects, but ICT is unique in affecting every function within the firm as well as every industry and service. Scientific and market research, design and development, machinery, instruments and process plant, production systems and delivery systems, marketing, distribution and general administration are all deeply affected by this revolutionary technology. Moreover the counter-inflationary effects of falling costs and prices in microelectronics, computers and telecommunications affect a widening range of products and services.

Table 17.1 *Estimates of increase in ICT capacity*

Area of change	(1) Late 1940s– early 1970s	(2) Early 1970s– mid-1990s	(3) Mid-1990s onwards 'optimistic' scenario
OECD installed computer base (number of machines)	30,000 (1965)	Millions (1985)	Hundred millions (2005)
OECD full-time software personnel	>200,000 (1965)	>2,000,000 (1985)	>10,000,000 (2005)
Components per microelectronic circuit	32 (1965)	1 mega-bit (1987)	256 mega-bit (late 1990s)
Leading representative computer: instructions per second	10^3 (1955)	10^7 (1989)	10^9 (2000)
Cost: computer thousand ops. per \$US	10^5 (1960s)	10^8 (1980s)	10^{10} (2005)

Source: Freeman and Soete (1994).

In attempting to assess the employment creation and destruction effects it is important to distinguish conceptually the direct from the indirect effects. The direct effects are the new jobs in producing and delivering new products and services. The indirect effects are the consequences elsewhere. Computer terminals are everywhere but it is not always clear whether they are displacing workers or adding additional new services and employment. The computer industry itself provided machines which displaced earlier types of electromechanical office equipment, while the microelectronic industry largely displaced the old valve (tube) industry. The new digital telephone exchanges require far less labour to manufacture and to maintain than the old electromechanical exchanges and the number of people working in the telephone switch industry has fallen in most industrial countries. Competitive restructuring of the old monopolistic networks has also resulted in a reduction of the number of employees, even though the number of firms and the number of lines and calls have increased enormously. However, the new telecommunications infrastructure provides the basis for numerous new information service industries and equipment, such as e-mail, fax, data banks and the new multimedia services of the future (Mansell and Silverstone, 1996; Dutton, 1996).

To compare the balance of gains and losses is thus a difficult undertaking as numerous empirical studies of the 1970s and 1980s confirm. The naïve view of ICT as simply a process of automation and job destruction has its counterpart in the equally naïve view of ICT as a purely positive source of new employment. Any sophisticated attempt to assess the employment effects must take into account both job destruction and job creation. Chapter 7 already emphasized these pervasive effects throughout the economic and social system, both in job creation and job destruction.

This approach (change of 'techno-economic paradigm') stresses both the direct and indirect effects of ICT. It points to the rise of entirely new industries in the second half of this century, such as the software industry, the electronic computer industry, the microelectronic industry, the VCR and TV industry. Each of these now employs millions of people but barely existed before 1950. Even more, however, it stresses the indirect effects of the ICT revolution, following Schumpeter's analysis of the bandwagon effects generated by the opening of new markets and numerous new opportunities for profitable investment. Keynes (1930) accepted unreservedly this Schumpeterian concept of new technology as the most potent stimulus to waves of new investment, subject only to the compliance of the monetary authorities.

Whether the net balance is positive or negative in a given national economy cannot be assessed by simply counting new jobs gained and old jobs destroyed. It has to be recognized that the expansionary effects on any national economy or the world economy as a whole paradoxically depend on rapid increases in labour productivity. What a revolutionary new technology can do is to create the basis for a virtuous circle of growth in which investment is high, labour productivity grows fast but output grows even faster, so that there is a net growth of employment. Whether this virtuous circle can be sustained depends on macro-economic policies, employment policy and trade as well as on the new technologies. If there is a good match between technologies, policies and institutions prolonged periods of full employment can result. This was the happy situation in Europe, Japan and North America in the 1950s and 1960s, based on cheap oil, very rapid expansion of the automobile and other consumer durable industries, of steel and plastics and of many related services. Productivity growth was high but output grew even faster, so there was also considerable growth of employment and very low unemployment (Figure 17.1) (the 'Verdoorn' effect). An orthodox 'optimistic' view is that the world economy will return to this state of affairs after the phase of structural adjustment. This view is illustrated in Box 17.1, taken from an OECD (1994) report on unemployment.

The virtuous circle of the 1950s and 1960s in most OECD countries has again been achieved with the aid of new technologies, but this time in Asia. The 'four tigers' of Eastern Asia and more recently some other economies of South East Asia have achieved remarkably high output growth, productivity growth and employment growth (Figure 17.1), described by the World Bank (1993) as the 'East Asian miracle' and already discussed in Chapters 12 and 13. The high labour productivity and falling costs of the ICT producers themselves facilitate diffusion of ICT throughout the area. Labour productivity increases in the ICT manufacturing sector have been extremely high because of the technical changes summarized in Table 17.1. For the other industrial countries the general evidence for the positive employment effect of advanced technologies and employment generation were analysed by the OECD (1991a, b, 1992) in their TEP report and other studies. In the case of manufacturing they are illustrated in Figure 17.2.

However, there are some reasons for taking a more cautious view of the trends in the world economy than those in the OECD study. The simultaneity between the rapid rise in unemployment in the developed countries and the rapid catching up growth of a number of East Asian countries, both in production and exports, combined with rapid diffusion of a pervasive technology throughout most of the developed countries, has led to a revival of academic and policy interest in the nature of the technology–employment

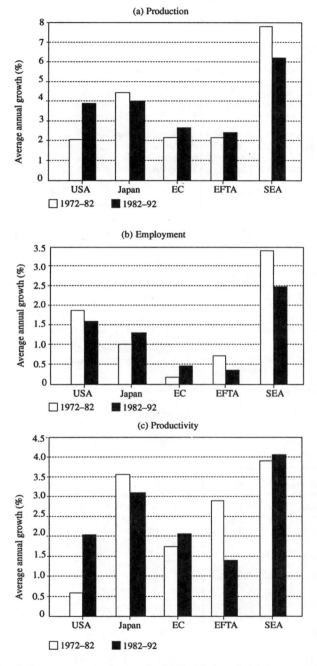

Fig. 17.1 (a–c) Average annual growth, 1972–92 (productivity growth = output growth per manhour)

Sources: ILO (1992); OECD, CRONOS Database.

Box 17.1   Technological change and employment: the lessons of history

Forecasts that the next wave of technological change will cause high unemployment and/or large declines in real wages have been frequently made in the past two hundred years. Today, these fears take a variety of forms; beliefs that computers and robots will soon take the place of unskilled labour or that the rate of structural change in the economy is too high and will lead to overwhelming dislocations through technological unemployment. So far, all such forecasts have been wrong. Increased productivity has been accompanied by rising demand for labour and rising real wages.

In the 1760s the French physiocrats were arguing that only agriculture was truly productive and that the shift of the French labour force into other sectors would reduce national wealth. As Adam Smith paraphrased their argument, the workings of the market cause the number of service and manufacturing workers to grow over time and 'encroach upon the share [of employment] which ought properly to belong to this productive [agricultural] class. Every such encroachment must . . . necessarily occasion a gradual declension in the real wealth and revenue of society.'

Over the following half century the agriculture share of the French labour force continued to fall. But the productivity increases led to average wages in France growing by nearly one-fourth, while unemployment showed no appreciable rise.

In the 1820s the 'Ricardian socialists' were arguing that increased productivity from the introduction of machinery would reduce employment, and put downward pressure on wages and on populations. They were relying on an extended numerical example from Chapter 30 of David Ricardo's revised *Principles of Political Economy*, after which Ricardo concludes 'that the opinion entertained by the labouring class, that the employment of machinery is frequently detrimental to their interests, is not founded on prejudice and error but is conformable to the correct principles of political economy'.

Over the following half century, capital accumulation and technical change caused average wages in the United Kingdom to more than double, while unemployment showed no appreciable rise.

In the 1860s Karl Marx was writing in *Capital*, 'the greater the social wealth, the functioning capital . . . the productivity of labour . . . the greater is the industrial reserve army [of the unemployed] . . . the more extensive [are] the pauperized sections of the working class'.

Over the following half century, as capital intensity and labour productivity continued to increase, average wages in the United Kingdom once again roughly doubled, and unemployment did not appreciably rise.

In the 1940s, after World War II, cyberneticist Norbert Wiener was forecasting that the invention of the computer would, as Paul Samuelson reports it, create 'technological unemployment that will make the Great Depression look like a picnic'. Over the following forty years, average hourly wages in the United States more than doubled, while unemployment increased by an average of 1 or 2 percentage points.

Source: OECD (1994).

---

relationship in the developed countries.[2] However, the latter relationship is no longer discussed within the traditional 'closed economy' framework of classical, neoclassical or Schumpeterian economic thinking, but within the framework of the rapid internationalization of production, liberalization of international trade and investment, and globalization of information and communication, as portrayed in Part Three and as discussed, for example, by Adrian Wood (1994).

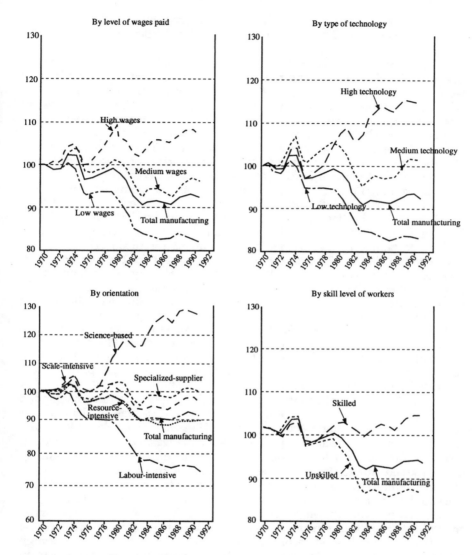

Fig. 17.2 Manufacturing employment by type of industry, averages for 13 OECD countries* (1970 = 100)

* Australia, Canada, Denmark, Finland, France, Germany, Italy, Japan, Netherlands, Norway, Sweden, UK, USA

Source: OECD (1994).

As before, this debate is dominated by policy concerns that the employment 'compensation' following from the gains in efficiency linked to new technologies and the (international) relocation and more efficient use of production factors, will not be an immediate and instantaneous process. In contrast to previous debates these concerns are now broadened to include more explicitly the direct and indirect effects of international trade and relocation. Two features have received particular attention in the recent literature.

At the employment level, the economic debate has become focused on the possible 'skill-biased' nature of recent technical change, particularly in relation to ICT (Berman *et al.*, 1994). The latter is now recognized by an increasing number of economists to represent a pervasive, so-called 'general purpose' technology which could imply significant increases in the demand for skilled labour and major time-consuming processes of structural adjustment as individuals, firms, industries, governments and all other institutions learn, largely by trial and error, how to use the new information and communication technologies more efficiently.[3]

At the aggregate level, this accentuation of the skill implications of ICT has led to a further shift from skill 'mismatch' adjustment issues and their possible growth bottleneck impact to growing concern, both in Europe, the USA and Japan about the skill distributional aspects of technical change, and in particular to the decline in the demand for unskilled labour. As suggested in Figure 17.2, the rate of unemployment for the less educated, low skilled groups of workers has been on average twice as high as the rate for the higher educated. Taking into account also those 'out of employment', the non-employment rate for the low educated labour force rose in some countries to more than 30 per cent in the 1990s.

The policy challenge behind this shift in focus is formidable: on the one hand, the adaptation towards an information society is likely to lead to substantial changes in the demand for various sorts of educational and skill requirements; on the other hand, the likelihood of large parts of the unskilled labour force becoming excluded is high.

Each of the countries reported in Figure 17.2 appears to have responded in the 1980s and 1990s in a different way to the decline in demand for unskilled labour in manufacturing. In the USA, labour market adjustment led to a substantial decline in real wages for the least educated and skilled workers. In Europe it led to much higher levels of unemployment among the unskilled labour force. In Canada much of the adjustment occurred through alterations in labour time. Whether this decline in demand for unskilled labour can be associated with technical change and ICT in particular has become one major issue of debate. This has raised the fundamental question: Can structural adjustment take place through an increase in skills and accelerated technical change or must it take place through wage reductions and decline in living standards for the low paid? In practice both tendencies are operating but clearly technology and training policies bear a major responsibility in association with macro-economic policies and regional policies for minimizing hardships.

17.2   INFORMATION SERVICE ACTIVITIES AND FUTURE EMPLOYMENT

The debate on the 'Information Society' highlights in particular the extraordinary importance of service industries. In one way services can be defined[4] as those activities (sectors) where output is essentially consumed when produced. While this might well be considered a rather narrow definition and one which covers only a limited number of sectors presently falling under the statistical definition of service sectors, it is an analytically useful definition because it highlights the intrinsic immaterial, intangible nature of many service activities whether they are personal services, such as hair-cutting; entertainment such as an opera performance; education such as teaching; health such as a doctor's visit; or public services such as welfare. It is the near-simultaneity of production and consumption which has generally limited productivity improvements in such activities. In parenthesis it is also worth noting that this provides an intellectual argument as to why the economic profession has generally tended to ignore the study and analysis of service activities.

Information and communication technologies, almost by definition, allow for the increased tradeability of service activities, particularly those which have been most constrained by the geographical or time proximity of production and consumption. By bringing in a space or time/storage dimension, information technology will make possible the separation of production from consumption in a large number of such activities, hence increasing the possible trade in them.[5]

In the case of the current 'new' information and communication technologies and their potential not just to collect, store, process and diffuse enormous quantities of information at minimal costs, but also to network, interact and communicate across the world – the world as a global village – both the time/storage and space dimensions of the new technology are likely to bring about the further opening up of many service activities, increasing their domestic and international tradeability. As in the case of the telephone, it is likely that the 'new' emerging computing and other electronics manufacturing sector will in the end remain relatively small compared to the growth and size of new information and communication service sectors. However, even the definition of the latter will become blurred as more and more of the traditional 'physically present' service activities will become 'info-type' service activities.

With regard to more traditional production processes, typical of industrial production, but also common in traditional service sectors concerned with the movement of goods such as transport, wholesale and retail trade, the impact of ICTs could well be characterized as exactly of the opposite kind. Rather than bringing time/storage between production and consumption as in services, ICTs will in the first instance aim at reducing the time/storage dimension between production and consumption. Many of the most distinctive characteristics of the new information and communication technologies are related directly to the potential of the new technology to link up networks of component and material suppliers, thus allowing for reductions in storage and production time costs – typified in the so-called just-in-time production system. At the same time, the increased flexibility associated with the new technology allows for a closer integration of production with demand, thus reducing the firm's own storage and inventory costs, which could be typified as Just-in-Time selling. Both features clearly work in the opposite direction from what was said above, i.e. they aim at reducing the time/shortage dimension between production and consumption. In doing so, they might well reduce the

Table 17.2 Industries' percentages of business employment of scientists and engineers, 1992

Field	Employment of scientists and engineers % (computer specialists %)
Manufacturing	48.1 (10.9)
Non-manufacturing	51.9 (23.7)
Engineering services	9.1 (3.2)
Computer services	8.3 (51.8)
Financial services	6.1 (58.5)
Trade	5.2 (25.5)

Source: NSF (1993) in Pavitt (1996).

'tradeability' of a number of those intermediary storage and inventory activities. In essence therefore, ICTs are making services more tradeable and more like manufacturing, leading to a further convergence of industrial and service activities.

Pavitt (1996) in particular has pointed to another of the fundamental changes in service industries with respect to the employment of scientists and engineers. For a long time service industries employed rather few of these highly qualified specialists but figures for 1992 showed that (for the first time in the history of the United States) 'non-manufacturing' industries employed more scientists and engineers than manufacturing (Table 17.2).

Computer services and financial services showed a particularly rapid growth and in both cases this was clearly due to the impact of ICT. In fact whereas computer specialists accounted for only just over 10 per cent of the total number of engineers and scientists in manufacturing in 1992, they accounted for nearly a quarter of that total in non-manufacturing industries. In an earlier publication (Freeman and Soete, 1987) we pointed to the fact that software R&D often went unrecorded in R&D statistics because of conceptual and measurement problems, and the same is of course even more true of patent statistics. Even in 1983, if total software expenditures were combined with electronics R&D, they accounted for over 50 per cent of the combined 'research, development and software' total in the UK. Much of the technical change associated with computerization takes the form of new software design and applications by users as well as producers so that the combined total (by now much larger) is a good indicator of the scope of this change related to ICT diffusion.

With the growing convergence between manufacturing and services, service activities, corresponding on average to over two-thirds of total economic activity in the EU countries, appear increasingly as important in their own right and in an increasing number of areas, as dominating manufacturing rather than the other way round (see, for example, the case of General Electric and other 'manufacturing firms' described in Parts One and Two). Particularly since the emergence of ICTs and their 'tradeability' impact on many service activities, the latter activities have emerged as 'core' value added activities. It is essential in this context to distinguish between data, information and knowledge.

ICTs play an essential role in the 'codification' of knowledge. The latter implies that knowledge is transformed into 'information' which can either be embodied in new material goods (machines, new consumer goods) or easily transmitted through information infrastructures. It is a process of reduction and conversion which renders the embodiment or transmission, verification, storage and reproduction of knowledge especially easy.[6] In contrast with codified knowledge, tacit knowledge refers to that which cannot be easily transferred because it has not been stated or measured in an explicit form. One important kind of tacit knowledge is skills. The skilled person follows rules which are not fully or formally known as such by the person following them. They are linked to activities acquired through learning but often of a non-routine kind.[7] An important impact of new ICTs is that they move the border between tacit and codified knowledge. They make it technically possible and economically attractive to codify kinds of knowledge which so far have remained in a tacit form. At the same time they generate entirely new forms of tacit knowledge, for example, in relation to software design.

The embodiment of codified knowledge in material goods has been typical of the dramatically improved performance of many new capital and consumer goods, incorporating many new electronic information and communication devices. The latter in turn have been at the core of the continuous productivity, investment and consumer demand growth in Western societies. As emphasized by authors criticizing the early 'post-industrial' literature,[8] this process could also be described as one of 'industrialization' of services: the continuous replacement of particular service activities by household material goods, embodying at least the 'codified' knowledge part (washing machines, television, dryers, etc.). The more recent electronic improvement in these products has further increased the 'household' performance of these products, freeing further household time. While the quality of these new material goods will not always completely substitute for the service activity they replace (a dishwasher is a good example), the codification process will be to some extent complete. The product might lack user friendliness (the typical example being the video player), but the user is not required to possess, or to understand, the knowledge embodied in the machine.

In services, by contrast and following the arguments set out above, while the codification of knowledge will have made it more accessible than before to all sectors and agents in the economy linked to information networks or with the knowledge of how to access such networks, its immaterial nature will imply that the codification will never be complete. The codification process will even rarely reduce the relative importance of tacit knowledge in the form of skills, competencies and other elements of tacit knowledge, rather the contrary. It is these latter activities which will become the main value of the service activity: the 'content'. While part of the latter might be based on pure tacitness, such as talent or creativity, the largest part will be greatly dependent on continuous new knowledge accumulation – learning – which will typically be based on the spiral movement whereby tacit knowledge is transformed into codified knowledge, followed by a movement back where new kinds of tacit knowledge are developed in close interaction with the new piece of codified knowledge. Such a spiral movement is at the very core of individual as well as organizational learning.

The implications for this continuous shift in value from manufactured goods embodying increasing amounts of 'codifiable' knowledge towards service-based 'tacit' knowledge activities is typical of the new emerging Information Society. It partly explains the attempts of electronic and computing manufacturing firms to enter information content activities. It also partly explains the change in structure and strategy of some firms which were once dominated by manufacturing activities (e.g. General Electric, Ericsson, etc., Chapter 11). Within services, it explains the move of 'carrier' operating firms being most directly confronted with the codification of knowledge and its distribution, to enter content sectors (media, education, culture).

The emerging Information Society will in other words become dependent on user demand for new information products and services. The 'demand articulation' of the latter depends crucially on the existing and new regulatory institutional framework as well as the overall macro-economic climate. In many cases of information or communication services, their commercial success depends on reaping the substantial economies of scale involved in typical network activities, with marginal costs being a fraction of fixed costs (one may think of movies, software, financial and insurance services, etc.). Europe with its fragmented national markets and cultural diversity is clearly at a disadvantage in reaping such economies of scale. The various proposals for liberalizing and harmonizing some of these markets more quickly, and in particular the national telecommunications operators' markets as, for example, in the Bangemann Action Plan or the G-7 Industrialists' Action Plan, corresponded to the urgency felt in addressing at least a part of this problem.

17.3 THE NEW EMPLOYMENT CHALLENGES

The employment concerns associated with the emerging Information Society relate in the first instance to the likely impact of ICT on productivity growth and output growth, particularly with respect to some of the new information service sectors described above. While much will depend on the new needs and markets that the new information service sectors will be capable of addressing; much will also depend on the way the productivity gains will be redistributed throughout the economy. These latter gains are as much based on the overall counter-inflationary effects of falling costs and prices in microelectronics, computers and telecommunications which are affecting a widening range of products and services, as on organizational improvements and other more dynamic learning efficiency improvements at the level of the shop floor, production planning or administrative activities.

As argued in Section 1 of this chapter, in attempting to assess the employment creation and destruction effects of ICT it is difficult to distinguish many of the direct negative and positive effects from the indirect effects. What brings employment concerns about the Information Society today back to the public policy forefront, despite the reassuring historical arguments and macro-economic compensation arguments (see Box 17.1), are the particular features and characteristics of the new information and communication technologies. Four features are especially noteworthy.

First, there is the particular impact of new ICTs on employment in the service sectors, particularly in those sectors and occupations hitherto largely 'protected' from automation or 'informationization'. Insofar as such 'sheltered' service employment has acted in the past as a main absorber of employment displacement in manufacturing and agriculture, there is increasing concern about whether new services will indeed be capable of providing sufficient new employment opportunities. Such new services are crucially dependent, as highlighted above, on the appropriate regulatory framework. The emergence of new markets for information services requires not just a more competitive framework, deregulation and open access, it also requires new institutions to set out the rules of such new markets, including those governing property rights, security, privacy, etc. At the same time, and as the case of Internet illustrates, the speed of change goes beyond the 'controlled' liberalization process pursued at the moment and involves a much more dramatic 'creative destruction' process, with a completely new communication pricing structure. Particularly in Europe, there is justified concern that the regulatory reform is too slow and that the development of new services in Europe is consequently lagging behind development in other parts of the world.

Second, insofar as the dramatic decrease in the cost for obtaining data and information with new ICTs can be compared with a macro-economic deflationary effect, the question can be raised whether our present statistical methods for assessing 'inflation' are still appropriate, and not increasingly measuring inflation 'illusion' rather than money illusion.[9] This point was endorsed by the *Economist* magazine in an editorial and feature article in its issue of 22 November 1996. It seems reasonable to assume that current European and United States estimates of inflation, ignoring many new products and services and many of the quality improvements typically associated with new ICTs, are overestimating inflation by a small percentage every year. As a result, and as one would expect on the basis of the widespread diffusion of ICTs and the emerging Information Society, many economies were no longer in a situation of inflation, but rather deflated in the mid-1990s. The strict monetary policies implemented in most EU member countries within the framework of the EMU convergence criteria, based on overestimated inflation estimates, were consequently unlikely to stimulate in the short term the positive output and employment growth aspects of the emerging IS. This is not to deny the possible longer term positive impact of budgetary fiscal consolidation policies on interest rates, private investment and employment growth – so-called crowding in – as argued, for example, in the Commission's European Employment Strategy. However, in the short term one may indeed wonder whether the priority given in the EU during the 1990s to strict macro-economic monetary policy was likely to be conducive to a rapid emergence of the information society in Europe, and even less to full employment.

Third, since ICT is an 'information' technology – the essence of which consists of increased memorization and storage, rapid manipulation and interpretation of data and information – it generally will further increase the possibilities to 'codify' large parts of human skills. As emphasized in the previous section, this is certainly not to deny the importance of the 'tacit' part of knowledge. On the contrary, as more knowledge becomes codifiable, the remaining non-codifiable part is likely to become even more crucial. Thus, the ability to codify relevant knowledge in creative ways as well as the competence to sort out relevant information and to use it efficiently has been and will continue to be of much greater importance. By the same token though, an increasing number of routine skills is becoming totally codifiable and their importance dramatically reduced.[10] As the largest part of employment in our societies does involve such routine tasks, there is concern about the distributional employment impact of the Information Society. Furthermore, confronted with the accompanying widespread use of various forms of information and computer technologies, 'skill mismatches' are likely to be of a much more pervasive and general nature, raising questions about the inherent 'skill bias' of new information and communication technologies. While these distributional concerns point to the crucial need for the broadening of education and training to all groups in society, they also raise fundamental questions with respect to possibly excluded groups, such as unskilled or routine skilled labour.

Fourth, as a consequence of the increased potential for further international codification and transferability, ICT could to some extent be considered as the first truly global technology. The possibility of ICT to codify information and knowledge over both distance and time, not only brings about more global access, it also enables firms/organizations to relocate the sort of routine activities which can be codified and thus also internationally traded. In other words ICT contributes to economic transparency and, generally brings to the forefront the cost advantages of alternative locations, to international capital mobility and international 'outsourcing' of particular activities. While the benefits to the world as a whole of a more transparent, borderless world are undisputed, there is again justified concern about the worldwide distribution of those benefits. As argued in Part Three, for the poorest, most peripheral countries/regions there is concern of becoming excluded; for the richer, technologically leading countries/regions, there is concern about the increasing erosion of the monopoly rents associated with innovation, and their implications for employment, wages and the built-up social security systems in Europe. In most EU countries, the financing of the national social security system and more generally the welfare state has been closely linked to employment, both through contributions from the employer and employee. In an increasingly global and economically transparent world, such a national link is becoming undermined.

It is difficult to predict the precise employment impact, whether in terms of volume, sectoral or occupational composition of the new information and communication technologies and the emerging Information Society. Both direct and indirect employment effects are likely to be substantial as a result of the structural transformation of society. Thus as the rise of entirely new industries in the second half of this century has already brought about new employment opportunities, such as the software industry, the electronic computer industry itself and the microelectronic industry, so will other new information and communication service sectors emerge and provide new employment opportunities. However, to do so will depend crucially on whether the appropriate regulatory institutional environment exists for the emergence of new information and communication markets;

and on the appropriate macro-economic climate. On both accounts there is concern that present-day conditions may not sufficiently favour the rapid emergence of such new markets.

At the same time, there is again, particularly in Europe, apprehension that both the direct and indirect employment displacement effects of the information society will be substantial. The potential offered by the new ICT to increase efficiency in some of the most typical information and communication dependent services sectors, such as finance, insurance and other business services, is high; the number of jobs displaced, particularly of the routine kind, is expected to be considerable in those sectors. Furthermore, there is concern that many traditional, 'non-tradeable' service activities might become internationally tradeable and that those activities will be relocated in low wage countries or regions. Again those jobs likely to be displaced are the sort of simple routine jobs, not necessarily of the unskilled, manual type, but rather those in which the information or knowledge content appears most easily subject to codification through ICT.

Whether the net balance of these direct and indirect employment effects will in the end be positive or negative cannot be assessed by simply counting new jobs gained and old jobs destroyed. It has to be recognized that the expansionary effects of the Information Society on any national economy or the world economy as a whole will paradoxically depend on the way the new information and communication technology will create the basis for a virtuous circle of growth in which investment is high, labour productivity grows fast, but output grows even faster, so that there remains a net growth in employment. That is why it is possible to be fundamentally optimistic about the inherent, long-term welfare and new employment opportunities associated with the emerging Information Society. However, these welfare gains and new opportunities are not given automatically. Whether the virtuous circle described above can be sustained will depend on macro-economic policies, regulatory and institutional reform in labour and product markets, more user friendly and focused technology policies, new distributional policies, policies aimed at boosting productivity growth as well as policies aimed at 'integrating' the new information and communication technologies in society. If there is a good match between technologies, policies and institutions prolonged periods of full employment are much more likely to result.

17.4 UNEMPLOYMENT AND INEQUALITY IN THE INFORMATION SOCIETY

It is not possible in a short chapter to address all the complex social problems associated with the coming of the Information Society.[11] We have chosen to address the problems of employment and unemployment because they have been in the forefront of concern. They were described, for example, by the Secretary-General of the OECD (1993) as 'disturbing, perhaps alarming'. Moreover, they are closely related to policies for technology, education and training. However, in concluding this chapter we briefly draw attention to a much wider problem associated with unemployment which is likely to become ultimately a source of great political and social tension: the growth of social inequality within and between countries.

The incidence of heavy structural unemployment in the 1980s and 1990s has combined with the increased dispersion of earnings for those in work to generate growing social inequality worldwide. In the former centrally planned economies the pursuit of market reforms and privatization programmes has led to polarization between rich and poor. Finally, as Schumpeter pointed out, the growth of new industries and firms also leads to exceptionally high rates of profit in some sectors side by side with the erosion or disappearance of profit in others.

The advent of the information society has thus been accompanied by a reversal of all those trends towards social justice and improved welfare services, which were such a characteristic feature of the quarter century following the Second World War. It seems probable that some features of the information society will exacerbate these trends even further. In particular a division is taking place between the 'information rich' and the 'information poor'. A fairly large number of people, even in the richest countries are unable or unwilling to use the new technologies or to gain access to those facilities where they might be used. 'Information poverty corresponds fairly closely to material (income) poverty but it is not identical. However, it can easily lead to material poverty in the labour market conditions generally prevailing. Social exclusion and the growth of a large underclass are thus becoming characteristic features of the information society, reinforced by the decline of the welfare state and the growth of regressive taxation. These tendencies have been particularly strong in the United Kingdom and the United States but they are clearly apparent in most countries. They are unlikely to disappear without deliberate policies designed to reverse this trend. Some such policies have been proposed by small groups of economists and social scientists, for example, in the report on *Constructing the Information Society for us All* published by the European Union (1996).

Somewhat surprisingly for a group looking at Information Technology, this 'Expert Group' recommended research on a new type of taxation, namely a 'Bit Tax' to raise sufficient revenues for government policies to tackle the serious problems of social inequality, which are everywhere apparent (Soete, 1996). This type of tax was proposed for several reasons but in the first place to broaden the revenue base in keeping with the vast structural transformation of the economy brought about by information technology. A second reason for serious research on a 'Bit Tax' or 'Information Transmission Tax' is paradoxically to improve the efficiency of the information industries. Information overload has been a problem in science and technology for a long time (see, for example, Bernal, 1939 or Ziman, 1994) and is also a growing problem for managers, politicians and for many consumers. The conversion of data and information into useful knowledge involves thought and preparation, leading almost always to a condensed presentation, codification and transmission. Overload very often leads to a closely related problem, information pollution: the unconsidered transmission of worthless information to those who have no desire to receive it. An appropriate tax regime would be a simple way to curb if not to eradicate these tendencies.

Finally, a related proposal has been made by James Tobin to deal with the growth of international inequality. He has pointed out that even a low rate of tax on the transactions of the international financial markets could easily yield far more than all the UN and other international aid programmes for the Third World. For reasons discussed in Part Three, the problems of international divergence in growth rates may well be more extreme than in the twentieth century so that those programmes for 'excluded' countries and regions will be even more important.

In considering these long-term social problems of growth and inequality, Kuznets argued that 'one might assume a long swing in the inequality characterizing the secular income structure: widening in the early phases of economic growth . . . becoming stabilized for a while; and then narrowing in the later phases' (Kuznets, 1953, p. 18). He viewed this long swing as 'part of the wider process of economic growth, interrelated with similar movements in other elements' including the 'rate of growth of population, rate or urbanization and internal migration, the proportion of savings or capital formation to national product . . . the ratio of foreign trade to domestic activities [and] in the aspects, if we could only measure them properly, of government activity that bear on market forces (there must have been a phase of increasing freedom of market forces, giving way to greater intervention by government). . . . The long swing in income inequality is also probably closely associated with the swing in capital formation proportions – in so far as wider inequality makes for higher, and narrower inequality for lower, country-wide savings proportions' (Kuznets, 1955, p. 20).

He concluded:

A final comment relates to the directions in which further explanation of the subject is likely to lead us. . . . If we are to deal adequately with processes of economic growth, processes of long-term change in which the very technological, demographic, and social frameworks are also changing – and in ways that decidedly affect the operation of economic forces proper – it is inevitable that we venture into fields beyond . . . economics proper. . . . [It] is imperative that we become more familiar with findings in those related social disciplines that can help us understand population growth patterns, the nature and forces in technological change, the factors that determine the characteristics and trends in political institutions. . . . Effective work in this field necessarily calls for a shift from market economics to political and social economy.

(Kuznets, 1955, p. 28)

NOTES

1. The latter rate, which is less subject to definitional differences in the measurement of unemployment between countries, is to some extent a more correct measure of the 'unused labour potential' in an economy.
2. There exists now a substantial empirical literature on this subject. For recent literature reviews see de Wit (1990), Vivarelli (1994).
3. Various authors have pointed to the historical analogy of the diffusion of electricity and in particular electric unit drive and ICT (Freeman and Soete, 1985, 1987; David, 1991. See also the discussion in Chapter 3).
4. For early analyses along these lines see Quinn (1986) and Soete (1987). See also Miles (1996). Another definition of services is that services are what you cannot drop on your foot. The fact that both definitions are becoming blurred

illustrates both the complexity of the issues and the urgent need for reclassi-
fication of economic activities in the so-called 'tertiary' sector.

5. This was certainly the case with regard to the invention of printing in the
Middle Ages and the impact this first new information technology had on the
limited tradeable 'service' activity of monks copying manuscripts by hand. It
was the time/storage dimension of the new printing technology which opened
up access to information in the most dramatic and pervasive way and led, to
use Marx's words, to the 'renaissance of science', the growth of universities,
education, libraries, the spreading of culture, etc. This opening up, 'tradeability' effect would become of far more importance to the future growth and
development of Western society than the emergence of a new, in this case
purely manufactured based, printing industry.

6. See in particular David and Foray (1995).

7. One might think of such activities as gardening, cycling or housekeeping. For a
thorough discussion of tacit knowledge from the standpoint of philosophy, see
Searle (1995). For a discussion of tacit knowledge in firms, see Nelson and
Winter (1982) and Senker (1995).

8. See among others Gershuny (1983) and Gershuny and Miles (1983).

9. See Soete (1996).

10. It is interesting to observe that by contrast, relatively simple human tasks (such
as gardening) might never become codifiable. This explains why the idea of
ICT as a 'skill-biased' technical change does not really capture the complexities
of the de- and reskilling processes. See in more detail OECD (1996).

11. For a more comprehensive discussion of numerous social and economic
aspects of the information society, see Castells (1996); Dutton, (1996); Gates
(1996); Whiston (1991); and for the Third World, see Nasthakken and Akhtar
(1994).

CHAPTER 18

TECHNOLOGY AND THE ENVIRONMENT

18.1 INTRODUCTION

Following the discussion of employment in Chapter 17, the central issue which is addressed in this chapter is the use of innovation, technology and other complementary policies to support the goal of environmentally sustainable development. Innovation and technology policy have an essential role in achieving this goal, because of the need for innovations to replace current unsustainable production methods and consumption patterns and because of the need for the more rapid development and diffusion of a wide range of alternative more environmentally friendly technologies. There are three additional reasons why focusing on environmental issues is a useful way of highlighting the particular policy challenges raised by technological change and innovation.

First, environmentally sustainable development is typically a long-term policy goal; one which is urgently in need of being put much higher on the agenda of science and technology policy ministries, agencies and other policy-makers. It requires a wide range of innovations and the gradual change of existing institutions and production and consumption technologies to meet new goals. Recent studies estimate that these changes will require, even with concerted effort, perhaps thirty to fifty years (Jansen and Vergragt, 1992). In comparison, many other European policy goals that involve technology policy are short- or medium-term goals, as discussed in the previous chapter, such as a reduction in the historically high unemployment rates of the last decade or steps to ensure a skilled workforce.

A second reason which warrants particular policy attention to sustainable development is the complexity of the issue. The goal of environmental sustainability, and in particular the development and diffusion of innovations to help meet this goal, cannot be reached without a wide range of policies, the involvement of many different economic actors, and changes in existing economic, social, and cultural institutions. The complexity of the problem provides a useful test of the merits of a systems approach in guiding the development of policy.

A final reason which warrants particular interest in environmental technologies and sustainable development is the close interaction between public and private goals. The goal of environmental sustainability is primarily a public goal, though it cannot be reached without ensuring that the private sector is both viable and capable of adjusting to change. Examining both of these issues permits a balanced application of a systems approach to both public and private goals.

18.2   ENVIRONMENTALLY SUSTAINABLE DEVELOPMENT: FROM OLD TO NEW 'MISSIONS' FOR SCIENCE AND TECHNOLOGY POLICY

There are many definitions of environmentally sustainable development. We use the term here in the sense of an economic system that can meet the needs of the present generation without irreplaceably reducing the resources available to future generations and without irreversibly damaging the environment.[1] This definition recognizes that the ability of an economic system to fulfil human needs over the long term is dependent on the viability of the ecological system and requires production and consumption technologies to meet two criteria. First, the production of waste products that cause environmental damage must be gradually eliminated through methods that prevent the creation of non-recyclable waste products. This is a more ambitious goal than the reduction of pollution, which can lead to policies which control pollution by shifting pollutants from one place to another, for example, from disposal in rivers to disposal in on-land waste disposal sites. Second, the total stock of available non-renewable resources must not be depleted. To meet this requirement, either production technologies will need to use renewable resources or substitutes will continuously need to be found to replace non-renewable resources as they are used up. In each case, a vital part of the solution is technological change through both the rapid diffusion of existing best-practice technologies and innovative activities that develop new technologies.

The use of science and technology policies to achieve environmental goals constitutes a new focus for technology policy. Superficially, this requires a return to the emphasis in the 1950s and 1960s on public goals that were met through mission-oriented projects. However, there is a fundamental difference between older mission-oriented projects, for example, nuclear defence, aerospace programmes, and new projects to support environmentally sustainable development. The older projects, discussed in Chapter 16, developed radically new technologies through government procurement activities that were largely isolated from the rest of the economy, though they frequently affected the structure of related industries and could lead to new spin-off technologies that had widespread effects on other sectors. In contrast, mission-oriented environmental projects will need to combine procurement with many other policies in order to have pervasive effects on the entire structure of production and consumption within an economy.

The pervasive character of new mission-oriented projects to meet environmental goals calls for a more systemic approach to policy. This approach results in substantial changes to the mission-oriented projects of the past.[2] Table 18.1 summarizes the key characteristics and differences between the old and new models of mission-oriented projects. We highlight a number of major characteristics of the systems approach to innovation.

The large number of organizations that will need to be involved in research, development and procurement programmes suggests using a network approach to support the rapid diffusion and appropriation of information. This can be achieved through research and procurement policies which encourage co-operative development projects and parallel

Table 18.1 Characteristics of old and new 'mission-oriented' projects

Old: defence, nuclear and aerospace	*New: environmental technologies*
The mission is defined in terms of the number and type of technical achievements with little regard to their economic feasibility.	The mission is defined in terms of economically feasible technical solutions to particular environmental problems.
• The goals and the direction of technological development are defined in advance by a small group of experts.	• The direction of technical change is influenced by a wide range of actors including government, private firms and consumer groups.
• Centralized control within a government administration.	• Decentralized control with a large number of involved agents.
• Diffusion of the results outside the core of participants is of minor importance or actively discouraged.	• Diffusion of the results is a central goal and is actively encouraged.
• Limited to a small group of firms that can participate owing to the emphasis on a small number of radical technologies.	• An emphasis on the development of both radical and incremental innovations in order to permit a large number of firms to participate.
• Self-contained projects with little need for complementary policies and scant attention paid to coherence.	• Complementary policies vital for success and close attention paid to coherence with other goals.

research by several organizations, though there is also a need for lead institutes that can fulfil both a leading research role and assist in the co-ordination of co-operative projects based on contracting out to private and public organizations. Co-operative projects should also include non-governmental organizations (NGOs) and scientists and organizations from the developing nations. NGOs not only bring valuable expertise to scientific and technical research projects, they can also help to articulate public demand and mediate between the goals of consumers and of industry. The participation of researchers from developing nations is needed to ensure the global diffusion of environmental technologies. The large number of participants in co-operative projects also requires an institutionalized means of evaluating each project. This could be built into the process through periodic technology assessments to influence the direction of technical development.[3]

18.3 TECHNICAL CHANGE AS A CUMULATIVE PROCESS: TOWARDS AN ENVIRONMENTALLY SUSTAINABLE TECHNOLOGICAL TRAJECTORY

A major difficulty for environmental policy is how to promote sustainable technologies within a market economy that selects products and processes, not on the basis of environmental criteria, but on the basis of profitability, which in turn is influenced by demand. A way through this difficulty is to develop policies that can take advantage of the cumulative and self-reinforcing characteristics of technical change. This can be done by developing policies which guide the continual search by industry for innovations and technologies towards environmentally beneficial directions. The goal of these policies is to start a self-reinforcing process in which additional searches for new technical solutions follow within the same technical pathway. For example, experience gained from generating electricity from photovoltaic cells in the few situations where this technology is economically competitive should lead to learning effects that gradually improve the cost effectiveness of photovoltaic cells and increase their competitiveness. Increasing competitiveness should then attract additional investment in this technology, leading to further technical improvements and cost reductions and a widening of the number of economically feasible applications. These effects have already been demonstrated in the falling costs of wind power and other sources of renewable energy in the British NFFO policy (Mitchell, 1995, 1996).

There are four main policy tools to guide private firms towards investing in the development of environmentally sustainable technologies: direct regulation, economic instruments, procurement, and policies to alter the social nexus of technical change. Each is briefly discussed in turn.

Direct regulation

Direct regulation, for example through air, water, soil and product quality standards, or limitations on the conditions and use of a product, is the most common method for reducing pollution or exposure to hazardous substances. Regulations have been extensively criticized, on theoretical grounds, as less efficient than economic instruments as a means of promoting innovation in less polluting technologies.[4] However, regulations are needed to set minimum standards and to prevent exposure to hazardous substances.

Economic instruments

Economic instruments include tradeable emission permits, emission and product taxes and in some cases subsidies, though subsidies are discussed below as part of procurement. Economic instruments differ from direct regulation by not setting standards for emissions. Pollution is permitted, but at a direct cost to the polluter ('the polluter must pay'). Economic instruments function through policies that estimate the externality costs of pollution and attach these costs to the inputs or outputs of production. For example, high-sulphur fuels could be taxed to include the estimated costs of environmental damage from acid rain. The additional cost to industry should attract innovative activity to find alternative fuels or new technologies that can reduce sulphur emissions.

The efficient use of economic instruments is dependent upon better accounting practices to estimate environmental costs and on technologies to measure emissions accurately. These have proved to be major obstacles, though the latter problem could be partly solved by investment in developing real-time monitoring systems for a range of substances. An additional need is to identify and remove perverse economic instruments such as tax credits and subsidies in agriculture that support environmentally harmful practices.

Procurement

Procurement policies, either through the direct support of research and development or through subsidizing the use or development of environmentally beneficial technologies within private firms, play an important role in achieving the goal of environmentally sustainable development. Direct procurement is probably most appropriate for the development of new technologies for use in infrastructural systems, such as energy, transportation and waste disposal systems. Subsidies, for example, through tax credits or co-operative research projects between industry and public research institutes may be the most effective means of developing cleaner process technologies where much of the existing knowledge is held by private firms and has not been codified or made publicly available.

Social nexus

The social nexus of a technology consists of a wide range of influences that limit the type of technologies that are both socially and economically feasible (Schot, 1992). This includes factors which affect consumer demand and the importance given by the management of firms to environmental issues. Both can be affected through educational programmes, for example, by including environmental courses in programmes for managers and engineers so that they learn to consider automatically the environmental effects of their business. Another goal is to use policy to support organizations that can increase the external and internal pressure on firms 'to promote the integration of environmental aspects into the general business-strategy of a firm' (Cramer et al., 1990), for example, through internal review and evaluation methods that include environmental objectives. There is some evidence that an increasing number of firms has begun to act in this way and to appoint senior managers with responsibilities in this field.

Organizations that can increase the external pressure on firms include, in addition to government, research institutes, the environmental technology sector, and non-governmental organizations (NGOs) such as consumer, public health, and environmental organizations. NGOs can increase pressure on firms by influencing demand through raising consumer awareness of environmental issues and of poor corporate practices. German consumer organizations appear to have been particularly effective in exerting this type of pressure.

Internal pressure can be placed on firms through trade and industry organizations and by the environmental and marketing departments within private firms. These organizations can alter acceptable practice by ensuring that environmental issues are included as an important factor in long-term strategy.[5] There is also a need here for more social science research to increase our understanding of the capabilities of organizations and individuals to adapt to environmental goals and on the design of appropriate policies to encourage this process.

One method which uses the social nexus of a technology to guide its development in beneficial directions is Constructive Technology Assessment (CTA).[6] We discuss this method in more detail in the final chapter.

CTA differs from other methods of technology assessment by its emphasis on developing a dialogue among producers, users and social groups which will be affected, either beneficially or negatively, by new technology. This widens the range of interest groups that can influence the development of a new technology and consequently improves the ability to guide innovative activities in directions that are environmentally and socially useful.

18.4   FROM LOCK-IN TO RADICAL BREAKTHROUGH

The future benefits and hazards of a developing technology can rarely be foreseen, given the uncertainty that characterizes technical change. Some technologies could be environmentally harmful during the early stages of their development but evolve into very beneficial technologies, whereas other technologies that at first appear to be beneficial or environmentally neutral could prove to be hazardous.[7] The inability to predict future developments suggests taking a cautious approach to regulations or other policies that could inhibit experimentation or lead to premature lock-in into an inferior technology.

Lock-in to existing technologies frequently leads to a preference for end-of-pipe technologies, which can simply transfer a pollutant from one area to another,[8] over cleaner process technologies. An example is the use of catalytic converters to reduce noxious gases in the exhaust of internal combustion engines, instead of developing an alternative type of engine that does not produce noxious gases in the first place.

End-of-pipe technologies offer definite advantages to private firms because of their greater short-term flexibility in comparison with some cleaner process innovations. They can be added to existing capital plant and they can require less learning than major process innovations. Under many conditions, they are a rational solution to the need to control pollution emissions from existing physical plant. In addition, end-of-pipe solutions can be applied to a wide range of production technologies and consequently they offer a larger market than cleaner process innovations for firms that develop and manufacture pollution control equipment. As a result of these advantages, policies to encourage environmental innovation can unintentionally support the development of inferior technical solutions based on end-of-pipe technologies that obstruct the long-term goal for cleaner process technologies that avoid pollution at the source. A reliance on end-of-pipe solutions could also be reinforced by the polluter pays principle, unless it is applied carefully so that it also encourages organizations to search for cleaner process technologies. This may not be possible, particularly for smaller firms, without some form of subsidy such as technical assistance in developing or applying cleaner processes.

One long-term solution is to encourage innovations which reduce the amount of all inputs needed to make a unit of output. This will not only reduce the resources needed to produce a given level of output, resulting in savings for the firm, but in most cases will also result in cleaner process technologies because the amount of all inputs, including those which end up as waste, will be reduced.[9] Lower input/output ratios are clearly technically feasible, as shown by the average 20% fall in energy intensity per 1000 ECU of GDP in the EC between the first major increase in energy

Table 18.2 Final energy intensity[a] of GDP (tons oil equivalent per 1,000 ECU) 1973–93

Country	1973	1988	1993
Belgium	0.64	0.49	0.47
Denmark	0.46	0.31	0.32
Germany	0.52	0.40	0.39
Greece	0.51	0.60	0.52
France	0.44	0.36	0.38
Ireland	0.59	0.57	0.43
Italy	0.46	0.36	0.37
Luxemburg	1.51	0.74	0.65
Netherlands	0.59	0.47	0.44
UK	0.62	0.45	0.44
EC	0.52	0.41	0.41

[a] Energy intensity – gross inland consumption divided by GDP at 1980 prices and 1980 exchange rates.

Sources: Eurostat (CEC, 1994a); updated from Eurostat (CEC, 1990, 1995), Eurostat Energy Statistics, EC, Eurostat, Luxemburg.

costs in 1973 and 1988. Energy intensity in Denmark fell by 33% in this time period. However, this improvement slowed down markedly in the 1990s (Table 18.2), possibly because relative energy prices no longer provided the same incentive for energy-saving and because of the failure to introduce carbon taxes.

18.5 THE ROLE OF PUBLIC POLICIES FOR ENVIRONMENTALLY SUSTAINABLE DEVELOPMENT

An essential goal of policies to support sustainable development is to encourage the rapid diffusion of environmentally beneficial technologies. Diffusion can be enhanced by programmes that increase the number of people that are knowledgeable about the innovation and capable of applying it to the needs of industry. Policies to support demonstration projects and technology transfer programmes can help by increasing exposure to a range of potential users.[10] Procurement programmes based on incrementalist principles (as discussed below) can also encourage the rapid diffusion of best-practice technologies by increasing the number of individuals with direct knowledge of the technology and the ability to use it within the private sector.

The goal of environmentally sustainable development requires the development and application of a wide range of technologies, each of which follow a unique pattern of development and application. The large number of technologies, needs and applications means that governments lack the necessary information to decide the details of what type of innovation is needed to solve particular problems. Instead of defining solutions, policies should be designed to influence the incentives that firms face, as discussed above, and to support research programmes that build upon the enormous diversity of the sources and applications of innovations by ensuring that the widest possible range of potentially beneficial

Table 8.3 Selected examples of the potential benefits from R&D investment in electronic sensor technology*

Technical measure	Effects
Cleaner processes	
Reduction of combustion air in rotary cement kilns.	20% reduction of NO emissions.
Improved control techniques for process heat and electricity generation.	10–20% reduction of SO_2 and NO_x emissions.
Improved control techniques for domestic hot water heaters.	10% reduction of CO_2 emissions.
Reduction of input/output ratios	
Online control of chemical nickel baths in electroplating by measuring the pH value and nickel content.	2–3-fold increase in the lifetime of the bath.
Onboard control of following distance for vehicles on highways.	2–3-fold increase in vehicle throughput (reduces need for new highway infrastructure).
Automatic dosing of separate detergent ingredients for washing machines.	50% reduction of washing agents; 10% reduction of energy and water use.
Online measurement of thickness of plastic sheets.	2–10% reduction in consumption of raw materials.

*   adapted from Angerer (1992).

technologies are explored and developed. This can be achieved through research and procurement programmes that include innovating firms, the potential users of new products and processes, public research institutions, and non-governmental organizations. In addition, to make sure that a wide range of potential applications are also explored, government procurement programmes to develop new technologies should also include a large number of firms that could potentially benefit from them. This last point is particularly important, given the importance of knowledge and the unique ways in which technologies are used by different firms.

Procurement programmes to support environmental sustainability will need to explore both traditional and new technologies. In particular, new technologies with many potential environmental advantages such as information, communications, new materials, and bio-engineering technologies will need to be thoroughly explored and applied to a wide range of production and consumption technologies. For example, information technologies can reduce energy and material inputs through improved control systems in industry and in home-heating. Table 18.3 provides several examples of the environmental benefits from cleaner processes and lower input/output ratios that could result from using information technology to develop real-time sensor devices. Another example is the use of information and communications technologies to reduce energy use due to travel through teleshopping, computers to permit working at home, and video links for business meetings.[11]

Both radical breakthrough technologies and incremental improvements to existing technologies are needed. An example of a future breakthrough innovation is solar or another renewable energy technology that could acquire a central role in an environmentally sustainable economy. An example of an incremental innovation with environmentally beneficial effects is a technical adaptation to jet aircraft engines to increase fuel efficiency and reduce NO_x emissions.

It should be noted that the term incrementalism is used by several researchers to describe a specific approach to the innovation process rather than to the taxonomic type of innovation. An incrementalist innovation process can produce both incremental innovations and radical innovations. The basic principle of incrementalism is that the innovation process should be designed to encourage frequent evaluations of a developing technology by a large number of researchers or potential users. This can be achieved through the use of relatively short development times for each technical advance, small project sizes, low capital investment levels for each project, and a minimum need for a dedicated infrastructure with no other uses.[12] The latter three criteria are designed to permit a range of firms and research institutions to conduct parallel research projects in a given technology in order to increase the opportunities for a large number of people to evaluate the technology critically, either through the rigours of the market or through non-market means, and to use the results of these evaluations to guide further technical developments.

The goal of an incrementalist approach to the innovation process is to avoid overinvestment in a limited range of expensive technologies that are later found to be unworkable, overly expensive, or environmentally more harmful than the technologies they were designed to replace. This is a potential problem with mission-oriented projects to develop radical new technologies through large, expensive, and long-term projects that limit the number of participants to a few technically advanced firms or research institutes. However, one or more incrementalist principles may need to be waived in cases where they would completely preclude the exploration of a potentially beneficial technology.[13]

18.6 A SYSTEMS APPROACH: COMPLEMENTARITY AND COHERENCE

The goal of environmentally sustainable development requires a wide range of complementary policies to support investment in new environmental technologies and the rapid diffusion of successful applications. These policies have been discussed in the preceding section and essentially fall into two main policy areas, as summarized in Table 18.4. The two areas consist of policies to guide investment towards a search for environmentally beneficial technologies and policies to promote the development and diffusion of these technologies. Policies to support the competitiveness and capacity of industry to change are a prerequisite for environmentally sustainable development. The opposite condition also applies: policies to promote sustainable development should not decrease the competitiveness of industry. International agreements are needed in some areas so that no nation is disadvantaged.

Policies to develop environmental technologies can help to improve the competitiveness of industry in two ways. First, new technologies which

Table 18.4 Main policy options to support the goal of environmentally sustainable development

---

1. **Policies that can be used to guide innovation, particularly towards cleaner process technologies and those with lower input/output ratios**

- Direct regulation such as air, water, soil, and product quality standards.
- Economic instruments such as emission and product taxes or tradeable emission permits.
- Procurement, either through the direct support of R&D or through subsidies.
- Policies to alter the social nexus, including social persuasion, demand factors and constructive technology assessment.

2. **Policies to influence the innovation process and to ensure the diffusion of new knowledge**

- Wherever possible use incrementalist principles based on short development times and small project sizes to allow a large number of organizations and firms to participate in co-operative R&D projects.
- Decentralized control of innovation projects using a network approach to link lead research institutes, private firms, and other organizations.
- Demonstration project and technology transfer programmes.

---

reduce the amount of material and energy inputs per unit of output will, on average, also reduce costs. For example, a study in the Netherlands of 45 pollution prevention projects in the public transport, food, electroplating, metal working and chemical industries found that only three projects increased costs to the firm and 19 had no effect on costs. Twenty projects reduced costs and 16 of them had a payback period of less than one year.[14] Second, policies that guide innovation towards the development of products and processes that meet stringent standards for environmental performance and safety can also increase the global competitiveness of industry if these standards are likely to be demanded in the future by many different countries.[15]

Environmental problems rarely respect national borders. The waste products created by industry or consumers in one country can be spread via air or water across many other countries or even globally, as with CFCs. Alternatively, the waste products produced by a large number of countries can be deposited in a single country or region. For example, most of the heavy metal contamination of the river deltas in the Netherlands is due to industry in Belgium, France, Switzerland, and Germany. Similar cross-border problems arise in the Great Lakes districts between Canada and the United States and in many other parts of the world. Given the international character of environmental problems, the goal of environmentally sustainable development is important to all of the different regions and nations of the world and requires the widespread diffusion of new technologies and supporting institutions. The multinational character of both the problems and the solutions suggests a strong role for supranational organizations such as the UN, while at the same time the localized nature of many of the sources of pollution and differences in the institutions and solutions that have developed to solve environmental problems, for example, to dispose of or recycle household and industrial waste, requires the extensive involvement of regional and national governments. For this reason, environmentally sustainable development can only be attained by the active involvement of all levels of government. This requires the careful application of the subsidiarity principle in order to determine the responsibilities at each level. The 'subsidiarity principle' may be defined as leaving decision-making to the lowest possible level consistent with the efficient fulfilment of the responsibility in each case.

Obviously in the case of a federal government, such as the United States or Canada, the division of responsibilities would depend on the specific constitutional, legal and political circumstances but the principle of subsidiarity is a useful guideline in order to maximize local grassroots participation and to minimize centralized bureaucracy. However, consideration of the case of the European Union or of the federal government in the United States shows that there are many responsibilities which are best undertaken at higher levels, or jointly by supranational, federal, national and local authorities. Hopefully in the future, worldwide intercontinental agencies will be able to assume greater responsibility for global standards in order to avoid the problem of loss of national or regional competitive advantage through strong environmental policies, for example, carbon tax.

18.7 CONCLUDING COMMENTS

In this chapter we have outlined the policy options and the general goal of environmentally sustainable development, but we have not discussed in depth which policy options are most effective in guiding technical change for specific goals. Both of these are political questions that will need to be debated by society and decided democratically.

The choice of which policy options to use depends less on their theoretical strengths and weaknesses, for example, between direct regulation and tradeable emission permits, than on pragmatic matters such as the ability to build a political consensus around the use of particular options. There is a need to use several options simultaneously, particularly procurement programmes, policies to alter the social nexus, and either direct regulation or economic instruments, but the importance attached to each particular option, and the best means of implementation, are political decisions.

Defining and developing a consensus around specific environmental goals is an even thornier problem, particularly when environmental goals require substantial changes to systemic and interlocked technologies. For example, agriculture is a production and consumption system that includes not only farmers and consumers, but government subsidy and income support programmes and the suppliers of equipment, pesticides and fertilizers. The solution to the environmental problems caused by agriculture could require not only minor changes, such as the development of less toxic pesticides, but a change in the entire structure of agricultural production. Similarly, the goal of reducing energy consumption could require systemic changes in the complex transport infrastructures that have developed over the last century in order to increase the attractiveness and flexibility of public transport in comparison with the private car. These types of major changes to the techno-economic system cannot be achieved without political debate to reach public consensus. It is to these political aspects of economics that we devote the concluding chapter.

NOTES

1. Some reduction of resources and some damage to the environment is inevitable. The problem is to constrain this reduction and damage so that resources can be replaced by recycling and substitution and damage contained or reversed by countervailing policies.
2. See Ergas (1987, 1992) for a critical discussion of the predominant features of mission-oriented projects.
3. For a German example of continual monitoring of a basic research programme for biological production of hydrogen, see Meyer-Krahmer and Reiss (1992).
4. A discussion of the benefits and disadvantages of different types of regulatory approaches, including the use of standards or economic instruments such as emission or product taxes, is provided by Pearce *et al.* (1989) and Milliman and Prince (1989). See also Howes *et al.* (1997).
5. There is some evidence, based on case studies, to show that firms are integrating environmental issues into corporate strategic planning. See Groenewegen and Vergragt (1991).
6. For a discussion of the theory and application of CTA, see van Boxsel (1989) and Schot (1992).
7. The latter case includes well-known examples such as chlorinated fluorocarbons (CFCs) used in refrigeration and which reduced food wastage, but damaged the ozone layer, DDT which provided protection from mosquito-borne diseases but killed beneficial insects, and irrigation schemes which increased agricultural land but in some cases led to salination problems and, paradoxically, to desertification. The former case is more difficult to document because examples are often controversial. Possible candidates include computers, which were originally developed for military use, and recombinant DNA techniques, which have raised concerns about the accidental development of new pathogens.
8. End-of-pipe technologies are designed to clean the waste stream of industrial processes. An example is scrubbers to remove sulphur dioxide from coal-fired electrical plants. Some end-of-pipe technologies can function as cleaner process technologies if the retrieved compound can be recycled, thereby reducing inputs.
9. This conclusion is illustrated mathematically by Pearce and Turner (1984), but in some cases a reduction in inputs will not be environmentally beneficial. This cannot be determined without considering the direct and indirect effects of reducing input requirements, because technologies to reduce inputs may increase environmental problems elsewhere. For example, energy-saving processes that require elaborate computer control systems could increase the use of toxic substances in the production of integrated circuits. The difficulty in assessing the environmental effect of a technology is compounded by international trade in materials and waste products.
10. The effectiveness of these programmes may depend upon the structure of the industry that supplies the new technologies. In general, the subsidization of demonstration projects and the diffusion of information will be less likely to have perverse or problematic effects on the speed of diffusion where the technology supplying industry is competitive.
11. See Freeman (1992) and Kemp and Soete (1990). Although the potential benefits of information technologies in reducing travel to work could be

counteracted by an increase in leisure travel (Cramer and Zegveld, 1991), this trend could also be counteracted by the increased use of sailing ships, airships and other energy-saving forms of public transport.

12. For an extensive discussion of incrementalism and several historical examples of failure due to overinvestment in large, capital intensive projects, see Collingridge (1981, 1990) and Collingridge and James (1989).

13. For example, technical restraints limit fusion research to large-scales mission-oriented projects. Fusion research within the EC has so far cost approximately 4.5 billion ECUs and could cost an additional 15 to 35 billion to build a prototype fusion electrical generator (Ford and Lake, 1991). The fusion reactor would be a successful example of a large-scale, mission-oriented project if it turns out to be economically and environmentally feasible. On the other hand, even if fusion generated electricity does meet these two criteria, it is possible that far fewer funds spent on other sources of electricity that are suited to small-scale units and incrementalist principles could attain the same or better level of commercial feasibility without the risk attached to investing so much in one project.

14. The effect on costs was not known for three projects at the time of the evaluation. See Dieleman and de Hoo (1993).

15. This point is emphasized by Porter (1990), Irwin and Vergragt (1989) and by Bill Clinton in an article outlining the environmental policy of the Democratic administration in the USA (*Guardian*, 9 November 1992). Jochem and Hohmeyer (1992) illustrate the argument with data for Germany which show that German exports of energy saving products increased in the 1980s at twice the rate of all other industrial exports.

CHAPTER 19

CONCLUSIONS: BEYOND THE ECONOMICS OF INDUSTRIAL INNOVATION

19.1 ECONOMICS AND TECHNOLOGICAL PROGRESS: WHAT IS THERE TO ASSESS?

In this last chapter a number of concluding arguments are put forward with respect to the need to broaden the policy debate beyond the narrow economic 'material welfare' perspective on science and technology. In the previous chapter we highlighted the need for a substantial reassessment of the role of science and technology in relation to long-term sustainability and welfare. In this chapter we focus on the broader social integration issues of science and technology. In the first section we briefly review some of the traditional economic misconceptions of technology, including its measurement. In the second part the analysis is broadened to focus on the endogenous nature of technological change. Finally, in the last section some fundamental problems of 'constructive' technology assessment are presented.

For most economists assessing technological change appears something of a puzzle, far removed from economic reality. The main reason for this goes back to the traditional economic framework, within which technology is reduced to an 'exogenous' external factor whose impact on, for example, economic growth can be best described – as in the case of population growth – in terms of a particular parametric value: a 'black box' variable, not to be opened except by scientists and engineers. This awareness of the limitations of the contribution of economics to society would be laudable if it did not also imply a particular economic vision and interpretation of the contribution of technology to economic development and growth. In this particular vision of economics, technological change is (and practically by definition can only be) associated with material welfare increasing aspects of economic growth, thanks to the assumed allocative efficiency of the market.

This is also reflected in the way technological change is generally measured. From an economic point of view, measurement of technological change is generally reduced to those new technologies which have a well-defined economic impact, either in terms of productivity growth or in terms of new product demand. Even with respect to the latter, the methodological problems raised in incorporating new products into the production function framework have generally led to a further reduction of the economic contribution of technological change to productivity growth

– typically expressed in terms of some weighted average of labour and capital productivity called total factor productivity. We raised this measurement issue already in the first chapter and again more systematically in Chapter 13. We raise it again here because it illustrates well the common perception in economic analyses of technology, that technological change and its social impact can be correctly assessed only in economic terms.

To economists it often comes as a surprise that there are many innovations which have very widespread *social* effects, but whose measurable economic effects are small or indirect in terms of macro-economic growth and efficiency. The innovation of an oral contraception device had a major impact on sexual behaviour in the 1960s and 1970s in most Western countries, giving rise to some fundamental debates about medical and social ethics. Its macro-economic impact was negligible and at best indirect through greater participation of women in the labour market. Genetic fingerprinting – a more recent technological advance in biotechnology – is said to be of great importance in forensic medicine, crime detection and the judicial process, especially in cases of rape, assault and murder. It could also have major implications for medical prognosis and life insurance, which will also raise some fundamental questions of medical and social ethics. Again though, the economic significance of this technical advance is difficult to predict, but probably rather small. For many other innovations the social impact may be great even though the direct measurable economic impact is insignificant.

Another distinction which is also insufficiently introduced in economic analyses and which at first sight might appear to have little to do with technology assessment is the difference between innovations which find applications in only one sector and those which affect many or all sectors of the economy. In the technological taxonomies suggested by a number of authors in the science and technology field (Freeman, 1982; Pavitt, 1984; Rosenberg, 1982), technological advances are often identified as either 'localized' or 'pervasive' in terms of their impact. An illustrative example will serve to clarify this distinction. The float glass process introduced by Pilkington's in the 1960s was certainly of enormous economic importance for that firm and for the glass industry generally, as it was licensed to almost all the major glass manufacturers in the world over the next few years. However, it has no applications outside the glass industry and its macro-economic significance is therefore relatively small. The microprocessor or the electronic computer, as we discussed in Chapters 7 and 17, have found applications in practically every single sector of the economy, with, one suspects, major economic impact on the efficiency, growth and even employment performance of the economy.

Economists in other words are not only rarely aware of the social impact of technological change, they are also insufficiently aware of the wide variance in economic impact of technological change. The problem has undoubtedly become more severe over the last decades. Whereas in the 1960s and 1970s most empirical economic studies in this area have limited the analysis to the industrial sector, either as purveyor of technological advances or as funding such advances, such an approach becomes, as we saw in Chapter 17, increasingly problematic when dealing with the increasing number of service sectors as major initiators of technological change. Already in Chapter 17 we pointed to the increasing problems of measurement and of indexation in relation to ICT in services and more generally quality improvements. A systematic inclusion of service activities means that one is increasingly confronted with questions about the actual direct economic impact of such technological advances. In many service sectors the separation between economic measurable impact and social quality of life impact of technological change is more difficult to make.

19.2   ASSESSING ECONOMIC PROGRESS AND ENDOGENOUS TECHNOLOGY

There is, however, increasing recognition, particularly among policy-makers but also among economists,[1] that the traditional economic 'exogenous' approach to technology and the related measurement biases is becoming a hindrance rather than a useful conceptual framework within which some of the recent policy concerns about the technology economy interaction can be discussed. With respect to the so-called Solow paradox of slow productivity growth it raises the more general issue of the assessment of economic 'progress'. It could be argued that the post-war period up to the mid-1970s – Jean Fourastié's 'Les trente glorieuses' – was a period in which there was general agreement that the quantitative economic data on the post-war increases in production and in productivity did provide a consistent rough picture of 'progress' over this same period. As Fourastié put it, 'the great hope of the twentieth century' has been fulfilled, based on 'the facts of production, consumption, length of working hours, health care and life expectancy'.

Over the more recent period, however, and in line with the productivity puzzle evidence, increasing discrepancies are emerging among such indicators and also in the perception of the weights allocated to material growth indicators and the many material and non-material 'externalities' of such 'growth'. There is far greater awareness that economic indicators do not measure correctly many of the social and environmental costs of economic growth. Similar arguments can be made with respect to the other concerns raised above. At the same time, there is growing awareness that much of the increased 'consumer welfare' associated with new and better products and services is not being measured in current aggregate economic indicators.[2]

Such arguments have gradually brought about a shift in the way technology has been analysed over the 1980s and 1990s. The OECD Brooks Report (1971b) already discussed environmental quality in relation to economic growth, but the Sundqvist Report (OECD, 1990) was probably the first OECD economic policy report which emphasized so strongly the importance of technological change as a wider process of social change. In the words of the Report:

> Technology can be defined as a social process which by meeting real or imagined needs changes those needs just as it is changed by them. Society is shaped by technical change, and technical change is shaped by society. Technical innovation – sometimes impelled by scientific discovery,[3] at other times induced by demand

– stems from within the economic and social system and is not merely an adjustment to transformations brought about by causes outside that system.

(OECD, 1990, p. 117)

In other words, technological change, if it is to have beneficial effects on society, will need to be 'embedded', integrated in society.

From such a perspective technological change is of course much less an exogenous 'manna from heaven' factor, superimposed from the outside through the activities of scientists and technologists, but rather an endogenous process whereby it will be continuously adapted and selected to the broad needs and requirements of society. An interesting illustrative example can be found in the lack of development and growth which took place in most East European countries over the last twenty years, despite massive investments in science and technology and higher education (Gomulka, 1990) in the 1970s and 1980s. While the lack of adequate market interaction pushed the science and technology system into isolation, the 'market' failure of the science and technology system in East European countries is clearly only one facet of this isolation. One of the greatest paradoxes of so-called socialist countries past development has been the lack of social integration of technical change (lack of safety standards, neglect of environmental pollution, higher health risks, lack of ergonomic considerations, etc.). Thus, in contrast to the so-called capitalist societies, science and technology became far more imposed on society and workers in particular, with the resulting lack of efficiency and quality improvements at the shop floor. Technological disenfranchisement accompanied political disenfranchisement.

While the political debate about the environment and other technology policy issues has certainly been much wider and thorough-going in political democracies than it was in the former socialist countries, it is still in many respects a debate among experts. In capitalist economies, the endogeneity of technology arises of course at all levels of technological development. At the level of technology 'creation', technological innovation is not only impelled by scientific discovery, but is also induced by demand. The development of a potential economic idea into new products and processes requires many stages of experimentation in which market possibilities interact with the original idea. The process of invention and innovation interacts in many loops before attempts to market the product or process are made. There are further interactions and feedback loops between the initial successful marketing of the product and its wider diffusion nationally and globally. The acceptability of a product or process will of course also be conditioned by social attitudes, standards, cultural norms and politics. Thus, in broader terms, technological change stems from within the economic and social system and is not merely an adjustment to transformations brought about by causes outside that system. Societies have, in other words, a say in the shape technology is likely to take. Hence, the importance of technology assessment for the policy choices which need to be made.

This rather obvious point seems to have been least recognized in the purpose and aims of most technology policies as they have been developed in most Western economies over the last decades.

19.3   ENDOGENOUS TECHNOLOGY POLICY: HOW
TO ASSESS TECHNOLOGY CONSTRUCTIVELY

The previous policy discussion has focused primarily on the economic context of technology. However, it is clear that the economic feasibility of a new process or product is only one, although often decisive, part of the 'social' integration of technology. Other contexts, social, ethical and socio-political also play important roles. Once the argument is accepted that the creation of technological capabilities does involve a complex, endogenous process of change, negotiated and mediated by society at large, it is obvious that policies in the area of science and technology cannot, nor should be limited to the economic integration of technological change, but must include all aspects of the broader social integration of such change. It is in its broader interaction with society that technological change is adapted and selected and the further realization of technological progress enhanced. From this perspective, technological renewal is by definition a broad concept encompassing research and development, diffusion, imitation and modification of new technologies, as well as the associated social and organizational changes and innovations.

It is within this broad conceptual framework that (constructive) technology assessment emerged as an institution and instrument of government policy aimed at improving on the one hand the social integration of technology and on the other the use and further development of technology itself (for the history and meaning of the term 'constructive' (see Rip *et al.*, 1995)) Technology assessment in most of its forms understandably developed separately from economics and in particular economic policy. The recognition of the importance of both the economic and broader social context conditions opens up the possibility of a far more coherent and complete conceptual framework for the development of technology policy, which in essence is and should be nothing other than continuous constructive technology assessment. Both the old (and what could be called static) TA issues have their place here, as well as the more dynamic (constructive) TA concerns. Already in the 1960s some types of cost–benefit analysis attempted to cover wider issues which were usually neglected in conventional investment appraisal (see, for example, the survey by Prest and Turvey (1966) for the American Economic Association and the Royal Economic Society).

Static TA issues have been discussed quite extensively for many years now and could be said to fall in economic language under the heading of the unequal distribution of positive and negative externalities of technological change. As emphasized, for example, by authors such as Harvey Brooks (OECD, 1971b), there is an apparent paradox in the distributional impact of technology. The costs or risks of a new technology frequently fall on a limited group of the population, whereas the benefits are widely diffused, often so much so that the benefits to any restricted group are barely perceptible even though the aggregate benefit to a large population amounts to considerably more than the total cost to the limited adversely affected group. Examples abound. Automation, for example, benefited consumers of a product by lowering its relative price, but the costs in worker displacement were borne by a small number of people and could be traumatic. A large electrical generating station may adversely affect the local environment, while providing widely diffused benefits to the population served by the electricity produced. Workers in an unusually dangerous occupation, such as mining, carry a disproportionate share of the costs associated with the resultant materials, which may bring wide benefits throughout a national economy.

This disproportion between costs and benefits can also work the other way as in many cases of environmental pollution and emissions. The effluents from a concentrated industrial region such as the Ruhr Valley or the American Great Lakes industrial complex may diffuse acid sulphates over a very large area which derives little benefit from the industrial activity, but may have its quality of life as well as agricultural productivity seriously degraded thereby.

The issue of sharing costs and benefits of technical change shows how important it is, both from the national and international point of view, to draw up some 'rules of the game' to ensure that adverse effects are less harmful than they would be if everything was left to free competition and, second, to establish such rules fairly early on, before vested interests acquire privilege and the fierceness of competition jeopardizes their compulsory application. The word 'static' is of course inappropriate to describe all such distributional issues. With the increase in the complexity of technology, possible risks not only threaten larger areas, they also might have impacts for long periods lasting several generations.

Such traditional TA or cost–benefit analysis should therefore be complemented with a more dynamic approach, whereby the broad social integration of technology as well as the adaptation of technology to society's needs are the central issues at stake. The issue is thus clearly more than one of dynamic externalities. However, from a practical economic policy perspective it seems useful, at least in the first instance, to elaborate further upon the externality terminology, rather than the constructive active TA one. One of the reasons for this has precisely to do with the dynamics of externalities, which themselves will not be susceptible to definitive categorization once and for all and are more intimately related to particular historical and institutional contexts. As Nelson and Winter put it:

> It could be said that technical change is continually tossing up new 'externalities' that must be dealt with in some manner or other. In a regime in which technical advance is occurring and organisational structure is evolving in response to changing patterns of demand and supply, new nonmarket interactions that are not contained adequately by prevailing laws and policies are almost certain to appear, and old ones may disappear. Long-lasting chemical insecticides were not a problem eighty years ago. Horse manure polluted the cities but automotive emissions did not. The canonical 'externality' problem of evolutionary theory is the generation by new technologies of benefits and costs that old institutional structures ignore.
>
> (Nelson and Winter, 1982, p. 120).

From this perspective the concept of a 'constructive' or 'socially optimal' way of assessing long-term impacts of technological change loses much of its meaning. Conflicts and uncertainty are inescapable. Occupying a more central place in technology policy analysis is now the notion that society ought to be engaging in experimentation and that the information and feedback from that experimentation will be of central concern in guiding the present evolution of the economic and technological system. As in the case of 'market failure', the complexity and subtlety of the dynamic interaction between technology and society suggest that simple normative rules will not be very helpful in the design of 'constructive' technology (assessment) policies. From this perspective, as Nelson and Winter (1982) suggested with respect to the market failure argument, such TA 'policies should focus on problems of dealing with and adjusting to change. This involves in the first instance abandonment of the traditional normative goal of trying to define an *optimum* and the institutional structure that will achieve it, and an acceptance of the more modest objectives of identifying problems and possible improvements' (Nelson and Winter, 1982, p. 394). This might seem a modest goal but it is a realistic one in terms of human history and the account of technical innovation and its uncertainties which has been given in this book. Some of the problems are both pervasive and complex,[4] as in the case of sustainable development or the reduction of inequality. They will be enough to keep us busy in the next century and, as Camus reminded us in his beautiful description of the labours of Sisyphus, 'Sisyphus was smiling.'

NOTES

1. Economists have of course always insisted that GNP was never a measure of 'happiness'. See, for example, Tibor Scitovsky's fascinating *The Joyless Economy* (1977) on the difference between material progress and happiness.
2. See, for example, Nakamura (1995); US Senate Committee (1995); Soete (1996).
3. For the socio-political element in the debate on choice of technology and in shaping technology, see especially Stirling (1994) and Molina (1995, 1996).
4. For a thorough discussion of the epistemological problems of complexity in economic theory and models, see Louçã (1997).

PART FOUR REVIEW ARTICLES, LITERATURE SURVEYS AND KEY REFERENCES

Ausubel, J.H. and Sladovic, H. (eds.) (1989) *Technology and the Environment*, Washington, National Academy Press.

Barras, R. (1990), 'Interactive innovation in financial and business services: the vanguard of the service revolution', *Research Policy*, vol. 19, no. 3, pp. 215–39.

Bell, M. and Pavitt, K. (1996) 'Development of Technological Capabilities', in Haque, I. (ed.), *Trade, Technology and International Competitiveness*, Washington, World Bank.

Bessant, J. and Dodgson, M. (1996) *Effective Innovation Policy*, London, Routledge.

Bijker, W.E. and Law, J. (eds) (1992) *Shaping Technology, Building Society, Studies in Socio-Technical Change*, Cambridge, MA, MIT Press.

Castells, M. (1996) *The Rise of the Network Society*, Oxford, Blackwell.

Clark, N., Perez-Trejo, F. and Allen, P. (1995) *Evolutionary Dynamics and Sustainable Development: A Systems Approach*, Aldershot, Elgar.

Daedalus (1996) 'The liberation of the environment', Special Issue, Summer.

De Witt, G.R. (1990) *A Review of the Literature on Technical Change and Employment*, European Commission, DGs V and XIII, Brussels.

Dutton, W.H. (ed.) (1996) *Information and Communication Technologies*, Oxford, Oxford University Press.

European Commission, High Level Expert Group on the Information Society (1996, 1997) *Interim Report* (1996), *Final Report* (1997), Brussels, EU.

Foray, D. and Grübler, A. (eds) (1996) *Technological Forecasting and Social Change*, special issue, Technology and the Environment, Amsterdam, North Holland.

Fransman, M. (1990) *The Market and Beyond: Cooperation and Competition in IT in the Japanese System*, Cambridge, Cambridge University Press.

Freeman, C. and Soete, L. (1994) *Work for All or Mass Unemployment: Computerised Technical Change into the 21st Century*, London, Pinter.

Gummett, P. (1992) 'Science and technology policy', in Hawkesworth, M. and Kogan, M. (eds) *Encyclopaedia of Government and Politics*, vol. 2, London, Routledge.

Howes, R., Skea, J.F. and Whelan, R. (1997) *Clean and Competitive? Motivating Environmental Performance in Industry*, London, Earthscan.

Kemp, R. (1995) *Environmental Policy and Technical Change: A Comparison of the Technical Impact of Policy Instruments*, Maastricht, University of Limburg.

Mansell, R. (1990) 'Rethinking the telecommunications infrastructure: the new "black box"', *Research Policy*, vol. 19, no. 6, pp. 501–17.

Mansell, R.M. (1995) *The New Telecommunications: A Political Economy of Network Evolution*, London, Sage.

Meadows, D.H., Meadows, D.L. and Randers, J. (1992) *Beyond the Limits: Global Collapse or a Sustainable Future?*, London, Earthscan.

Metcalfe, J.S. (1994) 'Evolutionary economics and technology policy', *Economic Journal*, vol. 104, pp. 931–41.

Metcalfe, J.S. (1995) 'The economic foundations of technology policy, equilibrium and evolutionary perspectives', in Stoneman, P. (ed.) *Handbook of the Economics of Innovation and Technological Change*, Oxford, Blackwell.

Meyer, W.B. (1996) *Human Impact on the Earth*, Cambridge, Cambridge University Press.

Miles, I. (1989) *Home Informatics: Information Technology and the Transformation of Everyday Life*, London, Pinter.

Mitchell, C. (1995) 'The renewables NFFO', *Energy Policy*, vol. 28, pp. 1077–91.

Mitchell, C. (1996) *Renewable Energy in the UK*, London, Council for the Preservation of Rural England.

Mowery, D. (1995) 'The practice of technology policy', in Stoneman, P. (ed.) *Handbook of the Economics of Innovation and Technological Change*, Oxford, Blackwell.

Pavitt, K. (1996) 'National policies for technical change: where are the increasing returns to economic research?', *Proceedings of the National Academy of Science, USA*, vol. 93, pp. 12693–700.

Pavitt, K. and Walker, W. (1976) 'Government policies towards industrial innovation', *Research Policy*, vol. 5, pp. 1–90.

Rip, A., Misa, T.J. and Schot, J. (eds) (1995) *Managing Technology in Society: The Approach of Constructive Technology Assessment*, London, Pinter.

Rothwell, R. and Zegveld, W. (1982) *Industrial Innovation and Public Policy*, London, Pinter.

Rush, H. and Bessant, J. (1996) 'Building bridges for innovation: the role of consultants in technology transfer', *Research Policy*, vol. 24, pp. 97–114.

Salomon, J-J. (1985) *Le Gauloise, Le Cowboy et Le Samurai*, Paris, CNAM.

Soete, L.L.G. (1991) *Synthesis Report, TEP, Technology in a Changing World*, Paris, OECD.

Stoneman, P. (ed.) (1995) *Handbook of the Economics of Innovation and Technological Change*, Oxford, Blackwell.

Teubal, M. (1987) *Innovation Performance, Learning and Government Policy, Selected Essays*, Madison, University of Wisconsin Press.

Tisdell, C. (1981) *Science and Technology Policy: Priorities of Governments*, London, Chapman & Hall.

Vivarelli, M. (1995) *The Economics of Technology and Employment: Theory and Empirical Evidence*, Aldershot, Elgar.

Williams, R. and Edge, D. (1996) 'The social shaping of technology', *Research Policy*, vol. 25, no. 6, pp. 865–901.

Ziman, J. (1994) *Prometheus Bound: Science in a Dynamic Steady State*, Cambridge, Cambridge University Press.

REFERENCES

Abernathy, W.J. and Utterback, J.M. (1978) 'Patterns of industrial innovation', *Technology Review*, 80, 2–9.

Abramovitz, M.A. (1979) 'Rapid growth potential and its realisation: the experience of capitalist economies in the postwar period, in Malinvaud, E. (ed.) *Economic Growth and Resources*, vol. 1, *The Major Issues*, Proceedings of the Fifth World Congress of the IEA, London, Macmillan, pp. 1–51.

Abramovitz, M.A. (1986) 'Catching up, forging ahead and falling behind', *Journal of Economic History*, vol. 46, pp. 385–406.

Abramovitz, M.A. and David, P.A. (1994) *Convergence and Deferred Catch-up: Productivity Leadership and the Waning of American Exceptionalism*, CEPR Publication no. 401, Stanford, Stanford University Press.

Achilladelis, B.G. (1973) 'Process innovation in the chemical industry', DPhil thesis, University of Sussex.

Achilladelis, B.G., Schwarzkopf, A. and Lines, M. (1987) 'A study of innovation in the pesticide industry', *Research Policy*, vol. 16, no. 2, pp. 175–212.

Achilladelis, B.G., Schwarzkopf, A. and Lines, M. (1990) 'The dynamics of technological innovation: the case of the chemical industry', *Research Policy*, vol. 19, no. 1, pp. 1–35.

Acs, Z.J. and Andretsch, D.B. (1988) 'Innovation in large and small firms: an empirical analysis', *American Economic Review*, vol. 78, no. 4, September.

Afuah, A. (1997) *Innovation Management: Strategies, Implementation and Profits*, Ann Arbor, University of Michigan Press.

Aghion, P. and Howitt, P. (1990) *A Model of Growth Through Creative Destruction*, NBER Working Paper no. 3223, Cambridge, MA.

Aghion, P. and Howitt, P. (1994) 'Endogenous Technical Change: the Schumpeterian Perspective', *Economic Growth and the Structure of Long-Term Development*, New York, St. Martin's Press.

Aghion, P. and Howitt, P. (1992) 'Model of growth through creative destruction', *Econometrica*, vol. 60, pp. 323–51.

Allen, D.H. (1968) 'Credibility forecasts and their application to the economic assessment of novel R and D projects', *Operations Research Quarterly*, p. 25.

Allen, D.H. (1972) 'Credibility and the assessment of R and D projects', *Long Range Planning*, vol. 5, no. 2, pp. 53–64.

Allen, G.C. (1981) 'Industrial policy and innovation in Japan', in Carter, C. (ed.) *Industrial Policy and Innovation*, London, Heinemann, pp. 68–87.

Allen, J.A. (1967) *Studies in Innovation in the Steel and Chemical Industries*, Manchester, Manchester University Press.

Allen, J.M. and Norris, K.P. (1970) 'Project estimates and outcomes in electricity generation research', *Journal of Management Studies*, vol. 7, no. 3, pp. 271–87.

Altshuler, A., Anderson, M., Jones, D.T., Roos, D. and Womack, J. (1985) *The Future of the Automobile*, London, Allen & Unwin.

Amable, B. (1993) 'National effects of learning, international specialization and growth paths', in Foray, D. and Freeman, C. (eds) *Technology and the Wealth of Nations*, London, Pinter.

Amable, B. (1994) 'Endogenous growth theory, convergence and divergence', in Silverberg, G. and Soete, L.L.G. (eds) *The Economics of Growth and Technical Change*, Aldershot, Elgar.

Amable, B. and Verspagen, B. (1995) 'The role of technology in market shares dynamics', *Applied Economics*, vol. 27, pp. 197–204.

Amann, R., Berry, M. and Davies, R.W. (1979) *Industrial Innovation in the Soviet Union*, Newhaven, Yale University Press.

Amendola, G., Dosi, G. and Papagni, E. (1993) 'The dynamics of international competitiveness', *Weltwirtschaftliches Archiv*, vol. 129, pp. 451–71.

Ames, E. (1961) 'Research, invention, development and innovation', *American Economic Review*, vol. 51, no. 3, pp. 370–81.

Ames, E. and Rosenberg, N. (1963) 'Changing technological leadership and industrial growth', *Economic Journal*, vol. 73, pp. 13–31.

Andress, J.F. (1954) 'The learning curve as a production tool', *Harvard Business Review*, January.

Andrews, K. and Poole, J.B. (1972) *Government of Science in Britain*, London, Weidenfeld & Nicolson.

Angerer, E. (1992) 'The role of electronics in environmental technology and its impacts on the environment, paper presented to the international conference Eco World '92, Washington, DC, 14–17 June.

Antonelli, C. (1994) 'Technological districts, localized spillovers and productivity growth: the Italian evidence on technological externalities in the core regions', *International Review of Applied Economics*, vol. 8, no. 1, pp. 18–30.

Antonelli, C. (1995) *The Economics of Localised Technological Change and Industrial Dynamics*, Dordrecht, Kluwer.

Aoki, M. (1986) 'Horizontal versus vertical information: structure of the firm', *American Economic Review*, vol. 76, no. 5, pp. 971–83.

Appleyard, M.M., Hatch, N.W. and Mowery, D.C. (1996) *Managing the Development and Transfer of Process Technologies in the Semi-conductor Manufacturing Industry*, Laxenburg, IIASA.

Arcangeli, F. (1993) 'Local and global features of the learning process', in Humbert, M. (ed.) *The Impact of Globalisation on Europe's Firms and Industries*, London, Pinter.

Archibugi, D. and Pinta, M. (1992) *The Technological Specialisation of Advanced Countries*, Report to the EC on International Science and Technology Activities, Dordrecht, Kluwer.

Archibugi, D., Cesaretto, S. and Sirilli, G. (1987) 'Innovative activity, R&D and patenting: the evidence of the survey on innovation diffusion in Italy', *STI Review*, no. 2, pp. 135–50.

Archibugi, D. and Michie, J. (eds) (1997) *Technology, Globalisation and Economic Performance*, Cambridge, Cambridge University Press.

Ariffin, N. and Bell, M. (1997) 'Patterns of subsidiary–parent linkages and technology capability-building in TNC subsidiaries: the electronics industry in Malaysia', in Jama, K.S. and Felker, G. (eds) *Malaysia's Industrial Technology Development, Political Economy, Policies and Institutions*, Oxford, Oxford University Press.

Armytage, W.H.G. (1965) *The Rise of the Technocrats*, London, Routledge & Kegan Paul.

Arrow, K.J. (1962a) 'Economic welfare and the allocation of resources for invention', in National Bureau of Economic Research, *The Rate and Direction of Inventive Activity*, Princeton, Princeton University Press.

Arrow, K.J. (1962b) 'The economic implications of learning by doing', *Review of Economic Studies*, no. 29, pp. 155–73.

Arthur, W.B. (1988a) 'Competing technologies: an overview', in Dosi, G., Freeman, C., Nelson, R., Silverberg, G. and Soete, L. (eds) *Technical Change and Economic Theory*, London, Pinter.

Arthur, W.B. (1988b) 'Self-reinforcing mechanisms in economics', in Anderson, P.W., Arrow, K.J. and Pines, D. (eds) *The Economy as an Evolving Complex System*, Reading, MA, Addison-Wesley.

Arthur, W.B. (1989) 'Competing technologies, increasing returns and lock-in by historical events', *Economic Journal*, vol. 99, no. 1, pp. 116–31, March.

Arundel, A. (1995) *PACE Report*, Maastricht, University of Limburg.

Ashton, T.S. (1948) *The Industrial Revolution, 1760–1830*, Oxford, Oxford University Press.

Ashton, T.S. (1963) 'The industrial revolution in Great Britain', in Supple, B. (ed.) *The Experience of Economic Growth*, New York, Random House.

Augsdorfer, P. (1994) 'Taxonomy of management attitudes towards bootlegging and uncertainty', in Oakey, R. (ed.) *New Technology-based Firms in the 1990s*, London, Chapman & Hall.

Augsdorfer, P. (1996) *Forbidden Fruit: An Analysis of Bootlegging, Uncertainty and Learning in Corporate R&D*, Aldershot, Avebury.

Autio, E. (1994) 'New technology-based firms as agents of R&D and innovation: an empirical study', *Technovation*, vol. 14, no. 4, pp. 259–73.

Ayres, R.U. (1988) *Technological Transformation and Long Waves*, Laxenburg, IIASA.

Ayres, R.U. (1991) *Computer-integrated Manufacturing*, vol. 1, London, Chapman & Hall.

Baba, Y. (1985) 'Japanese colour TV firms: decision-making from the 1950s to the 1980s', DPhil dissertation, Brighton, University of Sussex.

Babbage, C. (1834) *On the Economy of Machines and Manufactures*, London, [publisher not known].

Bacon, F. (1605) *The Advancement of Learning*.

Bacon, F. (1627) *The New Atlantis*.

Baines, E. (1835) *History of the Cotton Manufacture*, London, quoted in Rostow, W. (1963) and von Tunzelmann (1995).

Bairoch, P. (1981) 'The main trends in national economic disparities since the industrial revolution', in Bairoch, P. and Levy-Loboyen, M. *Disparities in Economic Development since the Industrial Revolution*, London, Macmillan.

Baker, R. (1976) *New and Improved: Inventors and Inventions that have Changed the Modern World*, London, British Museum Publications.

Baker, N.R. and Pound, W.H. (1964) 'R and D project selection: where we stand', *IEEE Trans Engin Manag*, vol. Em-11, no. 4, p. 124.

Ball, N.R. (1987) *Professional Engineering in Canada, 1857–1897*, Ottawa, National Museum of Canada.

Barker, B. (1990) 'Engineering ceramics and high technology superconductivity: two case studies in the innovation and diffusion of new materials', PhD thesis, University of Manchester.

Barker, G.R. and Davies, R.W. (1965) 'The research and development effort of the Soviet Union', in Freeman, C. and Young, A. *The Research and Development Effort in Western Europe, North America and the Soviet Union*, Paris, OECD.

Barna, T. (1962) *Investment and Growth Policies in British Industrial Firms*, NIESR Occasional Paper XX, Cambridge University Press.

Barnett, C. (1988) *The Audit of War*, Cambridge, Cambridge University Press.

Barras, R. (1990), 'Interactive innovation in financial and business services: the vanguard of the service revolution', *Research Policy*, vol. 19, no. 3, pp. 215–39.

Barro, R. (1991) 'Economic Growth in a Cross Section of Countries', *Quarterly Journal of Economics*, vol. 106, no. 2, pp 407–43.

BASF (1965) *Im Reiche der Chemie*, Düsseldorf, Econ-Verlag.

Baumler, E. (1968) *A Century of Chemistry*, Düsseldorf, Econ-Verlag.

Baumol, W.J. (1986) 'Productivity growth, convergence and welfare: what the long run data show', *American Economic Review*, vol. 76, pp. 1072–85.

Beattie, C.J. and Reader, R.D. (1971) *Quantitative Management in R and D*, London, Chapman & Hall.

Beer, J.J. (1959) *The Emergence of the German Dye Industry*, Chicago University of Illinois Press.

Bejar, L.E.M. (1994) 'Essays on the study of technological change and international trade', PhD thesis, University of Manchester.

Belden, T.G. and Belden, M.R. (1962) *The Lengthening Shadow: The Life of Thomas J Watson*, Boston, Little, Brown.

Bell, M. (1984) 'Learning and accumulation of industrial and technological capability in developing countries', in King, K. and Fransman, M. (eds) *Technological Capacity in the Third World*, London, Macmillan.

Bell, M. (1991) *Science and Technology Policy Research in the 1990s: Key Issues for Developing Countries*, Brighton, SPRU.

Bell, M. and Pavitt, K. (1996) 'Development of Technological Capabilities', in Haque, I. (ed.), *Trade, Technology and International Competitiveness*, Washington, World Bank.

Beloff, N. (1966) 'The learning curve: some controversial issues', *Journal of International Economics*, June.

Belussi, F. (1993) 'Industrial innovation and firm development in Italy: the Benetton case', DPhil thesis, University of Sussex.

Benham, F. (1938) *Economics*, London, Pitman.

Berman, E., Bound, J. and Griliches, Z. (1994) 'Changes in the demand for skilled labour within

US manufacturing: evidence from the annual survey of manufactures', *Quarterly Journal of Economics*, vol. 109, no. 2, pp. 367–97.

Bernal, J.D. (1939) *The Social Function of Science*, London: Routledge & Kegan Paul.

Bernal, J.D. (1953) *Science and Industry in the Nineteenth Century*, London, Allen & Unwin.

Bernal, J.D. (1958) *World Without War*, Routledge & Kegan Paul.

Bessant, J. and Dodgson, M. (1996) *Effective Innovation Policy*, London, Routledge.

Bhagwati, J.N. (1970) 'Comment', in Vernon, R. (ed.) *The Technology Factor in International Trade*, New York, NBER/Columbia University Press.

Bijker, W.E. and Law, J. (eds) (1992) *Shaping Technology, Building Society, Studies in Socio-Technical Change*, Cambridge, MA, MIT Press.

Boyer, R. (1988) 'Technical change and the theory of "regulation"', in Dosi, G., Freeman, C., Nelson, R.R., Silverberg, G. and Soete, L.L.G. (eds) *Technical Change and Economic Theory*, London, Pinter.

Boyer, R. (1997) *Les Systèmes d'Innovation à l'Ère de la Globalisation*, Paris, Economica.

Brander, J.A. and Spencer, B. (1983) 'International R&D rivalry and industrial strategy', *Journal of International Economics*, vol. 14, pp. 225–35.

Brander, J.A. and Spencer, B. (1985) 'Export subsidies and international market share rivalry', *Journal of International Economics*, vol. 17, pp. 83–100.

Braun, E. and MacDonald, S. (1978) *Revolution in Miniature: The History and Impact of Semi-Conductor Electronics*, Cambridge, Cambridge University Press.

Briggs, A. (1961) *The History of Broadcasting in the UK*, Oxford, Oxford University Press.

Bright, J.F. (ed.) (1968) *Technological Forecasting for Industry and Government*, New York, Prentice Hall.

Brock, G.W. (1975) *The US Computer Industry – A Study of Market Power*, Cambridge, MA, Harvard University Press.

Bruland, K. (1989) *British Technology and European Industrialisation: The Norwegian Textile Industry in the Mid-nineteenth Century*, Cambridge, Cambridge University Press.

Buchanan, J.M. and Yoon, J. (eds) (1994) *The Return of Increasing Returns*, Ann Arbor, University of Michigan Press.

Burn, D.L. (1967) *The Political Economy of Nuclear Energy*, London, Institute of Economic Affairs.

Burns, T. and Stalker, G. (1961) *The Management of Innovation*, London, Tavistock.

Bush, V. (1946) *Science: The Endless Frontier*, republished by United States National Science Foundation (NSF), Washington (1960).

Callon, M. (1993) 'Variety and irreversibility in networks of technique conception and adoption', Chapter 11 in Foray, D. and Freeman, C. (eds) *Technology and the Wealth of Nations*, London, Pinter.

Cantwell, J. (1989) *Technological Innovation and Multinational Corporations*, Oxford, Blackwell.

Cantwell, J. (1995) 'The globalisation of technology: what remains of the product cycle model?', *Cambridge Journal of Economics*, vol. 19, pp. 155–74.

Cappeleu, A. and Fagerberg, J. (1996) *East Asian Growth: A Critical Assessment*, Oslo, NUPI.

Carlsson, B. and Jacobsson, S. (1993) 'Technological systems and economic performance: the diffusion of factory automation in Sweden', Chapter 4 in Foray, D. and Freeman, C. (eds) *Technology and the Wealth of Nations*, London, Pinter.

Carroll, G.R. and Teece, D.J. (eds)(1996) *Industrial and Corporate Change*, Special Issue on Firms, Markets and Organisations, vol. 5, no. 2, pp. 203–645.

Carter, C.F. and Williams, B.R. (1957) *Industry and Technical Progress*, Oxford, Oxford University Press.

Carter, C.F. and Williams, B.R. (1959a) 'The characteristics of technically progressive firms', *Journal of Industrial Economics*, vol. 7, no. 2, pp. 87–104.

Carter, C.F. and Williams, B.R. (1959b) *Science in Industry*, Oxford, Oxford University Press.

Carter, C.F. and Williams, B.R. (1964) 'Government scientific policy and the growth of the British economy', *Minerva*, vol. 3, no. 1, pp. 114–25.

Casson, M. (1991) *Global Research Strategy and International Competitiveness*, Oxford, Blackwell.

Castells, M. (1996) *The Rise of the Network Society*, Oxford, Blackwell.

CEC (Annual, 1990s) *Eurostat Energy Statistics*, Eurostat, Luxembourg.

Centre for the Study of Industrial Innovation (1971) *On the Shelf: A Survey of Industrial R and D Projects Abandoned for Non-technical Reasons*, London, CSII.

Chadwick, R. (1958) 'New extraction processes for metals', in *Oxford History of Technology*. volume 5, Oxford, Clarendon Press.

Chambers, J.D. (1961) *The Workshop of the World: British Economic History from 1820 to 1880*, Oxford, Oxford University Press.

Chandler, A.D. (1977) *The Visible Hand: The Managerial Revolution in American Business*, Cambridge, MA, Belknap Press.

Chandler, A.D. (1992), 'What is a firm?: a historical perspective', *European Economic Review*, vol. 36, pp. 483–494.

Chenery, H. (1959) 'The interdependence of investment decisions', in M. Abramovitzm (ed.) *Allocation of Economic Resources*, Stanford, Stanford University Press, pp. 82–120.

Chesnais, F. (1988) 'Multi-national enterprises and the international diffusion of technology', Chapter 23 in Dosi, G. *et al.* (eds) *Technical Change and Economic Theory*, London, Pinter.

Chesnais, F. (1992) 'National systems of innovation, foreign direct investment and the operations of multinational enterprises', in Lundvall, B.A. (ed.) *National Systems of Innovation*, London, Pinter.

Christensen, (1995) in Stoneman, P. (ed.) *Handbook of the Economics of Innovation and Technological Change*, Oxford, Blackwell.

Church, R.A. and Wrigley, E.D. (eds) (1994) *The Industrial Revolution* (11 volumes, but see especially volumes, 2, 3, 8, 9), The Economic History Society, Oxford, Blackwell.

Cimoli, M. and Soete, L. (1992) 'A generalized technology gap trade model', *Economie Appliquée*, no. 3, pp. 33–54.

Clark, N., Perez-Trejo, F. and Allen, P. (1995) *Evolutionary Dynamics and Sustainable Development: A Systems Approach*, Aldershot, Elgar.

Clark, J.A., Freeman, C. and Soete, L.L.G. (1981) 'Long waves, inventions and innovations', *Futures*, vol. 13, no. 4, pp. 308–22.

Clark, R. (1979) *The Japanese Company*, Newhaven, Yale University Press.

Cohen, W.M. (1995) 'Empirical studies of innovative activity', in Stoneman, P. (ed.) *Handbook of the Economics of Innovation and Technological Change*, Oxford, Blackwell.

Cohen, W.M. and Levin, R.C. (1987) 'Empirical studies of innovation and market structure', in Schmalensee, R. and Willig, R. (eds) *Handbook of Industrial Organisation*, Amsterdam, North Holland.

Cole, H.S.D. *et al.* (1973) *Thinking about the Future*, London, Chatto & Windus.

Collingridge, D. (1981) *The Social Control of Technology*, Milton Keynes, Open University Press.

Collingridge, D. (1990) 'Technology organizations and incrementalism: the space shuttle', *Technology Analysis and Strategic Management*, vol. 2, pp. 181–200.

Collingridge, D. and James, P. (1989) 'Technology, organizations and incrementalism: high rise system building in the UK', *Technology Analysis and Strategic Management*, vol. 1, pp. 79–97.

Cook, P.L. and Sharp, M. (1991) 'The chemical industry', in Freeman, C., Sharp, M. and Walker, W. (eds) *Technology and the Future of Europe*, London, Pinter.

Coombs, R., Saviotti, P. and Walsh, V. (1987) *Economics and Technological Change*, London, Macmillan.

Coombs, R., Saviotti, P. and Walsh, V. (eds) (1992) *Technological Change and Company Strategies*, London, Academic Press.

Coombs, R., Richards, A., Saviotti, P. and Walsh, V. (eds) (1996) *Technological Collaboration: The Dynamics of Industrial Innovation*, Cheltenham, Elgar.

Cooray, D.V.B.N. (1980) 'The technological factor and its relevance to the competition between natural and synthetic rubber in international trade', DPhil thesis, University of Sussex.

Cornwall, J. (1977) *Modern Capitalism Its Growth and Transformation*, London, Martin Robertson.

Corry, A.K. (1990) 'Engineering methods of manufacture and production', in McNeil, I. (ed.) *An Encyclopaedia of the History of Technology*, London, Routledge.

Crafts, N. (1994) 'British economic growth, 1700–1831: a review of the evidence', in Hoppit, J. and Wrigley, E.A. (eds) *The Industrial Revolution in Britain*, vol. 2, Oxford, Blackwell.

Cramer, J., Schot, J., van den Akker, F. and Maas Geesteranus, G. (1990) 'Stimulating cleaner technologies through economic instruments; possibilities and constraints', *UNEP Industry and Environment Review*, vol. 13, no. 2, pp. 46–53.

Cramer, J. and Zegveld, W.C.L. (1991) 'The future role of technology in environmental management', *Futures*, vol. 20, pp. 451–68.

Cringeley, R. (1994) *Accidental Empires: How the Boys of Silicon Valley Make their Millions, Battle Foreign Competition and Still Cannot Get a Date*, London, Penguin.

Daedalus (1996) 'The liberation of the environment', Special Issue, Summer.

David, P. (1997) 'Review of Kealey, T. (1996) The Economic Laws of Scientific Research', *Research Policy*, vol. 26, no. 2.

David, P.A. (1984) *The Economies of Locational Tournaments, Center for Economic Policy Research, Technological Innovation Project*, Working Paper, Stanford University.

David, P.A. (1985) 'Clio and the economics of QWERTY', *American Economic Review*, vol. 75, no. 2, May, pp. 332–7. (An extended version is published in Parker, W.N. (ed.) (1986) *Economic History and the Modern Economist*, Oxford: Blackwell.)

David, P.A. (1991) *The Dynamo and the Computer: An Historical Perspective on the Modern Productivity Paradox*, Paris, OECD.

David, P.A. (1993) 'Path-dependence and predictability in dynamic systems with local network externalities: a paradigm for historical economics', Chapter 10 in Foray, D. and Freeman, C. (eds) *Technology and the Wealth of Nations*, London, Pinter.

David, P.A. and Foray, D. (1995) 'Accessing and expanding the science and technology knowledge base, STI outlook', *STI Review*, 16. Paris, OECD.

Davies, A. (1996) 'Innovation in large technical systems: the case of telecommunications', *Industrial and Corporate Change*, vol. 5, no. 4.

Davies, G.B. (1972) 'Contingency planning: the neglected end of the planning cycle', *Process Engineering*, December, pp. 93–7.

De Long, J. and Summers, L. (1991) 'Equipment Investment and Economic Growth', *Quarterly Journal of Economics*, vol. 106, pp. 445–502.

Dean, B.V. (1968) *Evaluating, Selecting and Controlling R and D Projects*, New York, American Management Association, p. 49.

Deane, P. (1965) *The First Industrial Revolution*, Cambridge, Cambridge University Press.

De Bell, J.M. (1946) *German Plastics Practice*, Huddersfield, Springfield.

DeBresson, C. (1987) *Understanding Technological Change*. Montreal, Black Rose Books.

DeBresson, C. (1991) 'Technological innovation and long wave theory: two pieces of the puzzle, *Journal of Evolutionary Economics*, vol. 1, no. 4, pp. 241–72.

DeBresson, C. (1996) *Economic Interdependence and Innovation Activity*, Cheltenham, Elgar.

DeBresson, C. and Amesse, F. (1991) 'Networks of innovators: a review and introduction to the issue', *Research Policy*, vol. 20, no. 5, pp. 363–79.

Dedijer, S. (1964) 'International comparisons of science', *New Scientist*, vol. 21, no. 379, pp. 461–4.

de la Mothe, J. and Paquet, G. (eds) (1996) *Evolutionary Economics and the New International Political Economy*, London, Cassell.

Delbeke, J. (1982) *The Mechanisation of Flemish Industry, 1812–1930: The Case of Antwerp*, Louvain, Centre for Economic Studies, Catholic University of Louvain.

Delorme, J. (1962) *Anthologie des Brevets sur les Matières Plastiques*, vols 1–3, Amphora.

Dertouzos, M., Lester, R. and Solow, R. (eds) (1989) *Made in America*, Report of the MIT Commission on Industrial Productivity, Cambridge, MA, MIT Press.

De Tocqueville, A. (1836) *Democracy in America*.

Devine, W. (1983) 'From shafts to wires: historical perspective on electrification', *Journal of Economic History*, vol. 43, no. 2.

De Witt, G.R. (1990) *A Review of the Literature on Technological Change and Employment*, Brussels, European Commission, DGs V and XIII.

Diebold, J. (1952) *Automation: the Advent of the Automatic Factory*, New York, Van Nostrand.

Dieleman, H. and de Hoo, S. (1993) 'PRISMA: the development of a preventative, multi media strategy for government and industry, in Fischer, K. and Schot, J. (eds) *Environmental Strategies for Industry: International Perspectives on Research Needs and Policy Implications*, Washington, DC, Island Press, pp. 245–75.

Dixit, A. (1986) 'Trade policy: an agenda for research', in Krugman, P. (ed.) *Strategic Trade Policy and the New International Economics*, Brighton, Wheatsheaf.

Dodgson, M. (1991a) *The Management of Technological Learning: Lessons from a Biotechnology Company*, Berlin, De Gruyter.

Dodgson, M. (1991b) 'Technological learning, technology strategy and competitive pressures', *British Journal of Management*, vol. 2, no. 3.

Dodgson, M. and Rothwell, R. (eds) (1994) *The Handbook of Industrial Innovation*, Aldershot, Elgar.

Dore, R. (1987) *Taking Japan Seriously: A Confucian Perspective on Leading Economic Issues*, London, Athlone Press.

Dory, J.P. and Lord, R.J. (1970) 'Does TF really work?', *Harvard Business Review*, vol. 48, no. 6, November–December, pp. 16–28.

Dosi, G. (1981) *Technical Change, Industrial Transformation and Public Policies: The Case of the Semi-conductor Industry*, Brighton, Sussex European Research Centre, University of Sussex.

Dosi, G. (1982) 'Technical paradigms and technological trajectories – a suggested interpretation of the determinants and directions of technical change', *Research Policy*, vol. 11, no. 3, pp. 147–62.

Dosi, G. (1984) *Technical Change and Industrial Transformation*, London, Macmillan.

Dosi, G. (1988) 'Sources, procedures and micro-economic effects of innovation', *Journal of Economic Literature*, vol. 36, pp. 1126–71.

Dosi, G., Freeman, C., Nelson, R., Silverberg, G. and Soete, L. (eds) (1988) *Technical Change and Economic Theory*, London, Pinter.

Dosi, G., Freeman, C., Fabiani, S. and Aversi, R. (1992) 'The diversity of development patterns: on the processes of catching up, forging ahead and falling behind', paper presented at the International Economics Association, Varenna, Italy, October.

Dosi, G., Pavitt, K. and Soete, L. (1990) *The Economics of Technical Change and International Trade*, Brighton, Wheatsheaf.

Downie, J. (1958) *The Competitive Process*, London, Duckworth.

Dowrick, S. (1997) 'Trade and Growth: a survey', in Fagerberg, J., Hansson, P., Lundberg, L. and Melchior, A. (eds), *Technology and International Trade*, Aldershot, Edward Elgar, pp. 197–26.

Dowrick, S. and Nguygen, D.T. (1989) 'OECD comparative economic growth 1950–1985: catch-up and convergence', *American Economic Review*, vol. 79, pp. 1010–30.

Drucker, P. (1946) *The Concept of the Corporation*, New York, John Day.

Dunning, J. (1988) *Exploring International Production*, London, Unwin Hyman.

Dunning, J. (1989) 'The theory of international production', in Fatemi, K. (ed.) *International Trade*, New York, Taylor & Francis.

Dunsheath, P. (1962) *A History of Electrical Engineering*, London, Faber.

Durlauf, S. and Johnson, P.A. (1992) 'Local versus global convergence across national economies', mimeo, Stanford University.

Dutton, W.H. (ed.) (1996) *Information and Communication Technologies: Visions and Realities*, Oxford, Oxford University Press.

Duysters, G. (1995) *The Evolution of Complex Industrial Systems: The Dynamics of Major IT Sectors*, Maastricht, UPM.

Duysters, G. and Hagedoorn, J. (1993) *An Analysis of Economic and Technological Differences in the International Information Technology Industry*, ENCIP Working Paper.

Dyer, J.H. (1996) 'How Chrysler created an American Keiretsu', *Harvard Business Review*, July–August, pp. 42–61.

Eads, G. and Nelson, R.R. (1971) 'Government support of advanced civilian technology', *Public Policy*, vol. 19, no. 3, pp. 405–27.

Economist (1996) 'The mystery of economic growth', 25 May.

Edqvist, C. (1997) *Systems of Innovation*, London, Pinter.

Edqvist, C. and Lundvall, B.-Å. (1993) Chapter on Denmark and Sweden in Nelson, R. (ed.) *National Innovation Systems*, Oxford, Oxford University Press, pp. 265–99.

Ellis, C.H. (1958) 'The development of railway engineering', in *Oxford History of Technology*, vol. 5, Oxford, Clarendon Press, pp. 322–50.

Encel, S. *et al.* (1975) *The Art of Anticipation: Values and Methods in Forecasting*, London, Martin Robertson.

Encyclopaedia Britannica, 9th edn. vol. 2, p. 543.

Enos, J.L. (1962a) *Petroleum Progress and Profits: A History of Process Innovation*, Cambridge, MA, MIT Press.

Enos, J.L. (1962b) 'Invention and innovation in the petroleum refining industry', in National

Bureau of Economic Research, *The Rate and Direction of Inventive Activity*, Princeton, Princeton University Press.

Ergas, H. (1987) 'Does technology policy matter?', in Guile, B. and Brooks, H. (eds) *Technology and Global Competition*, Washington, National Academy Press.

Ergas, H. (1992) A future for mission-oriented industrial policies? A critical review of developments in Europe, mimeo, Paris, OECD.

European Commission, High Level Expert Group on the Information Society, *Interim Report* (1996); *Final Report* (1997) Brussels, EU.

European Commission (1994) *European Science and Technology Indicators*, Luxembourg, EC.

European Union (1996) *Building the Information Society for Us All*, Interim Report, January.

Eversley, D.E.C. (1994) 'The home market and economic growth in England, 1750–1780', in Hoppit, J. and Wrigley, E.A. (eds) *The Industrial Revolution in Britain*, Oxford, Blackwell.

Ezrahi, Y., Mendelsohn, E. and Segal, H.P. (1995) *Technology, Pessimism and Post-Modernism*, Amherst, University of Massachusetts Press.

Fabian, Y. (1963) *Measurement of Output of R and D*, Paris, OECD, DAS/PD/6348.

Fabre, J. (1983) 'Le pouvoir structurant de l'electricité, *Bulletin Histoire de L'Electricité*, June, pp. 23–36.

Fagerberg, J. (1987) 'A technology gap approach to why growth rates differ', *Research Policy*, vol. 16, no. 2, pp. 87–99.

Fagerberg, J. (1992) 'The home market hypothesis re-examined: the impact of domestic–user-producer interaction in exports', in Lundvall, B.-Å (ed.) *National Systems of Innovation*, London, Pinter.

Fagerberg, J. (1994) 'Technology and international differences in growth rates', *Journal of Economic Literature*, vol. 32, pp. 1147–75.

Fagerberg, J. (1995) *Convergence or Divergence: The Impact of Technology*, Working Paper no. 524, Oslo, NUPI.

Fagerberg, J., Verspagen, B. and von Tunzelmann, N. (eds) (1994) *The Dynamics of Technology, Trade and Growth*, Aldershot, Elgar.

Farrell, G. (1954) 'The demand for motor-cars in the United States', *Journal of the Royal Statistical Society*, Series A, vol. 117, pp. 171–90.

Faulkner, W. (1986) 'Linkage between academic and industrial research: the case of biotechnological research in the pharmaceutical industry', DPhil thesis, Brighton, University of Sussex.

Faulkner, W., Senker, J.M. and Velho, L. (1995) *Knowledge Frontiers: Public Sector Research and Industrial Innovation*, Oxford, Oxford University Press.

Federation of British Industries (1947) *Scientific and Technical Research in British Industry*, London, FBI.

Federation of British Industries (1961) *Industrial Research in Manufacturing Industry*, London, FBI.

Ferraz, J.C., Rush, H. and Miles, I. (1992) *Development, Technology and Flexibility*, London, Routledge.

Fisher, F.M., Griliches, Z. and Kaysen, C. (1962) 'The costs of automobile model changes since 1949', *Journal of Political Economy*, vol. 70, no. 5, pp. 433–51.

Fleck, J. (1988) *Innofusion or Diffusation?*, ESRC, PICT Working Paper Series, University of Edinburgh.

Fleck, J. (1993) 'Configurations: crystallising contingency', *International Journal of Human Factors in Manufacturing*, vol. 3, no. 1, pp. 15–36.

Fleck, J. and White, B. (1987) 'National policies and patterns of robot diffusion: UK, Japan, Sweden and United States', *Robotics*, vol. 3, no. 1, pp. 7–23.

Floud, R.C. and McCloskey, D. (eds) (1981) *Economic History of Britain since 1700*, Cambridge, Cambridge University Press.

Foray, D. (1991) 'The secrets of industry are in the air: industrial cooperation and the organisational dynamics of the innovative firm', *Research Policy*, vol. 20, no. 5, pp. 393–407.

Foray, D. and Freeman, C. (eds) (1993) *Technology and the Wealth of Nations*, London, Pinter.

Foray, D. and Grübler, A. (eds) (1996) *Technological Forecasting and Social Change*, special issue, Technology and the Environment, Amsterdam, North Holland.

Foray, D. and Lundvall, B-Å. (1996) *Employment and Growth in the Knowledge-based Economy*, Paris, OECD.

Ford, G. and Lake, G. (1991) 'Evolution of European science and technology policy', *Science and Public Policy*, vol. 18, pp. 38–50.

Forester, T. (ed.) (1988) *The Materials Revolution: Superconductors, New Materials and the Japanese Challenge*, Oxford, Oxford University Press.

Fransman, M. (1990) *The Market and Beyond: Cooperation and Competition in IT in the Japanese System*, Cambridge, Cambridge University Press.

Freeman, C. (1967) 'Science and economy at the national level', in *Problems in Science Policy*, Paris, OECD.

Freeman, C. (1971) *The Role of Small Firms in Innovation in the United Kingdom since 1945: Report to the Bolton Committee of Inquiry on Small Firms*, Research Report no. 6, London, HMSO.

Freeman, C. (1982) *The Economics of Industrial Innovation*, 2nd edn, London, Pinter.

Freeman, C. (1987) *Technology Policy and Economic Performance: Lessons from Japan*, London, Pinter.

Freeman, C. (1989) 'The Third Kondratieff Wave: age of steel, electrification and imperialism, in Bohlin *et al.* (eds) *Samhällsvetenskap, ekonomi och historia*, Göterborg, Daidalos.

Freeman, C. (ed.) (1990) *The Economics of Innovation*, International Library of Critical Writings in Economics, vol. 2, Aldershot, Elgar.

Freeman, C. (1991) 'Networks of innovators, a synthesis of research issues', *Research Policy*, vol. 20, no. 5, pp. 499–514.

Freeman, C. (1992) 'A green techno-economic paradigm for the world economy', in Freeman, C. *The Economics of Hope*, London, Pinter, pp. 190–211.

Freeman, C. (1994) 'The economics of technical change: a critical survey', *Cambridge Journal of Economics*, vol. 18, pp. 1–50.

Freeman, C. (ed.) (1996) *The Long Wave in the World Economy*, International Library of Critical Writings in Economics, Aldershot, Elgar.

Freeman, C. (1995) *History, Co-evolution and Economic Growth*, IIASA Working Paper 95–76, Laxenburg, IIASA.

Freeman, C., Clark, J. and Soete, L. (1982) *Unemployment and Technical Innovation: A Study of Long Waves and Economic Development*, London, Pinter.

Freeman, C., Curnow, R.C., Fuller, J.K., Robertson, A.B. and Whittaker, P.J. (1968) 'Chemical process plant: innovation and the world market', *Nat Inst. Econ Rev*, no. 45.

Freeman, C., Fuller, J.K. and Young, A. (1963) 'The plastics industry: a comparative study of research and innovation', *National Institute Economic Review*, vol. 26, pp. 22–62.

Freeman, C. and Hagedoorn, J. (1992) *Globalization of Technology*, MERIT Research Memorandum 92-013, Maastricht.

Freeman, C. and Perez, C. (1988) 'Structural crises of adjustment: business cycles and investment behaviour', in Dosi, G., Freeman, C., Nelson, R., Silverberg, G. and Soete, L. (eds) *Technical Change and Economic Theory*, London, Pinter, pp. 38–66.

Freeman, C. and Soete, L. (1985) *Information Technology and Employment: An Assessment*, Brussels, IBM.

Freeman, C. and Soete, L. (1987) *Technical Change and Full Employment*, London, Pinter.

Freeman, C. and Soete, L. (eds) (1992) *New Explorations in the Economics of Technical Change*,

Freeman, C. and Soete, L. (1994) *Work for All or Mass Unemployment: Computerised Technical Change into the 21st Century*, London, Pinter.

Freeman, C. and Young, A. (1965) *The Research and Development Effort in Western Europe, North America and the Soviet Union*, Paris, OECD.

Freeman, C., Young, A. and Fuller, J.K. (1963) 'The plastics industry: a comparative study of research and innovation', *Nat Inst Econ Rev*, no. 26, pp. 22–62.

Fukasaku, Y. (1987) 'Technology imports and R&D at Mitsubishi Nagasaki shipyard in the pre-war period', *Bonner Zeitschrift für Japanologie*, pp. 77–90.

Gabor, D. (1964) *Inventing the Future*, Harmondsworth, Penguin.

Galison, P. and Hevly, B. (1992) *Big Science: The Growth of Large-Scale Research*, Stanford, Stanford University Press.

Gartmann, H. (1959) *Sonst Stunde die Welt Still*, Dusseldorf, [publisher not known].

Gates, W. (1996) *The Road Ahead*, London, Penguin.

Gatrell, V.A. (1977) 'Labour, power and the size of firms in Lancashire cotton in the second quarter of the nineteenth century', *Economic History Review*, vol. 30, pp. 95–139.

Gaynor, G.H. (ed.) (1996) *Handbook of Technology Management*, New York, McGraw Hill.

Gazis, D.C. (1979) 'The influence of technology on science: some experience at IBM research', *Research Policy*, vol. 8, pp. 244–59.

Gerschenkron, A. (1962) *Economic Backwardness in Historical Perspective*, Cambridge, MA, Harvard University Press.

Gershuny, J. (1983) *Social Innovation and the Division of Labour*, Oxford, Oxford University Press.

Gershuny, J. and Miles, I. (1983) *The New Service Economy*, London, Pinter.

Gibbons, M. and Johnston, R.D. (1972) *The Interaction of Science and Technology*, mimeo, Department of Liberal Studies in Science, University of Manchester.

Gibbons, M. and Johnston, R.D. (1974) 'The roles of science in technological innovation', *Research Policy*, vol. 3, no. 3, pp. 220–42.

Gibbons, M. and Littler, D. (1979) 'The development of an innovation: the case of Porvair', *Research Policy*, vol. 8, no. 1, pp. 2–25.

Gilfillan, S.C. (1935) *The Sociology of Invention*, Chicago, Follet.

Gille, B. (1978) *Histoire des Techniques*, Paris, Gallimard.

Gjerding, A.N. (1996) *Technical Innovation and Organisational Change*, Aalborg, Aalborg University Press.

Gold, B. (1971) *Explorations in Managerial Economics*, New York, Basic Books.

Gold, B. (1979) *Productivity, Technology and Capital: Economic Analysis, Managerial Strategies and Government Policies*, Lexington, Lexington Books.

Golding, A.M. (1972) 'The semi-conductor industry in Britain and the United States: a case study in innovation, growth and the diffusion of technology', DPhil thesis, University of Sussex.

Gomulka, S. (1990) *The Theory of Technical Change and Economic Growth*, London, Routledge.

Granstrand, O. (ed.) (1994) *Economics of Technology*, Amsterdam, North Holland.

Granstrand, O., Hakanson, L. and Sjølander, S. (1992) *Technology Management and International Business*, Chichester, Wiley.

Granstrand, O. and Sjølander, S. (1992) 'Managing innovation in multi-technology corporations', *Research Policy*, vol. 19, no. 1, pp. 35–61.

Graves, A. (1991) 'International competitiveness and technological development in the world automobile industry', DPhil thesis, Brighton, University of Sussex.

Greenhalgh, C. (1990) 'Innovation and trade performance in the United Kingdom', *Economic Journal*, vol. 100, pp. 105–18.

Gregory, G. (1986) *Japanese Electronics Technology: Enterprise and Innovation*, 2nd edn, Chichester, Wiley.

Griliches, Z. (1980) 'R&D and productivity slowdown', *American Economic Review*, vol. 70, pp. 343–8.

Griliches, Z. (1984) *R&D, Patents and Productivity*, Chicago, University of Chicago Press.

Groenewegen, P. and Vergragt, P. (1991) 'Environmental issues as threats and opportunities for technological innovation', *Technology Analysis and Strategic Management*, vol. 3, no. 1, pp. 43–55.

Grossman, G. and Helpman, E. (1990) 'Comparative advantage and long run growth', *American Economic Review*, vol. 80, pp. 796–815.

Grossman, G. and Helpman, E. (1991) *Endogenous Growth Theory*, Cambridge, MA, MIT Press.

Grosvenor, W.M. (1929) 'The seeds of progress', *Chemical Markets*.

Grupp, H. (1997), *Messung und Erklärung des technischen Wandels: Grundzüge einer empirischen Innovationsökonomik*, Berlin, Springerverlag.

Grupp, H. and Hofmeyer, O. (1986) 'A technometric model for the assessment of technological standards and their application to selected technology comparisons', *Technological Forecasting and Social Change*, no. 30, pp. 123–37.

Grupp, H. and Soete, L. (1993) *Analysis of the Dynamic Relationship between Technical and Economic Performances in Information and Telecommunication Sectors*, Final Report in Fulfilment of Project no. 8862 for DG XIII of the European Commission.

Guardian (1992) 9 November.

Gummett, P.J. (1980) *Scientists in Whitehall*, Manchester, Manchester University Press.

Gummett, P. (1992) 'Science and technology policy', in Hawkesworth, M. and Kogan, M. (eds) *Encyclopaedia of Government and Politics*, vol. 2, London, Routledge.

Habakkuk, J.H. (1962) *American and British Technology in the Nineteenth Century: The Search for Labour-saving Inventions*, Cambridge, Cambridge University Press.

Haber, L.F. (1958) *The Chemical Industry during the Nineteenth Century*, Oxford, Oxford University Press.

Haber, L.F. (1971) *The Chemical Industry, 1900–1930*, Oxford, Oxford University Press.

Hagedoorn, J. (1991) 'Networks in research and production', *International Journal of Technology Management*, special issue on The Increasing Role of Technology in Corporate Policy, pp. 81–95.

Hagedoorn, J. and Schakenraad, J. (1990) 'Strategic partnering and technological cooperation', in Freeman, C. and Soete, L. (eds) *New Explorations in the Economics of Technical Change*, London, Pinter.

Hagedoorn, J. and Schakenraad, J. (1991) 'The internationalization of the economy, global strategies and strategic technology alliances', *Nouvelles de la Science et des Technologies*, vol. 9, no. 2, pp. 29–41.

Hagedoorn, J. and Schakenraad, J. (1992) 'Leading companies and networks of strategic alliances in information technologies', *Research Policy*, vol. 21, no. 2, pp. 163–91.

Hahn, F. (1987) 'Neo-classical growth theory', in Eatwell, J. (ed.) *The New Palgrave Dictionary of Economics*, vol. 3, London, Macmillan.

Håkansson, H. and Snehota, I (eds) (1995) *Developing Relationships in Business Networks*, London, Routledge.

Hall, P. and Preston, P. (1988) *The Carrier Wave: New Information Technology and the Geography of Innovation*, London, Unwin Hyman.

Hamberg, D. (1964) 'Size of firm, oligopoly and research: the evidence', *Canadian Journal of Economics and Political Science*, vol. 30, no. 1, pp. 62–75.

Hamberg, D. (1966) *R and D Essays in the Economics of Research and Development*, New York, Random House.

Hamilton, A. (1791) *Report on the Subject of Manufactures*, reprinted US Government Printing Office, Washington (1913).

Hamilton, S.B. (1958) 'Building materials and techniques', in *Oxford History of Technology*, vol. 5, Oxford, Clarendon Press.

Hannah, L. (1983) 'Entrepreneurs and the social sciences', Inaugural Lecture, London School of Economics.

Hardie, D.W.F. and Pratt, J.D. (1966) *A History of the Modern British Chemical Industry*, Oxford, Pergamon.

Harris, R. (1984) 'Applied General Equilibrium Analysis of Small Open Economies with Scale Economies and Imperfect Competition', *American Economic Review*, vol. 74, no. 5, pp. 1016–32.

Harrod, R.F. (1939) 'An essay in dynamic theory', *Economic Journal*, vol. 49, pp. 14–33.

Harrod, R. (1948) *Towards a Dynamic Economics*, London, Macmillan.

Hart, A. (1966) 'A chart for evaluating product R and D projects', *Operations Research Quarterly*, vol. 17, no. 4, pp. 347–58.

Hawke, C.R. and Higgins, J.P.P. (1981) 'Transport and social overhead capital', in Floud, R. and McCloskey, B. (eds) *The Economic History of Britain Since 1700*, Cambridge, Cambridge University Press.

Heath, J.B. (1971) *International Conference on Monopolies, Mergers and Restrictive Practices*, London, HMSO.

Hessen, B. (1931) 'The social and economic roots of Newton's Principia', in Bukharin, N. (ed.) (1971) *Science at the Cross-roads*, with an introduction by P.G. Werskey, Ilford, Cass.

Hills, R.L. (1994) 'Hargreaves, Arkwright and Crompton: why three inventors?', in Jenkins, D.T. (ed.) *The Textile Industries*, Oxford, Blackwell.

Hirsch, S. (1965) 'The United States electronics industry in international trade', *Nat Inst Econ Rev*, no. 34, pp. 92–107.

Hobday, M. (1991) 'The European semi-conductor industry: resurgence and rationalisation', in Freeman, C., Sharp, M. and Walker, W. (eds) *Technology and the Future of Europe*, London, Pinter.

Hobday, M. (1994) 'Innovation in semi-conductor technology: the limits of the Silicon Valley network model', in Dodgson, M. and Rothwell, R. (eds) *The Handbook of Industrial Innovation*, Cheltenham, Elgar.

Hobday, M. (1995) *Innovation in East Asia: The Challenge to Japan*, Aldershot, Elgar.

Hobday, M. (in press, 1997) 'Product complexity, innovation and industrial organisation', *Research Policy*.

Hobsbawm, E. (1968) *Industry and Empire: An Economic History of Britain from 1750*, London, Weidenfeld & Nicolson.

Hodgson, G.M. (1992) 'Optimisation and evolution: Winter's critique of Friedman revisited', Newcastle Polytechnic, Department of Economics.

Hodgson, G.M. (ed.) (1995) *Economics and Biology*, The International Library of Critical Writings in Economics, vol. 50, Aldershot, Elgar.

Hoffmann, W.D. (1976) 'Market structure and strategies of R and D behaviour in the data processing market', *Research Policy*, vol. 5, pp. 334–53.

Hollander, S. (1965) *The Sources of Increased Efficiency: A Study of Du Pont Rayon Plants*, Cambridge, MA, MIT Press.

Hollingdale, S.H. and Toothill, G.C. (1965) *Electronic Computers*, Harmondsworth, Penguin.

Holroyd, Sir R. (1964) 'Productivity of industrial research with particular reference to research in the chemical industry', Institute of Chemical Engineers, *Proceedings of Symposium on Productivity in Research*, p. 6.

Hounshell, D.A. (1984) *From the American System to Mass Production, 1800–1932: The Development of Manufacturing Technology in the United States*, Baltimore, Johns Hopkins University Press.

Hounshell, D.A. and Smith, J.K. (1988) *Science and Corporate Strategy: DuPont R&D 1902–1980*, Cambridge, Cambridge University Press.

Howes, R., Skea, J.F. and Whelan, R. (1997) *Clean and Competitive? Motivating Environmental Performance in Industry*, London, Earthscan.

Hu, Y.S. (1992) 'Global or transnational corporations are national firms with international operations', *Californian Management Review*, vol. 34, no. 2, pp. 107–26.

Hufbauer, G.C. (1966) *Synthetic Materials and the Theory of International Trade*, London, Duckworth.

Hufbauer, G. (1970) 'The impact of national characteristics and technology on the commodity composition of trade in manufactured goods', in Vernon, R. (ed.) *The Technology Factor in International Trade*, New York, Columbia University Press.

Hughes, K. (1990) *Exports and Technology*, Cambridge, Cambridge University Press.

Hughes, T.P. (1982) *Networks of Power: Electrification in Western Society, 1800–1930*, Baltimore, Johns Hopkins University Press.

Hughes, T.P. (1989) *American Genesis*, New York, Viking.

Hulsink, W. (1996) *Do Nations Matter in a Globalising Industry? A Comparative Analysis of Telecommunication Re-structuring in France, Netherlands and UK (1980–1994)*, Rotterdam, Erasmus University.

Humbert, M. (ed.) (1993) *The Impact of Globalisation on Europe's Firms and Industries*, London, Pinter.

International Development Research Centre (1972) *Annual Report 1971–72*, Ottawa, IDRC.

Industrial Research Institute Research Corporation (1979) *Contribution of Basic Research to Recent Successful Industrial Innovations*, prepared for National Science Foundation, St Louis, IRI/RC (PB 80–160179).

Integrated Circuit Engineering Corporation (1995) *Worldwide IC Industry Economic Update and Forecast*.

Irvine, J.H. and Martin, B.R. (1980) 'A methodology for assessing the scientific performance of research groups', *Scientia Yugoslavica*, vol. 6, nos. 1–4, pp. 83–95.

Irvine, J.H. and Martin, B.R. (1981) 'L'évaluation de la recherche fondamentale: est-elle possible?', *La Recherche*, no. 128, pp. 1406–16.

Irvine, J.H. and Martin, B.R. (1984) *Foresight in Science: Picking the Winners*, London, Pinter.

Irwin, A. and Vergragt, P. (1989) 'Re-thinking the relationship between environmental regulation and industrial innovation: the social negotiation of technical change', *Technology Analysis and Strategic Management*, vol. 1, no. 1, pp. 57–70.

Jacob, M. (1988) *The Cultural Meaning of the Scientific Revolution*, New York, McGraw Hill.

Jaikumar, R. (1988) *From Filing and Fitting to Flexible Manufacturing: A Study in the Evolution of Process Control*, Cambridge, MA, Harvard Business School.

Jang-Sup Shin (1996) *The Economics of Late-comers, Catching-up, Technology Transfer and Institutions in Germany, Japan and S. Korea*, London, Routledge.

Jansen, J.L.A. and Vergragt, P. (1992) *Sustainable Development: A Challenge to Technology!*, The Hague, Ministry of Housing, Physical Planning and the Environment, Department for Information and International Relations.

Jantsch, E. (1967) *Technological Forecasting in Perspective*, Paris, OECD.

Jenkins, D.T. (ed.) (1994) *The Textile Industries*, vol. 9, in Church, R.A. and Wrigley, E.A. (eds) *The Industrial Revolution in Britain*, Oxford, Blackwell.

Jewkes, J., Sawers, D. and Stillerman, R. (1958) *The Sources of Invention*, (rev. edn. 1969), London, Macmillan.

Jochem, E. and Hohmeyer, O. (1992) 'The economics of near-term reductions in greenhouse gases', in Mintzler, I. (ed.) *Confronting Climate Change Risks, Implications and Responses*, Cambridge, Cambridge University Press.

Johnson, H.G. (1975) *Technology and Economic Interdependence*, London, Macmillan.

Jones, D.T. (1985) 'Vehicles', in Freeman, C. (ed.) *Technological Trends and Employment*, vol. 4, *Engineering and Vehicles*, Aldershot, Gower.

Jones, R.V. (1978) *Most Secret War: British Scientific Intelligence, 1939–1945*, London, Hamish Hamilton.

Jones, R. (ed.) (1981) *Readings from 'Futures'*, Guildford, Westbury House.

Kaldor, N. (1957) 'A model of economic growth', in *Economic Journal* (page references are to the reprint in Kaldor, 1980).

Kaldor, N. (1980) *Essays on Economic Stability and Growth*, 2nd edn, London, Duckworth.

Kamien, M.I. and Schwartz, N.L. (1975) 'Market Structure and Innovation: a survey', *Journal of Economic Literature*, vol. 23, no. 1, pp. 1–37.

Kanz, J. and Lam, D. (1996) 'Technology, strategy and competitiveness', in Gaynor, G.H. (ed.) *Handbook of Technology Management*, New York, McGraw Hill.

Kaplinsky, R. (1983) 'Firm size and technical change in a dynamic context', *Journal of Industrial Economics*, no. 32, pp. 39–59.

Kaplinsky, R. (1990) *The Economics of Small: Appropriate Technology in a Changing World*, London, IT Publications.

Katz, B.G. and Phillips, A. (1982) 'Government, economies of scale and comparative advantage: the case of the computer industry', in Giersch, H. (ed.) *Proceedings of Conference on Emerging Technology*, Kiel Institute of World Economics, Tubingen, J C B Mohr.

Katz, J.M. (ed.) (1987) *Technology Generation in Latin American Manufacturing Industries*, London, Macmillan.

Kaufman, M. (1963) *First Century of Plastics*, London, Plastics Institute.

Kaufman, M. (1969) *The History of PVC: The Chemistry and Industrial Production of Polyvinyl Chloride*, London, Maclaren.

Kay, N.M. (1979) *The Innovating Firm: A Behavioural Theory of Corporate R&D*, London, Macmillan.

Kay, M.N. (1982) *The Evolving Firm*, London, Macmillan.

Kay, M.N. (1984) *The Emergent Firm: Knowledge, Ignorance and Surprise in Economic Organisation*, London, Macmillan.

Kealey, T. (1996) *Economic Laws of Scientific Research*, London, Macmillan.

Keck, O. (1977) 'Fast breeder reactor development in West Germany: an analysis of government policy', DPhil thesis, University of Sussex.

Keck, O. (1980) 'Government policy and technical choice in the West German reactor programme', *Research Policy*, vol. 9, no. 4, pp. 302–56.

Keck, O. (1982) *Policy-making in a Nuclear Reactor Programme: The Case of the West German Fast Breeder Reactor*, Lexington, Lexington Books.

Kemp, R. (1995) *Environmental Policy and Technical Change: A Comparison of The Technical Impact of Policy Instruments*, Maastricht, University of Limburg.

Kemp, R. and Soete, L. (1990) 'Inside the "green box": on the economics of technological change and the environment', in Freeman, C. and Soete, L. (eds) *New Explorations in the Economics of Technological Change*, London, Pinter, pp. 245–57.

Kennedy, C. and Thirlwall, A.P. (1971) 'Technical progress', *Surveys in Applied Economics 1*, London, Macmillan, pp. 115–77.

Keynes, J.M. (1930) *Treatise on Money*, London, Macmillan.

Keynes, J.M. (1936) *General Theory of Employment, Interest and Money*, New York, Harcourt Brace.

Klein, B.H. (1977) *Dynamic Economics*, Cambridge, MA, Harvard University Press.

Kleinknecht, A. and Reijnen, J.O.N. (1992a) 'Why do firms cooperate on R&D? An empirical study', *Research Policy*, vol. 21, no. 4, pp. 347–60.

Kleinknecht, A. and Reijnen, J.O.N. (1992b) 'The experience with new innovation data in the Netherlands', *STI Review*, no. 11, pp. 64–76.

Kleinman, H.S. (1975) *Indicators of the Output of New Technological Products from Industry*, report to US National Science Foundation, National Technical Information Service, US Department of Commerce.

Knight, F.H. (1965) *Risk, Uncertainty and Profit*, London, Harper.

Knox, F. (1969) *Consumers and the Economy*, London, Harrap.

Kodama, F. (1995) *Emerging Patterns of Innovation: Sources of Japan's Technological Edge*, Cambridge, MA, Harvard Business School Press.

Kondratieff, N. (1925) 'The long wave in economic life', English translation, *Review of Economic Statistics*, vol. 17, pp. 105–15.

Krugman, P. (1979) 'A model of innovation technology transfer and the world distribution of income', *Journal of Political Economy*, vol. 87.

Krugman, P. (1986) *Strategic Trade Policy and the New International Economics*, Cambridge, MA, MIT Press.

Krugman, P. (1990) *Rethinking International Trade*, Cambridge, MA, MIT Press.

Krugman, P. (1995) 'Technological change in international trade', in Stoneman, P. (ed.) *Handbook of the Economics of Innovation and Technological Change*, Oxford, Blackwell.

Kuznets, S. (1940) 'Schumpeter's business cycles', *American Economic Review*, vol. 30, no. 2, pp. 257–71.

Kuznets, S. (1930), *Secular Movements in Production and Prices*, Boston, Houghton Mifflin.

Kuznets, S. (1953) *Economic Change*, New York, Norton.

Lall, S. (1995) 'Employment and foreign investment: policy options for developing countries', *International Labour Review*, vol. 134, pp. 521–39.

Landau, R. and Rosenberg, N. (eds) (1986) *The Positive Sum Strategy: Harnessing Technology for Economic Growth*, Washington, National Academy Press.

Landes, D. (1965) 'Technological change and industrial development in Western Europe, 1750–1914, in *Cambridge Economic History of Europe*, Cambridge, Cambridge University Press.

Landes, M. (1969) *The Unbound Prometheus: Technological and Industrial Development in Western Europe from 1750 to the Present*, Cambridge, Cambridge University Press.

Landesmann, M. and Goodwin, R. (1994) 'Productivity, Growth, Structural Change and Macroeconomic Stability', *Economic Growth and the Structure of Long-Term Development*, New York, St. Martin's Press.

Langrish, J., Gibbons, M., Evans, P. and Jevons, F. (1972) *Wealth from Knowledge*, London, Macmillan.

Lastres, H. (1992) 'Advanced materials and the Japanese national system of innovation', DPhil thesis, Brighton, University of Sussex, SPRU.

Lawson, W.D., Lynch, C.A. and Richards, C.J. (1965) 'Corfam: research brings chemistry to footwear', *Research Management*, vol. 8, no. 1, pp. 5–26.

Lazonick, W. (1990) *Competitive Advantage on the Shop Floor*, Cambridge, MA, Harvard University Press.

Lazonick, W. (1991) *Business Organisation and the Myth of the Market Economy*, Cambridge, Cambridge University Press.

Lenin, V.I. (1915) *Imperialism: The Highest Stage of Capitalism*, English edn, (1935), London, Lawrence & Wishart.

Lenin, V.I. (1917) *State and Revolution*, English edn, London (1933). Martin Lawrence.

Leonard-Barton, D. (1995) *Wellsprings of Knowledge: Building and Sustaining the Sources of Innovation*, Cambridge, MA, Harvard Business School Press.

Leontieff, W. (1953) 'Domestic production and foreign trade: the American capital position re-examined, in *Proceedings of the American Philosophical Society*, vol. 97.

Levin, R.C. (1988) 'Appropriability, R&D spending and technological performance', *American Economic Review (PP)*, no. 78, 424–8.

Levin, R.C., Klevorick, A.K., Nelson, R.R. and Winter, S.G. (1987) 'Appropriating the returns from industrial research and development', *Brookings Paper on Economic Activity*, no. 3, pp. 783–820.

Liebermann, M.G. (1978) 'A literature citation study of science technology coupling in electronics', *Proceedings of the IEEE*, vol. 66, no. 1, pp. 4–13.

List, F. (1841) *The National System of Political Economy*, English edn, London, Longman (1904).

Little, A.D. (1963) *Patterns and Problems of Technical Innovation in American Industry*, Washington, DC, USGPO.

Litvak, I.A. and Maule, C.J. (1972) 'Managing the entrepreneurial enterprise', *Business Quarterly*, vol. 37, p. 47.

Lockett, M. (1987) *The Factors Behind Successful IT Innovation*, Oxford, Templeton College.

Louça, F. (1997) *Turbulence in Economics: An Evolutionary Appraisal of Cycles and Complexity in Historical Processes*, Cheltenham, Elgar.

Loveridge, R. and Pitt, M. (eds) (1990) *The Strategic Management of Technological Innovation*, New York, Wiley.

Lucas, R.E.B. (1988) 'On the mechanisms of economic development', *Journal of Monetary Economics*, vol. 22, pp. 3–42.

Lundgren, A. (1991) 'Technological innovation and industrial evolution: the emergence of industrial networks', DPhil dissertation, Stockholm School of Economics.

Lundvall, B-Å (1985) 'Product innovation and user-producer interaction, *Industrial Development Research Series 31*, Aalborg, Aalborg University Press.

Lundvall, B-Å (1988a) Chapter 1, in Freeman, C. and Lundvall, B-Å (eds) *Small Countries Facing the Technological Revolution*, London, Pinter.

Lundvall, B-Å (1988b) 'Innovation as an interactive process: from user-producer interaction to the national system of innovation', in Dosi, G. *et al.* (eds) *Technical Change and Economic Theory*, London, Pinter.

Lundvall, B-Å (ed.) (1992) *National Systems of Innovation: Towards a Theory of Innovation and Interactive Learning*, London, Pinter.

Lundvall, B-Å (1993) 'User–producer relationships, national systems of innovation and internationalisation', Chapter 12 in Foray, D. and Freeman, C. (eds) *Technology and the Wealth of Nations*, London, Pinter.

Machlup, F. (1962) *The Production and Distribution of Knowledge in the United States*, Princeton, Princeton University Press.

McCutcheon, R. (1972) 'High flats in Britain, 1945–71', D.Phil. dissertation, University of Sussex.

McKay, A.L. and Bernal, J.D. (1966) 'Towards a science of science', *Technologist*, vol. 2, no. 4, pp. 319–28.

McKendrick, N. (1960, 1994) 'Josiah Wedgwood, an eighteenth century entrepreneur in salesmanship and marketing techniques', *Economic History Review*, vol. 12, pp. 408–33, reproduced in Hoppit, J. and Wrigley, E.A. (eds) (1994).

Mackenzie, D. (1990) *Inventing Accuracy: A Historical Sociology of Nuclear Missile Guidance*, Cambridge, MA, MIT Press.

Mackenzie, D. (1992) *Economic and Sociological Exploration of Technical Change*, in Coombs, R., Saviotti, P. and Walsh, V. (eds), pp. 25–48.

Maclaurin, W.R. (1949) *Invention and Innovation in the Radio Industry*, London, Macmillan.

McLeod, R. and Andrews, K. (1970) 'The origins of DSIR: reflections on ideas and men, 1915–1916', *Public Administration*, vol. 48, 23–48.

Maddison, A. (1982) *Phases of Capitalist Growth*, Oxford, Oxford University Press.

Maddison, A. (1987) 'Growth and slowdown in advanced capitalist economies: techniques of quantitative assessment', *Journal of Economic Literature*, vol. 25, pp. 649–98.

Maddison, A. (1995), *Monitoring the World Economy, 1820–1992*, Development Centre, Paris, OECD, p. 452.

Magnier, A. and Toujas-Bernatte, J. (1994) *Technology and Trade: Empirical Evidence for the Five Major Industrialized Countries*, Working Paper INSEE-DEEE G, no. 9207.

Maidique, M.A. and Zirger, B.J. (1985) 'The new product learning cycle', *Research Policy*, vol. 14, December, pp. 299–313.

Maizels, A. (1963) *Industrial Growth and World Trade*, Cambridge, Cambridge University Press and NIESR.

Mankiw, N., Romer, D. and Weil, D. (1992) 'A Contribution to the Empirics of Economic Growth', *Quarterly Journal of Economics*, vol. 107, pp. 407–37.

Mann, J. de L. (1958) 'The textile industry: machinery for cotton, flax, wool, 1760–1850', in Singer, C. (ed.) *et al. A History of Technology*, vol. 4, Oxford, Clarendon Press.

Mansell, R.M. (1995) *The New Telecommunications: A Political Economy of Network Evolution*, London, Sage.

Mansell, R.M. and Silverstone, R. (eds) (1996) *Communication by Design*, Oxford, Oxford University Press.

Mansfield, E. (1962) 'Entry, Gibrat's Law, innovation and the growth of firms', *American Economic Review*, vol. 48, pp. 1023–51.

Mansfield, E. (1968a) *Industrial Research and Technological Innovation*, New York, Norton.

Mansfield, E. (1968b) *The Economics of Technological Change*, New York, Norton.

Mansfield, E. (1980) 'Basic research and productivity increases in manufacturing', *American Economic Review*, vol. 70, pp. 865–73.

Mansfield, E. (1988a) 'Industrial Innovation in Japan and in the United States', *Science*, no. 241, pp. 1760–64.

Mansfield, E. (1988b) 'Industrial R&D in Japan and the United States: a comparative study', *American Economic Review*, vol. 78, pp. 223–8.

Mansfield, E. (1989) 'The diffusion of industrial robots in Japan and in the United States', *Research Policy*, vol. 18, pp. 183–92.

Mansfield, E. (1991) 'Academic research and industrial innovation', *Research Policy*, vol. 20, no. 1, pp. 1–13.

Mansfield, E. (ed.) (1993) *The Economics of Technical Change*, International Library of Critical Writings in Economics, vol. 31, Aldershot, Elgar.

Mansfield, E. (1995) *Innovation, Technology and the Economy, Selected Essays*, 2 vols. Aldershot, Elgar.

Mansfield, E. *et al.* (1971) *Research and Innovation in the Modern Corporation*, New York, Norton.

Mansfield, E. *et al.* (1977) *The Production and Application of New Industrial Technology*, New York, Norton.

Marris, R. (1964) *The Economic Theory of Managerial Capitalism*, London, Macmillan.

Marschak, T., Glennan, T.K. and Summers, R. (1967) *Strategy for R and D*, Vienna, Springer-Verlag.

Marshall, A.W. and Meckling, W.H. (1962) 'Predictability of the costs, time and success of development', in National Bureau of Economic Research, *The Rate and Direction of Inventive Activity*, Princeton, Princeton University Press, pp. 461–77.

Marshall, A. (1890) *Principles of Economics*, London, Macmillan.

Martin, B.R. and Irvine, J.H. (1983) 'Assessing basic research: some partial indicators of scientific progress in radio astronomy', *Research Policy*, vol. 12, no. 2, pp. 61–90.

Martin, B.R. and Irvine, J. (1989) *Research Foresight: Priority-Setting in Science*, London, Pinter.

Martin, P. and Thomas, S. (1996) *The Development of Gene Therapy in Europe and the US: A Comparative Analysis*, STEEP Special Report, no. 5, Brighton, SPRU.

Marx, K. (1858) *Grundrisse*, London, Allen Lane (1973).

Mass, W. and Lazonick, W. (1990) *The British Cotton Industry and International Competitive Advantage: The State of the Debates*, New York, Department of Economics, Columbia University Working paper 90-06, April.

Mathias, P. (1969) *The First Industrial Nation*, London, Methuen.

Meadows, D. *et al.* (1972) *The Limits to Growth*, New York, Universe Books.

Meadows, D.H., Meadows, D.L. and Randers, J. (1992) *Beyond the Limits: Global Collapse or a Sustainable Future?*, London, Earthscan.

Melto, de D.P. *et al.* (1980) *Innovation and Technological Change in Five Canadian Industries*, Discussion Paper no. 176, Ottawa, Economic Council of Canada.

Metcalfe, J.S. (1970) *The Diffusion of Innovation in the Lancashire Textile Industry*, Manchester School of Economics and Social Studies, no. 2, pp. 145–62.

Metcalfe, J.S. (1981) 'Impulse and diffusion in the study of technical change', *Futures*, vol. 13, pp. 347–59.

Metcalfe, J.S. (1982) 'On the diffusion of innovation and the evolution of technology', London, Conference of Technical Change Centre, January.

Metcalfe, J.S. (1988) 'The diffusion of innovation: an interpretative survey', in Dosi, G., Freeman, C., Nelson, R., Silverberg, G. and Soete, L. (eds) *Technical Change and Economic Theory*, London, Pinter.

Metcalfe, J.S. (1994) 'Evolutionary economics and technology policy', *Economic Journal*, vol. 104, pp. 931–41.

Metcalfe, J.S. (1995) 'The economic foundations of technology policy, equilibrium and evolutionary perspectives', in Stoneman, P. (ed.) (1995) *Handbook of the Economics of Innovation and Technological Change*, Oxford, Blackwell.

Metcalfe, J.S. (1997) *Evolutionary Economics and Creative Destruction*, London, Routledge.

Meyer-Krahmer, F. and Reiss, T. (1992) *Ex Ante Evaluation and Technology Assessment: Two Emerging Elements of Technology Policy Evaluation*, Research Evaluation, Karlsruhe, ISI.

Miles, I. (1989) *Home Informatics: Information Technology and the Transformation of Everyday Life*, London, Pinter.

Miles, I. (1996) *Innovation in Services, Services in Innovation*, Manchester, Manchester Statistical Society.

Miles, I. (1996) 'The information society: competing perspectives', in Dutton, W.H. (ed.) *Information and Communication Technology: Visions and Realities*, Oxford, Oxford University Press.

Milliman, S.R. and Prince, R. (1989) 'Firm incentives to promote technological change in pollution control', *Journal of Environmental Economics and Management*, vol. 17, pp. 247–65.

Mishan, E. (1969) *Technology and Growth: The Price we Pay*, New York, Praeger.

Mitchell, C. (1995) 'The renewables NFFO', *Energy Policy*, vol. 28, pp. 1077–91.

Mitchell, C. (1996) *Renewable Energy in the UK*, London, Council for the Preservation of Rural England.

Miyazaki, K. (1995) *Building Competencies in the Firm: Lessons for Japanese and European Opto-electronics*, Basingstoke, Macmillan.

Mjøset, L. (1992) *The Irish Economy in a Comparative International Perspective*, Dublin, National Economic and Social Council.

Modelski, G. and Thompson, R. (1993) *Seapower in Global Politics, 1494–1993*, London, Macmillan.

Mokyr, J. (1990) *The Lever of Riches: Technological Creativity and Economic Progress*, New York, Oxford University Press.

Molina, A.H. (1989) *Transputers and Parallel Computers: Building Technological Competencies Through Socio-technical Constituencies*, PICT Paper 7, Edinburgh, Research Centre for Social Sciences.

Molina, A.H. (1995) 'Socio-technical constituencies as processes of alignment', *Technology in Society*, vol. 17, no. 4, pp. 385–412.

Molina, A.H. (1996) 'Socio-technical alignment in the intra-organisational diffusion of information technology', in Inzelt, A. and Coenen, R. (eds) *Knowledge, Technology Transfer and Foresight*, Amsterdam, Kluwer, pp. 125–48.

Morand, J.C. (1970) *Recherche et dimension des entreprises dans la Communauté* Nancy, Economique Européenne.

Morris, P.J.T. (1982) 'The development of acetylene chemistry and synthetic rubber by IG Farben, 1926–1945', DPhil thesis, Oxford University.

Morris, P.J.T. (1989) *The American Synthetic Rubber Research Programs*, Philadelphia, University of Pennsylvania Press.

Mowery, D.C. (1980) 'The emergence and growth of industrial research in American manufacturing 1899–1946', PhD dissertation, Stanford University.

Mowery, D.C. (1983) 'The relationship between intra-firm and contractual forms of industrial research in American manufacturing, 1900–1940', *Explorations in Economic History*, vol. 20, pp. 351–74.

Mowery, D.C. (1995) 'The practice of technology policy', in Stoneman, P. (ed.) *Handbook of the Economics of Innovation and Technological Change*, Oxford, Blackwell.

Mowery, D.C. and Rosenberg, N. (1979) 'The influence of market demand upon innovation: a critical review of some recent empirical studies', *Research Policy*, vol. 8, no. 2, pp. 102–50.

Mowery, D.C. and Rosenberg, N. (1989) *Technology and the Pursuit of Economic Growth*, Cambridge, Cambridge University Press.

Mueller, W.F. (1962) 'The origins of the basic inventions underlying Du Pont's major product and process innovations, 1920–1950', in National Bureau of Economic Research *The Rate and Direction of Inventive Activity*, Princeton, Princeton University Press.

Murphy, K., Shleifer, A. and Vishny, R. (1989) 'Industrialization and the big push', *Journal of Political Economy*, vol. 97, pp. 1003–26.

Musson, A.E. and Robinson, E. (1969) *Science and Technology in the Industrial Revolution*, Manchester, Manchester University Press.

Nakamura, L. (1995) *Is US Economic Performance really that Bad?*, Federal Reserve Bank of Philadelphia Working Paper no. 95–21.

Narula, R. (1996) *Multinational Investment and Economic Structure*, London, Routledge.

Naslund, B. and Sellstedt, B. (1972) *The Implementation and Use of Models for R and D Planning*, Stockholm, European Institute for Advanced Studies in Management.

Nasthakken, D. and Akhtar, A. (1994) *Does the Highway Go South?*, Ottawa, IDRC.

National Science Foundation (1945) *Science, the Endless Frontier*, a report to the President by Vannevar Bush, reprinted 1960, Washington, NSF.

National Science Foundation (1973) *Interactions of Science and Technology in the Innovative Process*, Final Report from the Battelle Columbus Laboratory, NSF-667, Washington, NSF.

Nelson, D. (1980) *Frederick W. Taylor and the Rise of Scientific Management*, Madison, University of Wisconsin Press.

Nelson, R.R. (1959) 'The simple economics of basic scientific research', *Journal of Political Economy*, vol. 67, pp. 297–306.

Nelson, R.R. (1962) 'The link between science and invention: the case of the transistor', in National Bureau of Economic Research *The Rate and Direction of Inventive Activity*, Princeton, Princeton University Press.

Nelson, R.R. (1973) 'Recent exercises in growth accounting: new understanding or dead end?', *American Economic Review*, vol. 63, pp. 462–8.

Nelson, R.R. (1977) *The Moon and the Ghetto: An Essay on Public Policy Analysis*, New York, Norton.

Nelson, R.R. (1981) 'Research on productivity growth and productivity differences: dead ends and new departures', *Journal of Economic Literature*, vol. 19, pp. 1029–64.

Nelson, R.R. (1991) 'Why do firms differ, and how does it matter?', *Strategic Management Journal*, vol. 12, no. 1.

Nelson, R.R. (ed.) (1993) *National Systems of Innovation: A Comparative Study*, Oxford, Oxford University Press.

Nelson, R.R. (1994) 'What has been the matter with neoclassical growth theory?', in Silverberg, G. and Soete, L.L.G. (eds) *The Economics of Growth and Technical Change*, Aldershot, Elgar.

Nelson, R. (1996) *The Sources of Economic Growth*, Cambridge, MA, Harvard University Press.

Nelson, R.R., Peck, J. and Kalachek, E. (1967) *Technology, Economic Growth and Public Policy*, Washington, DC, Brookings Institution.

Nelson, R.R. and Salomon, J-J (1979) *Science and Technology in the New Economic Context*, Paris, OECD.

Nelson, R.R. and Winter, S.G. (1974) 'Neoclassical versus evolutionary theories of economic growth', *Economic Journal*, no. 84, pp. 886–905.

Nelson, R.R. and Winter S.G. (1977) 'In search of a useful theory of innovation', *Research Policy*, vol. 6, no. 1, pp. 36–76.

Nelson, R.R. and Winter, S.G. (1982) *An Evolutionary Theory of Economic Change*, Cambridge, MA, Harvard University Press.

Nevins, A. and Hill, F.E. (1954) *Ford: The Times, the Man, the Company*, New York, Scribner.

Niosi, J. (ed.) (1991) *Technology and National Competitiveness: Oligopoly, Technological Innovation and International Competition*, Montreal, McGill University Press.

Nonaka, I. and Takeuchi, H. (1986) 'The new product development game', *Harvard Business Review*, January–February, pp. 285–305.

Norris, K.P. (1971) 'The accuracy of project cost and duration estimates in industrial R and D', *R and D Management*, vol. 2, no. 1, pp. 25–36.

Nye, D.E. (1990) *Electrifying America: Social Meanings of a New Technology*, Cambridge, MA, MIT Press.

OECD (1963a) *The Measurement of Scientific and Technical Activities* (Frascati Manual), Paris, OECD (new edn 1993).

OECD (1963b) *Science, Economic Growth and Government Policy*, Paris, OECD.

OECD (1967) *The Overall Level and Structure of R&D Efforts in Member Countries*, Paris, OECD.

OECD (1968) *Gaps in Technology: Electronic Components*, Paris, OECD.

OECD (1971a) *R and D in OECD Member Countries: Trends and Objectives*, Paris, OECD.

OECD (1971b) *Science, Growth and Society (Brooks Report)*, Paris, OECD.

OECD (1980a) *Technical Change and Economic Policy*, Paris, OECD.

OECD (1980b) 'The measurement of the output of R and D activities: the 1980 conference on science and technology indicators', mimeo, Paris, OECD.

OECD (1981) *R and D Statistics*, Paris, OECD.

OECD (1982) *Historical Patent Statistics*, Paris, OECD.

OECD (1988) *New Technologies in the 1990s: a Socio-Economic Strategy (Sundqvist Report)*, Paris, OECD.

OECD (1991) *Technology in a Changing World*, Paris, OECD.

OECD (1991) *Technology and Productivity: The Challenges for Economic Policy*, Paris, OECD.

OECD (1992) *Technology and the Economy: The Key Relationships*, Paris, OECD.

OECD (1993) *The OECD Response*, Interim Report by The Secretary-General, Paris, OECD.

OECD (1994, 1996) *The OECD Jobs Study*, Paris, OECD.

OECD (1995) *MSTI Database*, Paris, OECD.

OECD (1996) *Technology, Productivity and Job Creation*, Vol. I and II, The OECD Job Strategy, Paris, OECD.

Ohmae, K. (1990) *The Borderless World*, New York, Harper.

Olin, J. (1972) *R and D Management Practices: Chemical Industry in Europe*, Zurich, Stanford Research Institute.

Ostry, S. (1995) *Technology Issues in the International Trading System*, October, mimeo, Paris, OECD.

Pareto, V. (1913) 'Alcuni relazioni fra la stato sociale e la variazoni della prosperita economica', *Rivista Italiana di Sociologia*, September–December, pp. 501–48.

Pasinetti, L.L. (1981) *Structural Change and Economic Growth: A Theoretical Essay on the Dynamics of the Wealth of Nations*, Cambridge, Cambridge University Press.

Patel, P. (1995) 'Localised production of technology for global markets', *Cambridge Journal of Economics*, vol. 19, no. 1, pp. 141–53.

Patel, P. and Pavitt, K. (1991) 'Large firms in the production of the world's technology: an important case of "non-globalisation"', *Journal of International Business Studies*, vol. 22, pp. 1–21.

Patel, P. and Pavitt, K. (1992) 'The innovative performance of the world's largest firms: some new evidence, *Economics of Innovation and New Technologies*, vol. 2, pp. 91–102.

Patterson, M. (1996) 'Innovation: managing the process', in Gaynor, G.H. (ed.) *Handbook of Technology Management*, New York, McGraw Hill.

Paulinyi, A. (1982) 'Der technologietransfer für die Metallbearbeitung und die preussische Gewerbeförderung 1820–1850', in Blaich F. (ed.) *Die Rolle des Staates für die wirtschaftliche Entwicklung*, Berlin, Blaich, pp. 99–142.

Pavitt, K.L.R. (1971) *The Conditions for Success in Technological Innovation*, Paris, OECD.

Pavitt, K.L.R. (ed.) (1980) *Technical Innovation and British Economic Performance*, London, Macmillan.

Pavitt, K.L.R. (1982) 'R and D, patenting and innovative activities: a statistical exploration', *Research Policy*, vol. 11, no. 1, pp. 33–51.

Pavitt, K. (1984) 'Patterns of technical change: towards a taxonomy and a theory', *Research Policy*, vol. 13, no. 6, pp. 343–73.

Pavitt, K.L.R. (1985) 'Technology transfer among industrially advanced countries: an overview', in Rosenberg, N. and Frischtak, C. (eds) *International Technology Trends*, New York, Praeger.

Pavitt, K. (1993) 'What do firms learn from basic research?', in Foray, D. and Freeman, C. (eds) *Technology and the Wealth of Nation's*, London, Macmillan.

Pavitt, K. (1996) 'National policies for technical change: where are the increasing returns to economic research?', *Proceedings of the National Academy of Science, USA*, vol. 93, pp. 12693–700.

Pavitt, K. (1996) 'Road to ruin', *New Scientist*, no. 2041, 3 August, pp. 32–6.

Pavitt, K.L.R., Robson, M. and Townsend, J. (1987) 'The size distribution of innovative firms in the UK: 1945–1983', *Journal of Industrial Economics*, vol. 35, no. 3, March, pp. 297–319.

Pavitt, K.L.R. and Soete, L.L.G. (1980) 'Innovative activities and export shares', in Pavitt, K. (ed.) *Technical Innovation and British Economic Performance*, Macmillan, London.

Pavitt, K.L.R. and Soete, L.L.G. (1981) 'International differences in economic growth and the international location of innovation', in Giersch, H. (ed.) *Emerging Technologies: The Consequences for Economic Growth, Structural Change and Employment*, Tübingen, Mohr.

Pavitt, K.L.R. and Walker, W. (1976) 'Government policies towards industrial innovation: a review', *Research Policy*, vol. 5, no. 1, pp. 1–97.

Pearce, D. and Turner, R.K. (1984) 'The economic evaluation of low and non-waste technologies', *Resources and Conservation*, vol. 11, pp. 27–43.

Pearce, D., Markandya, A. and Barbier, E.B. (1989) *Blueprint for a Green Economy*, London, Earthscan.

Pearce, R. and Singh, S. (1991) *Globalising Research and Development*, London, Macmillan.

Pearl, M.L. (1978) 'The iron and steel industry', in Williams, T.L. (ed.) *History of Technology*, vol. VI, pp. 462–99, Oxford, Clarendon Press.

Pearson, A.W. (1990) 'Innovation strategy', *Technovation*, vol. 10, no. 3, pp. 185–92.

Peck, J. (1968) 'British science and technology: the costs of overcommitment and gains from selectivity', in Caves, R. (ed.) *British Economic Prospects*, Washington, DC, Brookings Institution.

Peck, J. and Goto, A. (1981) 'Technology and economic growth, the case of Japan', *Research Policy*, vol. 10, pp. 222–43.

Peck, J. and Scherer, F.M. (1962) *The Weapons Acquisitions Process: An Economic Analysis*, Cambridge, MA, Harvard University Press.

Peck, M.J. and Wilson, R. (1982) 'Innovation, imitation and comparative advantage: the case of the consumer electronics industry', in Giersch, H.

Penrose, E. (1959) *The Theory of the Growth of the Firm*, Oxford, Blackwell.

Perez, C. (1983) 'Structural change and the assimilation of new technologies in the economic and social system', *Futures*, vol. 15, no. 5, pp. 357–75.

Perez, C. (1985) 'Micro-electronics, long waves and world structural change: new perspectives for developing countries, *World Development*, vol. 13, no. 3, pp. 441–63.

Perez, C. (1988) 'New technologies and development', in Freeman C. and Lundvall, B-Å. (eds.), *Small Countries Facing the Technological Revolution*, London, Pinter, pp. 85–97.

Perez, C. (1989) 'Technical change, competitive restructuring and institutional reform in developing countries', *World Bank Strategic Planning and Review*, Discussion Paper 4, Washington, DC, World Bank, December.

Perez, C. and Soete, L. (1988) 'Catching up in technology: entry barriers and windows of opportunity', in Dosi, G. *et al.* (eds) *Technical Change and Economic Theory*, London, Pinter, pp. 458–79.

Petroski, H. (1989) 'H D Thoreau, engineer', *American Heritage of Invention and Technology*, New Haven, Yale University Press, pp. 8–16.

Phelps, F.S. (1966) 'Models of technical progress and the golden rule of research', *Review of Economic Studies*, vol. 33, pp. 133–45.

Phillips, A. (1971) *Technology and Market Structure*, Lexington, Lexington Books.

Piore, M. and Sabel, C. (1984) *The Second Industrial Divide: Possibilities for Prosperity*, New York, Basic Books.

Polanyi, M. (1962) 'The republic of science', *Minerva*, vol. 1, no. 1, pp. 54–72.

Poon, A. (1993) *Tourism, Technology and Competitive Strategies*, Wallingford, CAB International.

Porat, M.U. (1977) *The Information Economy: Definition and Measurement*, vols 1–9, Washington, USGPO.

Porter, M. (1990) *The Competitive Advantage of Nations*, New York, Free Press.

Posner, M. (1961) 'International trade and technical change', *Oxford Economic Papers*, vol. 13, pp. 323–41.

Postan, M.M., Hay, D. and Scott, J.D. (1964) 'Design and development of weapons', in *History of Second World War*, London, HMSO.

Prahalad, C.K. and Hamil, G. (1990) 'The core competence of the corporation', *Harvard Business Review*, May–June, pp. 79–91.

Prais, S.J. (1981) 'Vocational qualifications of the labour force in Britain and Germany', *National Institute Economic Review*, no. 98, pp. 47–59.

Prest, A.R. and Turvey, R. (1996) 'Cost–benefit analysis: a survey', in *Surveys of Economic Theory*, vol. 3, for the American Economic Association and the Royal Economic Society, New York, St Martin's Press.

Price, D. de Solla (1963) *Little Science, Big Science*, New York, Columbia University Press.

Price, D.J. de Solla (1965) 'Is technology historically independent of science?', *Technology and Culture*, vol. 6, no. 4, p. 553.

Price, W.J. and Bass, L.W. (1969) 'Scientific research and the innovative process', *Science*, vol. 164, no. 3881, pp. 802–6.

Pursell, C.W. (ed.) (1991) *Technology in America: A History of Individuals and Ideas*, Cambridge, MA, MIT Press.

Quinn, J.B. (1986) 'Innovation and corporate strategy: managed chaos, in Horwitch, M. (ed.) *Technology in the Modern Corporation: A Strategic Perspective*, Oxford, Pergamon Press.

Quintella, R.H. (1993) *The Strategic Management of Technology*, London, Pinter.

Radio Corporation of America (1963) *Three Historical Views*, New York, RCA.

Radosevic, S. (1996) *Transfer of Technology*, Maastricht, INTECH.

Reddy, A.S.P. and Sigurdson, J. (1994) 'Emerging patterns of globalisation of corporate R&D and scope for innovation capability building in developing countries', *Science and Public Policy*, vol. 21, pp. 283–99.

Reinert, E. (1993) 'Catching up from way behind: a Third World perspective on First World history', in Fagerberg, J., Verspagen, B. and von Tunzelmann, N. (eds) *The Dynamics of Technology, Trade and Growth*, Cheltenham, Elgar.

Reinert, E. (1994) 'Symptoms and causes of poverty: underdevelopment in a Schumpeterian system', *Forum for Development Studies*, nos. 1–2, pp. 71–109.

Reinert, E. (1997) 'The role of the State in economic growth', *Working Paper 1997–5*, Centre for Development and Environment, University of Oslo.

Reinert, E.S. (1980) *International Trade and the Economic Mechanism of Underdevelopment*, New York, Graduate School of Cornell University.

Research Policy (1987) special issue on Output Measurement, vol. 16, nos. 2–4.

Ricardo, D. (1821) *Principles of Political Economy and Taxation*, London.

Rip, A., Misa, T.J. and Schot, J. (eds) (1995) *Managing Technology in Society: The Approach of Constructive Technology Assessment*, London, Pinter.

Roberts, E.B. (1968) 'The myths of research management', *Science and Technology*, no. 80, pp. 40–46.

Roberts, E.B. (1991) 'The technological base of the new enterprise', *Research Policy*, vol. 20, no. 4, pp. 283–98.

Robertson, A.B. and Frost, M. (1978) 'Duopoly in the scientific instrument industry: the milk analyser case', *Research Policy*, vol. 7, no. 3, pp. 292–316.

Romer, P. (1986) 'Increasing returns and long-run growth', *Journal of Political Economy*, vol. 94, no. 5, pp. 1002–37.

Romer, P.M. (1990) 'Endogenous technological change', *Journal of Political Economy*, vol. 98, pp. 71–102.

Rosegger, G. (1996), *The Economics of Production and Innovation: An Industrial Perspective*, Third Edition, Oxford, Buttersworth.

Rosenberg, N. (1969) 'The direction of technological change: inducement mechanisms and focusing devices', *Economic Development and Cultural Change*, vol. 19, pp. 1–24.

Rosenberg, N. (1976) *Perspectives on Technology*, Cambridge, Cambridge University Press.

Rosenberg, N. (1982) *Inside the Black Box: Technology and Economics*, Cambridge, Cambridge University Press.

Rosenberg, N. (1990) 'Why do firms do basic research with their own money?', *Research Policy*, vol. 19, no. 2, pp. 165–75.

Rosenberg, N. (1994) *Exploring the Black Box: Technology, Economics and History*, Cambridge, Cambridge University Press.

Rosenberg, N., Landau, R. and Mowery, D.C. (eds) (1992) *Technology and the Wealth of Nations*, Stanford, Stanford University Press.

Rostow, W.W. (1960) *The Stages of Economic Growth: A Non-Communist Manifesto*, Cambridge, Cambridge University Press.

Rothschild Report (1971) *A Framework for Government Research and Development*, London, HMSO.

Rothwell, R. (1976) *Innovation in Textile Machinery: Some Significant Factors in Success and Failure*, SPRU Occasional Paper no. 2, Brighton, University of Sussex.

Rothwell, R. (1979) *Technical Change and Competitiveness in Agricultural Engineering: The Performance of the UK Industry*, SPRU Occasional Paper no. 9, Brighton, University of Sussex.

Rothwell, R. (1992) 'Successful industrial innovation: critical factors for the 1990s', SPRU 25th Anniversary, Brighton, University of Sussex, reprinted in *R&D Management*, vol. 22, no. 3, pp. 221–39.

Rothwell, R. (1994) 'Industrial innovation: success, strategy, trends', in Dodgson, M. and Rothwell, R. (eds) *Handbook of Industrial Innovation*, Cheltenham, Elgar.

Rothwell, R and Gardiner, P. (1988) 'Re-innovation and robust design: producer and user benefits' *Journal of Marketing Management*, vol. 3, no. 3, pp. 372–87.

Rothwell, R., Freeman, C., Horsley, A., Jervis, V.T.P., Robertson, A.R. and Townsend, J. (eds) (1974) 'SAPPHO updated – project SAPPHO phase 2', *Research Policy*, vol. 3, no. 3, pp. 258–91.

Rothwell, R. and Dodgson, D. (1994) 'Innovation and size of firms', in Dodgson, M. and Rothwell, R. (eds) *The Handbook of Industrial Innovation*, Cheltenham, Elgar.

Rothwell, R. and Zegveld, W. (1982a) *Innovation and the Small and Medium-sized Firm*, London, Pinter.

Rothwell, R. and Zegveld, W. (1982b) *Industrial Innovation and Public Policy*, London, Pinter.

Rubenstein, A. (1966) 'Economic evaluation of R and D: a brief survey of theory and practice', *Journal of Industrial Engineering*, vol. 17, no. 11, pp. 615–20.

Ruigrok, W. and van Tulder, R. (1995) *The Logic of International Re-structuring*, London, Routledge.

Rush, H. and Bessant, J. (1996) 'Building bridges for innovation: the role of consultants in technology transfer', *Research Policy*, vol. 24, pp. 97–114.

Saechtling, H. (1961) *Werkstoffe aus Menschenhand*, Munich.

Sako, M. (1992) *Contracts, Prices and Trust: How the Japanese and British Manage their Sub-contracting Relationships*, Oxford, Oxford University Press.

Sakus, H. (1987) Centenary Address to the Canadian Engineering Institute, Montreal.

Salomon, J-J (1973) *Science and Politics*, London, Macmillan.

Salomon, J-J (1985) *Le Gauloise, Le Cowboy et Le Samurai*, Paris, CNAM.

Samuelson, P.A. (1967) *Economics*, 7th edn, New York, McGraw-Hill.

Saviotti, P. and Metcalfe, J.S. (eds) (1991) *Evolutionary Theories of Economic and Technological Change*, Chur, Harwood.

Saxenian, A. (1991) 'The origins and dynamics of production networks in Silicon Valley', *Research Policy*, vol. 20, no. 5, pp. 423–9.

Scherer, F.M. (1965a) 'Size of firm, oligopoly and research: a comment', *Canadian Journal of Economics and Political Science*, vol. 31, no. 2, pp. 256–66.

Scherer, F.M. (1965b) 'Firm size, market structure, opportunity and the output of patented inventions', *American Economic Review*, pp. 1097–1123.

Scherer, F.M. (1992a) *International High Technology Competition*, Cambridge, MA, Harvard University Press.

Scherer, F.M. (1992b) 'Schumpeter and plausible capitalism', *Journal of Economic Literature*, vol. 30, pp. 1416–33.

Scherer, F.M. (ed.) (1992c) *Monopoly and Competition Policy*, Library of Critical Writings in Economics, Cheltenham, Elgar.

Scherer, F.M. (ed.) (1994) *Monopoly and Competition Policy*, Library of Critical Writings in Economics, Cheltenham, Elgar.

Scherer, F.M. (1997) 'The distribution of profits from invention', final report (mimeo), Cambridge, MA, John F. Kennedy School of Government, Harvard University.

Schmookler, J. (1965) 'Catastrophe and utilitarianism in the development of basic science', in Tybout, R. (ed.) *The Economics of Research and Development*, Columbus, Ohio State University Press.

Schmookler, J. (1966) *Invention and Economic Growth*, Cambridge, MA, Harvard University Press.

Schneider, J. and Ziesemer, T. (1995) 'What's New and What's Old in New Growth Theory, Endogenous Technology, Microfoundations and Growth Rate Predictions', *Zeitschrift für Weltwirtschafts- und Sozialwissenschaften*, vol. 115, pp. 425–72.

Schneider, J. and Ziesemer, T. (1996) 'What's new and what's old in new growth theory: endogenous technology, microfoundation and growth rate predictions', *Zeitschrift für Wirtschafts – und Sozialwirtschaften*, no. 115, pp. 429–72.

Schot, J. (1992) 'The policy relevance of the quasi-evolutionary model: the case of stimulating clean technologies', in Coombs, R., Saviotti, P. and Walsh, V. (eds) *Technological Change and Company Strategies*, London, Academic Press, pp. 185–200.

Schott, K. (1975) 'Industrial R and D expenditures in the UK: an econometric analysis', DPhil thesis, Oxford University.

Schott, K. (1976) 'Investment in private industrial R and D in Britain', *Journal of Industrial Economics*, vol. 25, no. 2, pp. 81–99.

Schott, B. and Muller, W. (1975) 'Process innovations and improvements as a determinant of the competitive position in the international plastics market', *Research Policy*, vol. 4, pp. 88–105.

Schumpeter, J.A. (1912) *Theorie der wirtschaftlichen Entwicklung*, Leipzig, Duncker & Humboldt. English translation *The Theory of Economic Development*, Harvard (1934).

Schumpeter, J.A. (1928) 'The instability of capitalism', *Economic Journal*, pp. 361–86.

Schumpeter, J.A. (1939) *Business Cycles: A Theoretical, Historical and Statistical Analysis of the Capitalist Process*, 2 vols, New York, McGraw-Hill.

Schumpeter, J.A. (1942) *Capitalism, Socialism and Democracy*, New York, Harper & Row.

Sciberras, E. (1977) *Multinational Electronic Companies and National Economic Policies*, Greenwich, JAI Press.

Sciberras, E. (1981) 'Technical innovation and international competitiveness in the television industry', *Omega*, pp. 585–96.

Science Policy Research Unit (1972) *Success and Failure in Industrial Innovation*, London, Centre for the Study of Industrial Innovation.

Scitovsky, T. (1977) *The Joyless Economy: An Inquiry into Human Satisfaction and Consumers' Dissatisfaction*, Oxford, Oxford University Press.

Scott, A.J. (1991) 'The aerospace-electronics industrial complex of Southern California: the formative years 1940–1960', *Research Policy*, vol. 20, no. 5, pp. 439–57.

Scott, M.F. (1989) *A New View of Economic Growth*, Oxford, Clarendon Press.

Scoville, W. (1951) 'Minority migration and the diffusion of technology', *Journal of Economic History*, vol. 11, pp. 347–60.

Searle, J. (1995) *The Construction of Social Reality*, London, Penguin Books.

Seiler, R. (1965) *Improving the Effectiveness of Research and Development*, Maidenhead, McGraw-Hill.

Senker, J. (1991) 'Information technology and Japanese investment in Europe: implications for skills, employment and growth', *Futures*, vol. 20, pp. 36–56.

Senker, J. (1995) Tacit knowledge and models of innovation, *Industrial and Corporate Change*, vol. 4, no. 2, pp. 425–47.

Senker, J.M. and Faulkner, W. (1994) 'Making sense of diversity: public–private research linkage in three technologies', *Research Policy*, vol. 23, no. 6, 673–96.

Servan-Schreiber, J. (1965) *The American Challenge*, Harmondsworth, Penguin (English translation of *Le Défi Americain*, Paris, 1965).

Shackle, G.L.S. (1955) *Uncertainty in Economics and other Reflections*, Cambridge, Cambridge University Press.

Shackle, G.L.S. (1961) *Decision, Order and Time in Human Affairs*, Cambridge, Cambridge University Press.

Sharp, M. (1991) 'Pharmaceuticals and biotechnology', in Freeman, C., Sharp, M. and Walker, W. (eds) *Technology and the Future of Europe*, London, Pinter.

Sherwin, C.W. and Isenson, R.S. (1966) *First Interim Report on Project 'Hindsight'*, Washington, Office of the Director of Defence Research and Engineering.

Shimshoni, D. (1966) 'Aspects of scientific entrepreneurship', DPhil thesis, Harvard.

Shimshoni, D. (1970) 'The mobile scientist in the American instrument industry', *Minerva*, vol. 8, no. 1, pp. 59–89.

Silverberg, G., Dosi, G. and Orsenigo, L. (1988) 'Innovation, Diversity and Diffusion: A Self-Organisation Model', *Economic Journal*, vol. 98, pp. 1032–54.

Silverberg, G. and Lehnert, D. (1994) 'Growth fluctuations in an evolutionary model of creative destruction', in Silverberg, G. and Soete, L. (eds), *The Economics of Growth and Technical Change*, Aldershot, Edward Elgar.

Singer, C., Holmyard, E.J., Hall, A.R. and Williams, T.J. (eds) (1954–78) *A History of Technology* (6 vols), Oxford, Clarendon Press.

Sirilli, G. (1982) 'The researcher in Italian industry', mimeo, Science Policy Research Unit, University of Sussex.

Skoie, H. (1996) 'Basic research: a new funding climate?', *Science and Public Policy*, vol. 23, no. 2, pp. 66–75.

Sloan, A. (1963) *My Years with General Motors*, New York, Doubleday.

Smart, T. (1996) 'Jack Welch's encore: how GE's chairman is remaking his company – again', *Business Week*, 28 October, pp. 43–50.

Smith, A. (1776) *An Inquiry into the Nature and Causes of the Wealth of Nations*, Dent edn (1910).

Smith, A. and Venables, T. (1988) 'Completing the Internal Market in the European Community', *European Economic Review*, vol. 32 (7), pp. 1501–25.

Soete, L.L.G. (1979) 'Firm size and inventive activity: the evidence reconsidered', *European Economic Review*, vol. 12, pp. 319–40.

Soete, L.L.G. (1981) 'A general test of technological gap trade theory', *Weltwirtschaftliches Archiv*, vol. 117, no. 4, pp. 638–66.

Soete, L. (1987) 'The newly emerging information technology sector', in Freeman, C. and Soete, L. (eds) *Technical Change and Full Employment*, Oxford, Blackwell.

Soete, L.L.G. (1989) 'The impact of technological innovation on international trade patterns: the evidence reconsidered', *Research Policy*, vol. 16, nos. 2–4, pp. 101–30.

Soete, L.L.G. (1991) *Synthesis Report, TEP, Technology in a Changing World*, Paris, OECD.

Soete, L.L.G. (1994) *Technology, Economy and Productivity*, TEP Report, Paris, OECD.

Soete, L.L.G. (1996) 'New technologies and measuring the real economy: the challenges ahead', paper prepared for the ISTAT Conference on 'Economic and Social Challenges in the 21st Century', Bologna, 5–7 February.

Soete, L.L.G. and Arundel, A. (1993) *An Integrated Approach to European Innovation and Technology Diffusion Policy: A Maastricht Memorandum*, Commission of the European Communities.

Soete, L.L.G. and Kamp, K. (1996) 'The "Bit Tax": the case for further research', *Science and Public Policy*, vol. 23, no. 6, pp. 353–60.

Soete, L. and Turner, R. (1984) 'Technology Diffusion and the Rate of Technical Change', *Economic Growth and the Structure of Long-Term Development*, New York, St. Martin's Press.

Solo, R. (1980) *Across the High Technology Threshold: The Case of Synthetic Rubber*, Norwood PA, Norwood Editions.

Solow, R.M. (1956) 'A contribution to the theory of economic growth', *Quarterly Journal of Economics*, vol. 70, pp. 65–94.

Solow, R.M. (1957) 'Technical progress and the aggregate production function', *Review of Economics and Statistics*, vol. 39, pp. 312–20.

Steinmuller, E. (1994) 'Basic research and industrial innovation', in Dodgson, M. and Rothwell, R. (eds) *Handbook of Industrial Innovation*, Cheltenham, Elgar.

Stirling, A. (1994) 'The act of technology choice', DPhil thesis, University of Sussex.

Stokes, D.E. (1993) 'Pasteur's quadrant: a study in policy science ideas', mimeo draft, Princeton University.

Stoneman, P. (1987) *The Economic Analysis of Technological Change*, Oxford, Oxford University Press.

Stoneman, P. (ed.) (1995) *Handbook of the Economics of Innovation and Technological Change*, Oxford, Blackwell.

Storper, M. and Harrison, B. (1991) 'Flexibility, hierarchy and regional development: the changing structure of industrial production systems and their form of governance in the 1990s, *Research Policy*, vol. 20, no. 5, pp. 407–23.

Strange, S. (1988) *States and Markets*, New York, Blackwell.

Sturmey, S.G. (1958) *The Economic Development of Radio*, London, Duckworth.

Supple, B. (1963) *The Experience of Economic Growth: Case Studies in Economic History*, New York, Random House.

Surrey, A.J. and Thomas, S.D. (1980) *Worldwide Nuclear Plant Performance Lessons for Technology Policy*, SPRU Occasional Paper no. 10, University of Sussex.

Swann, P. (ed.) (1993) *New Technologies and the Firm*, London, Routledge.

Symeonidis, G. (1996) *Innovation, Firm Size and Market Structure: Schumpeterian Hypotheses and Some New Themes*, OECD Economics Department Working Paper no. 161, Paris, OECD.

Szakasits, G. (1974) 'The adoption of the SAPPHO method in the Hungarian electronics industry', *Research Policy*, vol. 3, no. 1, pp. 18–28.

Takeuchi, H. and Nonaka, I. (1986) 'The new product development game', *Harvard Business Review*, January–February, pp. 285–305.

Tamura, S. (1986) *Reverse Engineering, A Characteristic of Japanese Industrial R&D Activities*, Tokyo, Saitama University.

Taylor, C. and Silberston, A. (1973) *The Economic Impact of the Patent System*, Cambridge, Cambridge University Press.

Teece, D.J. (1986) 'Profiting from technological innovation: implications for integration, collaboration, licensing and public policy', *Research Policy*, vol. 15, no. 6, pp. 285–305.

Teece, D. (ed.) (1987) *The Competitive Challenge: Strategies for Industrial Innovation and Renewal*, Cambridge, MA, Ballinger.

Telefunken (1928) *25 Jahre Telefunken*.

Telefunken (1953) *50 Jahre Telefunken*.

Ter Meer, F. (1953) *Die IG Farben*, Dusseldorf [publisher not known].

Teubal, M. (1987) *Innovation Performance, Learning and Government Policy: Selected Essays*, Madison, University of Wisconsin Press.

Teubal, M., Amon, N. and Trachtenberg, M. (1976) 'Performance in innovation in the Israeli electronics industry', *Research Policy*, vol. 5, no. 4, pp. 354–79.

Teubal, M. and Steinmueller, E. (1982) 'Government policy, innovation and economic growth, lessons from a study of satellite telecommunications', *Research Policy*, vol. 11, pp. 271–87.

Thomas, H. (1970) 'Econometric and decisions analysis: studies in R and D in the electronics industry', PhD thesis, University of Edinburgh.

Tidd, J., Bessant, J. and Pavitt, K. (eds) (1997) *Managing Innovation: Integrating Technological, Market and Organizational Change*, Chichester, Wiley.

Tilton, J. (1971) *International Diffusion of Technology: The Case of Semi-Conductors*, Washington, DC, Brookings Institution.

Tisdell, C. (1981) *Science and Technology Policy: Priorities of Governments*, London, Chapman & Hall.

Tobin, S. (1955) *The Application of the Multi-variate Probit Analysis to Economic Survey Data*, Cowles Foundation Discussion Paper no. 1, New Haven.

Townsend, E.C. (1969) *Investment and Uncertainty*, Harlow, Oliver & Boyd.

Townsend, J. (1976) *Innovation in Coal-mining Machinery – the Anderton Shearer loader and the Role of the NCB and Supply Industry in its Development*, SPRU Occasional Paper no. 3, University of Sussex.

Townsend, J., Henwood, F., Thomas, G., Pavitt, K. and Wyatt, S. (1976) *Science and Technology Indicators for the UK: Innovations in Britain since 1945*, SPRU Occasional Paper no. 16, University of Sussex.

Turner, D.F. and Williamson, O.E. (1971) 'Market structure in relation to technical and organizational innovation', in Heath, J.B. (ed.) *International Conference on Monopolies, Mergers and Restrictive Practices*, London, HMSO.

Tweedale, G. (1987) *Sheffield Steel and America: A Century of Technical and Commercial Interdependence, 1830–1930*, Cambridge, Cambridge University Press.

UNESCO (1970) *Measurement of Output of Research and Experimental Development*, Paris, UNESCO.

US Department of Commerce (1967) *Technological Innovation: Its Environment and Management,* Washington, DC, USGPO.

US National Research Council (1989) *Materials Science and Engineering for the 1990s,* Washington, National Academy Press.

US Senate Committee (1995) KG Abraham, Commissioner Bureau of Labour Statistics before the Senate Finance Committee, 13 March 1995.

Utterback, J.M. (1993) *Mastering the Dynamics of Innovation,* Boston, Harvard Business School Press.

Uzawa, H. (1961) 'On a two-sector model of economic growth', *Review of Economic Studies,* vol. 29, no. 1, pp. 40–47.

Uzawa, H. (1965) 'Optimum technical change in an aggregative model of economic growth, *International Economic Review,* vol. 6, pp. 18–31.

Van Boxsel, J.A.M. (1989) 'Constructive technology assessment in the Netherlands: building a bridge between R&D policy and TA', paper presented to the symposium 'Range and Possibilities of Technology Assessment', Stuttgart.

van Gelderen, J. (1913) 'Springvloed Beschouwingen over industrielle Ontwikkeling en prijsbeweging', *De Nieuwe Tijd,* vol. 184, nos. 5 and 6, English translation by B. Verspagen in (ed.) Freeman, C. (1996).

Verdoorn, P.J. (1949) 'Fattori che regolano lo sviluppo della produttività del lavoro', *L'Industria,* no. 1, pp. 45–53.

Verdoorn, P.J. (1980) 'Verdoorn's Law in retrospect: a comment', *Economic Journal,* vol. 90, June, pp. 382–5.

Vernon, R. (1966) 'International investment and international trade in the product cycle', *Quarterly Journal of Economics,* vol. 80, pp. 190–207.

Verspagen, B. (1992) 'Endogenous innovations in neo-classical growth models: a survey', *Journal of Macro-Economics,* vol. 14, no. 4, pp. 631–62.

Verspagen, B. (1993) *Uneven Growth between Interdependent Economies: The Evolutionary Dynamics of Growth and Technology,* Aldershot, Avebury.

Verspagen, B. (1994) 'Technology and growth: the complex dynamics of convergence and divergence, in Silverberg, G. and Soete, L. (eds) *The Economics of Growth and Technical Change,* Aldershot, Elgar.

Villaschi, A.F. (1993) 'The Brazilian national system of innovation: opportunities and constraints for transforming technological dependency', DPhil thesis, University of London.

Vivarelli, M. (1995) *The Economics of Technology and Employment: Theory and Empirical Evidence,* Aldershot, Elgar.

Volti, R. (1990) 'Why internal combustion?', *American Heritage of Invention and Technology,* Fall, pp. 42–7.

von Hippel, E. (1976) 'The dominant role of users in the scientific instrument innovation process', *Research Policy,* vol. 5, no. 3, pp. 212–39.

von Hippel, E. (1978) 'A customer-active paradigm for industrial product idea generation', *Research Policy,* vol. 7, no. 2, pp. 240–66.

von Hippel, E. (1988) *The Sources of Innovation,* Oxford, Oxford University Press.

von Tunzelmann, G.N. (1978) *Steam Power and British Industrialisation to 1860,* Oxford, Oxford University Press.

von Tunzelmann, G.N. (1993) 'Technology in the early nineteenth century', in Floud, R.C. and McCloskey, D.N. (eds) *The Economic History of Great Britain,* 2nd edn, vol. 1, Chapter 11, Cambridge, Cambridge University Press.

von Tunzelmann, G.N. (1995a) *Technology and Industrial Progress: The Foundations of Economic Growth,* Cheltenham, Elgar.

von Tunzelmann, G.N. (1995b) 'Time-saving technical change: the cotton industry in the English industrial revolution', *Explorations in Economic History,* vol. 32, pp. 1–27.

Wade, R. (1990) *Governing the Market: Economic Theory and the Role of Government in East Asian Industrialisation,* Princeton, Princeton University Press.

Wakelin, K. (1997) *Links between Trade and Technology: A Microeconomic and Macroeconomic Study,* Aldershot, Elgar.

Walsh, V. (1984) 'Invention and innovation in the chemical industry: demand-pull or discovery push?', *Research Policy,* vol. 13, pp. 211–34.

Walsh, V., Townsend, J., Achilladelis, B.G. and Freeman, C. (1979) 'Trends in invention and innovation in the chemical industry', mimeo, Science Policy Research Unit, University of Sussex.

Ward, W.H. (1967) 'The sailing ship effect', *Bull Inst Physics*, vol. 18, p. 169.

Watson, T.J. (1963) 'Meeting the challenge of growth', McKinsey Foundation Lecture, no. 2.

Weber, M. (1922) *Economy and Society*, New York, Bedminster Press (1968).

Whiston, T. (1991) 'Forecasting the world's problems: the last empire, the corporatisation of society and the diminution of self', *Futures*, March, pp. 163–78.

Whiston, T.G. (1992) *Managerial and Organisational Integration: The Integrative Enterprise*, London, Springer.

Wiener, N. (1949) *The Human Use of Human Beings: A Cybernetic Approach*, New York, Houghton Mifflin.

Wilkins, G. (1967) 'A record of innovation and exports', in Teeling-Smith, G. (ed.) *Innovation and the Balance of Payments*, London, Office of Health Economics.

Williams, R. and Edge, D. (1996) 'The social shaping of technology', *Research Policy*, vol. 25, no. 6, pp. 865–901.

Williamson, O.E. (1975) *Markets and Hierarchies: Analysis and Antitrust Implications. A Study in the Economics of Internal Organization*, New York, Free Press.

Williamson, O.E. (1985) *The Economic Institutions of Capitalism*, New York, Free Press.

Williamson, O.E. and Winter, S.G. (eds) (1993) *The Nature of the Firm: Origins, Evolution and Development*, New York, Oxford University Press.

Wilson, C. (1955) 'The entrepreneur in the industrial revolution in Britain', *Explorations in Economic History*, vol. 7, no. 3, pp. 129–45.

Winter, S.G. (1986) 'Comments on Arrow and Lucas', *Journal of Economics*, no. 54, pp. 427–34.

Winter, S.G. (1988) 'On Coase, competence and the corporation', *Journal of Law, Economics and Organisation*, vol. 4, pp. 163–80.

Wise, T.A. (1966) 'IBM's $5,000,000,000 gamble', *Fortune*, September.

Witt, U. (ed.) (1991) *Evolutionary Economics, The International Library of Critical Writings in Econonomics*, Vol. 25, Aldershot, Elgar.

Womack, J.T., Jones, D.T. and Roos, D. (1990) *The Machine that Changed the World*, New York, Rawson Associates.

Wood, A. (1994) *North–South Trade, Employment and Inequality: Changing Fortunes in a Skill-driven World*, Oxford, Clarendon Press.

World Bank (1991) *World Development Report, 1991*, New York, Oxford University Press.

World Bank (1993) *East Asian Miracle: Economic Growth and Public Policy*, Washington, DC, World Bank.

Wynn, M.R. and Rutherford, G.H. (1964) 'Ethylene's unlimited horizons', *European Chemical News: Large Plants Supplement*, 16 October.

Yakushiji, T. (1986) 'Technological emulation and industrial development', DAEST/Stanford Conference on Innovation Diffusion, Venice.

Yarsley, V.E. and Couzens, E.G. (1956) *Plastics in the Service of Man*, Harmondsworth, Penguin.

Young, A.A. (1928) 'Increasing returns and economic progress', *Economic Journal*, vol. 38, pp. 527–42.

Ziman, J. (1994) *Prometheus Bound: Science in a Dynamic Steady State*, Cambridge, Cambridge University Press.

Zuse, K. (1961) *25 Jahre Entwicklung Programmgesteuerter Rechenanlagen*, Hersfeld.

INDEX